Geophysical Monograph Series

Including
IUGG Volumes
Maurice Ewing volumes
Mineral Physics Volumes

Geophysical Monograph 103

Measurement Techniques in Space Plasmas

Fields

Robert F. Pfaff
Joseph E. Borovsky
David T. Young

Editors

American Geophysical Union
Washington, DC

Published under the aegis of the AGU Books Board

Library of Congress Cataloging-in-Publication Data
Measurement techniques in space plasmas : fields / Robert F. Pfaff,
    Joseph E. Borovsky, David T. Young, editors.
        p. cm. -- (Geophysical Monograph ; 103)
    Includes bibliographical references.
    ISBN 0-87590-086-0
    1. Space plasmas--measurement.. 2. Electric fields--Measurement.
3. Cosmic magnetic fields--measurement. I. Pfaff, Robert F., 1953- .
II. Borovsky, Joseph E., 1954- . III. Young, David T. IV. Series
    QC809.P5M4   1998
    523.01--dc21   98-18385
                                                           CIP

ISBN 0-87590-086-0
ISSN 0065-8448

Printed in the United States of America.

Contents

Wave Measurements

Spacecraft-Charging Measurements and Effects

New Areas

Neutral-Particle Imaging

Multipoint Measurements

PREFACE

Space plasma measurements are conducted in a hostile, remote environment. The art and science of measurements gathered in space depend therefore on unique instrument designs and fabrication methods to an extent perhaps unprecedented in experimental physics. In-situ measurement of space plasmas constitutes an expensive, unforgiving, and highly visible form of scientific endeavor.

The demands on instrument performance, including survival during launch, far-flung journeys to distant reaches of the solar system, and years of unattended operation, all place severe constraints on the design, materials, electronic components, and other factors that directly relate to the success of a given scientific experiment. Furthermore, resources needed to build and operate the instruments, such as mass, power, memory, telemetry rate, and money, are always at a premium. In the past few years, the call has gone out for "faster, better, cheaper" spacecraft, instruments, and missions. This very worthy desire for more science per dollar has increased the demand on instrument performance while reducing resources available to the designer. The only resource not in scarce supply is the ingenuity of the space physics community, although the laws of physics and the rules of engineering do limit what even the most inventive can accomplish. This ingenuity is abundantly evident in this two-volume monograph.

This monograph brings together a unique combination of expertise concerning state-of-the art techniques which are designed to gather accurate measurements of space plasma physics phenomena. Topics include all of the major space plasma physics areas, such as thermal and energetic plasmas, including spectrometers that discriminate by energy, pitch angle and species; electric field detectors and magnetometers, including plasma wave interferometers, and several new areas, such as wave-particle correlators, multipoint measurements, neutral atom imaging, and miniaturization. One volume is devoted mainly to measurement techniques of plasma particles and to correlators, whereas the other concentrates on fields measurements and new areas of research. In order to keep the present work manageable, techniques to measure neutral gas parameters (i.e., to explore thermospheric physics) and "remote" techniques, such as cameras and photometers are not included here. Exceptions include neutral atom imaging techniques and radio remote sensing, which are used to study magnetospheric plasma dynamics.

The two volumes include 12 core chapters from recognized experts in the field. These papers are not reviews or tutorials per se, although some authors chose to provide such a synopsis. Rather, they provide an assessment of the current state-of-the art of each technique area, describing what constitutes the most successful instruments and where such techniques might still be lacking. The additional 72 contributions discuss other features related to individual measurement techniques.

Many previously published descriptions of instruments are inadequate because of omission of underlying theory of measurements, skimpy instrument descriptions, or omission of the intuitive aspects of development. Futhermore, as many of the old guard in the field of space physics are approaching retirement, a great deal of their art and know-how is rapidly disappearing with only a faint legacy preserved in the literature. One of the goals of the monograph is to fill this void. Although these volumes are not designed to provide a history, many of the techniques discussed here represent the fruits of much hard fought, trial-and-error development and space flight experience.

As space physics research is largely driven by space-based measurements, the correct interpretation of these measurements requires not only an understanding of the physics of what is being measured, but also an understanding of the experimental techniques used to obtain the measurements. Typically, issues about the validity of measurement techniques are not addressed in the literature, which mostly concentrates on results. Thus, an important goal of the present work is to critically assess the current capabilities of space physics instrument techniques and to provide a resource for scientists carrying out detailed analysis of space plasma physics using in-situ data. Such was the theme of a conference on instrument techniques held in Santa Fe, New Mexico on April 3-7, 1995, where scientists from a broad range of sub-disciplines gathered to critically assess space plasma measurement techniques, to discuss available alternatives, and to delineate the areas where additional methods are particularly needed. Indeed, the conference sub-title was "What works, what doesn't." The meeting was attended by 149 scientists from 16 countries and formed the springboard for the decision to produce the present two-volume work.

The editors thank all of the referees, a list of whom is attached, who provided insightful, critical reviews. We also thank the authors themselves, without whose concerted efforts this monograph would not have been possible.

Finally, we acknowledge financial assistance from the Space Physics Division of the National Aeronautics and Space Administration as well as from the Los Alamos National Laboratory. The editors also acknowledge, with thanks, the help of their assistants, Ms. Brenda Valette at the NASA/God-dard Space Flight Center and Ms. Eloisa Michel at the Los Alamos National Laboratory.

Robert F. Pfaff
NASA/Goddard Space Flight Center

Joseph E. Borovsky
Los Alamos National Laboratory

David T. Young
Southwest Research Institute

Editors

Reviewers

M. Acuña	R. Goldstein	J. Larsen	D. Potter
C. Alsop	P. Gough	M. Lessard	F. Primdahl
R. Arnoldy	M. Grande	A. Lazarus	J. Quinn
S. Barabash	R. Grard	G. Le	J. Raitt
B. Barraclough	J. Green	R. Lepping	P. Rodriguez
R. Belian	M.Gruntman	P.-A. Lindqvist	C. Russell
R. Benson	H. Hayakawa	H. Luhr	E. Scime
J.-J. Berthelier	R. Heelis	E. Lund	J. Scudder
M. Boehm	F. Herrero	R. Manning	E. Sittler
J. Borovsky	N. Hershkowitz	G. Marklund	J. Slavin
L. Brace	M. Hesse	N. Maynard	E. Smith
J. Burch	R. Holzworth	M. McCarthy	M. Smith
W. Burke	K. Hsieh	D. McComas	R. Snare
C. Carlson	J. Jahn	R. McEntire	R. Srama
P. Carter	A. James	J. McFadden	R. Stone
S. Chapman	G. James	C. McIlwain	O. Storey
D. Chornay	A. Johnstone	R. Merlino	D. Suszcynsky
J. Clemmons	M Kaiser	N. Meyer-Vernet	C. Swenson
V. Coffey	I. Katz	D. Mitchell	L. Tan
C. Curtis	J. Keller	E. Möbius	R. Torbert
S. Curtis	P. Kellogg	T. Moore	K. Tsuruda
R. Elphic	K. Khuran	F. Mozer	H. Vaith
R. Ergun	P. Kintner	T. Mukai	A. Vampola
A. Eriksson	E. Kirsch	J. Nordholt	D. Walton
D. Evans	C. Kletzing	K. Ogilvie	B. Wilken
J. Fainberg	D. Knudsen	S. Orsini	M. Wüest
W. Farrell	H. Koons	G. Papatheodorou	P. Wurz
T. Fritz	H. Laakso	G. Paschmann	J. Wygant
H. Funsten	J. LaBelle	A. Pedersen	A. Yau
D. Gallegher	J. Laframboise	R. Pfaff	D. Young
B. Gilchrist	S. Lai	C. Pollock	L. Zanetti

Electric Field Measurements in a Tenuous Plasma with Spherical Double Probes

A. Pedersen

Space Science Department, ESTEC, Noordwijk, The Netherlands

F. Mozer

University of California, Space Science Laboratory, Berkeley, California

G. Gustafsson

Swedish Institute of Space Physics, Uppsala Division, Sweden

Experiences with spherical double probes for measurements of quasistatic and wave electric fields in the Earth's *magnetosphere* were gained by experiments which were part of the GEOS, ISEE-1, Viking, Geotail and CRRES missions. These experiments were built with active control of the spherical probes by forcing a current from the probes to the spacecraft and thereby bringing the probes to a smaller and more comfortable probe-plasma impedance close the plasma potential. A spacecraft (and probes) will come to a positive potential in a magnetospheric plasma due to emission of photoelectrons: photoelectrons are in fact providing the necessary "contact" between spacecraft/plasma and probe/plasma. The understanding of these processes is a necessary condition for understanding electric field measurements in a tenuous plasma and also the spurious effects which influence measurements. Based on past experience it will be possible to measure quasistatic spacecraft spin plane components of the electric field perpendicular to the spacecraft-Sun direction with an accuracy of the order of 0.25 mVm^{-1} and with somewhat less accuracy in the spacecraft-Sun direction for double probe spin planes close to the ecliptic. When the spacecraft spin axis is close to the spacecraft-Sun direction, the accuracy can be considerably improved. The time resolution of such measurements is mainly limited by telemetry; probes and associated electronics permit measurements of electric fields up to a few MHz. A natural bi-product of electric field double probe measurements is the determination of the spacecraft/plasma potential which in turn provides high time resolution information about plasma density variations.

1. INTRODUCTION

A double probe for measurements of electric fields in a tenuous plasma must have a large probe separation and the only practical way is to deploy the probes radially from a

Measurement Techniques in Space Plasmas: Fields
Geophysical Monograph 103
Copyright 1998 by the American Geophysical Union

spinning spacecraft. Another reason for using a spinning spacecraft is that small electric fields can only be derived with confidence from the double probe spin modulation signal. This paper will describe spherical double probes; the advantage with this geometry is that the photoemission does not vary significantly during the spin, an important condition for measuring electric fields in a tenuous plasma.

Cylindrical double probes are technically less demanding; each probe consists of a radically deployed insulated wire with a non-insulated section at the tip. Photoemission will in this case vary with the spin and cause a spurious spin modulation signal. This technique has been used successfully in more dense magnetospheric plasmas and will be described in more detail by *Maynard* (this issue).

Spherical double probes for measurements of electric fields in a tenuous magnetospheric plasma were first launched on the S3-3 spacecraft in 1976 (Mozer et al. 1979) and on the GEOS-1 and ISEE-1 spacecraft in 1977. The experience gained from these experiments was used on other missions: GEOS-2, Viking, Geotail and CRRES. The performance of the GEOS-1, GEOS-2 and ISEE-1 experiments has been described by *Pedersen et al.* [1984] and that of the Viking experiment by *Block et al.* [1987]. The collected knowledge about the double probe technique from these missions has in turn formed the basis for the electric field double probe experiments on Polar and Cluster and has also provided useful inputs for the design of the Freja topside ionosphere electric field experiment [*Marklund et al., 1994; Eriksson, 1995*].

All the experiments mentioned above were designed based on the principle of using spinning wire booms with spherical sensors at the ends of coaxial wires. Each probe must have a comfortable impedance to its local plasma environment; the electric field can then be measured as the voltage difference between two radially opposite spherical sensors divided by their "effective separation", which can be smaller than the physical separation of 40 - 100 m for the above cases. This is caused by a partial short-circuit of the ambient electric field by the wire booms. As explained later, this effect can be calibrated. Work function differences between probes of the order 0.1 volt, may cause probe surface voltage differences of the same order corresponding to 1 mVm^{-1} for a 100m double probe. The spin modulation voltage signal is not influenced by work function differences and is used for slowly varying quasistatic electric field measurements. This is also a matter to be discussed in some detail.

The electric field measured is the spin plane component of the electric field vector. Figure 1a shows one double probe with a magnetic field at an angle α relative to the spin plane. For such a single double probe it takes one spin (a few seconds) to determine the spin plane component of the electric field. Rapid variations, over less than a few seconds, can therefore not be measured correctly in amplitude. For quasistatic electric fields, which do not vary significantly during one spin, it is a good assumption to write $E_\perp \gg E_\parallel$ or $\mathbf{E} \cdot \mathbf{B} = 0$ and the full $\mathbf{E}$ vector can then be constructed if $\mathbf{B}$ is not too close to the spin plane; in practice α should be at least $10° - 20°$. When α is smaller only the spin plane component of $\mathbf{E}$ perpendicular to $\mathbf{B}$ can be obtained. Shortlived electric field "spikes", of durations less than a spin period, can only be measured correctly if the "spike" is aligned with the double probe. Measurements of an arbitrary "spike" therefore show less than the full amplitude and the spin plane phase is not determined.

GEOS-1 and GEOS-2 carried a single double probe which only permitted quasistatic electric fields to be resolved with the spin period of 6 seconds, and higher frequency wave signals were spin modulated. ISEE-1 had a hybrid solution (like CRRES) with one double probe using two spherical sensors and another with cylindrical sensors.

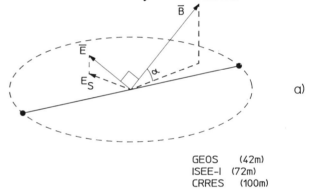

GEOS (42m)
ISEE-I (72m)
CRRES (100m)

a)

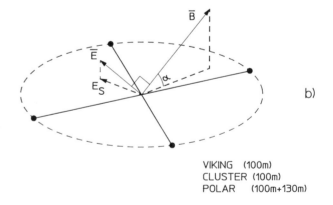

VIKING (100m)
CLUSTER (100m)
POLAR (100m+130m)

b)

Figure 1. a) One single double probe can be used to determine the spin plane component of the electric field, provided it does not change during one spin period, **E** is assumed to be perpendicular to **B**. b) Two crossed double probes can determine the spin plane component of the electric field up to high frequencies, only limited by probe functioning or telemetry.

Two crossed double probes (Figure 1b) were first used on S3-3 and later on Viking and will be used on Cluster and Polar. (Cluster was lost on the first Ariane 5 flight in June 1996 and future lessons from it will depend on the approval for reflight.) This makes it possible to obtain an electric field vector component in the spacecraft spin plane with high time resolution; in practice the time resolution is determined by telemetry and is limited to typically $10^{-2} - 10^{-1}$s. This permits the determination of the spin plane component of shortlived 'spikes'. For waves the full waveform electric field data can in practice only be transmitted up to approximately 100 Hz and occasionally in shorter bursts up to higher frequencies depending on on-board data storage. Electric field measurements with a double probe system, designed for magnetospheric conditions, are feasible up to approximately 1 MHZ. Due to telemetry limitations, it is in this case necessary to use on-board data analysis or transmit electric field power as a function of frequency with limitations on frequency resolution. Electric field measurements, as described here, assume that the dimensions of potential structures in space are much larger than the double probe dimension, i.e. the electric field is very close to being constant over the double probe. The spatial potential structures of some plasma waves are smaller than the double probe and the interpretation of data must take this into account.

This paper will recapitulate some of the basic principles for carrying out electric field measurements in a tenuous plasma, with emphasis on quasistatic and low frequency measurements, and will sum up the accumulated experience from past experiments. This will include descriptions of probe-plasma interactions under the influence of probe photoelectron emission and collection of ambient electrons. Influences from probe supports and the spacecraft must also be considered. The use of the probe for determining the spacecraft potential and its relation to electron density or electron flux will also be mentioned, and finally examples will be given of ways to check the correct operation of a double probe in different plasma environments.

2. PROBE-PLASMA COUPLING

A probe in a dense ionospheric plasma is coupled to the local plasma by ion and electron currents as illustrated in Figure 2a; these currents will balance each other at a negative floating potential of approximately $3 kT_e/e$ (T_e = electron temperature). The dynamic resistance between the probe and the plasma $R = \delta V/\delta I$ is of the order $10^6\Omega$ or smaller for a probe of a few cm. diameter, and a Debye shielding distance of a cm or less guarantees that the probe is connected to a small plasma volume surrounding the probe. It is necessary

to drive probe supporting elements (see Figures 2 and 4) at the same potential as the probe by connecting the low impedance output of the preamplifier to these elements (bootstrapping). Because ionosphere double probes require dimensions of a few meters, and the probe-to-plasma impedance is small, it is feasible and more comfortable, to place all amplifiers inside the spacecraft.

Sufficient coupling of an electric field probe to a tenuous magnetospheric plasma (electron density N_e as low as 10^4-$10^5 m^{-3}$) can only be achieved by photoemission, which means that the probe must be in sunlight to function. Figure 2b illustrates the situation in a tenuous plasma by the current voltage relation for an 8 cm diameter sphere (used on GEOS, ISEE-1 and POLAR). A high impedance preamplifier is placed inside the spherical electric field probe in order to effectively bootstrap probe support elements and in order to achieve low impedance output signals compared to input signals, with the typical probe-plasma impedance shown in Figure 2b. Other advantages, for wave measurements will be explained later. This arrangement requires that power must be supplied to the preamplifier from a floating power supply on the spacecraft. Photoemission will generate a photoelectron cloud around a probe; inside a radius r_a most photoelectrons are orbiting back to the positive probe and therefore do not contribute to any probe current. Outside r_a most photoelectrons escape to constitute the current l_a. Photoelectrons generated by the Sun have a near Maxwellian energy distribution, peaking near 2eV and with a smaller higher energy component [Grard, 1973, Pedersen, 1995]. I_e is the collected electron current; the ion current is completely negligible in comparison. I_a has been determined from detailed diagnostics with electric field probes [Pedersen, 1995]; this will be explained in more detail in the following section. For the purpose of explaining the principle of the measurement we will at this point only mention that l_a has an e-folding voltage of 2 volt for small positive probe potentials, changing to approximately 7.5 volt for probe potentials above 10 volt. This corresponds to photoelectron e-folding energies of respectively 2eV and 7.5 eV. In a thin plasma with an electron population corresponding to 10^5 m^{-3} and I keV in density and energy, the probe and the spacecraft will both float at a positive potential of +15 volt, and the probe will have a probe-plasma dynamic resistance $R \approx 5 \cdot 10^9 \Omega$ which is uncomfortably large and requires that spurious currents (e.g. from supports) be extremely well balanced on the two opposite probes.

The potential near a V_p probe will in vacuum fall off as $V(r) = V_p r_p/r$ where r_p is the probe radius and V_p its potential. Photoelectron space charge will cause a more rapid reduction of potential with r. For a probe with $r_p = 4-5$ cm r_a will range from a fraction of a meter to well above one meter for a very

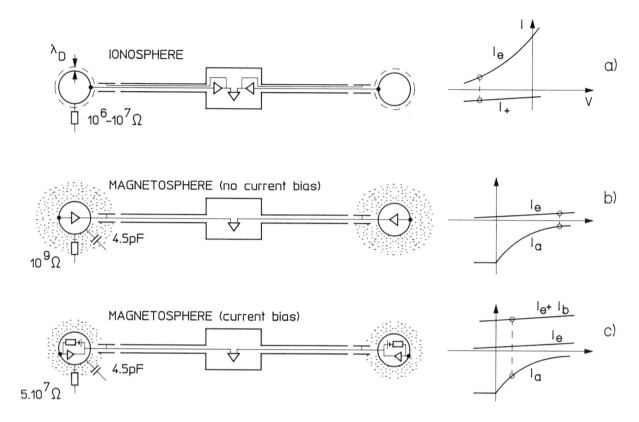

Figure 2. a) A double probe in the ionosphere will have each probe connected to the local plasma with a comfortable impedance. Each probe will float at a negative potential resulting from electron and ion currents to the probe (I_e and I_+) when the photoelectron current in comparison is small. b) A floating probe in the magnetosphere will be at a positive potential resulting from a balance of collected electrons (I_e) and escaping photoelectrons (I_a). The dotted regions around the probes indicate the extent of orbiting photoelectrons. A typical probe-plasma impedance is indicated. c) A bias current (I_b) from the probe to the spacecraft (using a high impedance current source) can bring a probe closer to the local plasma potential and can shrink the radius of orbiting photoelectrons. The resistive coupling ($\sim 5 \cdot 10^7 \Omega$) makes it possible to carry out reliable quasistatic and low frequency electric field measurements.

tenuous plasma. This is the typical distance for coupling between the probe and the plasma near the probe. We will next explain that it is possible to reduce r_a.

A spherical probe, in a plasma with a long Debye length, will collect electrons according to the formulae [*Whipple*, 1965]:

$$I_e = I_{eo}(1+V_p/V_e) \qquad (1)$$

where $V_e = kT_e/e$.

It is necessary to find ways to make the probe-plasma dynamic resistance smaller than that of a floating probe. By drawing a bias current from the probe via a high impedance current source referred to the preamplifier output, as indicated in Figure 2c (equivalent to a current of high energy electrons to the probe), it is possible to move the probe closer

to its local plasma potential where R is close to $5 \cdot 10^7 \Omega$. Furthermore this value for R is nearly independent of ambient plasma conditions. An additional benefit of the bias current to the probe is that the value of r_a will be reduced, making the coupling to the plasma via the photoelectrons better defined in spatial extent.

In parallel with the dynamic resistance R the probe is coupled to the plasma capacitively. For a tenuous plasma with a long Debye length the probe capacitance is that of the probe in free space to a very good approximation. For GEOS, ISEE-1, CRRES, Cluster and POLAR, with probes of 8 cm diameter, $C_p = 4.5$ pF. Figure 3a is a schematic drawing of a voltage preamplifier in an electric field probe with typical values for R and C_p as described. The transformation of a high impedance input signal V_{in} to a low impedance output signal V_{out}, can easily be achieved with an amplification A =

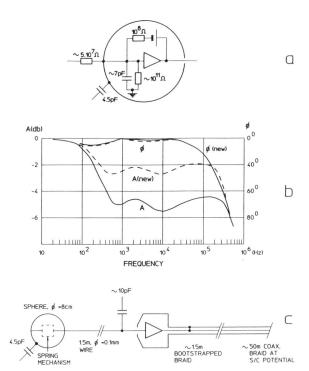

Figure 3. a) A spherical probe with preamplifier with schematic representation of probe/plasma impedance, amplifier input impedance and bias current source. b) Signal amplification, A, and phase, ϕ, for probe in test set up simulating space conditions. c) Modified probe design for Cluster/Phoenix; A and ϕ values are given in Figure 3b as A(new) and ϕ(new).

$V_{out}/V_{in} = 1$, and no significant phase shift, ϕ, up to approximately 100 Hz. The A and ϕ values for POLAR from 10 Hz to approximately 1 MHZ (Figure 3b) were obtained in a specially constructed test set-up consisting of a 1m diameter, 6m long metal tube. A drive signal was applied to the tube, and the probe and its bootstrapped wire elements were placed in the middle of the tube. This test is possible because the input signal is capacitively coupled to the probe at higher frequencies. At the time of submission of this paper, Polar has been launched, and for certain plasma conditions in the plasmasphere the amplifiers were found to be unstable over time intervals of 15-30 minutes. A thorough investigation, using the test set-up described above, has resulted in the understanding that the coupling between bootstrapped wire element and the probe, combined with an amplifier driven at higher frequencies than on earlier flights, caused the problem.

Based on this lesson, a modified electric field probe was developed for the Cluster spare spacecraft, called Phoenix, and was successfully tested and subsequently integrated on the spacecraft in early 1997. The new probe, shown in Figure

3c has a preamplifier in a separate bootstrapped box and the probe is connected to a 1.5m, 0.1mm diameter metal wire, and is deployed by the centrifugal force. The thin metal wire, in this case, part of the probe, will cause a small spin modulated photoemission, which can be calibrated for. The larger probe to plasma capacitor coupling (~ 14 pF) results in a better value for A at higher frequencies; ϕ is practically unchanged, as shown in Figure 3b.

3. INFLUENCES FROM PROBE SUPPORTS AND SPACECRAFT

The spacecraft must have 'sufficiently conductive' surfaces to serve as an electric potential reference for an electric field double probe. 'Sufficiently conductive' in practice means that all surfaces must have a surface skin conductivity of approx. $10^5 \Omega$ per square in order to conduct currents from the shadow side to the sunlit side of the spacecraft to limit any differential charge build-up on the spacecraft. The spacecraft, having a total surface approximately 500 times that of a probe will have a typical spacecraft-plasma dynamic resistance of 500 times less than that of a floating (not biased) probe, i.e. $10^7 \Omega$ for the plasma conditions described in connection with Figure 2b. Photoelectrons are emitted on sunlit areas and there will therefore be a sunward concentration of orbiting and escaping photoelectrons. This is more pronounced for the spacecraft than the probes because the distance of orbiting photoelectrons is comparable to the spacecraft dimensions and few photoelectrons will orbit to the shadowed side of the spacecraft. For the smaller probes the orbiting photoelectrons will be more evenly distributed.

Figure 4 is a schematic representation of a double probe aligned in the satellite-sun direction. Asymmetric emission of probe photoelectrons tend to give an apparent sunward electric field because photoelectrons on a probe positioned away from the Sun will lose more photoelectrons to the wire boom which is at the same potential as the spacecraft. Longer bootstrapped wire braid elements and negatively biased guards at the outer ends of the wire booms on Cluster and POLAR will minimise this effect (guards were missing on GEOS and ISEE-1 and an apparent sunward electric field of some mVm^{-1} had to be corrected for). Cluster will have an ion emitter to keep the spacecraft and the wire booms at a reduced positive potential of a few volts. This will further reduce what will remain of the asymmetric probe-wire boom photoelectron coupling.

Spacecraft photoelectrons will form a sunward cloud near the spacecraft, and a probe position sunward of the spacecraft will tend to attract more of them than the opposite probe anti-sunward of the spacecraft. This effect is however negligible

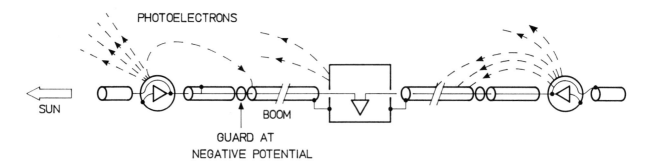

Figure 4. Each probe has bootstrapped supports close to the probe and a negatively biased guard close to the tip of the braid of the wire boom. This guard (not used onGEOS and ISEE-1) stops probe photoelectrons going to the wire boom in an asymmetric fashion. For no guard (or guard potential off) more photoelectrons will leave the probe positioned away from the Sun. This will cause a spurious sunward electric field.

for the probe design of Figure 4 because spacecraft photoelectrons will tend to return to the wire booms (at spacecraft potential) before reaching the probes, and the guards (at a negative potential relative to the spacecraft) will also help to deter spacecraft photoelectrons.

The spacecraft and the coaxial wire boom, with the outer braid electrically connected to the spacecraft, will tend to short out the ambient electric field so that the effective length of the double probe is less than the physical separation between two probes (L). Figure 5a illustrates this situation for half a double probe where ambient equipotential surfaces separated by L_e (probe effective length) are connected to the probes. Figure 5b shows that with longer bootstrapped supports, and negatively biased guards, the partial short-circuit effect will be reduced. The exact equipotential pattern will depend on the plasma shielding distance which can be several meters in a tenuous plasma.

It was possible to determine this effect on ISEE-1 in the solar wind; Figure 6 shows the measured electric field in the GSE y direction (E_{ym}) compared with that computed (E_{yc}) from solar wind speeds and magnetic fields. It can be seen in Figure 6 that E_{ym} has a fixed offset of ~ 1mVm^{-1} probably due to a small difference in photoemission between the probes on ISEE-1. Figure 6 (upper panel) illustrates that $L_e/L = 0.6$ whereas full bootstrapping of the wire booms on ISEE-1 (lower panel) result in $L_e/L \approx 1.0$. This latter arrangement has drawbacks which are not discussed here, however it is a useful demonstration that long bootstrapped probe supports on Cluster and POLAR (Figure 5b) will help with respect to shorting out electric fields.

4. PHOTOEMISSION FROM SURFACES IN SPACE

The double probe electric field experiments on GEOS, ISEE-1, CRRES and Viking have made it possible to study the long-term changes of photoemission properties of surfaces in space. The current-voltage curve of a probe is determined by photoelectrons, ambient electrons and the probe bias current. In a plasma environment where the electron population is known it is possible to determine the photoelectron current-voltage relation by either stepping the bias current and observing the change in probe potential or by varying the probe-spacecraft potential difference and observing the probe-spacecraft current. Details of such measurements are given in *Pedersen* (1995). The main results of this study are:

• The spacecraft conductive surfaces develop photoemission characteristics very similar to that of an electric field probe; this can be concluded from the fact that a floating probe (no bias current) has the same potential as the spacecraft, within a fraction of a volt, for a wide range of plasma densities.

• The full photoemission current density (current per unit area projected to sunlight) increases to approximately 80 μAm^{-2} over the first year in free space above the atmosphere and then seems to stabilize at this value. This is approximately four times more than what can be calculated from laboratory measurements [*Grard*,1973]. This could be due to ion implantation or UV modifications of surface absorbed gases.

• Spacecraft with perigees at altitudes of less than 500-600 km will be influenced by the atmosphere (probably atomic oxygen) which will reduce the photoemission to values much smaller than the long term free space value. In the case of highly eccentric orbits, this value stays during the whole orbit as long as perigee is below 500-600 km i.e. the orbital period is too short for recovery. This conclusion is based on measurements on ISEE-1, when perigee dropped to approximately 400 km for a few weeks, and from CRRES with perigee permanently near 300 km. The e-folding energy

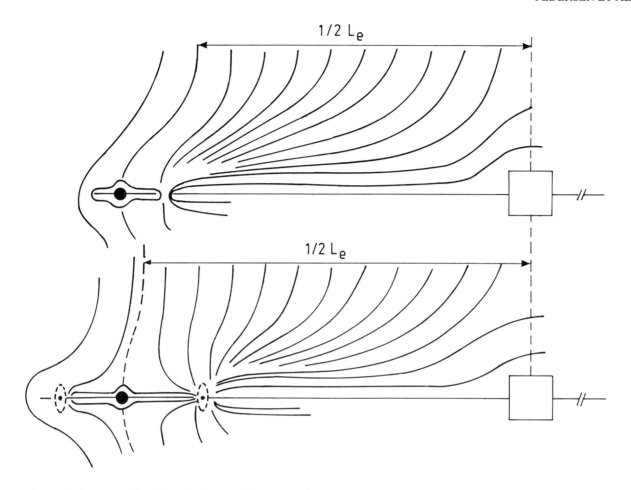

Figure 5. The spacecraft and the wire boom will cause a partial shorting out of the ambient electric field when the Debye length is comparable to or larger than the dimensions of one probe and its supports. The effective length of the double probe is L_e. By increasing the probe support dimensions and using negatively biased guards (shown as dotted rings) L_e can be increased to be closer to the probe to probe separation, L.

of photoelectrons is found to be close to 2 eV for photoelectron energies up to about 10 eV, with an additional component with an e-folding energy of approximately 7eV dominating higher photoelectron energies.

The photoelectron current density to the plasma from a spherical conductive body in free space for a long period (approximately 1 year) is shown in Figure 7 as a function of body potential and can be given by the analytic function:

$$J_a = 80 \ (\mu A m^{-2}) exp(-V/2) + 3(\mu A m^{-2}) exp(-V/7.5) \qquad (2)$$

An electric field probe with a current bias can be positioned at approximately +2 volt relative to its local plasma potential and its dynamic resistance which depends on the e-folding energy at low photoelectron energies (approx.2 volts) can be determined to be approx. $5 \cdot 10^7 \Omega$ for an 8 cm diameter current biased probe. The probe-plasma coupling resistance will be smaller and provide a better resistive coupling in a more dense plasma. The spacecraft will in a tenuous magnetospheric plasma be at potentials of +20 volt to +50 volt or more, determined by the balance of photoelectrons of e-folding energy of approx. 7.5 volt and ambient electrons (mostly above 100 eV for such plasma environments). The exact shape of the l_a curve for body potentials above approximately 10 volt requires further study and also must consider the influence from secondary electron emission.

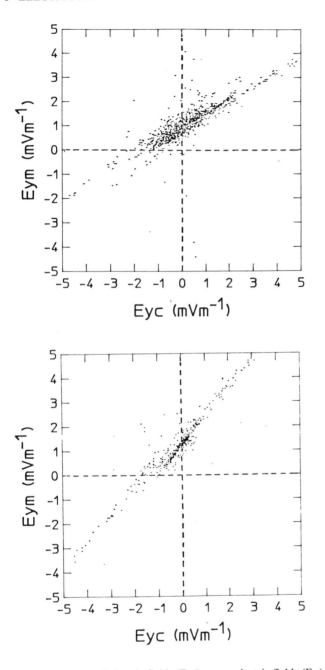

Figure 6. Measured electric fields (E_{ym}) versus electric fields (E_{yc}) which have been computed from solar wind speeds and magnetic fields. The upper panel is for grounded wire booms, and the lower panel is for bootstrapped wire booms.

plasma drift perpendicular to the magnetic field **B**, under conditions that **E** + **U** x **B** = 0. For such comparisons to be meaningful conditions must not vary significantly over several spacecraft spin periods. Complementary diagnostics can be carried out under long quiet magnetospheric conditions when the electric field in the frame of reference of the satellite is very small and instrument offsets, giving rise to spurious electric fields, are the only signals to be observed.

5.1 Probe Photoelectron Coupling to Supports

The GEOS and ISEE-1 experiments, with spin axis respectively perpendicular to the equator and close to the

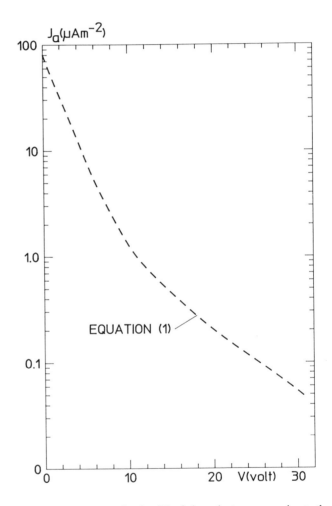

Figure 7. The current density (J_a) of photoelectroncs escaping to the ambient plasma (referred to the area projected to the Sun) versus spacecraft or probe potential relative to the local plasma potential. This curve has been established on the basis of comparisons between plasma and electric field data on GEOS, ISEE-1 and Viking. Spacecraft with low perigee or in low orbit will have a smaller J_a.

5. LESSONS FROM EXPERIMENTS IN SPACE

The performance of an electric field double probe for measurements of quasistatic electric fields in space can be established by comparison with measurements of **U**, the ecliptic plane, experienced a sinusoid-like spin modulation due to a stronger attraction of probe photoelectrons to the wire boom tips for a probe position away from the Sun. An offset (sunward apparent electric field) of several mVm^{-1} could be identified and mapped out in detail as a function of spacecraft potential and applied bias current for selected quiet periods. This effect could be reduced to approximately 2 mVm^{-1} on GEOS-2 by introducing a negative voltage barrier of -2 volt on part of the bootstrapped probe support. Smaller offsets were observed for the double probe axis perpendicular to the spacecraft - Sun direction; this is a symmetric geometry as far as supports are concerned and any offset therefore must be due to differences in photoemission. This can also be calibrated for during quiet periods. The Viking and CRRES spacecraft had their spin axes within a small deviation from the spacecraft-Sun direction and therefore had very small offsets due to varying coupling of probe photoelectrons to supports during a spin period.

5.2 Comparisons with Electron Drift Instrument on GEOS

Figure 8 shows data from the GEOS-2 double probe (translated into **E** x **B** velocities) and an electron drift instrument (during a PC-5 wave event) [*Pedersen et al.* 1984]. It was found that the double probe **E** x **B** velocity was phase shifted 20° - 40° relative to the electron drift direction. The explanation is most likely a wake effect caused by the positively charged boom tips approximately 1 m away from the probes. Cold ions convecting in the **E** x **B** direction will see this as a hindrance with the consequence that spurious electric fields are generated in the direction of the wake which has a potential minimum due to exclusion of convecting ions. This spurious electric field is, in the first approximation, proportional to the ambient **E** x **B** drift and is also proportional to the positive charging of the spacecraft which includes the boom tips. A tenuous plasma is therefore more critical for this influence.

The longer boom on ISEE reduced this wake effect to the point that it could not be detected (the ratio between spurious signal and probe differential signal became smaller). It will be totally negligible for experiments on Viking, CRRES and the future POLAR and Cluster spacecraft with even longer booms and several meters of bootstrapped cable between the probe and the boom tip connected to the spacecraft at a positive potential. Figure 8 also shows that the measured and subsequently corrected electric fields, translated to **E** x **B** velocities are approximately 70-80% of the electron **E** x **B** drifts measured by the GEOS-2 electron drift experiment, compared to 60% for ISEE-1 in the solar wind. The most likely explanation is that the more dense plasma on GEOS-2

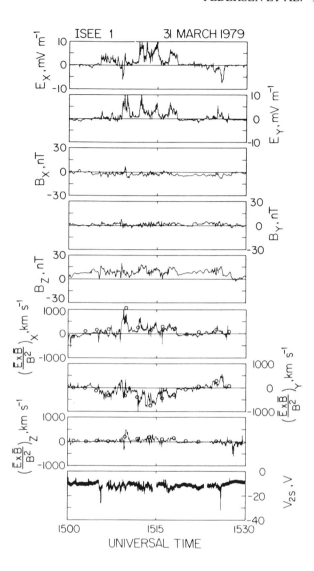

Figure 8. Comparison of plasma drift by three different techniques on GEOS-2. Data from the electron-beam experiment for measurement of gyro-centre drift is given as dots. **E** x **B**/B^2 calculated on the basis of electric and magnetic field data, is shown as thin lines, and thick lines for wake corrected data. The larger circles in the bottom panel show anisotropy directions derived from a low energy ion experiment [*G. Wrenn*, private communication].

in the inner magnetosphere, with a smaller Debye length, reduces the 'shorting out' effect.

5.3 Comparison with Ion Drifts at the Plasma Sheet Boundary Layer

Electric and magnetic fields in GSE coordinates, measured on ISEE-1 near the plasma sheet boundary layer at

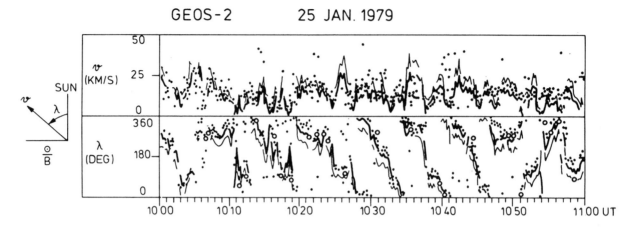

Figure 9. Electric and magnetic field data from ISEE-1 in the magnetotail (20.8 Earth radii, 0040 local time and ~4° GSM latitude). These data have been translated into **E** x **B** velocities and compared with plasma velocities perpendicular to **B** (L. Frank, private communication). The latter are shown as circles. The lower panel shows the negative of the potential between probe 2 and the spacecraft (V_{2s}). More negative values represent smaller densities.

approximately 20 R_e and near local midnight following substorms, are given in Figure 9a. The magnetic field is mainly in the Z direction and **E** and **E** x **B** are therefore both close to the XY-plane (the spin plane). The electric field components have been corrected for dc offsets and for the reduced effective length as described in paragraph 3. In Figure 9b are components of the **E** x **B** velocities compared with components of ion drift velocities [*L. Frank*, private communication]. The ion sampling period is much longer than for the electric field experiment; nevertheless the agreement between the two quantities is very good during this event which can be described as a 'bursty bulk flow event' [*Angelopoulos*, 1994]. A similar comparison between the double probe and another plasma experiment on ISEE-1 has been published by Mozer et al. (1983). The lower panel of Figure 9b shows that the potential difference between probe no. 2 and the satellite (V_{2s}) is of the order minus 10 to minus 15 volts. A drop in V_{2s} to more negative values means that a more tenuous plasma has been encountered; this can be seen near 1505 UT. This comparison carried out in the lobe and the plasma sheet is another demonstration that the spherical double probe performed correctly in these plasma environments.

5.4 Lower and Upper Limits of Measured Electric Fields

Diagnostics which permitted the determination of electric fields, and comparisons between plasma drift measurements and electric field measurements have made it possible to conclude that the spin plane component in the Sunward direction can be determined within ± 0.5 mVm⁻¹ and that the

component perpendicular to this direction can be determined within ± 0.25 mVm⁻¹; these values are somewhat better for Viking with the spin axis closer to the Sun. The smallest measurable electric fields on Cluster and POLAR are expected to be smaller.

Electric field 'spikes' of 100 mVm⁻¹, or more, and with a duration of a fraction of a second to some seconds are frequently observed near magnetospheric boundaries and on auroral field lines [*Mozer*, 1981]. An electric field double probe is probably the only experiment with sufficient dynamic range and time resolution to make measurements of these 'spikes'.

Wave measurements can be done in the same way as quasistatic electric field measurements up to approximately 100 Hz, and at higher frequencies keeping in mind that the probes are capacitively coupled to the plasma above this frequency. A condition for quantitative interpretation is that the wavelength is much larger than the probe separation so that the wave electric field is close to being linear over the double probe.

6. ELECTRON DENSITY OR FLUX MEASUREMENTS—A BI-PRODUCT

In Section 4 it was mentioned that a conductive spacecraft and an electric field probe would float at the same electric potential within a fraction of a volt in a wide range of plasma densities. This also means that the spacecraft can be approximated to be a sphere and that the balance of ambient electron and escaping photoelectron currents is nearly identical to that of the ideal sphere. This obviously does not hold any longer when the Debye length becomes comparable with the spacecraft radius. This situation will occur in a high density, low energy electron environment. The Debye length is 1m for 1eV electrons with a density of $6 \cdot 10^7 \text{m}^3$. This is therefore the approximate upper density limit for equation 1 to be used for the spacecraft. This upper limit of plasma density is less critical in the magnetosheath with more energetic electrons and consequently a larger Debye length for a given plasma density. Consider an electric field probe positioned approximately + 2 volt relative to its local plasma environment by a probe-spacecraft bias current, and the spacecraft floating at a potential determined by the ambient electron environment. The spacecraft-probe potential difference, ΔV, can be used for determination of electron density or flux (Pedersen 1995). Figure 10 demonstrates how ΔV is a function of electron density for densities in the range 10^6m^{-3} to $5 \cdot 10^7 \text{m}^{-3}$ and energies in the range 1 eV to 25 eV. For higher electron energies ΔV depends on electron flux, i.e. the product of electron density and velocity ($N_e \cdot v_e$). There is good agreement between calculated N_e-ΔV curves based on Equations 1 and 2 and measured values of N_e for different electron energies.

The usefulness of the "ΔV technique" for density measurements can be summed up as follows:
• It provides, in parallel with electric field measurements, a quick method for measuring the plasma density in the solar wind and in the regions of the inner magnetosphere with a low energy high density electron population.
Magnetosheath electron fluxes can be determined, and with knowledge of the electron energy distribution the plasma density can be determined.
• The plasmasheet and lobe electron fluxes can be determined to very small values.
• The method depends on the e-folding energy of photoelectrons and only to a lesser degree on the absolute values of the photoelectron current density.
• The method is complementary to other plasma experiments (plasma sounders and particle experiments) particularly by providing high time resolution data; 100 samples per s is probably possible from probe-plasma relaxation time considerations. This may also be the only way to diagnose a very tenuous plasma in the lobes of the magnetosphere.

Escoubet [1997] has used this tecnnique on ISEE-1 data to obtain a magnetospheric plasma density survey.

7. DISCUSSION AND CONCLUSIONS

Several lessons have been learned from double probe electric field experiments on a number of spacecraft, starting with S3-3 in 1976, GEOS and ISEE-1 in 1977. All probes

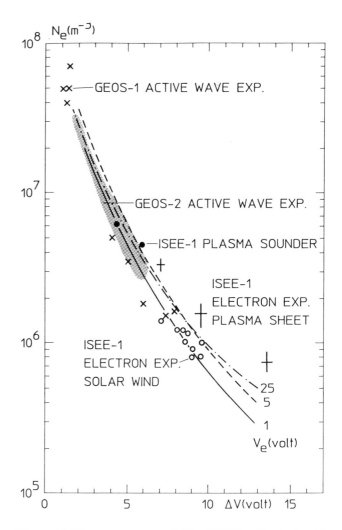

Figure 10. The measurements of ΔV = V(S/C) - V(probe) yields information on the ambient electron density for plasmas with electron energies up to several times 10 eV. The lines marked V_e = 1,5,25 volt (V_e = kT_e/e) are ΔV values resulting from the J_a curve in Figure 7 and electron populations with the above values for V_e. These curves are compared with different plasma density measurements.

are biased with a current to set the probe potential to be at approximately +2 volt relative to the ambient plasma for all plasma conditions in the magnetosphere. This is necessary for the optimum performance of the experiment, and at the same time permits to use the probes as a reference for measurements of the spacecraft potential which is ΔV+ approximately 2 volt. This is important information for interpretation of all low energy ion and electron measurements. The capability to measure electric fields in a tenuous magnetospheric plasma has been clearly demonstrated. The lower limit and accuracy of 0.25-0.5

mVm^{-1} may not suffice for all measurement needs. For higher accuracy needs, we note that the GEOS electron drift instrument could measure electric fields smaller than 0.1 mVm^{-1} via electron drift measurements. However, the strength of the double probe technique is the capability to measure electric fields over a large dynamic range with high time resolution. The two crossed double probes (4 spheres on four radial booms) on POLAR has been used to resolve the spin plane component of electric field 'spikes' with a time resolution of better than 0.01 seconds and allowing for amplitudes above 100 mVm^{-1}. The simultaneous measurements of the probe-spacecraft potential difference ΔV, provides electron density or flux information with similar high time resolution.

The spin plane component of wave electric fields can be measured up to frequencies in the 500 kHz - 1 MHz range. The amplitude can only be determined reliably for wavelengths larger than the effective double probe separation.

Laakso et al. [1995] have modelled the influence of plasma density gradients on a double probe. The concern is that the two probes are placed in different plasma densities and thereby a spurious signal is produced. This may in some cases be a concern for experiments on spacecraft in low orbits passing the auroral zone and possibly of concern at unusually steep gradients at the plasmapause. Plasma at magneto-spheric density gradients have gyro radii for ions as well as electrons much larger than the typical double probe dimension of 100 m and gradients, which must extend over many ion gyro radii, cannot therefore give rise to spurious electric field signals.

Experiments on POLAR and Cluster have been improved, based on previous experience, and it is expected that measurement uncertainties will be smaller. Comparison with new and improved plasma instruments, providing plasma drift information, will also provide good calibrations which will hopefully confirm this expectation. Cluster, if and when it is launched, will in this respect have the unique possibility to compare double probe electric fields with those derived from a sophisticated electron drift instrument.

REFERENCES

Angelopoulos, V., W. Baumjohann, C.F. Kennel, V. Coroniti, M.G. Kivelson, R. Pellat, R.J. Walker, H. Lühr and G. Paschmann, Bursty bulk flows in the inner central plasma sheet, *J. Geophys. Res.* 97, 4027-4039, 1992.

Block, L.P., C.-G. Fälthammar, P.-A. Lindqvist, G.T. Marklund, F.S. Mozer and A. Pedersen, Measurement of quasi-static and low frequency electric fields on the Viking satellite, *Technical report TRITA-EPP-87-02*, Royal Institute of Technology, Stockholm, 1987.

Eriksson, A.I. and R. Bostrom, IRF Scientific Report 220, 1995.

Escoubet, C.P., A. Pedersen, R. Schmidt and P.A. Lindqvist, Density in the magnetosphere inferred from ISEE-1 spacecraft potential, *J. Geophys. Res.,* in press, 1997.

Grard, R.J.L., Properties of the satellite photoelectron sheath derived from photoemission laboratory measurements, *J. Geophys. Res.* 78, 2885-2906, 1973.

Laakso, H., T. Aggson and R. Pfaff, Plasma gradient effects on double probe measurements in the magnetosphere, *Ann. Geophys.*, 13, 130-146, 1995.

Marklund, G.T., L.G. Blomberg, P.-A. Lindqvist, C.-G. Fälthammar, G. Haerendel, F. Mozer, A. Pedersen and P. Tanskanen, The double probe electric field experiment on Freja: Experiment description and first results, *Space. Sci. Rev.*, 70, 483-508, 1994.

Mozer, F.S., C.A. Cattell, M. Temerin, R.B. Torbert, S. von Glinski, M. Woldorff and J. Wygant, The dc and ac electric field, plasma density, plasma temperature, and field-aligned current experiments on the S3-3 satellite, *J. Geophys. Res.*, 84, 5875-5884, 1979.

Mozer, F.S., ISEE-1 Observations of electrostatic shocks on auroral field lines between 2.5 and 7 Earth radii, *Geophys. Res. Lett.* 8, 823-826, 1981.

Mozer, F.S., E.W. Hones, J. Birn, Comparisons of spherical double probe electric field measurements with plasma bulk flows in plasmas having densities less than 1 cm^{-3}, *Geophys. Res. Lett*, 8, 737-740, 1983.

Pedersen, A., C.A. Cattell, C.-G. Fälthammar, V. Formisano, P.-A. Lindqvist, F. Mozer and R. Torbert, Quasistatic electric field measurements with spherical double probes on the GEOS and ISEE satellites, *Space Sci. Rev.*, 37, 269-312, 1984.

Pedersen, A., Solar wind and magnetosphere plasma diagnostics by spacecraft electrostatic potential measurements, *Ann. Geophys.*, 13, 118-129, 1995.

Whipple, E.G.Jr., The equilibriam electric potential of a body in the upper atmosphere and in interplanetary space, *NASA Report X-615-65-296*, 1965.

Electric Field Measurements in Moderate to High Density Space Plasmas with Passive Double Probes

Nelson C. Maynard

Mission Research Corporation, Nashua, New Hampshire

The passive double floating probe instrument for measuring electric fields has, over the past twenty five years, been proven to be a reliable technique for the moderate to high density plasmas of the ionosphere and inner magnetosphere. Both cylindrical and spherical sensors have been successfully used in the high-density space plasmas in the E and F regions of the ionosphere. The technique has been extended to the lower density plasmas of the D region of the ionosphere and of the inner magnetosphere. While simple in concept, double floating probes have many pitfalls. Potential problems related to spacecraft attitude, sheaths, magnetic and vehicle-velocity wakes, photoemission, asymmetries, and capacitive coupling are discussed in the context of data from sounding rockets, DE-2, San Marco D, ISEE-1 and CRRES. At plasma densities below 10 cm^{-3} the passive technique may not reliably function as errors can become unmanageable. CRRES data show that the actively biased spherical probe technique can also be applied to cylindrical sensors to improve measurements at these densities.

INTRODUCTION

Over three decades have passed since the original proposal by *Aggson and Heppner* [1964] to use double probes for electric field measurements in space plasmas. Double-probe measurements have been extensively used on sounding rockets and satellites in the intervening years. The technique has evolved as the technology of deployable appendages improved and with the desire to make measurements in low density collisional and collisionless regimes. Success of the technique has not been without lessons learned. The purpose of this paper is to provide a primer for the design of a double-probe system, coupling some of the "oral tradition" to that which is in the literature. It will concentrate on passive double probes. Active biasing of the sensors for measurements in low density collisionless plasmas is covered elsewhere in this volume [*Pedersen*, 1997]. This paper is not meant to be a comprehensive review. Instead, it will highlight potential pitfalls, suggest mitigation techniques, and provide design trade-offs. Further considerations can be found in technique papers [e. g., *Fahleson*, 1967; *Aggson*, 1969; *Pedersen et al.*, 1984; *Maynard*, 1986] and instrument papers from various satellite experiments [e. g., *Heppner et al.*, 1978; *Maynard et al.*, 1981; *Wygant et al.*, 1992].

Passive double-probe instruments have been successfully used in the D, E and F regions of the ionosphere and in the inner magnetosphere on both satellites and sounding rockets. Measurements have also been made at the magnetopause and at the bow shock that have been tested with conformance to theory [*Aggson et al.*, 1983; *Scudder et al.*, 1986]. Both spherical and cylindrical sensors have been successfully used. The validity of double probes for measurements of electric fields in the moderate to high density E and F regions of the ionosphere is generally unquestioned. Papers, too numerous to mention, on results from Injun-5, OGO-6, S3-2, DE-2, San Marco D and many sounding rockets speak for themselves. *Aggson* [1969] showed that electric field measurements from a sounding rocket matched **v** x **B**

Measurement Techniques in Space Plasmas: Fields
Geophysical Monograph 103

when expected ambient electric fields were near zero (where **v** is the vehicle velocity). *Maynard et al.* [1970] found excellent agreement with Appleton-Hartree theory for AC electric field measurements in the VLF frequency range. *Hanson et al.* [1993] established that simultaneous double-probe electric field measurements and electric fields inferred from ion drift measurements from the DE-2 satellite generally agreed very closely. Occasionally occurring differences could be attributed to known technique deficiencies of both instruments. Plasmaspheric electric fields have been measured with long (200 m tip-to-tip) cylindrical passive double probes confirming that the technique functions with an accuracy of a fraction of a mV m^{-1} in collisionless plasmas at densities less than 10 cm^{-3}. Figure 1a shows close agreement between the azimuthal electric field in the plasmasphere measured by ISEE-1 and electric fields inferred from ground based whistler measurements, while Figure 1b compares the ISEE-1 results, projected along the magnetic field into the ionosphere, with measurements from the Millstone Hill radar [*Maynard et al.*, 1983]. Outside the plasmasphere, where densities continue to decrease and temperatures increase, biased probes are more accurate.

Most of the sources of error that will be discussed below affect the dc accuracy. AC electric field measurements can be carried out for wavelengths that are larger than the probe separation. Hence, while an accuracy of a fraction of a mV m^{-1} is possible for dc measurements, the ac measurement accuracy can be extended down to the system noise level of μV m^{-1}. Figure 2 shows an example from DE-2 of a pass through the dayside cusp with dc electric fields reaching nearly 150 mV m^{-1} [*Maynard et al.*, 1982]. The comb filter spectrometer measurements illustrate broadband electrostatic noise below 1 kHz combined with Alfvén waves at the lower frequencies, and also VLF hiss all associated with the cusp near 0420 UT. At lower latitudes ELF hiss is seen in the frequency range from 256 to 4000 Hz. In this case the separation distance was 21.4 m. Waves with a wavelength approaching that value, or harmonics of that value, will be underestimated in amplitude by the measurements. *Temerin* [1979] identified this artifact of a reduction at wavelengths that are harmonically related to the antenna length as a "fingerprint" effect in the electric field wave spectra [see also *Feng et al.* 1992]. This effect is not seen in the course resolution of the comb filter spectrometer, but is a feature of broadband spectrograms.

In the remainder of the paper we will explore the design of a double-probe electric field instrument, concentrating on the dc measurements while discussing the basic technique, sources of error and design trade-offs.

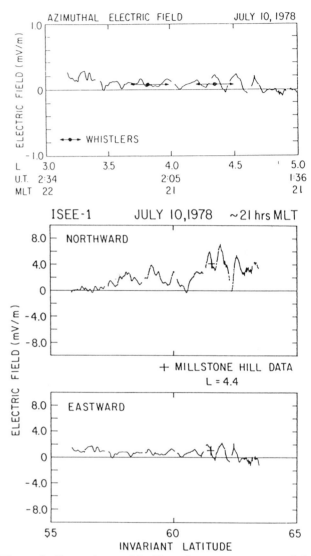

Figure 1. Comparisons of double probe measurements made by ISEE-1 in the plasmasphere with measurements made by ground based techniques. The top panel shows the comparison of the azimuthal component with electric fields derived from whistler measurements. The whistler values are an average over time indicated by the length of the attached horizontal arrows. The bottom panel shows the ISEE-1 electric field projected along equipotential magnetic field lines into ionosphere northward and eastward components and compared with electric field measurements obtained with the Millstone Hill radar [*Maynard et al.* 1983].

PASSIVE DOUBLE PROBE TECHNIQUE

The double-probe technique utilizes two identical floating sensors separated along an axis by a separation

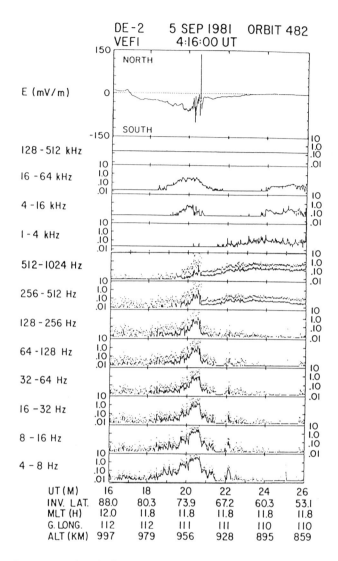

Figure 2. Electric fields observed by the DE-2 satellite from a southern hemisphere polar cusp crossing on September 5, 1981. The dc data (upper panel) are 1/2 s averages and represent the electric field in the horizontal direction along the orbit track. The ac data are from a comb filter spectrometer. The ac measurement axis is at 45° to the horizontal and as near perpendicular to B as possible. The solid curve represents the RMS average value within the passband while the dots are peak values within the sampling interval [*Maynard et al.* 1982].

distance **d** (Figure 3). Ideally, the difference in the local plasma potential at these two sensors is related to the sum of the ambient electric field and the **v x B** = electric field from the motion of the probe system across the Earth's magnetic field.

$$(\Phi_1 - \Phi_2)/|d| = (E + v \times B) \bullet d \qquad (1)$$

Hence, we only need to measure the potential difference between the two sensors, assuming that the sensor system motion relative to the magnetic field is known and assuming that we have not perturbed the ambient medium by the presence of our sensor system. A key word in the above is "identical".

Any probe or body immersed in a plasma will acquire a potential such that the net current to that body is zero. The floating potential of the probe is the sum of the local plasma potential and the current balance potential. The primary currents from the collection of ambient electrons (I_e) and positive ions (I_i) dominate the current balance in the high density plasmas of the ionosphere. Since electrons are more mobile than ions, the probe must acquire a negative potential to repel some of the electrons to equalize the currents. That potential will be a fraction of the temperature of the electrons and is typically a few volts or less in the ionosphere. The current collection equations, developed by *Mott-Smith and Langmuir* [1926], describe the normal Langmuir probe I-V characteristic and are functions of the shape as well as the sensor potential. If the probes are identical and the plasma uniform, the current balance potential will be the same for both probes, and Equation 1 applies directly.

The problems come from the presence of the sensor system in the plasma medium. Additional sources of current exist because of the sensors themselves (from photoemission) and because of the support structure (spacecraft and booms) for the sensor. As ambient densities decrease, photoelectron currents become significant and eventually dominate, driving the sensors positive. In most cases there is a large spacecraft or long rocket body in between the two sensors which is also photoemitting. Additional currents include photoelectrons that are emitted from the sensor (I_p), sensor emitted photoelectrons that return to the sensor (I_{pr}), photoelectrons that escape from the support structure or the spacecraft which are intercepted by the sensor (I_{sp}), and currents drawn by the voltmeter to make the measurement (I_v). Thus, the current balance equation for passive double probes becomes

$$I_e + I_{pr} + I_{sp} + I_v = I_i + I_p \qquad (2)$$

Ideally, all currents except the first on each side of the equation should be minimized. In practice photoemission can not be eliminated. It is essential to make it as unvarying as possible through symmetry of the probes and uniformity of the surfaces. The effect of photoemission in low density plasmas is to bias the sensor along the Langmuir probe current voltage characteristic to where small

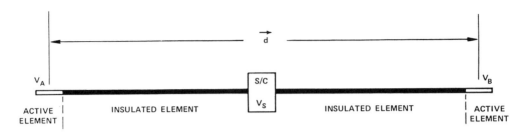

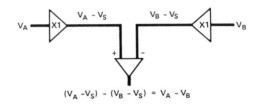

Figure 3. Schematic representation of the double probe technique for electric field measurements.

changes in current produce large voltage changes. This increases errors by amplifying asymmetries in the system. I_v is a function of the floating potential, but it can be made negligibly small by using high-input-impedance preamplifiers. However, active biasing of double probes adds a fixed current to the input of the preamplifier to counteract the bias of photoemission [see *Pedersen* 1997]. One other source of current, secondary emission, has been neglected in the above. This will only be a factor in regions of low cold plasma density and high energetic particle fluxes.

The ambient potential distribution can also be modified by the presence of the support structures and spacecraft. All surfaces will charge to some degree in the plasma to maintain their own current balance. These potentials cause a plasma sheath to form which shields the potentials from the plasma. The characteristic dimension of the sheath is the Debye length (λ_d) which is expressed as

$$\lambda_d = [\varepsilon_o kT/ne^2]^{1/2} \qquad (3)$$

where ε_o is the permitivity of free space, k is the Boltzman constant, T is the electron temperature, n is the electron density and e is the electron charge. To make sure that the potential of the main body is not seen by the sensors, the sensors should be many Debye lengths away. Photoemission distorts the sheath around the spacecraft body by emitting electrons only on the sunlit side, creating a cloud of electrons and excess negative charge in front of the spacecraft. The resulting sheath distortion can create

an apparent sunward electric field in low density plasmas when the sensors are not well outside of the sheath [*Cauffman and Maynard*, 1974]. The important criterion is to keep the sensors many Debye lengths from any source of potential that may distort the ambient distribution.

Sensors must be symmetric. They can be either spherical or cylindrical. Other shapes will create problems in maintaining symmetry. The most symmetric sensor is spherical. No matter what orientation, the same amount of the sensor is sunlit, making photoemission nearly constant. Cylindrical sensors change their orientation relative to the Sun with spacecraft spin, resulting in a spin-variable photocurrent; however, they have other advantages. The electrons photoemitted from the spacecraft and collected by a probe depend on cross-section presented by the sensor and the potential distribution. Cylindrical antennas minimize the cross-section. They are easier to deploy to long lengths and have less an impact on the moment of inertia for a given length. Cylindrical sensors integrate the potential over the sensor length which is usually meters, while spherical sensors provide more of a point measurement.

The surface work function of the sensor adds a contact potential to the mix of what is measured. If that work function is uniform over the whole sensor surface, both sensors will be the same and the potential difference will not be affected. The most stable relative to surface work function or contact potential are the noble metals. Note that in later sections of the paper the term contact potential will be used, as it normally is in the community, to refer to a collection of dc offset errors which includes, but is not restricted to, the work-function-related contact potential.

Many surfaces or surface coatings have been tried. Gold or silver plated sensors provide surface stability, but a fingerprint can destroy the uniformity. Metals, especially noble metals, are large photoemitters, and fingerprints as well as non-uniform surface crystal structure also affect the uniformity of photoemission. Graphite or amorphous carbon has been used to reduce photoemission and obtain more uniform surface properties. The technique most often used to obtain a graphite surface is to paint the sensors with a graphite paint such as Electrofilm 4306 or DAG 213. It is important in ionospheric applications to be sure that the binder is stable in the presence of bombardment by energetic oxygen ions. Early shuttle missions brought back spherical sensors that were scrubbed to bare aluminum by the oxygen ion dominated plasma environment [*Shawhan et al.*, 1984]. A vitreous carbon sphere was also tried in the GEOS experiments, but results were no better than the graphite films, and the sensors were more expensive. In addition to being a low photoemitter, the graphite creates a "uniformly dirty" sensor. A uniformly dirty sensor is to be preferred over a plated surface that has surface variations or fingerprints. Using this concept, beryllium-copper and stainless steel sensor surfaces have been left "dirty" and bare for many sounding rocket flights where cleanliness control may be difficult.

To get the sensors away from the spacecraft requires a deployable appendage or support structure. The easiest support structures are fold-out hinged booms; however they are limited in length by where they can fit within the launch shroud. Articulation of the booms increases the length by a factor of two with added complexity and chance of failure. Beryllium-copper and some forms of stainless steel can be fabricated with a built-in memory. Stored flat on a spool, they form into a cylindrical shape as they are deployed. They can be made more rigid by interlocking the edges or by building in a helical twist which creates a tight overlap with length. A spherical sensor can be supported by these structures, or the outer part of the elements can be used as a cylindrical sensor. These antennas with trade names of "Stacer" (Weitzman), "Interlocked Tee" (Fairchild) and "Stem" (Spar-Astro Research) can be deployed up to 20 meters from three-axis stabilized vehicles, subject to design and spacecraft constraints. Long antennas impact the spacecraft attitude control system through large increases in spacecraft

moments of inertia. The antennas are subject to thermal bending and in some configurations can oscillate. The 11 m antennas on DE-2 were thermally compensated to prevent thermal bending [*Maynard et al.*, 1981]. On spinning spacecraft wire antennas can be reeled out in the spin plane to very long lengths. The ISEE-1 cylindrical wire antennas were 200 meters tip-to-tip. Those with spherical sensors were required to be shorter because of increased moment of inertia. Centrifugal force keeps the antennas straight. Space qualified wire antenna deployers have been built by both Weitzman and Fairchild. In the spinning spacecraft configuration, spin axis antennas must be of the tubular variety and are limited in length by spacecraft stability criteria.

In the cylindrical sensor configuration the extendible device is used both as the support and the sensor. The inner part of the wire or tube is coated with an insulator such as Teflon or kapton while the outer few meters is left bare to provide the sensor. The high-input-impedance preamp is located at the base of the antenna. The sensor separation distance or baseline for the measurement is the distance between the midpoints of the bare elements. However, at higher frequencies, signals capacitively couple across the insulator, leading to integration of the potential over the whole length, thus reducing the baseline. Using spherical sensors on the end of the antenna allows the preamps to be located in the spheres. This steps down the impedance, limiting capacitive coupling to that of the sensor only, and keeps the point measurement intact for ac signals. Signals are reduced in amplitude only when the baseline becomes comparable to the wavelength as noted above. The outside of the support structure can be made conductive for potential control.

The outputs of the preamps are differentially subtracted and processed. The electronics become tailored to the experiment goals and are limited only by spacecraft constraints, imagination and money.

SOURCES OF ERROR

A number of sources of errors have been discussed or hinted at above. Table 1 provides a summary list of problems that create errors in the measurements and standard mitigation techniques to minimize the effects. The following paragraphs provide further details and examples.

Symmetry, or lack there of, is one of the largest factors in electric field measurement accuracy. Symmetry of sensor surface properties has been addressed in the previous section. Symmetry is also aspect sensitive. Shadowing of all or a portion of a sensor changes the

Table 1. Double Probe Error Sources

Problem	Mitigation
Current balance problems	
Voltmeter current	High input impedance
Sensor asymmetry (slowly varying)	*Minimize supporting hardware*
Photoemission	Lower photoemission; Use spherical sensors
Surface work function	Maximize uniformity of surface properties
Sensor orientation	Improve thermal design; Use spherical sensors
(mechanical or thermal distortion)	
Collection of body photoelectrons	Repel with guard electrodes
Sensor asymmetry (aspect sensitive)	*Minimize supporting hardware*
Shadowing	Use outboard stub booms; Lengthen baseline; Cull data
Magnetic wake	Avoid; Lengthen baseline; Cull data
Velocity wake	Avoid; Lengthen baseline; Cull data
Plasma gradients	Shorten baseline
Potentials from the presence of hardware	*Minimize supporting hardware*
Differential charging	Maximize electrostatic cleanliness
Velocity wake	Avoid; Lengthen baseline; Cull data
Asymmetric photoelectron sheath	Lengthen baseline
Capacitive coupling	Increase sensor size; Put preamps in sensor
v x B electric fields	
Spacecraft attitude knowledge accuracy	Design and test improved attitude determination system
Sensor attitude knowledge	Improve testing and thermal design

photoemission current from the sensor and affects the current balance. With spherical sensors, the support boom provides a shadow on the sensors for some aspect angles. To provide the same amount of shadowing on both sensors for a given aspect angle, a stub boom is extended outward from each sensor having the same diameter as the support boom. For shadowing of a sensor by the central body, there is no mitigation technique other than to ignore the data for the time that only one sensor is in shadow or in the "solar wake". Just as there are solar wake effects that affect the current balance, there are also velocity wake effects and magnetic wake effects. The density is decreased in the velocity wake, which changes the current collection. The greater reduction of ions compared to electrons in the near wake further complicates the current balance. Long antennas or support structures can be used to minimize wake effects by placing the sensors outside the highly disturbed region. Magnetic wake effects result when there are asymmetries in the flow along magnetic field lines. The central body then shields the flow from one sensor more than the other creating the asymmetric currents. Magnetic wake effects can manifest themselves as an apparent parallel electric field as observed by *Bering* [1983] with sounding rocket data. Wake effects of all types that remain after the best

possible mitigation are best removed by culling the affected data. Lengthening antennas to get the sensors outside of the wake or to cut down the solid angle intercepted by the central body is the best means for mitigation of wake effects.

At lower densities wake effects not only affect the current balance but can also create potential perturbations that can affect the measured potential. Figure 4 shows a calculation for the potential around a large central body moving to the left that is charged to a vehicle potential of 0.5 V. Calculations were done with the Phillips Laboratory POLAR code for spacecraft charging and wakes (D. Cook, private communication, 1992). The left panel is for an oxygen dominated medium, while the right panel is for a hydrogen dominated medium. Antennas extending 6 m in length from the low-earth orbiting spacecraft traveling approximately 7.5 km s^{-1} are shown. In this case the Debye length is of the order of a meter. Ideally, $d \gg \lambda_d$. The potential is disturbed slightly even at 45° from the wake axis in the hydrogen medium and is significantly disturbed along the wake axis for both mediums. The booms must also be long enough to get the sensors well outside of both potential and current balance effects of wakes. Potential distortions are also seen at low densities from the photoelectron sheath on the sunlit side,

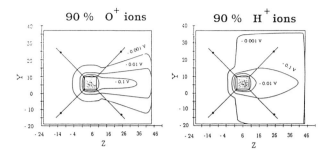

Figure 4. Examples of the spacecraft sheath distortion by the velocity wake. The left panel is for an oxygen atmosphere which would be typical of nighttime low latitude E region conditions. The right panel is for a hydrogen dominated atmosphere and would be typical of nighttime conditions above the F region with no particle precipitation. The vehicle potential is set a 0.5 V and the density is 10^2 cm^{-3}. The simulation was run with the POLAR code of Phillips Laboratory (D. Cook, private communication, 1992).

as noted in the previous section. Both potential distortions from the photosheath and current balance effects from body photoelectrons lead to an apparent sunward electric field [*Cauffman and Maynard*, 1974].

A mitigation technique to minimize the collection of body photoelectrons by the sensors in a low density plasma is to use a guard electrode located along the support structure inboard of the sensor and biased to a negative voltage above the maximum energy of the photoemitted electrons. This repels those from the central body traveling along the boom back toward the center and some of those from the sphere back to the sphere. The net gain is to reduce the interaction between bodies, limiting the current balance errors.

In low density plasmas in the presence of significant energetic electrons, spacecraft charging can occur up to a fraction of the energetic component temperature. Since the level of charging is a function of the surface characteristics, some areas of a spacecraft charge more than others. The differential charging distorts the sheath and, if strong enough, can cause breakdown. Maximizing the conductive surface area while minimizing exposed potentials keeps the surface potential more uniform. Conductive coatings over solar cells are expensive and generally are not cost-effective in low-earth orbit where spacecraft charging only occurs rarely. Electrostatic cleanliness to as great a degree as practical is a basic criteria in any space electric field experiment.

Long wire antennas that have the preamplifier at the base must consider an additional error source. It was earlier noted that capacitive coupling across the insulator

at high frequencies shortened the effective base line for ac measurements. It was also found on the San Marco D experiment that capacitive coupling at the spin frequency of a large **v** x **B** signal can provide a significant current into the antenna that is out of phase from the main signal [*Aggson et al.*, 1992]. If the area of the sensor is small, that error current can become a significant fraction of the currents collected by the sensor and cause a small phase shift and error in the results. This limited the San Marco measurements to regions where the plasma densities were above 10^4 cm^{-3}. San Marco was especially sensitive to this effect because the ambient mV m^{-1} electric fields are a very small fraction of the 250 to 300 mV m^{-1} **v** x **B** signals. A small phase error becomes important when subtracting two large vectors in order to obtain a small resultant ambient vector. This effect was not a significant problem on ISEE because of much smaller **v** x **B** signals and larger sensor area. Figure 5 shows an example of high-resolution despun data from San Marco with the data popping in and out of the phase shift error problem. The expanded scale plot at the bottom covers the blackened area in the top panel. The variations in the electric field in the region of the problem are at the fourth harmonic of the spin frequency resulting from vectorially combining errors on each of the two axes at the second harmonic of the spin frequency. The decrease in effective length and phase shift change and causes are shown schematically at the left. This effect could be mitigated by increasing the sensor size (i. e. decreasing the resistive coupling imped-ance to the plasma), decreasing the insulated portion (i. e., decreasing the capacitive coupling), or using sensors with internal preamps (eliminates the capacitive coupling problem from the support structure).

A mitigation technique that appears in Table 1 several times relative to general categories is to "minimize supporting hardware". In developing a payload primarily for D region measurements, this concept was employed in conjunction with maximizing symmetry [*Maynard* 1986]. The resulting daughter payload was flown on top of a mother payload with its 1 m booms folded into the nosecone. The booms were deployed, and the daughter was separated from the mother payload. The central body was a 12 inch diameter sphere. The X and Y axes were at 45° to the spin axis, the Z axis was in the spin plane orthogonal to the XY plane, and the I axis was a shorter axis in the spin plane (included for inertial stability). Data taken, after the booms had unfolded, just before and just after separation are shown in Figure 6 along with a configuration drawing of the attached structures. In the left set of panels (still attached) the distorted "sinusoidal" variations are the result of variations in the sheath around

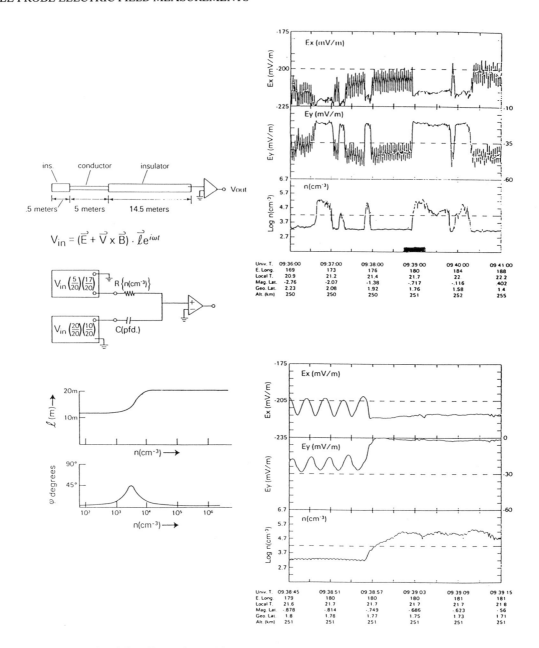

Figure 5. An example of the effects of capacitive coupling of the large **v x B** electric field across the insulator on the San Marco satellite. At the left is a schematic explanation of the resulting effective length change and phase shift. The lower panel at the right is and expansion of the area with the black bar in the upper panel, and the sinusoidal variation is an artifact (see text). The data are in error whenever the density drops below 10^4 cm^{-3} [*Aggson et al.* 1992].

the large elongated rocket-body-payload-configuration as it spins. Note that the largest amplitude signal is from the I axis which had the shortest baseline. The downward spikes in the X and Y axes data are from shadowing of lower sensor from the Sun by the rocket body. The spikes at the middle of the wave in the I axis data going in both directions are from shadowing of both sensors as they spin behind the payload. Note that once the payload is separated, all of these error signals go away except the shadowing of the I axis sensors.

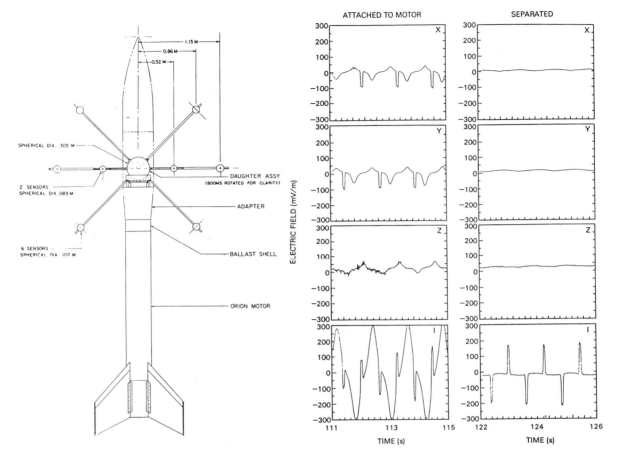

Figure 6. An example of sheath potentials affecting the measurements and the effects of shadowing of sensors. At the left is a schematic of the daughter payload attached to the rocket motor. The booms in a three-axis orthogonal configuration are rotated into the plane of the paper for clarity. Note that while the daughter is attached the central body is irregular and large compared to the length of the sensors. After the daughter is separated, the opposite is true. Data is shown just before and just after separation (see text) [*Maynard* 1986].

Data taken on the downleg of a flight just before dawn (Figure 7) illustrate the effects of not having long enough antennas. Three different effective sensor separation lengths were flown for spin plane measurements: 2.3 m for the Z axis, 1.7 m for the *X* and *Y* axes (2.3 m / cos 45°), and 1.0 m for the *I* axis. The amplitude variations of **E** + **v x B** as recorded for the *Z* axis are shown in the top left panel. The ratio of the *Y* to *Z* axes and *I* to *Z* axes are shown in the top center and top right panels. respectively. The unity ratio in the center panel down to 75 km shows that both axes are recording the same electric field while the deviation from unity in the right panel indicates that the I booms are too short, allowing sheath potentials to influence the result. The differences between the axes are plotted in the bottom three panels. The first two indicate agreement within less than 1 mV m⁻¹, while the I-Z difference shows 4 to 5 mV m⁻¹ errors in the D region. In a similar flight after sunrise, errors were detected higher - below 85 km. The higher altitude of the cut off of good measurements in sunlight is a result of photoemission from both the sensors and the support structures distorting both the sheath potentials and the current balance as the density decreases in the lower D region.

One of the easiest problems conceptually, and one of the most challenging, is the vector subtraction of the motional **v x B** electric field from the measured value to obtain the ambient electric field. Grouping dc offset errors into one term with equivalent electric field units which is generally referred to as contact potential (E_{c12}), it follows from Equation 1 that for an axis defined by sensors 1 and 2

$$(\Phi_1 - \Phi_2)/|d_{12}| = E_{m12} = [\mathbf{E} + \mathbf{v} \ \mathbf{x} \ \mathbf{B}] \bullet \mathbf{d_{12}} + E_{c12} \quad (4)$$

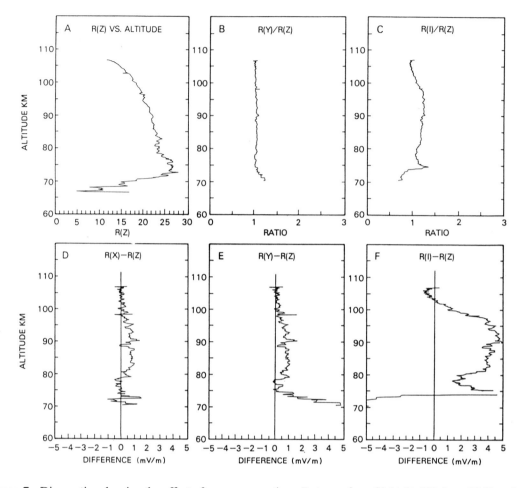

Figure 7. Diagnostics showing the effect of sensor separation. Data are from flight 31.039 from Wallops Island, VA, on November 19, 1983, at 1046 UT. The 1.7 m separation (X and Y axes) is shown to be adequate to make the measurement down to 75 km with this payload configuration, while the 1.0 m separation (I axis) data show the effect of sheath potentials (see text) [*Maynard* 1986].

For a three axis measurement similar equations exist for the other two axes. The problem is that each of these quantities is known in a different frame of reference. For instance the vehicle velocity is usually known in the inertial reference frame for spacecraft, but may be in the Earth's co-rotating frame for sounding rockets. Electric fields are usually expressed in the co-rotating frame in the ionosphere. The co-rotation velocity is used in a $\mathbf{v} \times \mathbf{B}$ correction to translate electric fields between the inertial and co-rotating reference frames. Attitude information of the orientation of the spacecraft generally has limited accuracy; however this is the primary information for finding a common system to do the vector subtraction. That information may be derived in the inertial frame or in a local co-rotating frame. Attitude accuracy of 0.2° is difficult. In the presence of a 450 mV m^{-1} $\mathbf{v} \times \mathbf{B}$ electric

field (typical for a polar orbiting satellite in the ionosphere traveling nearly 8 km/s across a 56,000 nT magnetic field) every 0.1° in attitude accuracy translates to an accuracy of 0.8 mV m^{-1} from the vector subtraction. The measurements are made in the frame of reference of the sensors. An additional problem comes from the knowledge of the position of the actual sensor axis in the satellite reference frame. Mechanical alignment of the sensors is never perfect. Thermal distortion of the sensor support system can introduce further non-orthogonalities in the actual measurement axes. To place all quantities in a common reference frame in the satellite system for the vector subtraction requires a non-orthogonal rotation of the measurements and a series of orthogonal transformations of the velocity vector from inertial space. If the measured magnetic field is used, that also must be rotated from the sensor axes into the satellite system. In the ionosphere where variations in the magnetic field rarely change the vector orientation by 0.1° from the average, it is often possible to use a calculated model magnetic field rotated into the satellite coordinates without a significant loss in accuracy.

Improving the accuracy of the attitude determination is the primary means to minimize $\mathbf{v} \times \mathbf{B}$ subtraction errors. Accurate mechanical alignment of the sensors coupled with alignment testing and calibration is also necessary. Long deployable antennas used in a three axis stabilized system must be calibrated for straightness. Thermal bending of these antennas or other support structures can be reduced by proper design. In the case of on board magnetic field measurements or magnetic aspect determination, calibration of the system in a controlled magnetic environment is also needed. Magnetic cleanliness of the payload minimizes errors.

The last term in the above equation, E_{c12} or contact potential, groups a number of current balance errors together into a dc offset term. In a spinning configuration, this is easily removed as the constant offset of the sinusoidal variation with spin phase. In a three-axis stabilized configuration the dc offset is not as easily determined. Sometimes it is possible to use unique features of the orbital variation in a region to extract information about the offset errors. In the DE-2 experiment the two axes were located in the orbit plane at ± 45° to the horizontal. The integrated potential along the spacecraft track (calculated from the horizontal electric field component) is very sensitive to constant offset errors. Assuming conditions do not change during the 20 minutes necessary for a polar pass, the integrated potential should be zero at each mid-latitude end of the polar pass. This assumption allowed accurate automatic adjustment of the contact potentials by changing the values for each axis in the same direction (as explained in *Hanson et al.*, [1993]). For instance a 2 mV m^{-1} adjustment of each contact potential changes the cross polar cap potential at the end point by 30 kV. In the slowly varying conditions across the magnetic equator, a more complete adjustment could be made by minimizing the resultant electric field along B [*Maynard et al.*, 1988].

In regions where the orientations of the sensors are stable (i. e., thermal bending is not variably changing the nonorthogonality of the system) and the contact potentials are constant, it is possible to analytically determine over an extended data sample the nonorthogonality correction matrix and the contact potentials for a three-axis stabilized system. Nine independent data samples are needed in which the orientation of the magnetic field and

the electric field significantly change. In the context of Equation 4 we can express the orthogonality correction as

$$E_{mi}^{\ k} - E_{ci} = \sum_j \Omega_{ij} [E_{aj}^{\ k} + (\mathbf{v} \times \mathbf{B})_j^{\ k}] \quad (5)$$

$$\sum_j \Omega_{ij}^{\ 2} = 1 \quad (6)$$

both for i = 1, 2, 3, and

$$\mathbf{E_a}^k \bullet \mathbf{B}^k = 0. \quad (7)$$

This is a system of 7 equations with 3 unknown components of E_a, 3 unknown contact potentials (E_c) and 9 unknown components of the nonorthogonality correction matrix (Ω_{ij}). Assuming that the contact potentials and the correction matrix remain the same, each independent data set (or k iteration) adds 4 equations but only 3 new unknown components of E_a. Nine independent data sets results in 39 equations and 39 unknowns which can be solved by numerical techniques. This technique was planned for regular use with the DE-2 data, but it could not be implemented because of the failure of one of the axes to deploy. A truncated version of this technique was used on a number of carefully picked data sets to determine an average nonorthogonality correction matrix that was applied to all DE-2 data. However, the contact potentials could not be analytically determined within the two axis system nor could the technique be routinely used.

All of the above discussion has assumed that the plasma conditions are identical at each sensor outside of the ambient potential variation. An additional error is possible from plasma gradients in density and temperature affecting the current balance. *Laakso et al.* [1995; 1997] have derived expressions for plasma gradient effects, finding them to be small when the plasma currents dominate over the photoemission currents, the ambient temperature is low, and the sensors float negative. This is typical of most ionospheric conditions. Plasma gradient errors are also small in very tenuous plasmas. In between in plasmas typical of the outer plasmasphere where the density drops below 10^2 cm^{-3}, the error can be significant. Here photoemission drives the sensors positive. The largest problems occur when the ambient temperature exceeds 100 V. Driving a bias current (its magnitude is not so important) to the sensors reduces the problem significantly. These studies neglected the effects of space charge. *Diebold et al.* [1994] determined that the maximum space-charge-enhanced plasma-gradient error is a function of the escaping photoelectron current. Reducing the total photoemission current as well as lengthening the baseline reduces the maximum error when space charge is considered. However, lengthening the baseline increases the chances that significant gradients can be encountered.

Temperature gradient effects are very small in tenuous plasmas for biased probes but are relatively large for non-biased probes [*Laakso et al.*, 1997]. The largest errors from temperature gradients occur when the probes are floating negative relative to the ambient plasma and in a high temperature plasma. For the typical ionospheric plasma with temperatures less than 1 eV, temperature gradient effects are small [*Laakso et al.*, 1997]. Under conditions where the large, high-temperature auroral fluxes could dominate the low temperature ambient environment and spacecraft charging could occur (as has occasionally been observed on the DMSP satellite [*Gussenhoven et al.* 1985]), measurements could be affected by the temperature gradients. Note that DMSP is at 840 km where the ambient density is significantly less than the F region peak, making it possible for extremely intense auroral fluxes to dominate and charge the spacecraft. Under these conditions, preamplifiers are generally driven out of their range, making measurements impossible because of limitations of the electronics. Two such instances in the lifetime of DE-2 were observed (DE-2 spent a large percentage of its time at lower altitudes).

Passive double-probe instruments have been extensively and successfully used in the ionosphere and even out into the inner magnetosphere. In the tenuous and more energetic plasmas of the magnetosphere the addition of bias currents to the sensors improves the accuracy, overcoming errors that would be fatal to dc measurements with passive probes (see *Pedersen* [1997]). Figure 8 shows a sample of data from the CRRES satellite in the near-Earth magnetosphere at L = 4 and a density of 10^2 cm^{-3} on which both spherical and cylindrical sensors were flown [*Wygant et al.*, 1992]. Both types of sensors were biased. The agreement of the two axes, which have different baselines and different sensors, to within 5% testifies that the errors have been minimized. Biasing the cylindrical sensors, which were similar to those on ISEE-1, extended the range of valid measurements to regions where the ISEE-1 passive double probe measurements could not be used because of the errors discussed above.

DESIGN TRADE-OFF CONSIDERATIONS

The previous two sections have discussed the basic double probe technique and sources of errors. Designing an electric field experiment requires a number of factors or properties of the instrument and the vehicle to be considered. Many factors may be predetermined; however others may be up for choice. The choices must be melded

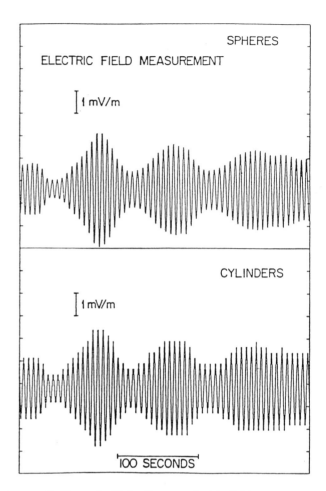

Figure 8. Example of simultaneous electric field measurements from spherical sensors and cylindrical sensors on the CRRES satellite [*Wygant et al.*, 1992].

with the mitigation techniques for reducing errors in order to maximize the accuracy of the measurements. Tables 2 and 3 attempt to focus these considerations by listing a number of properties that must be considered, the common choices, and the principal advantages and disadvantages (or concerns) of each choice. Table 2 concentrates on factors that affect spacecraft or vehicle design, while Table 3 is in general more instrument related. Both are intertwined and a decision in one area may limit choices in other areas. Specific criteria are not provided, since many combinations of choices are possible and criteria change dependent on the combination. The advantages and disadvantages of each choice, provide pointers back into the discussion of the technique and errors. Further elaboration can be found in the given references.

In considering instrument choices, it is necessary to remember that extendible elements of all kinds

Table 2. Passive Double-Probe Electric Field Measurements - Trade-offs: Spacecraft

Property/Choice	Advantages	Disadvantages/Concerns
Central body		
Long cylinder	Natural shape of rockets	Larger distortion of plasma
	Stable - large moment of inertia ratio	Deployables decrease stability
Large, short cylinder	Natural configuration of satellites	Needs longer antennas
	Stable - large moment of inertia ratio	
Small spherical	Minimizes distortion of plasma	Difficult to instrument
	Minimizes wake errors	Stability must come from antennas
Stabilization		
Spinning	Accuracy - easy removal of offsets	Spin axis component difficult
	Long antenna lengths easier	
Non-spinning	Time history clearer	Ambiguity in determining dc offsets
	Interpretation of ac E easier	Requires self-supporting antennas
Attitude determination		
Accuracy requirement		Large $\mathbf{v} \times \mathbf{B} \Rightarrow$ increased accuracy
		$300 mVm^{-1} \Rightarrow 0.1°$ for $\pm 0.5 mVm^{-1}$
Translation from sensor		Mechanical straightness
		Mechanical alignment
		Thermal bending
Attitude sensor types		
(need two vector directions)		
Gyro	Good accuracy, 2 vectors	Drift limits duration of use
Star	Excellent accuracy	Difficult except at night
Solar	Good accuracy	Daytime
Horizon	Moderate accuracy	Emission layers vary in height
Magnetic	Fair accuracy	$\mathbf{B}$ varies with activity

significantly affect spacecraft design in ways that are not immediately obvious at first glance. Because of large increases in moments of inertia, the spacecraft stabilization systems must be significantly more robust. Spacecraft maneuvers, sometimes needed for thermal control become more difficult. Magnetic and electrostatic cleanliness procedures necessary for good measurements also drive spacecraft design and costs. Electric field experimenters are viewed warily by spacecraft project managers. It is important to understand spacecraft constraints and the impacts of design decisions on those constraints. Mitigation techniques for error avoidance can be expensive.

SUMMARY

Double-floating probes are a viable technique for accurate electric field measurements for both ac and dc electric fields. Passive double-probe instruments have been successfully used in the D, E and F regions of the ionosphere and in the inner magnetosphere on both satellites and sounding rockets. The concept is simple; however, the plasma environment presents a number of pitfalls that affect the accuracy. Careful attention to the design of both the instrument and the spacecraft can minimize or mitigate most of the errors. The choices of sensor properties, antenna baselines, spacecraft stabilization, and attitude determination must be made based on the plasma characteristics in region of the measurements, mission constraints, and a factor not explicitly discussed, but always in the background, cost. It is relatively easy to make accurate measurements at densities $\geq 10^3$ cm^{-3}. Extending down to densities $\geq 10^2$ cm^{-3} requires careful consideration of sources of error. At densities below 10^2 cm^{-3} passive double probes will provide accurate measurements only in low temperature plasmas. Active biasing alleviates this problem.

Table 3. Passive Double-Probe Electric Fields - Trade-offs: Instrument

Property/Choice	Advantages	Disadvantages/Concerns
Antenna type		
Hinged boom	Simplicity + low cost	Limited length
Extendible - tubular	Short to intermediate lengths	Thermal bending
		Large MI for intermediate lengths
Extendible - wire	Very long lengths possible (200 m t-t)	Spin plane deployment only
		Large MI for long lengths
Antenna length		
Long	Removes sensor away from sheath	ac baseline varies with wavelength
	Better signal to noise	
Short	Clear baseline at high frequencies	Sheath errors
		Lower signal to noise
Sensor type		
Cylindrical	Longer lengths with less impact	Photoemission varies with angle to Sun
Spherical	"Point" measurement	Higher moment of inertia
	Symmetry relative to Sun	More complicated to make and deploy
Sensor surface		
Noble metal	Lower work function,	Higher photoemission
	Thermal design easier	Sensitive to handling
Plain metal	Symmetry by uniformly dirty	Higher photoemission
Graphite coated	Lower photoemission,	Sensitive to handling
	Lower dc offsets	Care with bonding
Sensor preamp location		
In sensors	Better control of frequency response	More complicated deployer
	Easiest for spherical sensors	
At base of deployer	Simplicity	Insulate boom inside of sensor
	Easiest for cylindrical sensors	Capacitive coupling across insulator
Sensor axis orientation		
Perpendicular to Sun line	Direction of minimum sheath extent	
	Eliminates shadowing	
Perpendicular to B	Elimates magnetic wake errors	
Perpendicular to velocity	Elimates velocity wake errors	

Acknowledgments. I especially thank Drs. T. L. Aggson and J. P. Heppner who pioneered the development of the passive double probe technique and have worked with me through many experiments. Drs. F. S. Mozer and A. Pedersen have lead the extension of the technique into tenuous plasmas with the concept of biasing of the sensors. Sensor surface properties have been researched by Drs. A. Pedersen and R. J. L. Grard at ESTEC and by W. Viehmann at NASA/GSFC. Contributions to understanding the dynamics of extendible mechanisms have been made by Drs. J. Fedor and H. Hoffman of NASA/GSFC. Many others, too numerous to mention have supported or helped the development of this technique and associated hardware. This compilation was supported in part by the Air Force Office of Scientific Research task 2310 under a contract with Phillips Laboratory.

REFERENCES

Aggson, T. L., Measurements of electric fields in space, in *Atmospheric Emissions*, edited by B. M. McCormac and A. Omholt, p. 305, Van Norstrand Reinhold, New York, 1969.

Aggson, T. L., and J. P. Heppner, A proposal for electric field measurements on the ATS-4 satellite, submitted to NASA, August 1964.

Aggson, T. L., P. J. Gambardella, and N. C. Maynard, Electric Field Measurements at the Magnetopause: 1. Observation of Large Convection Velocities at Rotational Magnetopause Discontinuities, *J. Geophys. Res., 88*, 10,000, 1983.

Aggson, T. L., N. C. Maynard, W. B. Hanson, and J. L. Saba, Electric Field Observations of Equatorial Bubbles, *J. Geophys. Res., 97*, 2997, 1992.

Bering, E. A., A sounding rocket observation of an apparent wake generated parallel electric field, *J. Geophys. Res., 88*, 961, 1983.

Cauffman, D. P., and N. C. Maynard, A model of the effect of the satellite photosheath on a double floating probe system, *J. Geophys. Res., 79*, 2427, 1974.

Diebold, D. A., N. Hershkowitz, J. R. DeKock, T. P. Intrator, S.-G. Lee, and M.-K. Hsieh, Space charge enhanced, plasma gradient induced error in satellite electric field measurements, *J. Geophys. Res., 99*, 449, 1994.

Fahleson, U., Theory of electric field measurements conducted in the magnetosphere with electric field probes, *Space Sci. Rev., 7*, 238, 1967.

Feng, W., D. A. Gurnett, and I. H. Cairns, Interference patterns in the Spacelab 2 plasma wave data: Oblique electrostatic waves generated by the electron beam, *J. Geophys. Res., 97*, 17,005, 1992.

Gussenhoven, M. S., D. A. Hardy, F. Rich, W. J. Burke, and H.-C. Yeh, High-level spacecraft charging in the low-altitude polar auroral environment, *J. Geophys. Res., 90*, 11,009, 1985.

Hanson, W. B., W. R. Coley, R. A. Heelis, N. C. Maynard, and T. L. Aggson, A comparison of in situ measurements of **E** and **-v x B** from Dynamics Explorer 2, *J. Geophys. Res., 98*, 21,501, 1993.

Heppner, J. P., E. A. Bielecki, T. L. Aggson and N. C. Maynard, Instrumentation for dc and low frequency electric field measurements on ISEE-A, *IEEE Trans. on Geosci. Electronics, GE-16*, 253, 1978.

Laakso, H., T. Aggson, and R. Pfaff, Plasma gradient effects on double probe measurements in the magnetosphere, *Ann. Geophys., 13*, 130, 1995.

Laakso, H.,T. L. Aggson , and R. F. Pfaff, Plasma gradient effects on double probe measurements, this issue, 1997.

Maynard, N. C., Measurement techniques for middle atmosphere electric fields, *Handbook for MAP, 19*, edited by R. A. Goldberg, p. 188, SCOSTEP Secretariat, Urbana, Ill, 1986.

Maynard, N. C., T. L. Aggson, F. A. Herrero, and M. C. Liebrecht, Average Low-Latitude Meridional Electric Fields from DE-2 During Solar Maximum, *J. Geophys. Res., 93*, 4021, 1988.

Maynard, N. C., T. L. Aggson, and J. P. Heppner, The plasmaspheric electric field as measured by ISEE-1", *J. Geophys. Res., 88*, 3991, 1983.

Maynard, N. C., J. P. Heppner, and A. Egeland, Intense, variable electric fields at ionospheric altitudes in the high latitude regions as observed by DE-2", *Geophys. Res. Lett., 9*, 981, 1982.

Maynard, N. C., E. A. Bielecki, and H. F. Burdick, Instrumentation for vector electric field measurements from DE-B, *Space Sci. Instr., 5*, 523, 1981.

Maynard, N. C., T. L. Aggson, and J. P. Heppner, Electric field observations of ionospheric whistlers, *Radio Sci., 5*, 1049, 1970.

Mott-Smith, H. M., and I. Langmuir, The theory of collectors in gaseous discharges, *Phys. Rev., 28*, 727, 1926.

Pedersen, A., C. A. Cattell, C.-G. F H{a}lthammar, V. Formisano, P.-A. Lindqvist, F. Mozer, and R. Torbert, Quasistatic electric field measurements with spherical double probes on the GEOS and ISEE satellites, *Space Sci. Rev., 11*, 77, 1884.

Scudder, J. D., A. Mangeney, C. Lacombe, C. C. Harvey, and T. L. Aggson, The resolved layer of a collisionless, high β, supercritical, quasi-perpendicular shock wave: 2. Dissipative fluid electrodynamics, *J. Geophys. Res., 91*, 11,053, 1986.

Shawhan, S. D., G. B. Murphy, and J. S. Pickett, Plasma diagnostics package initial assessment of the shuttle orbiter plasma environment, *J. Spacecr. Rockets, 21*, 387, 1984.

Temerin, M., Doppler shift effects on double-probe measured electric field power spectra, *J. Geophys. Res., 84*, 5929, 1979.

Wygant, J. R., P. R. Harvey, F. S. Mozer, N. C. Maynard, H. Singer, M. Smiddy, W. Sullivan, and P. Anderson, The CRRES electric field/Langmuir probe instrument, *J. Spacecraft and Rockets, 29*, 601, 1992.

Nelson C. Maynard, Mission Research Corporation, One Tara Boulevard, Suite 302, Nashua, NH 03062. (e-mail: maynard@zircon.plh.af.mil)}

The Electron Drift Technique for Measuring Electric and Magnetic Fields

G. Paschmann[1], C. E. McIlwain[2], J. M. Quinn[3,4], R. B. Torbert[4], and E. C. Whipple[5]

The electron drift technique is based on sensing the drift of a weak beam of test electrons that is caused by electric fields and/or gradients in the magnetic field. These quantities can, by use of different electron energies, in principle be determined separately. Depending on the ratio of drift speed to magnetic field strength, the drift velocity can be determined either from the two emission directions that cause the electrons to gyrate back to detectors placed some distance from the emitting guns, or from measurements of the time of flight of the electrons. As a by-product of the time-of-flight measurements, the magnetic field strength is also determined. The paper describes strengths and weaknesses of the method as well as technical constraints.

1. INTRODUCTION

The electric field is an essential quantity in space plasmas, yet it is one of the most difficult to measure. This is because in many important circumstances the electric fields are very small (less than 1 mV/m) and the plasma is very dilute. Under such circumstances it is often difficult for the conventional double-probe technique to distinguish natural fields from those induced by spacecraft wakes, photoelectrons, and sheaths. The electron drift technique has been developed to check and complement the double-probe technique. The drift method involves sensing the drift of a weak beam of test electrons emitted from small guns mounted on the spacecraft. This drift is related to the electric field, but gradients in the magnetic field can contribute to the drift. Comparing the drifts at different electron energies, the electric and magnetic drifts can be separated.

When emitted in the proper directions, the electron beam returns to dedicated detectors on the spacecraft after one or more gyrations. During these gyrations, the beam probes the ambient electric field at a distance from the spacecraft that for sufficiently small magnetic fields is essentially outside the latter's influence. In this paper we describe the basis of the method and the constraints imposed by the magnetic and electric field strengths to be encountered. We emphasize the criteria that led to the design of the Electron Drift Instrument (EDI) for the Cluster mission. EDI employs two electron guns, each of which can be aimed electronically in any direction over more than a hemisphere. A servo loop continuously re-aims the electron guns so that the beams return to dedicated detectors. The electron drift can be calculated by triangulation of the two emission directions. For small magnetic fields, the triangulation method becomes inaccurate, and the drift will instead be calculated from the measured differences in the time of flight of the electrons in the two nearly oppositely

[1]Max-Planck-Institut für extraterrestrische Physik, 85740 Garching, Germany

[2]University of California at San Diego, La Jolla, CA 94304, USA

[3]Lockheed Space Science Laboratory, Palo Alto, CA 92093, USA

[4]University of New Hampshire, Durham, NH 03824, USA

[5]University of Washington, Seattle, WA 98195, USA

Measurement Techniques in Space Plasmas: Fields
Geophysical Monograph 103
Copyright 1998 by the American Geophysical Union

directed beams. The time-of-flight measurements also yield an accurate determination of the magnetic field strength.

The electron drift technique has a number of limitations. First, performance is strongly affected by the magnitudes of the fluxes of returning beam electrons and of ambient electrons. Second, measurements will be interrupted whenever the beam is strongly scattered by instabilities or interactions with ambient fluctuations. Third, beam tracking will be interrupted by very rapid changes in either the magnetic or the electric field. Fourth, accurate separation of the electric and magnetic components of the drift may not always be possible with only a limited range of electron energies.

2. PRINCIPLE OF OPERATION

2.1. Drift Velocity from Beam Direction Measurements

The basis of the electron drift technique is the injection of test electrons and the registration of their gyrocenter displacements after one or more gyrations in the magnetic field, **B**. The displacement, **d**, referred to as the drift step, is related to the drift velocity, $\mathbf{v}_D$, by:

$$\mathbf{d} = \mathbf{v}_D \cdot N \cdot T_g, \tag{1}$$

where T_g is the gyroperiod and N denotes the number of such periods after which the electrons are captured. If the drift is solely due to an electric field, $\mathbf{E}_\perp$, transverse to **B**, then (using MKSA units)

$$\mathbf{d} = \frac{\mathbf{E} \times \mathbf{B}}{B^2} \cdot N \cdot T_g. \tag{2}$$

Or, numerically, for $N = 1$

$$d\,(\mathrm{m}) = 3.57 \times 10^4 \, \frac{E_\perp \,(\mathrm{mV/m})}{B^2 \,(\mathrm{nT})}. \tag{3}$$

Values of the drift step d as a function of magnetic field strength and drift velocity are shown in Figure 1.

The B^{-2} scaling implies that for a given electric field (1 mV/m, say), the drift step varies between 0.06 mm at low altitudes (25000 nT), and 1428 m in the solar wind or the central plasma sheet (5 nT), i.e., by a factor of $2.5 \cdot 10^7$. For small drift steps, the electrons gyrate nearly back to their origin and can be intercepted by a detector essentially collocated with the electron source. This is the scheme chosen for the Freja mission and described in another article in this monograph [Kletzing et al., 1997]. The first application of the electron-drift technique was designed for the few-hundred nT fields

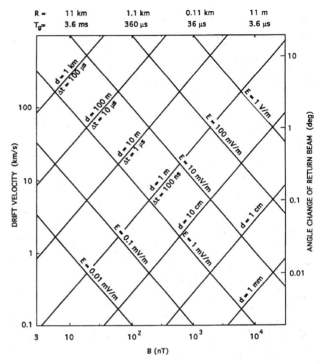

Figure 1. Magnitudes of the quantities directly measured by the electron drift technique, in terms of lines of constant drift step d and time-of-flight difference Δt, as a function of magnetic field strength and electron drift velocity, for 1 keV electrons. Lines of constant electric field are also provided. The electron gyroradius R and gyrotime T_g are indicated along the top, the angle change of the return beam along the vertical axis on the right.

at synchronous altitude [Melzner et al., 1978] where the drift step can become much larger than the spacecraft dimensions. This is even more true for the Cluster (and Phoenix) missions where magnetic fields range from <1000 nT at perigee to only a few nT at apogee. These large drift steps require a totally different measurement concept, as discussed in the next subsection.

That electrons emitted by an electron gun mounted on a spacecraft can gyrate back to a detector on the same spacecraft, even if the drift step is much larger than the spacecraft dimensions, can be understood in two ways. Consider first the electron motion in a moving frame where there is no electric field (Figure 2). In this frame all electron trajectories are circles and return to the origin regardless of their emission direction. The spacecraft, on the other hand, now moves with the electron drift speed along a straight line that intersects the possible electron orbits in varying phases of their gyration. Now it is easy to see which electrons will hit the spacecraft: those that arrive at the intersec-

ELECTRON DRIFT TECHNIQUE
IN CO-MOVING FRAME

(E=0)

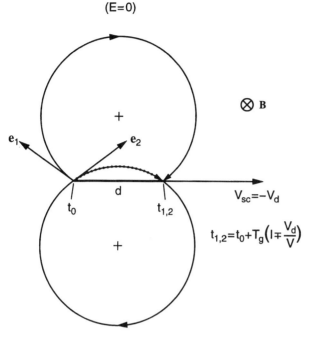

$$t_{1,2}=t_0+T_g\left(1\mp\frac{V_d}{V}\right)$$

Figure 2. Principle of drift step determination in a coordinate system where there is no electric field. The arrows denoted e_1 and e_2 are the two gun pointing directions that cause the beams to hit the detector when it has moved to the point denoted $t_{1,2}$.

tion of their orbits with the spacecraft trajectory at the same time the spacecraft does. Two trajectories meet this requirement. To first order, these are the two circles (shown in the figure) whose secant is the distance (equal to the drift step) that the spacecraft moves in a gyroperiod. This is not quite correct; the upper trajectory has completed less than a full circle when it intersects the spacecraft trajectory, thus needing less than a gyroperiod, while the lower one needs more than a full gyroperiod. As a consequence, the true solutions are circles with emission directions that differ slightly from those shown, intersecting the spacecraft trajectory slightly left and right, respectively, of the single intersection shown in the figure. Note that the figure is for an unrealistically high ratio of drift step to gyroradius, and thus grossly exaggerates the difference between the two trajectories. For realistic ratios, the two have more nearly equal flight times, and emission directions that are closer to 180° apart. For the realistic case, the beam return directions are also more nearly parallel to the emission directions than shown in Figure 2.

While the co-moving frame is useful for explaining why there always are two trajectories that hit the space-

craft, considerations of the effects of the actual electron gun and detector geometry requires treatment in the spacecraft frame. For this we turn to Figure 3. First one notes that all electrons emitted from a common source S in a plane normal to $\mathbf{B}$ are focussed after one gyration onto a single point that is displaced from S by the drift step, $\mathbf{d}$. The variability and size of the drift step makes it impossible to have an electron gun at S, and at the same time a detector D at the focus. But one does not really need a gun at S: a beam from a gun at an arbitrary location will also hit the detector at D, provided the beam is directed towards or away from S. In the first case the gun can be thought of as supplying electrons to the source at S, in the second the gun furbishes electrons emanating from the source. If two guns are used, as shown in the figure, determination of the beam emission directions that return a beam onto the detector yields the displacement, $\mathbf{d}$, and thus the drift velocity, $\mathbf{v}_D$. This is a classical triangulation problem.

Of course, noting that the position of S is constantly changing in response to the varying electric and magnetic fields, finding the direction from each gun to the

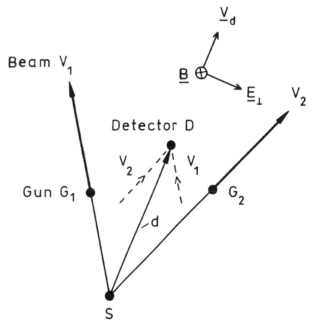

Figure 3. Principle of drift step triangulation in the spacecraft frame. If as a result of the drift all electrons emitted from S reach the detector location D after one gyration, then beams V_1 and V_2 directed along lines through S will also strike the detector. The drift step $\mathbf{d}$ is therefore the vector from the intersection point of the two beam directions to the detector D, once the beams are steered such that an 'echo' is received by the detector. Trajectories are straight lines in this figure because the drift step is assumed small compared with the gyroradius.

'target' at S (or away from it), and then keeping the beams on-target requires an active control of the beam pointing based on information from the detector.

As is true for any triangulation problem, the size of the baseline, b, and the beam pointing uncertainty determine the accuracy with which the displacement, d, can be measured. If the pointing accuracy of either beam is $\delta\alpha$, the error in the drift step is

$$\delta d = 2b \left(1 + \frac{d^2}{b^2}\right) \delta\alpha \qquad (4)$$

The baseline is defined as the distance transverse to $\mathbf{v}_D$ between a gun and its associated detector when projected into the plane perpendicular to $\mathbf{B}$. The maximum baseline is determined by the spacecraft diameter which in the case of Cluster is 3 m. As discussed below, for the gun-detector configuration used for the Electron Drift Instrument on Cluster, the effective baseline is actually twice as large, i.e., 6 m. Assuming a pointing accuracy $\delta\alpha$ of 0.5°, accuracies are better than 15% for drift steps <100 m. To determine larger drift steps one has to rely on measurements of the electron time of flight that are discussed in the next section.

The two beams return to the detector generally with substantially different directions. As detectors cannot view both these directions at the same time, one therefore needs a dedicated detector for each beam. Ideally, guns and detectors should not be co-planar, but rather form a tetrahedron. Otherwise there will be situations where the triangulation baseline vanishes, i.e., when $\mathbf{B}$ and $\mathbf{v}_D$ are in the gun/detector plane. Technical constraints ruled out such a solution on Cluster, and one gun and one detector were therefore combined in a single unit and two such gun-detector units mounted on opposite sides of the spacecraft. As a consequence, the triangulation baseline will vanish each time the projection of the two packages in the plane perpendicular to $\mathbf{B}$ is aligned with $\mathbf{v}_D$. Even though this will cause a spin-modulation of the accuracy with which the drift step is triangulated, the electron guns will stay on track. Note that in the worst case of a spin axis perpendicular to both $\mathbf{B}$ and $\mathbf{v}_D$, the baseline is always zero. Note also that with two detectors the triangulation scheme is modified as shown at the top of Figure 4, and becomes equivalent to one where the baseline is twice as large (bottom part of Figure 4).

With regard to Figures 3 and 4, note that when guns outside of S are used, D is no longer a focal point of the beams, nor do the travel times precisely equal the gyrotime, T_g. If the beam is directed towards (away from) the target S, the travel time will be longer (shorter) than T_g. This difference is the topic of the next subsec-

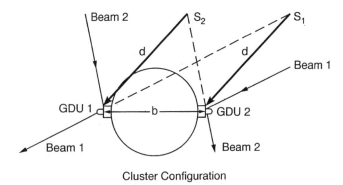

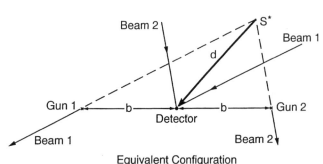

Figure 4. Triangulation scheme for two gun/detector units, GDU1 and GDU2, located on opposite sides of the spacecraft. S_1 and S_2 are the virtual source points for the two detectors (top). The problem is equivalent to one where a single detector is placed inbetween two guns at a distance b from each gun, and the virtual source becomes a single point, S^* (bottom). Note that the effective baseline is doubled this way.

tion. Note also that electron trajectories are straight lines only if, as assumed in Figures 3 and 4, the drift step is small compared to the gyroradius. In the same approximation, the returning beams are parallel to the outgoing ones. Both approximations are no longer true for large drift steps. In this case the curvature of the electron orbits must be taken into account. The resulting effects were already discussed in conjunction with Figure 2.

Figure 5 shows the electron orbits in the spacecraft frame when one beam is directed towards the target, the other away from it. The differences in beam emission and arriving directions are clearly visible, as are the different trajectory shapes that imply differences in electron time of flight. The effects are highly exaggerated because the figure is drawn to scale for the case of a gyroradius of 9 m and a drift step of 3 m that imply magnetic and electric fields of 11.8 μT and 12 V/m, respectively.

The angle between outgoing and returning beam (in radians) is given by

Cluster EDI

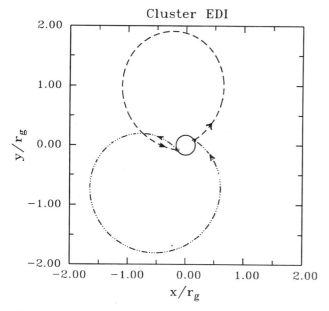

Figure 5. Electron orbits that return to detectors on the opposite side of a 3 m-diameter spacecraft, for a drift step of 4.5 m and a gyroradius of 9 m. These parameters were chosen so that spacecraft and orbits could be drawn to scale, but in fact are for unrealistically large magnetic and electric fields, namely 11.8 μT and 12 V/m, respectively. The magnetic field is pointing out of the page and the drift velocity is directed along X. The X and Y axes are scaled in gyroradii r_g.

$$\delta \approx 2\pi \frac{v_D}{v}, \qquad (5)$$

where v is the electron speed which for 1 keV electrons is 18742 km/s. For drift speeds of 100 km/s and 1 keV electrons, δ is 1.9°, increasing to almost 10° at 500 km/s. This effect can complicate operation, because if δ exceeds the width of the detector angular acceptance, the detector viewing direction must be offset from the direction anti-parallel to the gun firing direction by an amount dependent on the quantity to be measured.

Before turning to the time-of-flight measurements, we should mention that in the original application of the electron-drift technique on the Geos spacecraft, a much simpler triangulation scheme was used that required only a single gun and no active control, at the expense of providing only a single drift-step measurement per spacecraft revolution. The Geos scheme is illustrated in Figure 6. Electrons emitted in a fixed direction are displaced by $S = d \sin \alpha$ after they have gyrated once, where α is the angle between beam and drift directions. Spacecraft spin causes α to cover the full range between 0° and 360° twice per spin. When α

is such that $S = a$, where a is the gun-detector spacing, the beam will hit the detector. This condition is met twice per spacecraft spin. From the spin-phases when this occurs, the drift direction and magnitude can be reconstructed. The analysis assumes that the drift stays constant over times the order of the satellite spin period which for Geos was 6 s.

2.2. Drift Velocities from Time-of-Flight Measurements

As already illustrated in the previous section, the electrons in the two beams returning to the detectors travel different path lengths. As a result their flight times differ by an amount given by

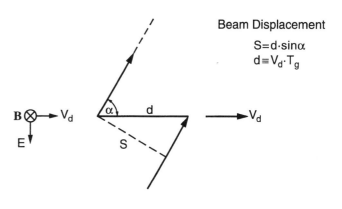

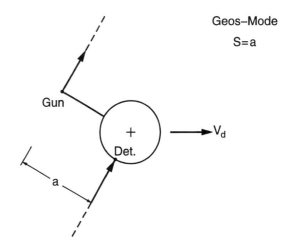

Figure 6. Beam displacement S as a function of the angle α between beam emission and drift velocity directions (top), and utilization of the variation of α with spacecraft rotation until S matches the gun-detector separation a, in the GEOS application of the electron drift technique (bottom).

$$\Delta T = T_{\text{to}} - T_{\text{aw}} = 2 \cdot T_g \cdot \frac{v_D}{v} \propto \frac{d}{v}, \qquad (6)$$

where T_{to} and T_{aw} are the flight times for the beam electrons aimed towards and away from the target, respectively. The idea to use electron times-of-flight to obtain the drift velocity is due to *Tsuruda et al.* [1985]. It has been successfully applied in the 'Boomerang' instrument on Geotail [*Tsuruda et al.,* 1994].

In the limit of very large drift steps, as depicted in Figure 7, the towards (away) beams are directed essentially anti-parallel (parallel) to the drift velocity. For simplicity it has been assumed here that guns and detectors are collocated, or in other words that the drift step is large compared to gun-detector separations.

The gyroperiod itself is obtained from the mean of the travel times:

$$T_g = \frac{T_{\text{to}} + T_{\text{aw}}}{2} \qquad (7)$$

Measuring T_{to} and T_{aw} permits determination of $\mathbf{v}_D$. ΔT scales directly as d/v. Hence, while the triangulation becomes increasingly less accurate, the time-of-flight method becomes more accurate with increasing drift step d, limited only by detector signal-to-noise effects. As shown in Figure 1, ΔT is many μs for those regions where the triangulation method starts to fail. Appropriate pulse-coding of the beams makes it possible to measure the time of flight of the electrons with a resolution better than $1\,\mu$s.

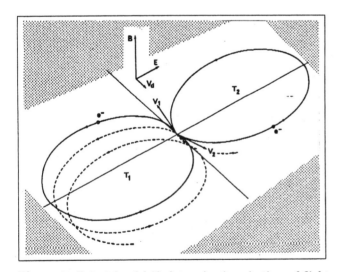

Figure 7. Principle of drift determination via time-of-flight measurements. Electrons emitted in a direction, v_1, opposing the drift, v_d, travel a longer path, and thus have a longer gyrotime, T_1, than electrons emitted along the drift, which take T_2 (from *Tsuruda et al.,* 1985).

2.3. Measurement of B

As discussed in the previous subsection, measurement of electron times of flight yields their gyroperiod, T_g, which in turn yields the magnetic field strength, B, via

$$T_g = \frac{2\pi m}{eB}. \qquad (8)$$

Values of T_g range from about 0.1 to 10 ms (see Figure 1) and are thus easily measurable. As a result, the magnetic field strength can be determined with very high accuracy, as demonstrated on Geotail [*Tsuruda et al.,* 1994].

2.4. Separation of Electric and Magnetic Gradient Drift

The beam electrons are subject not only to electric field drifts, but also to drifts caused by magnetic field gradients, $\nabla_\perp B$, directed perpendicular to the magnetic field. When the scale-length, ℓ, of such inhomogeneities becomes small (as it does at the bow shock, the magnetopause, or at the edges or the center of the plasma sheet), the gradient drift will make a significant contribution to the test electrons' displacement. The ratio of this drift, v_B, to that caused by the transverse electric field in the spacecraft frame of reference, v_E, is

$$\frac{v_B}{v_E} = 10^3 \, \frac{W_e \,(\text{keV})}{E_\perp \,(\text{mV/m})} \, \ell^{-1} \,(\text{km}), \qquad (9)$$

where W_e is the energy of the electron. For 1 keV test electrons and a field of $1\,\text{mV/m}$, v_B/v_E reaches unity if ℓ approaches 1000 km. The equation suggests that to separate the two drifts, one should employ electrons at different energies.

When the total drift is measured at two energies, W_1 and W_2, with $r = W_2/W_1$, the electric and magnetic drifts are obtained from the following expressions:

$$\mathbf{v}_E = \frac{(r\mathbf{v}_1 - \mathbf{v}_2)}{(r - 1)} \qquad (10)$$

$$\mathbf{v}_B(W1) = \frac{(\mathbf{v}_2 - \mathbf{v}_1)}{(r - 1)} \qquad (11)$$

where $\mathbf{v}_1$ and $\mathbf{v}_2$ are the (total) drifts that are inferred from the triangulation analysis applied to the two measurements at energies W_1 and W_2, respectively.

When the time-of-flight measurement technique is used in the presence of a significant gradient in $\mathbf{B}$, the analysis is more complicated. Since the gyrotime is defined in terms of the magnetic field at the center of the gyro circle, the two beams fired parallel and anti-parallel to the drift direction (Figure 7) will have different gyrotimes. If there were only the $\nabla_\perp B$ drift (i.e., no electric field), then the gyrotimes are given by:

$$T_g = T_0 \left(1 + \frac{R_g \sin \Phi_0}{\ell} \right) \qquad (12)$$

For the anti-parallel beam, $\Phi_0 = \pi/2$ and for the parallel beam $\Phi_0 = -\pi/2$, if one assumes that the gradient in $\mathbf{B}$ is in the same direction as $\mathbf{E}$ in Figure 7. Here T_0 is the gyrotime as given by the magnetic field at the spacecraft and R_g the corresponding gyroradius. The drift velocity and drift step are:

$$v_B = v_{B0} \left(1 + \frac{2R_g \sin \Phi_0}{\ell} \right) \qquad (13)$$

$$d_B = d_{B0} \left(1 + \frac{3R_g \sin \Phi_0}{\ell} \right) \qquad (14)$$

where again v_{B0} and d_{B0} are the values in terms of the magnetic field at the spacecraft. Use of the measured times now results in:

$$T_{\text{to}} + T_{\text{aw}} = 2T_0 \qquad (15)$$

as before, but

$$\Delta T = T_{\text{to}} - T_{\text{aw}} = 6T_0 \frac{v_{B0}}{v_0}. \qquad (16)$$

When both an electric field and a gradient in the magnetic field are present it can be shown that the difference in measured times for two beams fired anti-parallel and parallel to the net drift direction is given to leading order by:

$$\Delta T = T_0 \left(4 \frac{v_{B0}}{v_0} \sin \delta\Phi + \frac{2}{v_0} |\mathbf{v}_E + \mathbf{v}_{B0}| \right) \qquad (17)$$

Here, $\delta\Phi$ is the starting angle of the anti-parallel beam with respect to the direction of the magnetic field gradient.

In general there are four unknown quantities: the magnitude and direction (or equivalently, the x and y components) of both $\mathbf{v}_B$ and $\mathbf{v}_E$ in the plane perpendicular to $\mathbf{B}$. When two energies are used there will be six measured quantities: two net drift directions (one at each energy) and two pairs of flight times for the parallel and anti-parallel beams. However the flight times are not all independent since the sum of the pair for each energy is $2T_0$. Nevertheless the net drift directions and the differences in measured flight times for the two energies provide four relations that make it possible to obtain the four unknown drift components of $\mathbf{v}_B$ and $\mathbf{v}_E$.

3. CONCEPTUAL DESIGN CRITERIA

In this section we focus on the parameters that determine instrument conceptual design, with emphasis on the application for the Cluster mission. A technical description of the Electron Drift Instrument for Cluster has been published elsewhere [Paschmann et al., 1997].

3.1. Beam Return Fluxes

A key design criterion for the electron drift measurements is the magnitude of the flux of returning electrons incident on the detectors. This flux depends upon many factors, including the angular current distribution of the outgoing beam, the beam gyroradius, possible beam modification by electrostatic or wave-particle forces, and the geometrical arrangement of the gun and detector with respect to the drift step vector. The outgoing beam has an opening angle, β, of approximately $1°$. Thus the beam diverges along the magnetic field direction by a distance $s_\parallel = 2\pi R_g \beta / 57.3$, where R_g is the gyroradius, but is focussed in the plane perpendicular to $\mathbf{B}$ after one gyro orbit. By definition, this focus is located one 'drift step' from the gun. In general, those beam electrons with the proper firing direction encounter the detector either somewhat before, or somewhat after, this focus point. Thus the beam divergence leads to a large variation in the return beam flux with magnetic field strength and drift step. If gyro radius and drift step are as large as encountered on Cluster (see Figure 1), the detector intercepts only a small part of the emitted beam. Large detector sensitivity and particle counting techniques are necessary under such conditions. For the Freja application where gyro-radius and drift step are small, the return fluxes are large enough so that simple detection schemes are feasible [Kletzing et al., 1997].

Equation 18 gives the beam flux, F, in cm^{-2}s^{-1}, at the detector for an emitted beam with a flat current distribution over a square angular cross-section, where I is the gun current in nA, B the magnetic field strength in nT, W the beam energy in keV, x the distance from the gun's gyrofocus to the detector in m, β the beam full width parallel and perpendicular to $\mathbf{B}$ in degrees, and R_g the gyroradius in m.

$$F = 3.1 \cdot 10^3 \frac{IB}{x\beta^2 W^{1/2} (1 \pm x/2\pi R_g)} \qquad (18)$$

The $\pm$-term in the denominator accounts for divergence parallel to $\mathbf{B}$ between the gun's gyrofocus and the detector, depending upon whether the detector intercepts the beam before $(-)$ or after $(+)$ the gyrofocus.

Of course a square, uniform cross-section is not a realistic representation of the actual beam. However for that portion of the real beam that has the same angular current density I/β^2 as the uniform beam, the return flux at the detector would be the same.

Note that above equation is for electrons having gyrated once. For electrons having gyrated N times, the return flux scales as $1/N^2$.

3.2. Requirements on Gun and Detector Designs

For applications in weak magnetic fields, the electron drift measurements require electron guns capable of providing a beam that can be steered rapidly into any direction within more than a hemisphere. Electron energies must be variable in order to separate $\mathbf{E} \times \mathbf{B}$ and $\nabla_\perp B$ drifts. At the same time the energy dispersion must be small to restrict beam spreading in space and time. Beam currents must be variable between 0.1 nA and more than 1000 nA to account for the strong dependence of return fluxes on B and E. To maximize the return signal in the detectors and to provide accurate triangulation, the angular width of the beam must be kept small, but still large enough to account for uncertainties in pointing direction. The pointing accuracy must be 0.5° or better in order to extend the triangulation measurements to large enough drift steps that electron time-of-flight measurements can take over. The time-of-flight measurements in turn require that the beam be modulated and coded.

To detect the electrons from the two beams one needs two detectors. Each detector must be able to look in any direction within a region greater than a 2π steradian hemisphere. To follow the rapidly changing beam return directions, the detectors must be capable of changing their look directions in less than a millisecond. The detectors must have a large effective area, as much as two or three square centimeters, to guarantee adequate count rates under conditions when the return fluxes are weak. At the same time sensitivity to electron fluxes from the ambient plasma must be suppressed as much as possible to provide adequate signal-to-noise ratios. Unlike the ambient electrons, the returning beam is monoenergetic and unidirectional. Therefore, the signal-to-noise ratio is improved by designing the detector to be selective in velocity space. By changing the width of the accepted energy band and the focussing properties of the detector optics, one can choose different combinations of the sensitive area and angular acceptance width for beam electrons on the one hand, and the geometric factor for ambient electrons on the other hand. This is achieved by properly choosing the voltages on the electrodes in the detector optics. We refer to the different combinations as detector 'states'. Choice of the detector 'state' is to be based on beam fluxes and signal-to-noise ratio considerations.

3.3. Beam Recognition, Tracking and Coding

To detect the beam electrons in the presence of background counts from ambient electrons, and to measure their flight time, the electron beam is intensity modulated with a pseudo-noise code (PNC). The modulation frequency must reflect the expected time delays and the desired delay-time accuracy, and for EDI on Cluster can be chosen between 8 kHz and 4 MHz. The stream of electron event pulses received by the detectors are fed in parallel into an array of counters ('correlators'), each one gated with its individual copy of the PNC, shifted by one chip from counter to counter, and delayed as a whole (by a variable amount) against the PNC used to modulate the outgoing beam. The correlator that is controlled by the PNC matching the flight time will receive the most signal counts. For details concerning the correlator design, we refer the reader to the paper by Vaith et al. [1997].

We have experimented with several correlator schemes. The one presently implemented uses a 15-chip pseudo-noise code and 15 correlator channels. It guarantees that beam electrons are always counted in one of the channels, regardless of flight time. But it has the disadvantage that the flight time is determined only modulo the code duration and thus can be ambiguous. Electrons having gyrated more than once can also not be uniquely distinguished from the electrons having gyrated once. The ambiguity in flight time can usually be removed by starting out with a sufficiently low code-clock frequency such that the entire range of expected flight times fits within one code-length. This initial frequency is based on the gyrofrequency computed from the magnetometer data and an assumed 10% variation in flight-time to account for large electric fields.

In the standard mode of operation foreseen for the instrument on Cluster ('Windshield-Wiper Mode'), both beams are swept in the plane perpendicular to $\mathbf{B}$ in 0.25°-steps every ms. After every step, the signal-to-noise ratio is computed from the maximum and minimum counts in the set of 15 correlator channels associated with each detector. If that ratio exceeds some (selectable) limit, this is an indication that the beam is striking its detector. The angular sweep is continued until the signal is lost, i.e., the beam has completed its pass over the target. When this happens, the beam sweep direction for the gun in question is reversed and the process repeated. The same procedure is followed for the other gun. This way both beams are independently sweeping back and forth over the target. Beam currents and detector 'states' are continuously adjusted to maximize signal and/or signal-to noise.

If the beam were always returning parallel to its emission direction, it would be sufficient to simply steer each detector to look at a direction anti-parallel to the emission direction of the associated beam. But as we have seen above, there is an angle δ between the two directions. As long as δ is less than the acceptance angle of the detector, this does not cause any problem. We therefore use detector optics 'states' with large acceptance angles for target acquisition. Once signal has been acquired, δ can be computed from the time-of-flight differences between the two beams. This allows switching over to a detector state with better signal or signal-to-noise properties, but narrower acceptance angle by offsetting the detector look-direction according to the sign and magnitude of δ.

4. CAPABILITIES AND LIMITATIONS OF THE TECHNIQUE

The electron drift technique primarily measures the electron drift velocity, from which the electric field perpendicular to the magnetic field, $\mathbf{E}_\perp$, including its component along the spacecraft spin axis, can be derived. By contrast, the double-probe technique usually measures $\mathbf{E}_\perp$ in the plane of the wire booms only. Under favorable conditions, it may also be possible to infer $\mathbf{E}_\parallel$ from electron drift measurements. If local magnetic field gradients, $\nabla_\perp B$, contribute significantly to the electron drift velocity, the electron drift technique provides the unique capability of determining these local magnetic field gradients from a comparison of the electron drift at different energies. When electron time-of-flight measurements are made, the technique also yields accurate measurements of the magnetic field strength, B.

The quantities directly measured by the electron drift technique all scale with some power of the ambient magnetic field strength B. For a given electric field E, the drift step and the time-of-flight difference scale as B^{-2}, the return flux as B^3, and the angle between beam emission and return directions as B^{-1}. In addition, the electron gyroradius which determines the scale over which the measurements are made, scales as B^{-1}. Figure 8 summarizes the scaling relations in terms of B, E, and the electron speed v. Because of the strong B-dependence, the expected values of B very much determine whether triangulation or time-of-flight techniques

EDI SCALING

$$d \equiv v_d \cdot T_g \quad \propto \quad \frac{E}{B^2}$$

$$T_{1/2} \equiv T_g \left(1 \pm \frac{v_d}{v}\right)$$

$$T_2 - T_1 = 2\,T_g\,\frac{v_d}{v} \quad \propto \quad \frac{E}{B^2}$$

$$T_g = \frac{T_1 + T_2}{2}$$

$$Flux \quad \propto \quad I_B \cdot \frac{B^3}{E}\,\frac{1}{v}$$

$$\delta \quad \propto \quad \frac{E}{B}\,\frac{1}{v}$$

Figure 8. Scaling relations for the quantities that characterize the electron drift technique: drift step d; electron time-of-flights and their difference T_1, T_2, and $T_2 - T_1$; gyrotime T_g; return beam flux; and angle change δ of return beam.

are applicable and what complexity in the gun and detector designs is required.

For example, if B is high, such as on low-altitude spacecraft, the drift step is so small that the beam always returns very close to its origin and the return fluxes are so large that simple detection techniques are feasible and the drift step is directly measured. Electrons gyrating more than once are no concern because they can be easily intercepted. On the other hand, gyrotimes are so small that time-of-flight measurements are not feasible. As a result, B is not measured under these circumstances. There is also the problem that the gyroradius can become so small that spacecraft effects cannot be ignored.

If B covers the range from very small to medium, such as on Cluster, time-of-flight measurements of the drift velocity are feasible, and B is thus measured as well. Spacecraft effects are of little concern because the gyro-radius is large. On the other hand, because of the large size of the drift step, the gun-detector geometry must be carefully chosen, and the beam firing directions must be actively controlled. Large gyroradius and drift step conspire to make the beam return fluxes generally small even for the largest feasible beam emission currents. This together with the presence of large fluxes of background electrons from the ambient plasma, requires elaborate detectors and correlation techniques to compensate for the low signals and/or low signal-to-noise ratios. On the other hand, when B increases, and thus the gyro-radius and drift step become smaller, the fraction of the beam electrons that return becomes larger and larger, eventually requiring a reduction in beam current in order not to saturate the detectors. Thus beam emission currents must be constantly adjusted. The implied large variations in signal-to-noise ratio affect the achievable time resolution of the measurements and can lead to loss of track.

To instantaneously separate electric and magnetic components of the electron drift, one would ideally want two fully redundant gun-detector systems and associated control, operating at widely different energies. Because resource limitations ruled out such a solution on Cluster, we use the same system to sequentially operate at the two energies which the design supports, namely 0.5 and 1.0 kev. A factor of two in beam energy will not always be sufficient to separate the drifts. Furthermore, variations in the fields on the time-scale of the energy variation will also cause difficulties.

In addition to the constraints already noted, the electron drift measurements can be adversely affected by intrinsic beam instabilities, strong scattering of the beam by ambient fluctuations, large-amplitude 'spikes' in the electric field, and very rapid magnetic field variations. All these effects can cause a loss of beam track and thus a momentary loss of data.

While the electron drift measurements might at times be compromised, there is another measurement capability that enhances the scientific return from such an instrument at no extra cost: the ability to make unique measurements of ambient electron and ion distributions, thanks to the special properties of the detectors and their control. Each of the two detectors can be commanded to look in any direction over greater than a 2π steradian hemisphere, and since the two detectors are mounted on opposite sides of the spacecraft, full-sky surveys can be achieved without relying on spacecraft spin to complete the coverage of phase space. In addition, since much of the control is already based on the **B** field measured by the on-board magnetometer, specialized surveys can be performed in coordinates fixed with **B**, and they can be performed continuously.

REFERENCES

Kletzing, C.A., G. Paschmann, and M. Boehm, Electric field measurements using the electron beam technique at low altitudes, 1997 (this volume).

Melzner, F., G. Metzner, and D. Antrack, The Geos electron beam experiment, *Space Sci. Rev.*, **4**, 45, 1978.

Paschmann, G., F. Melzner, R. Frenzel, H. Vaith, P. Parigger, U. Pagel, O.H. Bauer, G. Haerendel, W. Baumjohann, N. Sckopke, R.B. Torbert, B. Briggs, J. Chan, K. Lynch, K. Morey, J.M. Quinn, D. Simpson, C. Young, C.E. McIlwain, W. Fillius, S.S. Kerr, R. Maheu, and E,C. Whipple, The electron drift instrument for Cluster, *Space Sci. Rev.*, **79**, 233, 1997.

Tsuruda, K., H. Hayakawa, and M. Nakamura, in A. Nishida (ed.), *Science Objectives of the Geotail Mission*, ISAS, Tokyo, p. 234, 1985.

Tsuruda, K., H. Hayakawa, M. Nakamura, T. Okada, A. Matsuoka, F.S. Mozer, and R. Schmidt, Electric field measurements on the Geotail satellite, *J. Geomag. Geoelectr.*, **46**, 693, 1994.

Vaith, H., R. Frenzel, G. Paschmann, and F. Melzner, Electron gyro time measurement techniques for determining electric and magnetic fields, 1997 (this volume).

C. E. McIlwain, Center for Astrophysics and Space Science, University of California at San Diego, La Jolla, CA 94304, USA (e-mail: cmcilwai@ucsd.edu)

G. Paschmann, Max-Planck-Institut für extraterrestrische Physik, P.O.Box 1603, D-85740 Garching, Germany (e-mail: gep@mpe-garching.mpg.de)

J. M. Quinn, Institute for the Study of Earth, Oceans and Space, University of New Hampshire, Durham, NH 03824,USA (e-mail: jack.quinn@unh.edu)

R. B. Torbert, Institute for the Study of Earth, Oceans and Space, University of New Hampshire, Durham, NH 03824, USA (e-mail:torbert@unhedi1.unh.edu)

E. C. Whipple, Geophysics Department, Box 351650, University of Washington, Seattle, WA 98195, USA (e-mail: whipple@geophys.washington.edu)

Electric Field Measurements in the Magnetosphere by the Electron Beam Boomerang Technique

K. Tsuruda and H. Hayakawa

Institute of Space and Astronautical Science, 3-1-1, Yoshinodai, Sagamihara, Kanagawa 229, Japan

M. Nakamura

Department of Earth and Planetary Science, University of Tokyo, Tokyo 113, Japan

This paper describes the outline of the electron beam boomerang experiment carried out on the Geotail spacecraft. The experiment uses an electron beam artificially emitted from the satellite to measure the three dimensional electric field as well as the magnitude of the magnetic field. Unlike the Geos electron beam experiment, The boomerang experiment measures the time of flight of the electron beam returning to the spacecraft, which improves the measuring accuracy significantly.

INTRODUCTION

The electron beam technique measures the drift motion of the test electrons emitted from a spacecraft to derive the electric field. The test electrons travel a few tens kilometers before they return to the spacecraft. Since most part of the orbit of the test electrons is outside the Debye sphere, the electron beam method is less affected by the local electric disturbances. The electron beam technique was first used to detect electric field on Geos [Melzner et al., 1978]. The Geos electron beam experiment measured the drift of test electrons directly by a set of electron gun and detector separated each other by a fraction of the drift step which is defined as the distance the test particle's guiding center moves during one gyro period. The Geos method is difficult to realize when the drift step is large. This drawback is removed by Tsuruda et. al [Tsuruda et. al, 1988] by measuring the time of flight of the test particle

Measurement Techniques in Space Plasmas: Fields
Geophysical Monograph 103
Copyright 1998 by the American Geophysical Union

instead of the drift step and tested by a sounding rocket[Nakamura et al., 1989]. This method is referred as 'boomerang method'. Geotail is the first satellite which has succeeded in measuring the electric field by the boomerang method in the magnetosphere.

THE MEASUREMENT PRINCIPLE

As is well known, a charged particle's guiding center drifts in the direction perpendicular to both the electric field and the magnetic field as;

$$\vec{V}_d = \frac{\vec{E} \times \vec{B}}{B^2} \qquad (1)$$

where V_d is the drift velocity of the guiding center. The electron(ion) beam method measures this drift velocity to derive the electric field. There are two ways to measure the drift velocity, one is to measure the drift distance during one gyro period as is done on Geos and the other is to measure the time of flight of the test particle which return to the original point. The name 'boomerang' comes from the resemblance of the orbit of the test particle to that of boomerang in Australia. We assume the parallel electric field is negligibly small in the following discussion. For the test particle to return to the original point it has to be ejected in the plane perpendicular to the magnetic field(condition 1) and at some angle which is nearly equal to the right angle from the direction of the electric field(condition 2). If these two conditions are met the test particle will return to the original point after approximately one gyro period. There are two directions which satisfy these two conditions as is shown in Figure 1. The times (T_1 and T_2) of flight of the test particle and the angles (ϕ_1 and ϕ_2) of the initial velocity are written as;

$$T_1 = \frac{2\pi}{\Omega}(1+\Delta), \qquad \phi_1 = \pi(1-\Delta) \qquad (2)$$

$$T_2 = \frac{2\pi}{\Omega}(1-\Delta), \qquad \phi_2 = \pi\Delta \qquad (3)$$

$$\Delta = \frac{V_d}{V_0} \qquad (4)$$

where the suffices 1 and 2 denotes a larger orbit and smaller orbit of the test particles. V_0 is the speed of the test particle which is assumed to be a constant.

The vector electric field and the magnitude of the magnetic field can be obtained from the relations below;

$$T_g = \frac{T_1+T_2}{2} \qquad B = \frac{2\pi m_e}{eT_g} \qquad (5)$$

$$\Delta = \frac{T_1-T_2}{2T_g} \qquad E = \Delta V_0 B \qquad (6)$$

$$\phi_E = \frac{\phi_1+\phi_2}{2} \qquad (7)$$

The accuracy of the time measurement depends on the count rate of the return particles. A simple geometrical consideration gives the following formula for the count rate of return particles F as;

$$F = \frac{S \times I}{V_0 V_d T_g^2 \delta^2} \propto \frac{B^3}{E} \qquad (8)$$

where I, S, δ, V_0, V_d, T_g, E are the beam current, detection area of the detector, and the beam divergence, beam velocity, drift velocity, gyro period and the magnitude of the electric field, respectively. Note that the count rate of the return flux is proportional to the third power of the magnetic field and inversely proportional to the magnitude of the electric field.

BOOMERANG INSTRUMENT ON GEOTAIL

The boomerang instrument onboard Geotail consists of 4 electron guns, two electron detectors, two ion emitters,

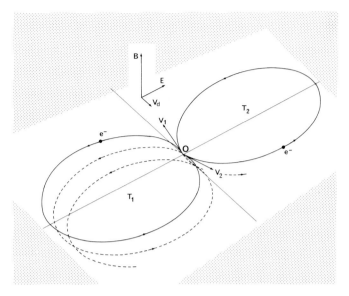

Figure 1. Principle of the boomerang experiment. Test electrons emitted from a spacecraft(center) in the directions V_1 and V_2 return to the center after T_1 and T_2. V_1 and V_2 are approximately perpendicular to both B and E and T_1 is a little larger than the electron gyro period Tg while T_2 is a little smaller than Tg. T_1-Tg or T_2-Tg is proportional to the magnitude of the electric field.

and control electronics. The two guns installed on the top panel shares the beam direction so that one(main gun) emits electron beam in the elevation angle from 0 to 90 degrees, and the other gun(sub gun) emits in the elevation angle from -45 to 45 degrees. The main gun is used mostly in the lobe while the sub gun is used mostly in the near earth magnetosphere. A similar set of guns is installed on the bottom panel of the spacecraft to cover the other half hemisphere. The field of view of the detector is fan shaped with -20 to 100 degrees in the elevation and approximately 10 degrees in the azimuth. One of the detectors is installed on the top panel and the other on the bottom panel. Two ion emitters are installed, one is of field emission type and emits Indium ion of 5 to 10 keV[Schmidt et al., 1993] and the other is of thermo ionic type which emits Lithium ions of about 500 eV, for the purpose to neutralize the charging of the spacecraft due to the emission of the electron beams. The install configuration of these active elements of the experiment is given in Figure 2. The detailed specifications are given elsewhere[Tsuruda et al., 1994].

The electronics do mainly three functions. One is to control the direction of the electron beam. In the near earth magnetosphere, the data from the magnetometer[Kokubun et al., 1994] are processed onboard to get the beam direction so that it is shot in the direction perpendicular to the magnetic field, while in the tail lobe, the beam

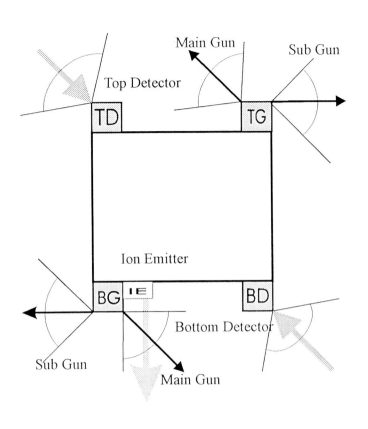

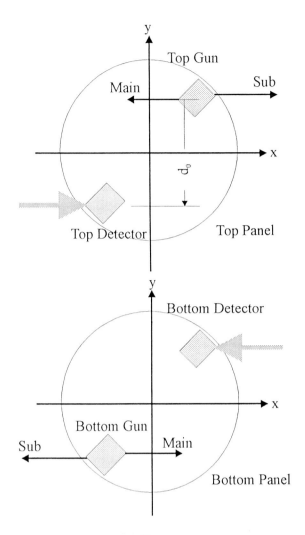

Figure 2. The installation and the filed of views of the elements of the boomerang experiment on Geotail.

direction is determined by the onboard software as is discussed in the later section. The second function is to put modulation on the electron beam for the purpose of flight time measurement. The boomerang experiment onboard Geotail employs simple on/off signal with the periods of one of 8,16,32,...,1024μsec. An appropriate modulation period is selected by the command. The third function is to make correlation of the return flux with the modulation signal. This function is performed by specially designed integrated circuits. The correlation of the return flux is made with modulation signal as well as with that shifted by 90 degrees in phase. The outputs from the onboard correlator are total count(tc), correlation coefficient with the modulation signal(s), and that with phase shifted modulation signal(c). The time of flight(*tof*) is obtained by the equation;

$$tof = (1 + \frac{s}{sig}) \times \frac{T_{\mathrm{mod}}}{4} \qquad (c \leq 0) \qquad (9)$$

$$tof = (3 - \frac{s}{sig}) \times \frac{T_{\mathrm{mod}}}{4} \qquad (c > 0) \qquad (10)$$

$$sig = |c| + |s| \qquad (11)$$

There are ambiguities in *tof* thus obtained of multiples of T_{mod} but this ambiguity can be removed by a proper selection of T_{mod} and by the use of Tg measured by the magnetometer.

BOOMERANG EXPERIMENT IN THE DAYSIDE MAGNETOSPHERE

Plate 1 shows an example of the boomerang experiment in the dayside magnetosphere where the magnetic field is nearly parallel to the satellite spin axis. The electron beam of 1 μA is used. The beam direction is controlled to be perpendicular to the magnetic field which hits the right direction to return twice per spacecraft spin period. The

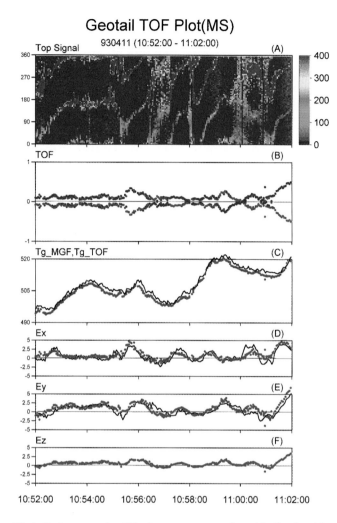

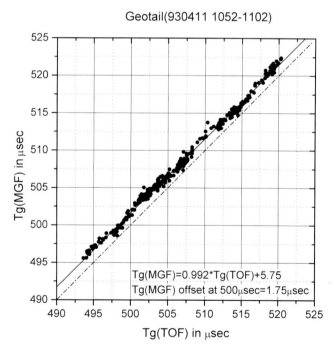

1, the bottom detector detected similar return electrons. The color code gives the count rate per 21 msec which is the time unit of the onboard correlator. The second panel (B) shows the time of flight of the return electron beams normalized to the modulation period of, for this case, 8 μ sec. The two circles of different colors correspond to the top detector(red) and the bottom one(blue). The separation distance between the two circles is proportional to the magnitude of the electric field. At around 10:56:30 and 11:00:00 the separation becomes small suggesting that the electric field is small. The return flux in the top panel at these time intervals does not show distinct two peaks but spreads over the spin phase. This can be explained by the fact that the electron beam ejected at any angle in the plane perpendicular to the magnetic field returns to the original point when the electric field is zero. The width of the phase of the return beam is determined by the magnitude of the electric field and the divergence of the electron beam. Although divergence of the electron beam at the electron gun is about 1 degree, the phase width of the return beam is much wider than the expected width. The space charge of the beam itself, the high frequency electromagnetic noise and some other unspecified effects

Plate 1. An example of the boomerang experiment in the dayside magnetosphere. Panel A: Correlated component of the return electron flux(color code) detected by the top detector in satellite phase and time plane. Note that the high return flux is detected twice per one satellite spin. Panel B: Time of flight of the return beam. Full scale is normalized to modulation periods of 8 μsec. Panel C: Comparison of the gyro period(in μsec) measured by the boomerang experiment(red dots) and the magnetometer(thick line). Panels D, E, F: x, y, and z components of the electric field derived from the boomerang experiment(red dots) and x, and y components of the electric field derived from the probe experiments. Note that the probe Ex is corrected in offset by -5 mV/m and Ey is multiplied by 1.3.

Figure 3. Comparison of the gyro periods measured by the onboard magnetometer(MGF) and those from the boomerang experiment(TOF). At around Tg of 500 μsec, the sensitivity of the magnetometer of 0.992(-0.8 %) with the offset of 1.75 μ sec(0.35 %) are obtained.

top panel (A) shows the return electron flux detected by the top detector and correlated to the modulation signal. The vertical axis is the satellite spin phase and the horizontal axis is the time. Except for the two periods around 10:56:30 and 11:00:00, the flux increases twice per spacecraft spin as expected. Though not displayed in Plate seem to exist and have enlarged the divergence of the beam.

The 3rd panel (C) shows the gyro period in μsec calculated from the magnetic field data obtained by the onboard magnetometer(black curve) and those obtained by the boomerang experiment(red circles). As is seen, the agreement is excellent. The small difference of about 2 μ sec is considered due to the offset in the spin axis sensor of the magnetometer which cannot be removed by the spacecraft spin motion. The two data are plotted in Figure 3 with the regression line. The regression analysis shows the sensitivity of the magnetometer is 0.992 of the true value and the offset is 1.75 μsec.

The two panels (D) and (E) show the x and y components of the electric field derived from the boomerang experiment(red circles) and the two components of the electric field obtained by the probe experiment(black curves). The panel (F) shows the z component of the electric field obtained by the boomerang experiment. The probe measurements are not done for z component electric field. The Ex measured by the probe on Geotail always shows some offset of the order of a few mV/m. In addition, the boomerang experiment often disturbs the probe experiment due to the emission of neutralizing ion beam. The ion beam changes the photoelectron cloud around the spacecraft. Considering these factors, corrections are made to both Ex and Ey in the plotted probe data. The two measurements are compared in detail in Figure 4. The correlation analysis shows the two measurements can be expressed as;

$$Ex(probe) = 0.92 * Ex(TOF) + 4.92 \qquad (12)$$

$$Ey(probe) = 0.545 * Ey(TOF) - 0.03 \qquad (13)$$

The data points scatters when the magnitude is small. This would be attributed to the inaccuracy of the time of flight because multiply gyrated electrons can reach the detector when the electric field is small.

BOOMERANG EXPERIMENT IN THE TAIL LOBE

The operation of the boomerang experiment in the tail lobe is more difficult than in the dayside magnetosphere by two reasons. One reason is simply due to the orientation of the Geotail spacecraft spin axis. As is discussed in the previous section, the electron beams can be emitted only in a plane parallel to the spacecraft y-z plane(Figure 2). This plane rotates with the spacecraft and if the spin axis is perpendicular to the magnetic field as is in the tail lobe, a vector in the plane can be perpendicular to the magnetic field only when the magnetic field direction becomes the

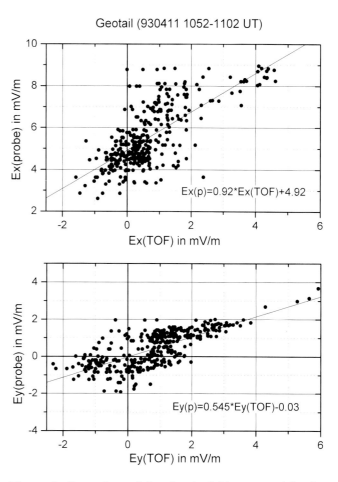

Figure 4. Comparison of the electric fields measured by the double probe and the boomerang experiment(TOF) both on Geotail.

normal to the plane. This condition is met for a short time (a few tens of millisecond) determined by the beam divergence once per spin period. By this reason, the spin motion of the spacecraft cannot be used to search the right direction of the beam, a special search logic is needed for the lobe operation. Even with such search logic, the chances of detecting the return electron beam are much limited in the lobe than in the dayside magnetosphere. The other reason is the weakness of the magnetic field. As is discussed in the previous section, the return flux is proportional to the cube of the magnetic field. In order to avoid undesirable charging effects, the electron beam current cannot be increased sufficiently high to compensate the dilution due to the weak lobe magnetic field and as a result the return flux become smaller than that in the dayside operation.

Figure 5 shows an example of the lobe operation. The top two panels show the correlated component of the return flux detected by the top and bottom detector, respectively. About 10 μA of beam current is emitted from the two electron guns. Unlike the operation in the dayside magnetosphere, the flux is detected when the spacecraft spin phase is around 90 degrees and 270 degrees where the magnetic field is perpendicular to the gun-detector plane. The bottom panel shows the elevation angle of the electron beam in the gun-detector plane. The searching logic currently used for Geotail lobe operation are as follows. The elevation angle is kept constant for one spin period. It is changed according to the following rules; 1) if the return signal is not detected during the preceding two spin periods the elevation angle is decreased by one step, 2) if the return signal is detected, the elevation angle is not changed until the signal is lost, and 3) when the signal is lost the elevation angle is increased by a few steps and return to 1). The black circles in the bottom panel indicate the detection of the signal while the open circles show that the return signal is not detected. The third panel shows the time of flight of the return beams(circles) and the two gyro periods, one obtained by the magnetometer(continuous curve) and the other(crosses) obtained by the time of flight. Note that the gyro period is displayed only in a fraction of 256 μsec which is the period of the modulating signal for this experiment. Though the times of flight and the gyro period derived from them track along the gyro period calculated from the magnetometer, the agreement is worse than those seen in the experiment in the dayside magnetosphere. One of the reasons for this is the smallness of the magnetic field in the lobe. A typical value of the difference between the two gyro periods is 30 μsec which corresponds to about 10^{-3}nT and is below the noise level of the instrument. The curve in the third panel is produced by making the average of the raw data over the spin period. The differences between the crosses and the corresponding circles (T_+ and T_-) are of the order of 30μsec which correspond to 0.6 mV/m for the measured magnetic field of about 6.8 nT.

DISCUSSIONS

Boomerang experiment has turned out to be a promising method to measure the electric field in a hot tenuous magnetospheric plasma. However, it is still in a very primitive stage and many technical problems are left unsolved. In the previous discussions we assumed smallness of the parallel electric field. If it is not the case, the beam has to be shot with a small offset angle

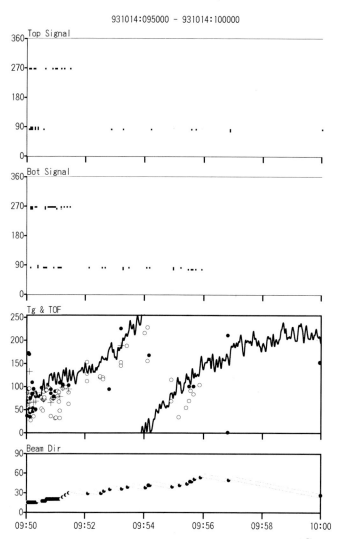

Figure 5. Example of the boomerang experiment in the tail lobe at X_{GSE}=-165R_E, Y_{GSE}=-7R_E, and Z_{GSE}=10R_E. The top two panels show the return fluxes detected by the top and bottom detectors. The 3rd panel shows the time of flight of the return beam(open circle:top detector, black circle: bottom detector) and the two gyro periods one measured by the onboard magnetometer(continuous curve) and those measured by the boomerang experiment(cross). The gyro period is about 5.2 msec and only a fraction of modulation period(256 μsec) is shown in the panel. Bottom panel shows the beam elevation angle in the spacecraft coordinate.

$$\alpha = \pi \frac{E_{\parallel}}{V_0 B} \qquad (14)$$

Usually this angle is small and it is difficult to derive the parallel electric field from the measurement of this angle.

The return electrons come to the detector as a parallel beam while the ambient plasma electrons are omni directional. High background count in the plasmasheet masked return beam in our experiment. A new detector design which is sensitive only to the parallel beam is necessary to apply the boomerang technique in the various region of the magnetosphere.

The current design uses the spin motion of the spacecraft to search the appropriate beam direction. This method limit the time resolution of the measurement to half the spin period. The intrinsic limitation on the time resolution of the method is the electron gyro period. Development of auto tracking technique will improve the time resolution of the measurements greatly.

Acknowledgments. We owe greatly to the Geotail MGF team they provided us with the onboard magnetic field data which are essential to operate the experiment

REFERENCES

Melzner, F., G. Metzner, and D. Antrack, The GEOS electron beam experiment, *Space Sci. Instrum.*, 4, 45-55, 1978.

K. Tsuruda, H. Hayakawa, and M. Nakamura, Ion beams as diagnostic tools -DC electric field measurement in the ionosphere, *Adv. Space Res.*, Vol.8, 165-174, 1988.

M. Nakamura, H. Hayakawa, and K. Tsuruda, Electric field measurement in the ionosphere using the time of flight technique. *J. Geophys. Res.*, vol 94(A5), 5283-5291, 1989.

K. Tsuruda, H. Hayakawa, M. Nakamura, T. Okada, A. Matsuoka, F. S. Mozer, and R. Schmidt, Electric Fied Measurements on the GEOTAIL Satellite, *J. Geomag. Geoelectr.*, 46, 693-711, 1994.

S. Kokubun, T. Yamamoto, M. H. Acuña, K. Hayashi, K.Shiokawa, and H. Kawano, The Geotail Magnetic Field Experiment, *J. Geomag. Geoelectr.*, 46, 7-21, 1994.

R. Schmidt, H. Arends, A. Pedersen, M. Fehringer, F. Rüdenauer, W. Steiger, B. T. Narheim, R. Svenes, K. Kvernsveen, K. Tsuruda, H. Hayakawa, M. Nakamura, W. Riedler, and K. Torkar, A novel medium-energy ion emitter for active spacecraft potential control, *Rev. Sci. Instrum.*, 64(8), 2293 - 2297, 1993.

Electron Gyro Time Measurement Technique for Determining Electric and Magnetic Fields

H. Vaith, R. Frenzel, G. Paschmann, and F. Melzner

Max-Planck-Institut für extraterrestrische Physik, 85740 Garching, Germany

The paper describes the technique the Electron Drift Instrument (EDI) on CLUSTER employs to measure the flight times of the electrons between the time they are emitted and the time they return after having completed a gyration in the ambient magnetic field. In the presence of an electric field two different trajectories exist which return to the detector, and thus there are two possible flight times. Their difference is a measure of the electric field, while their average is determined solely by the magnetic field. EDI uses beams which are modulated with selectable frequency and pseudo-noise coded. A set of correlators determines the time delay of the received electron counts. An auto-tracking feature allows to follow time variations caused by changes in magnetic and/or electric fields.

1. INTRODUCTION

The Electron Drift Instrument on CLUSTER measures the drift of two emitted electron beams that return to the spacecraft after one or more gyrations when emitted into certain directions. The drift is related to the perpendicular electric field $E_\perp$ and the perpendicular gradient in the magnetic field $\nabla_\perp B$. If the displacement of the beam trajectories after one gyration – the drift step d – is small, its magnitude and direction can be calculated by triangulation of the emission directions of the two beams. In case of large drift steps (> 30 m in case of EDI on CLUSTER) the direction of the drift step can still be found by triangulation, however, the accuracy of the magnitude deteriorates and is better determined from the difference in the times of flight

of the electrons in the two nearly oppositely directed beams. This technique was developed by Tsuruda at ISAS [*Tsuruda et al.*, 1985].

Figure 1 illustrates the operation of EDI in a frame of reference where the electric field vanishes. In this frame the S/C travels at the negative drift speed: $\mathbf{v}_{SC} = -\mathbf{v}_d = \frac{-\mathbf{E} \times \mathbf{B}}{B^2}$. Two electron beams launched into directions $\mathbf{e}_1$ and $\mathbf{e}_2$ at time t_0 return to the S/C at times t_1 and t_2, having performed slightly less and slightly more than one gyration, respectively. In the approximation of $v_d \ll v_e$ (v_e = electron velocity), the times of flight T_1 and T_2 are given by

$$T_{1/2} = t_{1/2} - t_0 = T_g(1 \mp \frac{v_d}{v_e}) \qquad (1)$$

where T_g is the gyro-time of the electrons and is equal to $\frac{2\pi m}{eB}$. By taking the sum and the difference of T_1 and T_2, one can calculate the gyro-time T_g and v_d, and thus $B = \frac{2\pi m}{eT_g}$ and $E_\perp = v_d B$.

The strategy to initially acquire a return signal on the detector and to subsequently launch the electron

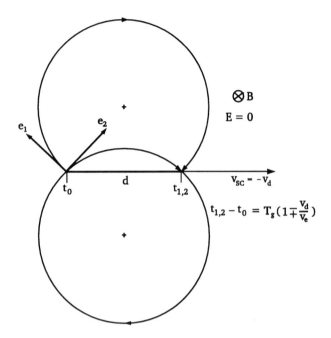

Figure 1. Principle of drift step determination in a coordinate system where there is no electric field

beam into the proper direction to keep the return signal is described in [*Paschmann et al.,* 1997a]. A more detailed analysis of the relation between the times of flight and the electric/magnetic fields, involving grad B effects, can also be found there.

For the task of measuring the flight times of the beam electrons a Correlator has been designed which provides a modulation waveform for the emitted beams and measures the phase delay of the return beam by correlation. It is equipped with a Delay-Lock Loop (DLL), also referred to as automatic track facility, which enables it to track changes in the time of flight (ToF), caused by variations in the magnetic and/or electric field, in a closed loop operation.

2. PSEUDO NOISE CODES

Pseudo-Noise (PN) Codes possess autocorrelation properties which make them ideal for use in DLL tracking systems. They can be generated in digital electronic circuits using shift registers some of whose outputs are combined logically and fed back into the input of the first stage. Using the proper logic function and output stages for the feed-back, a *maximal-length* PN sequence of period $M \cdot \Delta = (2^n - 1) \cdot \Delta$ can be generated, with n being the number of stages in the shift register and Δ the inverse of the clock frequency f_c. The employed

shift register consists of four stages and the generated PN Code sequence is $u_k = 111011001010000$, where k denotes the start phase of the code which is defined by the initial contents of the shift register. The single code elements (chips) are represented by specifying their position within the sequence in parentheses. Thus, $u_k(i)$ is the i-th chip of the code sequence u_k. Both i and k run from 0 to 14.

Maximal-length PN codes have several properties among which the two-valued autocorrelation function $R(j)$ is of particular interest:

$$R(j) = \frac{1}{M} \sum_{i=0}^{M-1} S_k(i) \cdot S_k(i+j) = \left\{ \begin{array}{l} \frac{-1}{M} : j \neq 0 \\ 1 : j = 0 \end{array} \right. \quad (2)$$

where $S_k(i) = 1$, if $u_k(i) = 0$ and $S_k(i) = -1$, if $u_k(i) = 1$. Replacing the sum by an integral yields R as a function of the (continuous) time shift ϵ between the two PN sequences, $R(\epsilon)$. Note, that $R(\epsilon)$ has only a single peak and no side lobes at all.

The task of a Delay-Lock Loop (DLL) is to obtain a correction (magnitude and sign) to an estimate $\hat{\tau}$ of the true delay τ of a signal $s(t + \tau)$ of known shape and thereby improve the estimate $\hat{\tau}$, or, equivalently, reduce the delay error $\epsilon = \tau - \hat{\tau}$. In case of a differentiable, periodic signal this can be done by multiplying the received signal $As(t + \tau)$ of amplitude A with the derivative of the signal, delayed by the estimate $\hat{\tau}$. The time average of this product, also called Delay Discriminator characteristic is

$$D(\epsilon) = \overline{As(t + \tau)\frac{d}{dt}s(t + \hat{\tau})} = -AR'(\epsilon) \quad (3)$$

where $R'(\epsilon)$ is the derivative of the autocorrelation function of the signal $s(t)$ with respect to ϵ [*Spilker,* 1977]. For the binary sequence of a PN code signal the differentiated signal consists of spikes at the transition points of the code and is inconvenient to handle in a digital electronic circuit. For this reason, the term $\delta s(t + \hat{\tau}) = s(t + \hat{\tau} + \frac{\Delta}{2}) - s(t + \hat{\tau} - \frac{\Delta}{2})$ is used instead of the derivative $\frac{d}{dt}s(t + \hat{\tau})$. The Delay Discriminator characteristic is then found to be

$$D(\epsilon) = \overline{s(t + \tau) \cdot \delta s(t + \hat{\tau})} = R(\epsilon - \frac{\Delta}{2}) - R(\epsilon + \frac{\Delta}{2}) \quad (4)$$

Figure 2 shows $D(\epsilon)$ for the employed 15-chip PN code. Note, that in the range $-\frac{\Delta}{2} \leq \epsilon \leq \frac{\Delta}{2}$ it provides an average correction which is directly proportional to the delay error ϵ. In the range $-\frac{3\Delta}{2} \leq \epsilon \leq -\frac{\Delta}{2}$ and $\frac{\Delta}{2} \leq \epsilon \leq \frac{3\Delta}{2}$ the correction falls off and is no longer

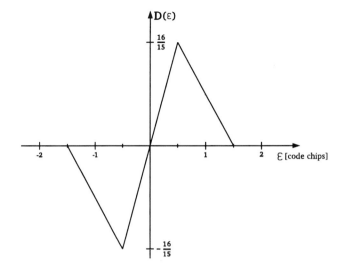

Figure 2. Delay Discriminator characteristic for a 15-chip maximal-length Pseudo-Noise Code

proportional to ϵ, although its sign is still right. For delay errors larger than $\frac{3\Delta}{2}$ there is no average correction at all.

3. IMPLEMENTATION AND FUNCTIONAL DETAILS

The EDI Time-of-Flight Correlator is implemented in a RAM-based Field Programmable Gate Array (FPGA) of the Xilinx type. The configuration data for the FPGA are held in EEPROM and loaded into the Xilinx upon instrument configuration. Other correlator schemes could be uploaded in flight, if necessary. Details about the hardware instrumentation can be found in [*Paschmann et al.*, 1997b].

3.1. Main Elements

The main elements of the EDI Correlator are a PN Code generator, 15 correlator counters and a DLL, which can be used in two different modes of operation. The clock frequency f_c of the PN Code generator is selectable in a wide range to account for the large range of times of flight to be encountered along the CLUSTER orbit. For the selected initial value the code period T_{PN} ($= 15 \cdot \Delta$) is typically much less than the ToF and therefore the measured delay is ambiguous. The ambiguity can be removed by taking into account the magnitude of the magnetic field as provided by the Flux-Gate Magnetometer. The relatively short period of only 15 code chips has, however, the advantage, that it is possible to cover the entire range of possible code phases equidis-

tantly and without gaps by creating 15 copies of the PN code, each one delayed by Δ relative to its neighbor code. Each of these codes, together with a counter, forms a simple correlator counting or ignoring the received pulses depending on the states − 1 or 0 − of the respective code at the instants the pulses arrive. The autocorrelation function for this type of correlation is

$$R_p(j) = \frac{1}{M} \sum_{i=0}^{M-1} u_k(i) \cdot u_k(i+j) = \left\{ \begin{array}{ll} \frac{M-1}{2M} & : \; j \neq 0 \\ 1 & : \; j = 0 \end{array} \right.$$

(5)

and yields a less optimal noise discrimination than the autocorrelation function $R(j)$. The reason to use this easier type of correlation is that it requires only simple up-counters rather than up/down counters and thus takes up less space in the FPGA, leaving room for other features.

For the PN code sequence shown earlier, which consists of seven '1's and eight '0's, all 15 correlator counters will on average accumulate 7/15 of the background events resulting from ambient plasma electrons in the energy range close to the beam energy. Signal event pulses from the received electron beam carry a PN-modulation the phase of which depends on the ToF. If this phase matches the phase of one of the 15 channels exactly, then all the signal event pulses will be accumulated in this channel and all the other channels will on average count only 3/7 of the signal pulses. In case the phases do not match exactly, the signal is spread over two adjacent channels.

As mentioned earlier the DLL can be operated in two modes. The operation in tracking-mode is schematically shown in Figures 3 and 4. Two PN codes are used which are phase-shifted by $-\frac{\Delta}{2}$ and $+\frac{\Delta}{2}$, respectively, relative to the code of correlator channel 7. Instead of performing the correlations of the received event pulses with these two codes separately and then taking the difference, as (4) would suggest, the two codes are used to form a phase comparator. The output of the phase comparator consists of an up/down signal (early/late) and an enable signal for the subsequent up/down counter. The up/down signal indicates whether a received pulse is considered as too early or too late for a PN modulation matching the phase of code 7. This decision can only be made if the states of the two codes differ at the instant of the pulse arrival. Therefore, the up/down counter can be disabled in order to ignore pulses arriving while the two codes are in the same state. If the magnitude of the counter contents exceeds a selectable threshold ("tracking sensitivity") the system will attempt to decrease the delay error by shifting the phase

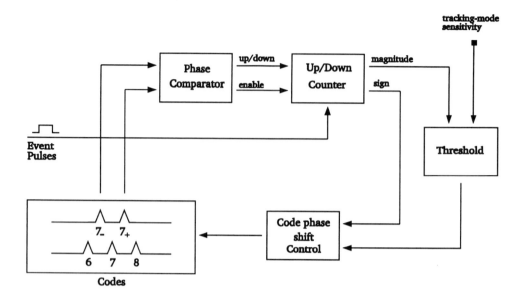

Figure 3. Block diagram of the Delay-Lock Loop circuit

of all 15 codes by a fraction of the code chip duration Δ, the code shift step DT. The balance of code phase shifts, kept in a second up/down counter, provides a measure of the change in the time of flight.

The large number of different situations in terms of variability in ToF, signal and background count rates which will be encountered with CLUSTER requires the ability to select the code clock frequency f_c, and thus the code chip length Δ, independent from the shift step DT. Therefore two parameters, n and m, are used together to set the chip length Δ (cf. Table 1). While the first also determines the size of the code phase shift step DT, the latter can be used to dynamically increase the measurement accuracy by means of decreasing Δ in factors of 2, once the DLL is in the locked state and working stably. Due to the delay of the return beam electrons relative to the emitted beam electrons this is not a straightforward process and requires some compensation circuitry to avoid unlocking of the DLL.

With the initially selected chip length, as outlined further below, the signal may appear initially in any of the 15 channels after the Correlator has been started. The automatic tracking, however, works only if the magnitude of the delay error ϵ is smaller than $\frac{3\Delta}{2}$, i.e. when the signal is in channels 6, 7, or 8. One could, of course, rely on variations in the ToF to move the signal into channels 6-8. However, in order to speed up the process, the DLL can be operated in a second mode, called drift-mode, to quickly shift the phase of the 15 codes so that the delay error becomes small enough to enable the tracking-mode. In this mode the code phase shift direction is not derived by the phase comparator but by a circuit that periodically evaluates which of the 15 (periodically resetted) correlator counters contains the largest number of counts. Depending on the channel number of this counter relative to channel 7 the drift direction is chosen so that the signal peak is shifted towards the auto-track channel 7. The rate of the drift is proportional to – beside the shift-step size DT – the received event rate and, as in the case of the tracking-mode, to a selectable threshold ("drift sensitivity") which determines the number of event pulses required to perform a single shift. The counter that is used in tracking-mode to keep the balance of code phase shifts is also used to count the number of shifts in drift-mode. This way, it is possible to switch back and forth between tracking- and drift-mode without invalidating the contents of the shift step counter.

In order to avoid time delays caused by the DPU analyzing the data provided by the Correlator and taking the appropriate action, the Correlator can, when it is commanded into drift-mode, switch automatically to tracking mode whenever it detects the largest number of counts in channel 7 or its next neighbors. The switching is reversible and can be made dependent upon the additional condition, that drift-mode is used only, if the largest number of counts of the 15 channels exceeds a selectable threshold. This way the system is more immune to false peaks caused by the background when the beam has been lost temporarily.

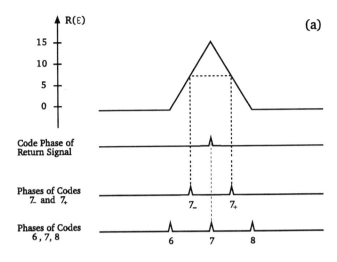

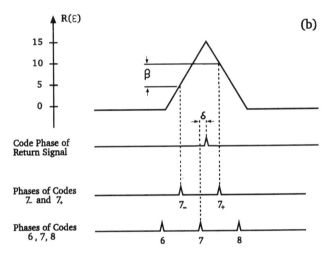

Figure 4. Illustration of DLL operation in tracking-mode. (a) Situation where the delay of channel 7 matches the delay of the return signal exactly. The correlator channels 7_- and 7_+ will on average accumulate the same number of counts. (b) A slight mismatch δ between the delays of channel 7 and the return signal, will result in different accumulation numbers in channels 7_- and 7_+. The sign and magnitude of the difference β can be used to correct the delay.

3.2. Setup and Operation

The parameters which have to be set for a start of the Correlator are n and m, i.e. the code phase shift step DT and the code chip length Δ, and the tracking-mode and drift-mode sensitivities. To remove the ambiguity caused by the code period T_{PN} being shorter than the time of flight, the chip length Δ has to be sufficiently large such that the range of expected flight times, based on the magnitude and accuracy of the magnetic field, as provided by the on-board magnetometers FGM and

Table 1. Shift step size DT as a function of n and Code chip length Δ ($= n \cdot m / 2^{23}$ seconds) vs. n and m. The code period T_{PN} is equal to $15 \cdot \Delta$.

n	1	2	4	8	16	32	64
$DT[\mu s]$	0.06	0.12	0.24	0.48	0.95	1.91	3.81

m			$\Delta[\mu s]$				
2	0.24	0.48	...	...	...	7.63	15.3
4	0.48	0.95	...	...	...	15.3	30.5
8	0.95	1.91	...	...	...	30.5	61.0
16	1.91	3.81	...	...	...	61.0	122

Table 2. Initial selection of code phase shift step DT, code chip length Δ and tracking-mode sensitivity SE for sample conditions $(B, E, dToF/dt)$ in some key regions of CLUSTER.

	Solar Wind	Cusp	Tail Lobe	Plasma Sheet
$B[nT]$	8	40	30	20
$E[mV/m]$	3.6	4.0	0.5	1.0
$T_g[ms]$	4.5	0.9	1.2	1.8
$\Delta T[\mu s]$ [a]	214	9	2	9
$\frac{dToF}{dt}$	10^{-3}	10^{-3}	10^{-5}	10^{-4}
$Sig[s^{-1}]$ [b]	$1.9 \cdot 10^4$	$4.2 \cdot 10^5$	$5.0 \cdot 10^5$	$3.5 \cdot 10^5$
$Bgd[s^{-1}]$ [c]	$3.0 \cdot 10^4$	$1.5 \cdot 10^5$	$3.0 \cdot 10^4$	$3.0 \cdot 10^5$
$DT[\mu s]$	1.91	0.48	0.95	0.95
$\Delta[\mu s]$	61.0	15.3	30.5	30.5
SE	3	17	17	9
m_{min} [d]	8	2(<)	2(<)	2(<)

[a] ΔT is the difference between the two times of flight.

[b] Signal count rates are calculated from typical electron beam currents based on the respective magnetic and electric field.

[c] Background count rates are taken from AMPTE data.

[d] Indicates how far the code chip length Δ can be reduced under the assumed conditions. ($<$) indicates that a further decrease in DT (n) is possible.

STAFF and on an assumed 10% reduction/increase in flight time to account for large electric fields, fits within the length of the code ($15 \cdot \Delta$). With m being initially set to 16 (in order to allow a later increase of accuracy, as described below) this determines the setting of n and thus DT (cf. Table 1). The initial value for the tracking mode sensitivity is based on the selected DT, the background and expected signal count rates and on the rate of change in the gyro time which the DPU finds by analyzing the variability of B. The drift-mode sensitivity is chosen such that under the given event rates the drift would shift the codes by approximately $\frac{\Delta}{2}$ in the time frame of 244 μs which is the period of the maximum-channel evaluation. This leads to a shift of more than half the code length (8 chips) in 4 ms, the time frame on which the DPU analyzes the data, and should thus cause the Correlator to switch from drift-mode to tracking-mode within this time frame, once a beam signal has been detected.

In addition, the remainder of the quotient of estimated gyro time and code period T_{PN} can be provided to the Correlator. This affects the initial phase shift of channel 7 relative to the gun code modulating the outgoing electron beam and is used to get the first signal peak as close to channel 7 as possible and thus reduce the amount of time required to shift the peak into channel 7 in drift-mode.

The tracking-mode and drift mode sensitivities are continuously updated (every 4 ms and 250 ms, respectively) to adapt to changes in the event rates and/or $\frac{dToF}{dt}$. Once the tracking loop is working stably, the DPU evaluates whether it is possible to switch to the next lower code chip length under the given conditions of signal/background event rates and $\frac{dToF}{dt}$, and which tracking-mode sensitivity to choose in this case. This information is stored in tables which are based on ground calibration results.

Table 2 shows the values initially selected for DT, Δ and the tracking-mode sensitivity for typical conditions in some of the key regions of CLUSTER.

REFERENCES

Paschmann, G., C.E. McIlwain, J.M. Quinn, R.B. Torbert, and E. C. Whipple, The Electron Drift Technique for Measuring Electric and Magnetic Fields, 1997 (this volume)

Paschmann, G., F. Melzner, R. Frenzel, H. Vaith, P. Parigger, U. Pagel, O.H. Bauer, G. Haerendel, W. Baumjohann, N. Sckopke, R.B. Torbert, B. Briggs, J. Chan, K. Lynch, K. Morey, J.M. Quinn, D. Simpson, C. Young, C.E. McIlwain, W. Fillius, S.S. Kerr, R. Maheu, and E,C. Whipple, The Electron Drift Instrument for Cluster, *Space Sci. Rev.*, **79**, 233, 1997.

Spilker, J.J., Digital communications by satellite, pp 528, Englewood Cliffs: Prentice-Hall 1977, Prentice-Hall information and system sciences series

Tsuruda, K., H. Hayakawa, and M. Nakamura, in A. Nishida (ed.), *Science Objectives of the Geotail Mission*, ISAS, Tokyo, p. 234, 1985.

R. Frenzel, Max-Planck-Institut für extraterrestrische Physik, P.O.Box 1603, D-85740 Garching, Germany (e-mail: rhf@mpe-garching.mpg.de)

F. Melzner, Max-Planck-Institut für extraterrestrische Physik, P.O.Box 1603, D-85740 Garching, Germany (e-mail: fhm@mpe-garching.mpg.de)

G. Paschmann, Max-Planck-Institut für extraterrestrische Physik, P.O.Box 1603, D-85740 Garching, Germany (e-mail: gep@mpe-garching.mpg.de)

H. Vaith, Max-Planck-Institut für extraterrestrische Physik, P.O.Box 1603, D-85740 Garching, Germany (e-mail: hav@mpe-garching.mpg.de)

Electric Field Measurements Using the Electron Beam Technique at Low Altitudes

C. A. Kletzing

Department of Physics and Astronomy, University of Iowa, Iowa City, IA, USA

G. Paschmann

Max-Planck-Institut für extraterrestrische Physik, Garching, FRG

M. Boehm

Jet Propulsion Laboratory, Pasadena, CA, USA

The electron beam technique determines the perpendicular electric field by using a weak beam of electrons to measure the $E \times B$ drift. An advantage of the technique is that it can measure the component of the electric field along the spin axis of a rotating spacecraft without the use of booms. Because the magnetic field is quite strong at low altitudes, the measured quantity, the displacement of the beam after one gyration, is quite small – a few mm for moderate electric fields. This limits the lowest fields that the technique can measure to about 20 mV/m. For such small fields, offsets and distortions of the signal the same size as or greater than the actual signal from the electric field must be subtracted and the data must be smoothed. Nonetheless, the technique does measure electric fields which compare favorably with the electric fields determined by the double probe technique.

INTRODUCTION

The electron beam technique has been developed as an alternative and complement to double probe measurements of electric fields in space plasmas. The technique is based on sensing the $E \times B$ drift of weak electron beams. The beam technique was originally developed at MPE and first proven on the ESA Geos spacecraft [*Melzner et al.*, 1978]. Instruments based on the same principle have been selected for the Geotail, Cluster, and Equator-S missions [*Tsuruda et al.*, 1985; *Paschmann et al.*, 1989]. These missions are all at high altitude and have small ambient magnetic fields. At low altitude, an ion beam version has been flown on sounding rockets [*Nakamura et al.*, 1989].

Measurement Techniques in Space Plasmas: Fields
Geophysical Monograph 103

We report here on an implementation of the technique that was developed for the Freja satellite. Because of its 63° inclination orbit, Freja was always near its apogee of ~1700 km while in the northern auroral zone. At this relatively low altitude, the Earth's magnetic field is quite strong and the drift velocity correspondingly small, allowing a somewhat simpler instrument than those used at higher altitudes.

PRINCIPLE OF OPERATION

Figure 1 schematically illustrates the principle of the technique. A beam of electrons, emitted perpendicular to the ambient magnetic field, B, would in the absence of an electric field, E, return to its origin after one gyration. A finite E-field will induce a drift V_d which causes a displacement L_d (the 'drift step') of the returning beam defined as $L_d = V_d \cdot T_g$, where T_g is the electron gyrotime. The emitted beam has a finite angular width. In the plane perpendicular to B, the beam is refocussed after one gyration. Along B the beam continuously disperses. As a result, the beam will appear as

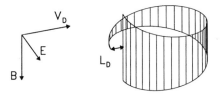

Figure 1. Principle of sensing electron drift in crossed electric and magnetic fields

a thin sheet at the time it returns. If one senses the drift step, one can calculate V_d, and from that, the electric field.

The gyro diameter of the electron orbit is not large (less than 15 m for 4 keV electrons), and the return beam is quite intense, in spite of the beam dispersion. For a 3 μA emitted beam and a 1 cm wide detector, the intercepted return current will be about 5 nA. At such high return beam intensities, a direct current measurement is possible and electron multiplier techniques are not required.

Because of the strong magnetic fields (26000 nT) at Freja apogee, the drift step is small, only 5 mm for an electric field of 100 mV/m. This limits the sensitivity of the instrument and is a difficulty the technique faces at low altitudes. The small drift step does, however, allow the beam detector to be collocated with the gun, simplifying the measurement. Using position-sensing anodes, the measurement system is capable of resolving displacements of about 1 mm, corresponding to electric fields of 20 mV/m at Freja apogee. The beam displacement for a given electric field scales as $1/B^2$ but is independent of electron energy. Thus the sensitivity worsens with decreasing altitude. For example, at 1000 km altitude it takes 40 mV/m to produce the same displacement of 1 mm.

In the Freja application, beam detection is further simplified by requiring no energy discrimination. Background fluxes from ambient cold plasma and energetic auroral primaries are suppressed by beam modulation and synchronous detection combined with suppressor grids. The intrinsic time resolution of the measurements is determined by the integration time of the detection system, typically below 1 ms.

The beam will return to the spacecraft only if emitted at right angles to the ambient magnetic field, to within the width of the beam. To achieve this for all orientations of the magnetic field relative to the spacecraft, the beam must be steerable within a half-plane. The actual emission direction within that plane is calculated from the on-board magnetometer data.

The emitted and returning electron beams can, in principle, cause interference with other measurements. The interaction of the beam with the ambient medium can create plasma waves detectable by the wave instruments and the returning beams can fall within the acceptance angle of electron spectrometers, causing spurious counts. On Freja, one of the electron spectrometers did measure electrons from the emitted beam. This appears to have occurred when the beam was blocked by a part of the spacecraft structure and scattered into the spectrometer's field of view. This problem was mostly eliminated by not pointing the gun in the blocked directions.

INSTRUMENT DESCRIPTION

Each Gun/Detector Unit (GDU) consists of an electron gun with magnetic deflection system, a position-sensing detector, analog detection electronics, gun and deflector high voltage supplies, and the interfaces to the central DPU which provides power and control, and handles commands and data. Three GDU's were used on Freja with total mass of 2.4 kg and power consumption of 4.5 W. For further technical information refer to the instrument paper [*Paschmann et al.*, 1994].

Electron Gun

The electron gun provides a narrow (3-4°) beam of electrons, with selectable energy up to 4 KeV and intensity up to 10 μA. The beam is steerable over 180° in a plane and can be angle modulated. A pointing accuracy of about 1° can be tolerated. A larger electron energy is desirable because it would increase the gyro radius, but was outside the capability of the electron source being used.

The electron source is a commercial electron gun designed for oscilloscopes which was modified by replacing the cathode with pure tungsten and retaining only one of the deflection-plate pairs allowing beam deflection within a plane. The initial electrostatic deflection controls where the beam enters a magnetic deflection amplifier which consists of two sections with oppositely directed, nearly homogeneous fields generated by ceramic magnets. The pole pieces are wedge-shaped such that the length of the electron path in the strong field between the pole pieces varies so as to amplify the initial injection direction. Maximum deflection (90°) is achieved for an initial injection at 18°. Because the magnetic deflector is a quadrupole configuration, the stray field falls off as the fourth power of the distance. Details of the design of the gun and deflection system are described in *Parl* [1991].

The beam is angle-modulated by adding a small AC voltage to the deflector voltage at a frequency between 1 and 133 kHz to allow synchronous detection. This produces a binary coding by causing the beam to hit the detector during the 1's in the code, and to miss it otherwise. Three different 8-bit codes are used to distinguish the three beams and avoid false associations between different GDU's.

Beam Detector and Mechanical Design

The beam detector is capable of determining the position of the returning electron sheet with an accuracy of about 1 mm over a displacement range of ±5 cm in the presence of uniform background currents. The detection technique is based on the division of the return beam current in two oppositely directed wedge-shaped discrete anodes arranged alongside each other with small gaps between them (cf. Figure 2). As the beam returns as a thin sheet, it cuts across the wedges and the current division changes linearly with position along the anodes. Position is obtained from the ratio of the currents in the two wedges.

To measure the electron drift under varying geometries requires three beams whose displacement is measured along three mutually perpendicular (or near-perpendicular) directions. This is achieved by having three identical GDU's with detectors that are inclined at 45° and mounted 120° apart. For each unit, ideally a full 360° must be kept unobstructed, 180° for the emitted beam, the opposite 180° for the return beam. This has not been possible in practice.

Figure 3 shows the gun/detector geometry. There are two separate detectors, inclined at an angle of 90° with respect to each other. The two detectors are needed to cover all possible return beam orientations. Each detector assembly consists of a metal anode printed on a PC board, and suitable baffles and grids for limiting the beam and rejecting background particles. The anodes are 10 cm long and 0.8 cm wide. Going inward, toward the anode, the grids are on ground, +9V, and -9V, respectively. A square metal shield extending outward from the GDU assembly prevents electrons which have made more than one gyration from reaching the detector.

Synchronous Detection

The scheme for synchronous detection is shown in functional form in Figure 2. The top half of the figure shows a returning beam slicing across the detector wedge anodes at approximately half way between the middle of the detector and the B end of the detector. At this return beam position the B anode receives three times the current that the A anode receives. To determine the return beam position, we calculate the anode signal ratio $(A - B)/(A + B)$. For the example shown in Figure 2 this gives a value of -1/2. The ratio varies from +1 for a beam return at the extreme end of the A anode to -1 at the opposite extreme end on the B anode (in practice, the ratio actually varies between ±2 because the $A + B$ am-

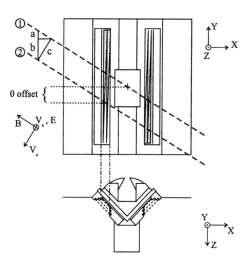

Figure 3. Gun/Detector geometry.

plifiers have half the gain of the $A - B$ amplifiers). In the presence of background current (assumed constant over the anode), the denominator becomes larger, but the numerator remains constant because of cancellation in the subtraction.

The panels in the bottom half of Figure 2 show how the synchronous detection system removes the background. The uppermost panel shows the gun code. This is a square wave which when zero causes the gun deflection system to steer the beam so that it is not perpendicular to B and therefore does not return to the spacecraft. When non-zero the beam returns to the detector strips. The next two panels (moving downward) show the current returned to the A and B anodes. A constant offset which is present at all times is shown (light shading) along with the gun signal which is present only for non-zero gun code (dark shading). The gun signal in B is three times that in A.

The next two panels show the signal after it has gone through the stages in the diagram in the upper half of the Figure 2 marked A (or B) sync. This is a unity-gain circuit when the gun code is non-zero, but is a unity gain inverter for zero gun code. The signals in these panels are the same as those in the upper two panels (discussed above) when the gun code is non-zero (unity gain) but are opposite in polarity when the gun code is zero (unity gain inverter). The dashed line shows the zero level in these two panels.

These signals are then fed to integrators as shown in the upper diagram. The output of the integrators is shown in the lower two panels in the bottom half of Figure 2. When the gun code is non-zero, the integrator integrates in the positive direction both the return beam and the background. When the gun code is zero (no return beam), the inverted signal of background only integrates in the negative direction, canceling the previous positively integrated background and leaving only the return beam signal. By integrating over several gun code periods, the gun signal is amplified to a sufficient level for sampling by the system's A/D converter. The asymptote for the signal is shown by the dashed line in this figure. Because the asymptote and the actual signal are the same only at the end of a complete gun code period, the signal is sampled by the A/D only at the end of gun code periods.

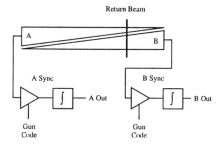

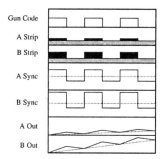

Figure 2. Schematic diagram showing the position sensing technique and how synchronous detection reduces background current effects.

Operational Aspects

In the nominal mode of operation, the instantaneous deflection angle of each beam is computed on the basis of the flux-gate magnetometer data available in real-time through the telemetry interface and is used to set the high-voltage supplies which deflect the beam. The parameters sent to the beam deflection supplies are interpolated from tables stored in memory. Initial deflection tables were derived from rough laboratory calibrations of the deflection systems. After launch, the instrument was operated in a calibration mode from which corrected deflection tables were generated. This procedure revealed some effects which were unanticipated.

First, the beam deflection tables should nominally depend only on the projection of the magnetic field into the beam emission plane. However, we have found that there is a small dependency of the deflection angle on the out-of-plane magnetic field component. Second, the two occurrences in each spin when the projection of the magnetic field into emission plane (in the rotating GDU reference frame) had precisely the opposite direction also required slightly different deflection voltages. After extensive testing, the deflection tables were enlarged and onboard software implemented so that both effects were taken into account. For two of the units, we were able to obtain good signals over the entire spacecraft rotation regardless of magnetic field orientation. The third had portions of its range of emission direction which were blocked by parts of the spacecraft structure, thereby preventing beam return for certain angles. For all unblocked angles we were able to get a good return signal.

ANALYSIS

Geometrical Considerations

The signal that is determined from synchronous detection is not simply the drift step along the measurement axis of the detector. It also contains effects due to the geometry of the injection and return of the beam to the detectors. There are two main effects. Both arise when the beam-magnetic field geometry is such that the returning beam plane does not intersect the detector strips perpendicular to the detector axis. The first effect produces a multiplicative distortion of the returned signal. The second effect produces an offset to the signal.

In Figure 3 we show two views of the gun-detector system. The upper half of the figure shows the view looking toward the two detector strips along the direction in which the beam is fired for 0° deflection angle. The lower half of the figure shows the view looking down along the detector measurement axis, that is, looking down the vertical axis of the upper half of the figure.

To understand how the geometry affects the beam signal, we illustrate a simple case with the two lines labeled 1 and 2 in the upper left half of Figure 3. The orientation of E, B, V_B, and V_D for this example is shown in the middle left of the figure. Line 1 represents intersection of the return beam plane and the plane of the figure when there is no electric field. The magnetic field points to the left and is parallel to the line because the return beam is a plane due to the spreading along the magnetic field. For this orientation of the magnetic

field, the beam firing direction is into the paper, indicated by V_B, which is 0° deflection angle direction. The drift step is perpendicular to both the electric and magnetic fields and points to the lower left is shown by Line 2 which represents the drift step displacement from Line 1.

The first of the geometrical effects is shown by the set of lines labeled a, b, and c on the upper left side of Figure 3. The drift step is the perpendicular distance between the lines 1 and 2 shown by the line labeled c. We wish to determine the component of this drift step along the vertical axis. This is the distance labeled b. However, the measured distance is $a + b$ which is the displacement of the beam along the vertical. As can be determined from the geometry shown, the relation between $a + b$, the measured quantity, and b, the component of the drift step along the vertical axis is $b = (a + b) \cos^2 \theta_B$, where θ_B is the angle the normal to the return plane makes to the detector axis. When the beam does not return perpendicular to the vertical axis, the quantity measured is always greater than the actual drift step component along this axis. This effect is a multiplicative one, and we multiply the measured quantity by the appropriate factor which is always less than or equal to one.

To determine θ_B for all orientations of beam firing direction and magnetic field, we must solve for the interception of the return beam plane and the detector in a more general way. We use the coordinate system shown in the upper and lower right for each view of Figure 3. The y-axis is along the detector axis, the x-axis points to the right, and the z-axis points anti-parallel to the 0° deflection direction and completes the right-handed system. In the general case, θ_B is the angle between the y-axis and the projection of the normal n to the returning beam plane. The detection point is sufficiently close to the beam emission point that the beam return direction is the smae as the emission direction V_B. The normal to the return beam plane n is defined by $n = b \times V_B$. This gives

$$\cos^2 \theta_B = \frac{n_y^2}{n_y^2 + n_z^2}$$

where b is the unit vector along the magnetic field. To get the beam to return at all, we must have $V_B \cdot b = 0$. Using this and the approximation that the beam is emitted only in the x-z plane allows to write the correction term as:

$$\cos^2 \theta_B = \frac{1}{1 + [b_y b_z / (b_x^2 + b_z^2)]^2}$$

which depends only on the components of b. In practice the correction factor is solved separately in the plane each detector using coordinate systems rotated around $\hat{y}$ by ±45°.

The second geometrical effect arises due to non-collocation of the beam emission point and the detectors and is also illustrated by Figure 3. For this effect we need only examine the line labeled 1. The + in the center of the figure marks the beam emission point as projected into this plane. However, the wedge anodes (one half of each set is shaded) can be seen to be displaced to the left and to the right of this point. Because the magnetic field has a $\hat{y}$ component, the point at which the beam intercepts the detector is offset from zero (shown as a dotted line labeled 0 offset). For the example shown the left-hand strip has positive offset and the right-hand strip has negative offset. This is an additive effect and is subtracted from the measured values before the multiplicative factor is used.

To solve for the general case of the additive offset, we use the same coordinates as before. The beam firing direction is in the $-\hat{z}$ direction. We define the beam emission point to be the origin of this coordinate system and assume that the zero point of the detector is at (A,0,B), that is, the zero point lies in the x-z plane but is removed from the emission point. The additive offset depends only the $\hat{x}$ component of the zero point position $L = A$ and is given by:

$$\Delta = L \tan \theta_B = L \left(\frac{b_y b_z}{b_x^2 + b_z^2} \right)$$

which is consistent with the definition of θ_B given above. Other gun firing directions require a rotation about the $\hat{y}$-axis and give the same result, but we then have $L = A \cos \theta_G + B \sin \theta_G$, where θ_G is the gun firing direction with positive sense for return beams which fall on the left detector as shown in Figure 3.

Due to the nature of the beam deflection system, the emission point is not fixed. As the beam emission direction is changed, the emission point moves. To zeroth order, this is the point at which the beam leaves the magnet system. However, the magnetic field of the deflection system is appreciably larger than the ambient magnetic field over a distance of 2-3 cm from the edge of the pole pieces, and the beam continues to bend in this fringing field. The in-situ field then becomes dominant and the emitted electrons begin to gyrate in that field. It is this point which must be used to determine the additive offset. Uncertainties in the knowledge of the fringing fields of the flight magnets introduce errors in the position of the source location on the order of a few mm. In addition, the above analysis assumes that the beam is emitted only in a plane. Slight variations out of this emission plane will also introduce errors into these formulae for the geometrical effects. Investigations of the fringe fields from a test unit show that deviations out of the plane of 1-2° may be present in the flight units. These errors can be corrected empirically, as described below.

Data

In Figure 4 we show an example of data from one of the GDU's (Unit 2) which displays the various quantities used to determine the electric field. In the bottom panel we show the deflection angle vs. time (the curved lines running from +90° to -90°) and the angle that the magnetic field makes with respect to the spin plane (the slowly varying line near zero). The next panel shows the $A + B$ current. Note that due to the 4 times lower sample rate, this data is shown repeated 4 times at the same value. The next panel is the $A - B$ current, and is shown at its full resolution. In the second panel from the top, we show the ratio $(A - B)/(A + B)$ (dark line). We also show (light dotted line) the correction due to the non-collocation offset. It is clear that the raw ratio closely resembles this offset.

The uppermost panel shows a comparison between the ratio signal corrected for the geometric effects described above (dark line) and the signal expected as calculated from $v \times B$ using the spacecraft velocity solution and the on-board magnetometer data (light dotted line). It is important to remember

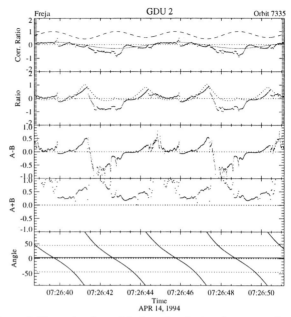

Figure 4. Example of raw data from the electron beam experiment.

that the rapid motion of the spacecraft induces an electric field of about 200 mV/m in the reference frame of the spacecraft which appears as a sinusoidal signal. This gives a quantitative measure of how well the experiment measures electric fields for regions along the orbit when we are reasonably certain there are no real electric fields of similar magnitude which would confuse the data. While the agreement with the computed $v \times B$ signal is, at times, quite good, at others there are substantial discrepancies which are attributed to imprecise knowledge of the exact location of the emission point and the emission direction. We also show the multiplicative correction (dashed line) to show how it varies.

Empirical Correction and Comparison with Double Probe

To understand how well the Electron Beam Instrument responds to electric field signals, we have developed an empirical correction to the data. The procedure takes the ratio which has been corrected for the geometrical effects (as shown in the top panel of Fig. 4) and determines the difference between this ratio and the expected $v \times B$ signal. This difference is determined for a single revolution of the spacecraft at a single, specified time, but is then used for the entire period of interest to correct the ratio signal.

By definition, there is perfect agreement between the empirically corrected signal and $v \times B$ for the period from which the correction was generated. As one moves away from the calibration period, the empirically corrected signal deviates increasingly from the expected signal. This is due to variations in required correction as the magnetic field varies with respect to the spacecraft. As shown in the preceding section, the two geometrical corrections depend on all three components of the magnetic field. By generating the correction over one spin, we account for the variation of the magnetic field in the spin plane with the spin axis component being effectively constant. As the angle between B and the spin plane varies, the spin axis component changes and the correction becomes imperfect. The empirical correction is generally good until the magnetic field orientation with respect to the spin plane varies by a few degrees (typically a few minutes) from that during calibration period. Because this calibration generally introduces some noise into the measurement, we typically perform smoothing on the resulting time series using a median filter.

A comparison between the GDU-measured perpendicular electric field and that measured by double probe on Freja is shown in Figure 5. The data are shown in a despun spacecraft coordinate system. The bottom two panels show a comparison of the component of electric field in the spin plane of the spacecraft which is perpendicular to the magnetic field and which is labeled E_2. The middle two panels show a comparison between the two techniques for the component which lies in the spin plane and is parallel to the projection of the B vector into the spin plane and is labeled E_1. The top panel shows the component of $E_\perp$ along the spin axis and is labeled E_z. Because the double probe does not measure this component, it is usually inferred by assuming $E \cdot B = 0$. However, in cases such as this one, in which the magnetic field lies very near the spin plane, such an inference is not possible, and only the electron beam measurements are shown. The calibration period is labeled in the top panel of the figure. The data from both measurements have been smoothed using a median filter with a width of 490 ms.

The comparison, is, in general quite good. The beam measurement shows some spin-periodic noise with amplitude of ~20-30 mV/m in the leftmost and rightmost parts of Figure 5. This an artifact of the correction procedure and is not seen in the double probe E_2 component. Perhaps the most significant disagreement is between 0727:35 and 0728:10. In fact, if one covers up the righthand portion of the plot from 0727:35 onward, other than the noise in the beam measurement, the agreement is quite striking. If one covers up the lefthand portion of the plot up to this time, then the agreement is rather poor.

The poor agreement in the E_1 component occurs because the angle that the ambient magnetic field makes with respect the spin plane (λ_B on the x-axis) is quite small with a value of 2-4°. When the two probes become aligned along the magnetic field with the spacecraft body in between, the upper and lower probes can experience different plasma conditions which may give rise to a spin-modulated signal [*Lindqvist and Marklund*, 1990]. However, the poor agreement of the E_2 components is not explained by this effect because the orthogonal set of probes which measure the E_2 component will not be influenced by shadowing. The cause of the disagreement of the two measurements of E_2 is not yet understood.

One of the advantages of the beam technique is that it directly determines the spin-axis component of the electric field. As can be seen in the example shown in Figure 5, this component can be quite large. Without this component it can be impossible to understand the orientation of the electric field. For the period 0727:40 to 0727:50 this component is dominant and without it, the orientation would be in error by more than 45°. For further examples of electron beam data, comparisons with double probe data, and relation of electric field to auroral precipitation, we refer the interested reader to *Kletzing et al.* [1994].

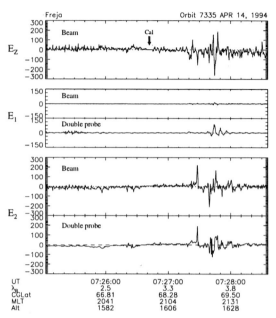

Figure 5. Comparison between double probe electric field measurements and the electron beam measurements.

Acknowledgments. We would like to acknowledge our Co-Investigators and others who have contributed to this work: G. Haerendel, R. Frenzel, H. Höfner, W. Baumjohann, N. Sckopke, E. Sartori, G. Marklund, and P.-A. Lindqvist.

REFERENCES

Kletzing, C. A., G. Paschmann, M. H. Boehm, G. Haerendel, N. Sckopke, W. Baumjohann, R. B. Torbert, G. Marklund, and P.-A. Lindqvist, Electric fields derived from electron drift measurements, *Geophys. Res. Lett.*, *21*, 1863, 1994.

Lindqvist, P.-A., and G. T. Marklund, A statistical study of high-altitude electric fields measured on the viking satellite, *J. Geophys. Res.*, *95*, 5867, 1990.

Melzner, F., G. Metzner, and D. Antrack, The Geos electron beam experiment, *Space Sci. Instr.*, *4*, 45, 1978.

Nakamura, M., H. Hayakawa, and K. Tsuruda, Electric field measurement in the ionosphere using the time-of-flight technique, *J. Geophys. Res.*, *94*, 5283, 1989.

Parl, M., Diplomarbeit, 1991.

Paschmann, G., F. Melzner, G. Haerendel, O. H. Bauer, W. Baumjohann, M. Boehm, M. Nakamura, N. Sckopke, R. Treumann, C. E. McIlwain, W. Fillius, E. C. Whipple, R. B. Torbert, J. M. Quinn, V. Formisano, A. Amata, A. Pedersen, K. Tsuruda, and H. Hayakawa, *The electron drift instrument*, p. 55, SP–1103. European Space Agency, 1989.

Paschmann, G., M. Boehm, H. Höfner, R. Frenzel, P. Parigger, F. Melzner, G. Haerendel, C. A. Kletzing, R. B. Torbert, and G. Sartori, The electron beam instrument (F6) on Freja, *Space Sci. Rev.*, *70*, 447, 1994.

Tsuruda, K., H. Hayakawa, and M. Nakamura, Electric field observations on Geotail, in *Science Objectives of the Geotail Mission*, edited by A. Nishida, p. 39, ISAS, Tokyo, 1985.

M. Boehm, Jet Propulsion Laboratory, Mail Stop 169-506, 4800 Oak Grove Drive, Pasadena, CA, 91109.

C. A. Kletzing, Department of Physics and Astronomy, The University of Iowa, 203 VAN, Iowa City, IA, 52242.

G. Paschmann, Max-Planck-Institut für extraterrestrische Physik, Giessenbachstrasse, Postfach 1603, D-85740 Garching, Germany.

Electric Field Instrument Using Radiated Electrons (E-FIRE): An Innovative Approach to the Measurement of Electric Fields in the Earth's Magnetosphere

A M Jorgensen, K L Hirsch, M J Alothman, S Braginsky, T A Fritz,

Center for Space Physics, Boston University

We are proposing a novel compact sensor system for measuring electric fields in space plasmas. The measurement of electric fields from earth orbiting satellites has been a difficult challenge for the large scale DC component of this field. Traditional techniques using double-probe style instruments are troubled by contact potentials to the plasma environment and photoelectric effects due to solar radiation. Newer techniques using low energy electron beams put requirements on costly spacecraft electric cleanliness, in order to avoid beam deflection by spacecraft potentials. E-FIRE uses energetic electrons of approximately one MeV in connection with recently developed ion-implanted solid-state detectors which have good position resolution and energy determination. The electrons are produced by a radioactive source with a line spectrum that tags the particles as being emitted from the satellite. These electrons will return to the spacecraft within a few microseconds having displacements of millimeters to centimeters due to the ambient electric (and in some cases magnetic) field, independent of any electric potentials that may have been produced by solar radiation or the interaction of the spacecraft with the local plasma environment.

1. INTRODUCTION

Electric fields in space are very difficult to measure. They are usually small: on the order of a few millivolts per meter. Such fields must be measured on a spacecraft whose motion through the magnetic field can induce electric fields that may be an order of magnitude greater than the ambient field.

Electric fields are traditionally measured with double-probe style instruments. These consist of long booms with good conductive contact to the ambient plasma at each end, and isolated everywhere else. The elec-

tric field component along the boom can thus be determined by measuring the potential difference between the ends of the booms, subtracting out the electric field induced by the motion of the spacecraft. Two possible source of error remain with the double-probe measurement. Firstly, there is the contact potential between the probes and the ambient plasma. If this is the same for both probes, it does not contribute. However, the contact potential depends in detail on the surface properties of the probes, and on the local plasma environments, and could therefore be different if the probes are not completely identical. Secondly, solar radiation may set up photo-electric potentials between the ambient plasma and the probes, a potential that could also differ between probes. If the photo-electric potential is the same for both probes, it does not contribute to the measurement. However, variations in surface properties or illumination can introduce this problem. In most cases there are solutions for both of these problems, but their implementation often contribute to making good electric field measurments expensive.

Interactions of the spacecraft with the plasma medium can induce large potential differences between different parts of the spacecraft and between the spacecraft and the ambient plasma medium. While this is not a problem for the double probe instrument, which does not measure the spacecraft potentials, but only the probe potentials difference, this can cause problems for low energy electron beam techniques. In low earth orbit, a typical low energy electron gyro radius is not much larger than the spacecraft, which implies that the spacecraft may disturb the beam [Kletzing, et al.,1994]. Spacecraft potentials can be be a significant fraction of the energy of the beam, and therefore cause significant deflection of the beam, corrupting the measurement of a small drift step caused by the ambient electric field. We present a new technique that has the potential for determining the ambient geoelectric field with minimal impact on spacecraft resources of weight, power, and special structures.

2. THEORY OF OPERATION

E-FIRE measures the drift of energetic electrons. The drift has the following three components: (a) The $\vec{E} \times \vec{B}$ drift, caused by the ambient electric field, (b) The motion of the spacecraft in the frame of the magnetic field, (c) Magnetic gradient drift. In the frame of the spacecraft, the total drift of the electron in one gyration (see Figure 1) can thus be written as:

$$\Delta \vec{r} = (\vec{v}_{\vec{E} \times \vec{B}} - \vec{v}_{s/c} + \vec{v}_{\vec{\nabla} B})$$

$$\Delta \vec{r} = (\frac{\vec{E} \times \vec{B}}{B^2} - \vec{v}_{s/c} + \frac{E_{kin,\perp}}{q} \frac{\vec{B} \times \vec{\nabla} B}{B^3}) \tau_g \quad (1)$$

Note that the $\vec{E}$ is the electric field in the frame stationary with respect to the magnetic field, and that the inclusion of $\vec{v}_{s/c}$ in the above expression is equivalent to the inclusion of a $\vec{v}_{s/c} \times \vec{B}$ term in an expression for the total electric field in the spacecraft frame. In a low altitude polar cap region, typical numbers are B=0.3-0.5 Gauss, E=100 mV/m, giving $v_{\vec{E} \times \vec{B}}$=2.0-3.3 km/s. $v_{s/c}$ is typically 8 km/s for a low altitude satellite. The field aligned value of $\vec{\nabla} B$ in the polar caps is 1.3 Gauss/R_E. Since most of the gradient in the polar caps is field aligned, it is reasonable to assume an upper limit on the perpendicular gradient of $\nabla B = 0.15$ Gauss/R_E.

Combined with an electron kinetic energy of 1 MeV, we can estimate a maximum gradient drift as $v_{\vec{\nabla} B} = 1.0$ km/s. If the instrument is capable of using particles of at least two different energies, the gradient of the magnetic field can be calculated by writing equation 1 for the two cases, subtracting, which eliminates the $\vec{E} \times \vec{B}$ and $\vec{v}_{s/c}$ terms, and then solve for $\vec{\nabla} B$ (Recall that all particles with the same relativistic mass have the same gyroperiod. For instance, all non-relativistic electrons have the same gyroperiod, but their gyroradii, and therefore the distance away from the spacecraft they sample varies with their energy - this is how the gradient can be separated). In this connection, one should note that the $\vec{\nabla} B$ compoent does not affect low energy electrons, which does present a small advantage to low energy electron beam experiments.

From the measurement of the drift displacement, the electric field can be determined, if the drift components due to spacecraft motion and magnetic field gradient can be determined and subtracted with sufficiently high accuracy. Revisiting equation 1, we see that three parameters must be well determined in order to calculate $\vec{E}$ from the $\Delta \vec{r}$: $\vec{B}$, $\vec{v}_{s/c}$, and $\vec{\nabla}_{\perp} B$. Using an onboard magnetometer in connection with at least two different particle energies, allows us to determine $\vec{B}$, and $\vec{\nabla}_{\perp} B$. These two parameters could also be determined from magnetic field models, or a combination of modelling and measurements could be used. If we use only measurments and assume negligible uncertainties on $\vec{B}$ (0.1% or better) and $\vec{v}_{s/c}$ (retrospective orbit fitting can reduce this to 0.05% or better), the dominant source of uncertainty comes from the measurement of the drift step, and the determination of $\vec{\nabla} B$. As mentioned above, $\vec{\nabla} B$ can be determined using two particles of different energies, and in a typical situation, one would expect the uncertainty on $\vec{\nabla} B$ to be about equal to the uncertainty on the determination of a drift step. When this value is then used in combination with measured drift steps to calculate $\vec{E}$, one arrives at an expected uncertainty on $\vec{E}$ that is a few times the electric field corresponding to the position resolution of the instrument.

3. IMPLEMENTATION

E-FIRE has three main components: a collimated radioactive source, a position and energy sensitive detector, and the processing electronics.

3.1 Source

There are several criteria to weigh when selecting the radioactive source for the experiment. It must be easy to determine whether a detected particle originates from the source or from the ambient plasma population. This criterion favors a line source, since particles are emitted with very specific energies, allowing for easy identification. The particles from the source must not have a to large gyroradius, since then only very few will return to the detector, making a very intense (and potentially dangerous) source necessary. Particles with very long gyroperiods (such as He^{2+}) will take too long to return resulting in a very large drift - of the order of meters for He^{2+}, as oposed to millimeters for electrons - which is impossible to measure with a small instrument. These criteria favor low energies, and electron sources, since heavier or more energetic particles will have too large gyroradii, too long gyroperiods, and too little returned flux. The third criterion is to select the line source with an energy where the background radiation is minimized. This criterion favors high energies. Weighing all these factors, we propose to use ^{207}Bi which has several β lines at around 1 MeV energy. A listing of the line energies and intensities of ^{207}Bi is given in table 1.

As discussed above, the source strength necessary depends critically on the gyroradius of the particles, and on the background radiation at energies corresponding to the beta radioactive line. When the electons are emitted, they will return to the focus point (displaced from the source by the drift vector) as a very extended sheet parallel to the magnetic field, because the perpendicular motion has refocused the particles in that direction, while the particles continue to spread out along the field line. Kletzing et al. [1994] contains an excellent illustration of this (see their Figure 1). Therefore, the particles that return are those that are emitted within an angle from the perpendicular direction, that is half the angular field aligned size of the detector as seen from a distance equal to the circumference of the particle orbit. In addition, one can reduce smearing of the return spot in the event that the drift places the focus off the surface of the detector, by introducing an exit aperture, which allows all pitch angles through, but restricts particles emitted perpendicular to the field line (the particles that will return to the detector) in only a narrow

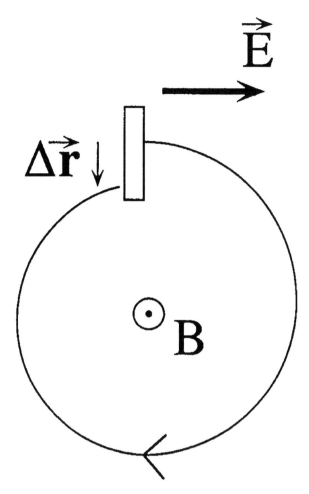

Figure 1. An electric field, $\vec{E}$ will cause a particle to drift in it's gyration orbit in a direction perpendicular to the electric field and to the magnetic field.

angular range. Table 2 demonstrates the number of particles expected to return to a 1 cm wide (field aligned dimension) detector from sources with different intensities. The table gives count rates for both a 1 degree and 10 degree wide exit aperture. Since 1 MeV electron fluxes in the polar cap are very low, the choice of source intensity depends more on the desired time resolution (time between particle detections), than on the efficiency of subtracting out an ambient background. The background can be further reduced by incorporating energy discrimination narrowly around the source line. Based on these consideratioins, we would consider a good choice of source to be in the range 10-100 mCi.

Flying a radioactive source may raise logistic issues, and restrictions may be imposed which require a rigorous hazards control program to be implemented. One concern is for those working on the manufacturing and

Table 1. The 4 prominent electron lines of the ^{207}Bi spectrum. The columns are: energy, gyro radius, gyro period and relative efficiency.

E/MeV	r_g/m	$\tau_g/\mu s$	R/%
0.482	66.3	1.39	22
0.554	70.3	1.49	6
0.972	99.0	2.07	55
1.044	103.8	2.17	17

Table 2. The number of particles to expect to return to a 1 cm wide detector. Aperture size is the angular spread of the electrons, perpendicular to the magnetic field, as they leave the source.

Source Strength	Aperture Size	Electron Energy / MeV	Count Rate / s^{-1}
1 Ci	1°/ 10°	0.482	542/5420
		0.554	140/1400
		0.972	909/9088
		1.044	268/2679
100 mCi	1°/ 10°	0.482	54/542
		0.554	14/140
		0.972	91/909
		1.044	27/268
10 mCi	1°/ 10°	0.482	5/54
		0.554	1/14
		0.972	9/91
		1.044	3/27
1 mCi	1°/ 10°	0.482	0.5/5
		0.554	0.1/1
		0.972	1/9
		1.044	0.3/3

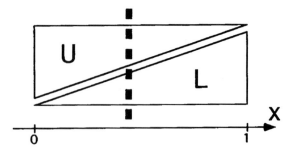

Figure 2. A wedge detector consists of two identical, but oppositely directed triangular strips. The returning electron beam will be smeared along the field line, drawing a strip across the detectors as indicated by the thick dashed line. The relative number of counts detected by the two strips determines the displacement of the returning 'sheet' of electrons.

testing of the instrument and spacecraft. This concern could be greatly reduced by working with a much weaker source during this phase, and then introduce the flight source only immediately before launch. Another concern has to do with the risk of the payload being destroyed during launch. If this were to happen, one could imagine the radioactive material being dispersed, creating a potential environmental hazard. The small size of the source will allow it to be encased in a rugged, heat resistant plastic, which would render it virtually indestructible. In case of an accident, the radioactive material could remain unrecovered for weeks, or even years without causing environmental problems. In fact, one of the authors (TAF) has flown multiple radioactive source on the ISEE and POLAR spacecrafts.

Another potential concern is that of having a strong radioactive source located close to the instrument electronics, as would be necessary in a compact design. With a very compact source, efficient shielding of the electrons from the electronics is possible with only a few grams of material. The source holder can be designed to attenuate the electron flux in the direction of the electronics to tolerable levels. Some X-rays will be produced, but they are less of a fluence problem.

3.2 Detector

For illustrative purposes, we have included a description of the wedge detector as well as the integrated solid state strip detector that we propose to use for E-FIRE.

3.3 Wedge Detector

The wedge detector determines position of particles from counting statistics on two oppositely oriented wedge strips. The position along the detector strip depends on the count rates of the two wedges, C1 and C2, as $x = x_{max}(C1/(C1 + C2))$ for a detector of length x_{max}. To achieve high precision, high count rates are necessary. In E-FIRE, high count rates are difficult to achieve without using an intense source. The wedge detector concept was used on the Swedish satellite Freja [*Paschmann et al.,*1994]. It has the advantage of being extremely simple in design, consisting of just two independent solid state detectors (see Figure 2). The advantage of this system is its simplicity. The disadvantage is the high count rates necessary for good position determination.

3.4 Strip Solid State Detector

The strip detector determines positions by registering which one of a large number of elements the particles hit. This detector concept has the advantage of being able to determine the drift step of a single particle - in addition to it's energy - allowing the drift to be sampled at the single particle count rate. The typical geometry of a strip detector is shown in Figure 3. The resolution of such a detector depends solely on the density with which the individual elements are spaced. Table 3 gives some typical resolution figures for using the ^{207}Bi source on a low altitude spacecraft. The example uses a detector of length 41 millimeters, which corresponds to a dynamic range of approximately $\pm$ 100 mV/m in addition to the dynamic range needed for $\vec{v}_{s/c}$, and $\vec{\nabla}B$ drifts. The dynamic range could easily be changed to suit a specific situation. To increase the dynamic range, a longer strip could be used, or two or more strips could be placed end-to-end. A solid state detector strip can contain thousands of individual detectors, and readout can be a problem if the electronics is not miniaturized. There are several ways of doing this: the detection electronics can be fabricated directly on the detector, making a single chip which both detects and processes. It is also possible to fabricate the processing electronics on separate dies, and wire-bond these to the detector. These multiple dies and detectors will then be mounted in a casing, making a single detection and processing unit as shown in Figure 4. Much work is being done in miniaturizing the analog portion of the front end electronics. These advances take place especially in the areas of experimental high-energy physics, in which very large numbers of detectors are placed in relatively small volumes in order to trace the results of nuclear reactions. One such effort at Rutherford Appleton Laboratories has produced numerous high density pre-amplifier circuits, such as the MX series [*Seller, et. al,* 1988], some of which have already been successfully flight tested on space physics instruments [*Levine,* 1994].

4. ELECTRONICS

Since the electronics of the detector will be contained in a very small package, in order to take maximum advantage of this miniaturization, the E-FIRE data processing unit (DPU) will be made as compact as possible. The processor will be a single chip computer (such as the Motorola 6811) with a small amount of external memory. A block-diagram of E-FIRE is shown in figure 5. Most communication with the detector and with the spacecraft will be implemented directly on-chip through

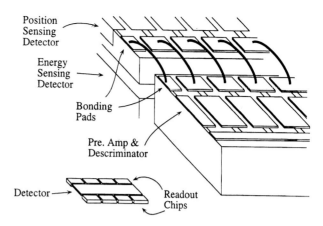

Figure 4. The figure demonstrates how the strip detector output can be acquired and processed in a compact manner by external amplifier/readout chips. These different chips can be packaged into a single compact unit with digital event readout to the instrument computer.

programmable I/O pins. The instrument can process the data in several different ways:

1) Single event mode. Selected detections (according to some criteria) are transmitted to the ground as a word containing energy and position information.

1) Histogram mode. Detections are accumulated into a histogram of spin phase, position, and energy. These histograms are then transmitted to the ground at regular intervals.

2) Fitting mode. Another possibility is that the onboard software can perform necessary fits to the histograms to calculate the average electric field vector over the accumulation interval.

All three modes could operate on the same data, and share the same telemetry stream to provide data useful for a wide variety of applications.

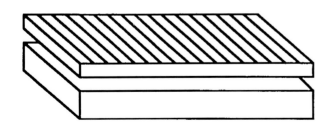

Figure 3. Basic geometry of the strip detector. On top is a layer of narrow parallel strips, and underneath is a thick back detector. The electron penetrates the thin strip detector, leaving a small amount of energy as the position signal, and the rest in the back detector. The back detector energy signal is used to identify the particle as coming from the ^{207}Bi source, or from the ambient plasma.

Table 3. The table demonstrates, for different field resolutions, the needed density of strips, and total number of strips that are needed for a 41 mm long detector, capable of measuring a low altitude spacecraft velocity plus a $\pm$ 100 mV/m electric field.

Field Resolution	Spatial Resolution	Number of Distinguishable Positions
100 mV/m	4.1 mm	10
10 mV/m	0.4 mm	100
1 mV/m	41 μm	1,000
0.5 mV/m	21 μm	2,000
0.1 mV/m	4 μm	10,000

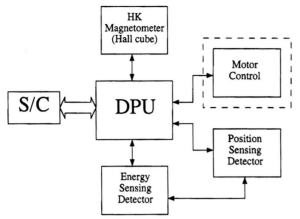

Figure 5. A schematic diagram of the different parts of E-FIRE. The energy signal from the thicker energy detector is used to determine whether the signal from the position sensing detector should be processed as an electric field drift. Possible extensions to the simple design could be a housekeeping magnetometer whose signal is used to control a motor to align the solid state strips along the magnetic field.

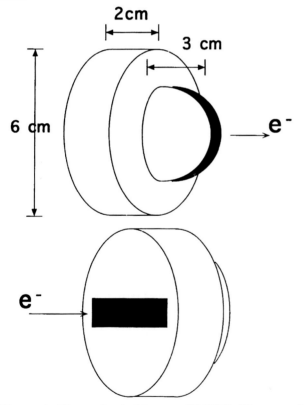

Figure 6. The mechanical layout of E-FIRE. The source is housed at the center of the dome to the right. A slit in the dome acts as an exit aperture which collimates the electron beam. On the left side of the instrument is the detector. The instrument works best when placed such that the individual strips (and the slit in the dome) are parallel to the magnetic field.

4.1 Packaging

All components of E-FIRE (except for power supply and spacecraft interface) can be packaged into a very small compact and lightweight package as shown in Figure 6. It will be approximately 6 centimeters in diameter, and 5 centimeters tall. The total weight (excluding power supply and spacecraft interface) will be less than 500 grams.

5. CONCLUSIONS

E-FIRE presents a very compact and inexpensive alternative for making routine measurements of the electric field in the low-altitude polar cap region. It wins over the double-probe style instrument both in size, simplicity and cost. It is not sensitive to spacecraft charging or sheath effects that can affect the low energy electron beam instrument. E-FIRE is the ideal piggyback instrument for providing routine, synoptic polar cap electric field measurements. It is compact, economical, independent of spacecraft resources, and agrees well with recent trends towards smaller, faster, cheaper missions.

REFERENCES

Kletzning, C A, G Paschmann, M H Boehm, G Haerendel, N Sckopke, W Baumjohan, R B Torbert, G Marklund, P-A Lindqvist. Electric fiels derived from electron drift measurments, *Geophys. Res. Lett.*, 21, 17, 1994

Levine, G L. Characterization of the RAL and Hugins Imaging Electron Spectrometers, *Master's Thesis in Electrical Engineering at Boston University*, 1994.

Paschmann, G, Boehm, M, Hoefner, H, Frenzel, R, Parigger, P, Melzner, F, Haerendel, G, Kletzing, C, A, Torbert, R, B, Sartori, G. The Electron Beam Instrument (F6) on Freja, *Space Science Reviews,* 70:447-463, 1994.

Paschmann, G, Melzner, F, Haerendel, G, Bauer, O, H, Baumjohan, W, Sckopke, N, Treumann, R, McIlwain, C, E, Fillius, W, Whipple, E, C, Torbert, R, B, Quinn, J, M. The Electron Drift Instrument for Cluster, *GGS Instrument Papers,* 1995

Seller, P, Allport, P, P, Tyndel, M. Results of Silicon Strip Detector Readout Using a CMOS Low Power Microplex (MX1). *IEEE Transactions on Nuclear Science,* Vol. 35, No. 1, February, 1988.

725 Commonwealth Ave, Boston, MA 02215, USA

Gradient-Induced Errors

N. Hershkowitz and D. A. Diebold

Department of Nuclear Engineering and Engineering Physics, University of Wisconsin, Madison, WI

Plasma gradients are a possible source of error in the measurement of weak double layers in the magnetosphere by satellite double probes. Space charge enhancement of Plasma Gradient Induced Error (PGIE) has been considered elsewhere [*Diebold et al.*, 1994] and is briefly reviewed here. Possible effects of presheaths on space charge enhancement of PGIE are discussed. Gradients in wave amplitude may lead to Wave Gradient Induced Error (WGIE) and this is also discussed.

1. INTRODUCTION

Double probes, i.e., two spatially separated probes [*Aggson and Heppner*, 1964, *Boyd*, 1967] which can be either spheres or wires, are a common diagnostic on satellites and rockets. Electric fields in space are often determined by taking the difference in potential between the two probes of a double probe and dividing by the distance between the probes [*Pedersen et al.*, 1978]. Our special concern has been double probe measurements of weak double layers in the magnetosphere, where probe photoemission is an important (and often dominate) source of probe current. Photoemission can be a source of double-probe electric-field-measurement error. The following techniques are among those that are employed to help ensure good magnetospheric double probe measurements in the presence of strong photoemission: double probe surfaces arc made of conducting materials so that they are equipotential [*Mozer et al.*, 1978], double probes are constructed so that both probes are equally shadowed at all spin angles [*Fehringer*, 1989], negatively biased guards are often placed between probes and satellites to minimize leakage currents [*Mozer et al.*, 1978], and current sources are used to bias probes when electric field measurements are

being made (this makes the probes less sensitive to spurious currents) [*Pedersen et al.*, 1978].

Another source of error are gradients. When the spacing of double probes are comparable to plasma gradient scale lengths, errors in the electric field measurements made by probes are possible [*Fahleson*, 1967]. Such errors are known as Plasma Gradient Induced Errors (PGIE). In addition, gradients in wave amplitude may lead to Wave Gradient Induced Errors (WGIE) which are similar to PGIE. We have previously [*Diebold et al.*, 1994] considered space charge (associated with strong photoemission) enhancement of PGIE and concluded that it does not account for typical double probe measurements of weak double layers (specifically, the calculated electric field measurement error associated with this effect was calculated to be less than Viking satellite electric field measurements of weak double layers [*Boström et al.*, 1988] by at least an order of magnitude). Here, we briefly review our previous work and then consider the possible effects of presheaths (which were not consider in our previous work) on space charge enhancement of PGIE. We also discuss WGIE.

2. SPACE CHARGE EFFECTS

Space charge associated with photoemitted electrons modifies sheaths in such a way as to tend to repel the photoemitted electrons. It is important to recognize that when strong photoemission is present, space charge effects can result in a plasma potential that undershoots near the emitting probe [*Guernsey and Fu*, 1970]. A qualitative

Measurement Techniques in Space Plasmas: Fields
Geophysical Monograph 103
Copyright 1998 by the American Geophysical Union

depiction of such a radial plasma potential profile (measured from the probe surface at R_P) is shown in Figure. 1. Here F_T, the trapping potential, is the potential difference between the probe bias voltage, V_P, and the potential minimum. The ambient plasma potential is shown as F_A. The difference in potential between Φ_A and the potential minimum is shown as Φ_M. The sheath potential, Φ_S, is $\Phi_S \int V_P - \Phi_A = \Phi_T - \Phi_M$.

When using double probe data to determine electric field, it is assumed that the F_S's of the two probes are equal. Error in the double probe method that is cause by plasma gradient induced differences in the Φ_S's is referred to as PGIE or plasma gradient induced error. When differences in the Φ_S's are caused by differences in the Φ_M's, we refer to the error as space charge enhanced PGIE. Relatively small amounts of plasma can have large effects on space charge because the conservation of angular momentum allows ions to spend relatively long periods of time orbiting probes and alleviating space charge [Kingdon, 1923]. A depiction of space charge enhanced PGIE is schematically shown in Figure 2. The Φ_S of the probe on the right is greater than that of the left probe because of the right probe's smaller Φ_M. For the situation shown in Figure 2, the electric field calculated from the voltages of the double probe would be greater than the ambient electric field because of the greater space charge near the left-hand probe. Such a situation could come about if there was less plasma density at the left probe (as compared to the right probe) and this smaller density resulted in less alleviation by ions

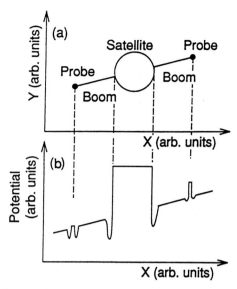

Figure 2. (a) Schematic of a satellite with double-probe. Probe and satellite size has been exaggerated for clarity. (b) The potential structure associated with the satellite/double-probe system when there is space charge enhanced PGIE.

of the space charge associated with the left probe (and, hence, a greater Φ_M for the left probe).

Boström et al. [1988] reported Viking data showing electric fields associated with large ($\delta n/n \leq 50\%$), local depletions of plasma density. The electric fields associated with these depletions were such that the voltage of the probe that was in a region of depletion at any given time was often 2 to 3 V ($\approx T_e/e$, where T_e is the electron temperature of the ambient plasma and e is the charge of an electron) less than the voltage of the other probe. This is qualitatively consistent with space charge enhanced PGIE, such as that depicted in Figure 2.

3. EXPERIMENTAL RESULTS

Laboratory data from Diebold et al. [1994] which illustrate space charge enhanced PGIE are shown in Figure 3. Two current-versus-voltage characteristics of one probe are shown. The probe was cylindrical, 55 cm in length and 1.3×10^{-2} cm in diameter. For convenience, the probe was thermionically emitting rather than photoemitting. The probe was situated in a stainless steel chamber that was 64 cm in length and 60 cm in diameter. Neutral pressure (air was bled into the machine) was in the 10^{-4} to 10^{-3} torr range. The probe was heated by applying a half-wave rectified voltage across it. This voltage was large enough that during the "on" part of the heating cycle electrons were emitted with enough energy to ionize the

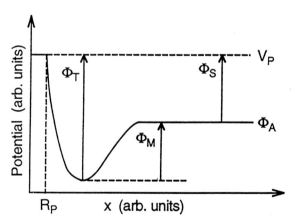

Figure 1. A qualitative illustration of plasma potential (in arbitrary units) versus radius (also in arbitrary units) near a photoemitting probe affected by space charge. The probe radius is R_P, the probe voltage is V_P, and the ambient plasma potential is Φ_A. The trapping potential Φ_T is the difference between V_P and the potential minimum, while Φ_M is the difference between Φ_A and the potential minimum. The sheath potential is $\Phi_S \equiv V_P - \Phi_A = \Phi_T - \Phi_M$.

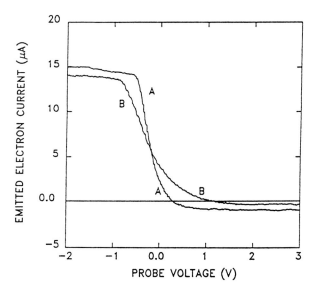

Figure 3. The data labeled A and B are two current (in microamperes) versus voltage (in volts) characteristics measured from the same laboratory probe under different plasma and probe conditions.

residual gas. In this way, plasma was created. The plasma electron temperature was approximately 2 eV and the plasma density was in the 10^3 to 10^5 cm^{-3} range.

Collected electron current is shown in Figure 3 as negative emitted-electron-current. The collected electron current corresponding to characteristic A of Figure 3 was greater than that corresponding to B, which indicates that the plasma density corresponding to characteristic A was greater than that corresponding to B. Consequently, there were more space-charge-alleviating ions present when characteristic A was taken. Characteristic A was found by Diebold et al. [1994] to be well described by theory in which it was assumed that the probe was not affected by space charge while characteristic B was found to be well described by theory in which was assumed that the probe was affected by space charge.

As can be seen in Figure 3, the emitted electron currents corresponding to characteristics A and B are approximately constant at probe voltages less than -0.5 V and -0.9 V, respectively. These constant currents are the saturated (limited by the temperature of the probe) emitted currents. The current of a characteristic such as A, which is not affected by space charge, begins to decrease from the saturated emitted current when the probe voltage becomes more positive than the local plasma potential. Hence, from inspection of Figure 3, the plasma potential near the probe when characteristic A was taken was approximately -0.5 V.

The current of characteristic B begins to deviate from its saturated emitted current at approximately -0.9 V, which is less than (not equal to) the ambient plasma potential that was near the probe when characteristic B was taken. Again, characteristic B is affected by space charge while characteristic A is not. Space charge makes it more difficult for photoemitted electrons to escape from the probe to the plasma. To achieve a given emitted electron current bias, a probe's voltage must be more negative when it is affected by space charge than when it is not. Qualitatively, the greater the emitted electron bias current, the greater the difference between the probe's voltage when it is affected by space charge as compared to when it is not affected by space charge. It follows that, in the absence of differences in plasma potential and in collected electron current, the characteristics of B and A would approach each other in the limit of zero bias current. As can be seen from inspection of Figure 3, at zero emitted electron current, the voltage of probe B is more positive than probe A by approximately 0.9 V. This suggests that the ambient plasma potential near the probe when characteristic B was taken was more positive than that when A was taken by roughly 1 V. Diebold et al. [1994] found that, when the differences in collected electron currents were taken into account, the (more accurate) difference between the plasma potentials was $\approx$ 0.5 V (rather than 1 V).

For the purpose for illustrating space charge enhanced PGIE, imagine that characteristics A and B are the characteristics of two probes on a satellite that are being used to measure electric field. For this situation, the ambient plasma density near probe A is greater than that near B (resulting in probe A not being affected by space charge and probe B being affected by space charge) and the ambient plasma potential near probe A is less than that near B. Depending on the current bias chosen, the probes would indicate positive, zero or negative electric field (positive electric field being defined here to be in the direction of the true electric field which points from B to A). If the probes were biased below, approximately at, or above 6 mA, they would indicate positive, zero, or negative electric field. For example, at a bias of 3 mA the voltage of probe A would be approximately 0.2 V greater than the voltage of probe B, whereas at a bias of 10 mA the voltage of probe A would be approximately 0.2 V less than the voltage of probe B. At a bias of 6 mA, the voltages of probes A and B would be approximately equal.

4. DISCUSSION

Diebold et al. [1994] quantitatively calculated the maximum space charge enhancement of PGIE, which occurs when space charge effects were present at one probe and absent at the other probe. The qualitative potential structure corresponding to such a situation is schematically shown in Figure 4. In Figure 4, it is assumed that $\nabla \Phi_A = 0$. The error in the double probe method is then due to the difference between the probe voltages. When, as depicted in Figure 4, the Φ_T's associated with the two probes are equal, the difference in the probe voltages (Δ) is equal to Φ_M (as can be easily seen from inspection of Figure 4).

Diebold et al. [1994] assumed Φ_M was such that the Child-Langmuir law for emitted electrons at the potential minimum (of the probe affected by space charge) was satisfied and derived the following expression for the maximum expected space charge enhanced PGIE:

$$E_M \text{ (mV/m)} \approx [I_U(100 \text{ nA})\gamma]^{2/3}/d(50 \text{ m}) \qquad (1)$$

when the probes and sheaths are cylindrical (I_U is the photoemitted current (in units of 100 nA) which passes through the potential minimum, γ is the ratio of sheath radius to probe length (and must be roughly ≤ 1 for the probe's sheath to be cylindrically symmetric), and d is the distance (in units of 50 m) between the two probes of a double probe) and

$$E_M \text{ (mV/m)} \approx 2[I_U(100 \text{ nA})(\alpha^2/5)]^{2/3}/d(50 \text{ m}).. \qquad (2)$$

when the probes and sheaths are spherical (α^2 is of the order of 5 and is very weakly dependent on the ratio of the probe radius to the probe's sheath thickness). Hence, for a roughly typical magnetospheric double probe with a probe separation of 50 m and a current bias ($\approx I_{II}$) of 100 nA,

E_M calculated from equations 1 and 2 is ≈ 1 to 2 mV/m depending on the geometry of the probes and probe sheaths.

The weak-double-layer electric fields observed by Viking were typically ≈ 2.5 V/80 m or 30 mV/m (where 2.5 V is typical of the voltage difference between the spherical probes and 80 m is the probe separation) and the current bias of the probe ($\approx I_U$) was 150 nA (R. Boström, private communication, 1991). Inputting $I_U \approx 150$ nA, d = 80 m and $\alpha^2 \approx 4$ into equation (2) yields $E_M \approx 1.4$ mV/m, which is more than order of magnitude less than the 30 mV/m associated with the weak double layers typically observed by Viking. The following four effects, however, were not considered by Diebold et al. [1994]: the possibility of an ion presheath associated with the probe affected by space charge; the possibility of an electron presheath associated with the probe not affected by space charge; geometrical (spherical and cylindrical) effects on Φ_T (as opposed to geometrical effects on Φ_M which were considered) and on the possible presheaths; asymmetric emission (photoemission only on the sunward half of a probe, not over the entire probe); and WGIE.

5. PRESHEATHS

Ion accelerating presheaths associated with planar, ion rich sheaths of boundaries (the voltages of these boundaries were below the plasma potential) have been observed in laboratory plasmas [Meassick et al., 1985, Meyer et al., 1992]. The $T_e/2e$ potential drop across such presheaths accelerates ions so that the ions enter the sheaths at the ion acoustic speed, which is necessary if Poisson's equation is valid in the sheaths. Likewise, there should be presheaths with a potential increases of $T_i/2e$ across them associated with the electron rich sheaths of nonemitting, planar objects biased above the plasma potential. If one probe of a double probe were affected by space charge and had an ion accelerating presheath with a voltage drop of $T_e/2e$ across it (the minimum in potential caused by the space charge is less than the ambient plasma potential) and the other probe (biased above the ambient plasma potential) were not affected by space charge and had an electron accelerating presheath with a voltage increase of $T_i/2e$ across it, then the error in the double probe method (of calculating electric field) due to these differences in presheaths (which in turn are due to differences in the effect of space charge on the two probes) would be $(T_e + T_i)/2ed$, where d is the probe separation. This error would be both qualitatively and quantitatively consistent with typical Viking electric field measurements.

Before concluding, however, that differences in probe presheaths could lead to errors consistent with typical

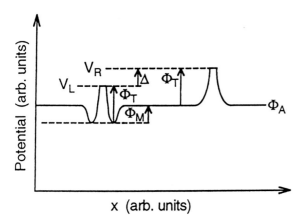

Figure 4. Qualitative potential structure associated with two probes in the absence of ambient electric field. The probe on the left is affected by space charge; the probe on the right is not affected by space charge.

Viking electric field measurements of weak double layers, much work would need to be done. The Viking probes are spherical, not planar, and the probes are strongly photoemitting, rather than nonemitting. To our knowledge no one has yet tried to measure the presheaths of nonplanar, emitting objects. Neither has there been theoretical work on this subject. Although Parrot et al. [1982] theoretically investigated the presheaths of cylindrical and spherical nonemitting probes in the limit of vanishing Debye length λ_D to probe radius R_P, not only are magnetospheric double probes strongly photoemitting but also $\lambda_D \sim 10$ m $\gg R_P$. Further, electron accelerating presheaths, although predicted by theory, have never been experimentally measured, not even for nonemitting, planar objects (again, to our knowledge, no one has ever tried). Lastly, the work of Diebold et al. [1994] suggests that for magnetospheric probes the potential drop Φ_M is much less than $T_e/2$, while experimental work has dealt with potential drops greater than $T_e/2e$. The theoretical work of Parrot et al. [1982] suggests that such small potential drops would happen almost entirely across the presheaths. In such a case, the potential drop across the presheath would be Φ_M rather than $T_e/2$.

Gradients in wave amplitude which result in greater oscillations in the ambient plasma potential near one probe of a double probe as compared to the other probe may lead to Wave Gradient Induced Error (WGIE) in electric field measurements made by the probes. For the purposes of illustrating WGIE, consider Figure 5. The plasma sheath of one probe of a double probe is represented in Figure 5 as a resistance and capacitance (R_{sh} and C_{sh}, respectively) in parallel. This is an appropriate representation when considering plasma potential oscillations of amplitude small compared to T_e/e. The finite resistive and capacitive

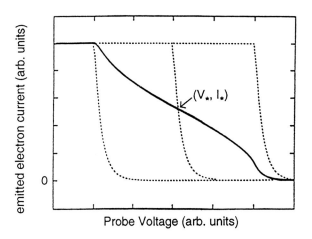

Figure 6. The left-hand, middle, and right-hand dotted curves are qualitative depictions of the current versus voltage characteristics of a probe in the presence of sinusoidal, low frequency, small amplitude plasma potential oscillations at times of minimum, average, and maximum potential, respectively. We refer to the middle dotted characteristic as the "voltage averaged" characteristic. The solid line is the "current averaged" characteristic of the probe.

impedances (R_{bi} and C_{bi}, respectively) across the current bias source of that probe are also explicitly shown in Figure 5. The resistance R_{bi} is generally made to be large enough that the current through R_{bi} is much less than the probe bias current I_B.

For oscillations of sufficiently low frequency, small amplitude or both, the currents through both C_{sh} and C_{bi} are small and can be ignored. This is how one typically thinks of probe operation in the absence of high frequency oscillations, with the bias current I_B flowing through a resistive sheath. In such a case, the voltage of the probe oscillates with the oscillation in plasma potential to ensure that the current through the probe is constant, i.e., that the current through the probe remains equal to I_B, at all times throughout the oscillation cycle. The left-hand, middle, and right-hand dotted curves of Figure 6 are qualitative depictions of the current versus voltage characteristics of a probe in the presence of sinusoidal, low frequency, small amplitude plasma potential oscillations at times of minimum, average, and maximum potential, respectively. Note that the average characteristic is the same as the characteristic of the probe in the absence of plasma potential oscillations. Hence, in the case of an absence of oscillations at one probe of a double probe, zero DC electric field between the probes, and sinusoidal, low frequency, small amplitude plasma potential oscillations at the other probe, the average voltage of the probe affected by oscillations is equal to the voltage of the other probe for all

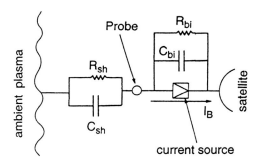

Figure 5. Qualitative circuit equivalent of one probe's sheath and current bias source. The plasma sheath is represented as a resistance and capacitance (R_{sh} and C_{sh}, respectively) in parallel. The finite resistive and capacitive impedances across the current bias source of that probe are shown as R_{bi} and C_{bi}, respectively.

current biases, i.e., there is no WGIE. We will refer to the middle dotted characteristic of Figure 6 as the "voltage averaged" characteristic because it is the characteristic one obtains under these conditions if the voltage assigned to each distinct current bias is the probe's <u>voltage</u> (when the probe is biased at that distinct current bias) <u>averaged</u> over time, i.e., averaged over an oscillation cycle.

WGIE may occur, however, at when there are relatively large amplitude oscillations of higher frequencies. For example, in a particular situation, there might be higher frequency plasma potential oscillations (at one probe but not the other) for which $(\omega C_{bi})^{-1} \ll R_{bi}$ and $(\omega C_{sh})^{-1} \gg R_{sh}$. Further, the amplitude of these oscillations might be large enough that the oscillating current through C_{bi} is (less than but) comparable to the saturated emitted electron current. In such a case, C_{bi} would essentially short-circuit the effect of the oscillation on the current source and probe, i.e., the voltage across the current source would be fairly constant over an oscillation cycle and, hence, the probe's voltage would also be fairly constant (if, as will be assumed, the satellite's voltage is fairly constant). Note that, in such a case, although the voltage across the current source is fairly constant over an oscillation cycle, if the current through C_{bi} has a DC component, then the voltage across C_{bi} (and, hence, the probe voltage also) can and will change substantial over many oscillation cycles until it reaches an equilibrium voltage for which there is no DC component to the current passing through C_{bi}. The solid curve in Figure 6 is what we refer to as the "current averaged" characteristic. This is the characteristic one obtains under the conditions $(\omega C_{bi})^{-1} \ll R_{bi}$ and $(\omega C_{sh})^{-1} \gg R_{sh}$ if the voltage assigned to each distinct current bias is the equilibrium voltage, i.e., the probe's voltage (when the probe is biased at that distinct current bias) for which there is zero <u>averaged</u> (over an oscillation cycle) <u>current</u> through C_{bi}. The reader may wish to note that at the equilibrium voltage, the averaged current through the probe is I_B. For a more detailed description of how the current versus voltage characteristics of probes in fluctuating potentials are calculated, the reader is referred to *Hershkowitz* [1989].

Inspection of Figure 6 shows that WGIE would be possible if there were plasma potential oscillations at higher frequencies (at one probe but not the other) for which $(\omega C_{bi})^{-1} \ll R_{bi}$ and $(\omega C_{sh})^{-1} \gg R_{sh}$. Specifically, the middle-dotted or voltage-averaged characteristic (corresponding to the probe at which the plasma potential is constant) and the solid or current-averaged characteristic (corresponding to the probe at which the plasma potential fluctuates) are not equal for current biases I_B such that $0 < I_B < I_*$ and $I_* < I_B <$ the saturated emitted electron current. For I_B such that $0 < I_B < I_*$, the voltage of the probe at which the plasma potential fluctuates would be greater than the other probe (even though, again, the ambient DC plasma potential at the two probes are equal); and for I_B such that $I_* < I_B <$ the saturated emitted electron current, the voltage of the probe at which the plasma potential is constant would be greater. WGIE is also expected to occur when large plasma potential oscillations exist at one probe (but not the other) at higher frequencies for which $(\omega C_{sh})^{-1} \ll R_{sh}$ (such that the oscillating current through C_{sh} is less than but comparable to the saturated emitted electron current).

REFERENCES

Aggson, T. L., Probe measurements of electric fields in space, in ATMOSPHERIC EMISSIONS, edited by B. M. M:cCormac, pp. 305-316, van Nostrand Reinhold, New York, 1969

Boström, R., G. Gustafsson, B. Holback, G. Holmgren, H. Koskinen, and P. Kintner, Characteristics of solitary waves and weak double layers in the magnetospheric plasma, Phys. Rev. Lett., 61, 82, 1988.

Boyd, R.,L.,F., Measurement of electric fields in the ionosphere and magnetosphere, Space Sci. Rev., 7, 230, 1967.

Diebold, D.A., N. Hershkowitz, J.R. DeKock, T.P. Intrator, S.-G. Lee, and M-K. Hsieh, Space Charge Enhanced, Plasma Gradient Induced Error in Satellite Electric Field Measurements, J. Geophys. Res., 99, 449, 1994.

Fahleson, U., Theory of electric field measurements conducted in the magnetosphere with electric probes, Space Sci. Rev., 7, 238, 1967.

Pedersen, A. C. A. Cattel, C.-G. Falthammar, VC. Formisano, P.-A. Lindqvist, F. Mozer, and R. Torbert, Quasistatic electric field measurements with spherical double probes on the GEOS and ISEE satellites, Space Sci. Rev., 37, 269-312, 1984

Guernsey, R.L., and J.H.M. Fu, Potential distribution surrounding a photo-emitting plate in a dilute plasma, J. Geophys. Res., 75, 3193, 1970.

Hershkowitz, N., How Langmuir probes work, PLASMA DIAGNOSTICS, ed. by O. Auciello and D.L. Flamm, Academic Press, p. 164, 1989.

Kingdon, K.H., A method for the neutralization of electron space charge by positive ionization at very low gas pressures, Phys. Rev., Ser. II, 21, 408, 1923.

Meassick, S., M-H. Cho, and N. Hershkowitz, Measurement of plasma presheath, IEEE Trans. Plasma Sci., PS-13, 115, 1985.

Meyer, J.A., G.-H. Kim, M.J. Goeckner, and N. Hershkowitz, Measurement of the presheath in an electron cyclotron resonance etching device, Plasma Sources Sci. Technol., 1, 147, 1992.

Mozer, F.S., R.B. Torbert, U.V. Fahleson, C.G. Falthammer, A. Gonfalone, and A. Pedersen, Measurements of quasi-static and low-frequency electric fields with spherical double probes on the ISEE-1 spacecraft, IEEE Trans. Geosci. Electron., GE-16, 258, 1978.

Parrot, M.J.M., L.R.O. Storcy, L.W. Parker, and J.G. Laframboise, Theory of cylindrical and spherical Langmuir probes in the limit of vanishing Debye number, Phys. Fluids, 25, 2388, 1982.

Pedersen, A., R. Grard, K. Knott, D. Jones, A. Gonfalone, and U. Fahleson, Measurements of quasi-static electric fields between 3 and 7 radii on GEOS-1, Space Sci. Rev, 22, 333, 1978.

Daniel. A. Diebold and Noah Hershkowitz, Department of Nuclear Engineering and Engineering Physics, University of Wisconsin, 1500 Engineering Dr., Madison, WI 53706

Plasma Gradient Effects on Double Probe Electric Field Measurements

Harri Laakso

Geophysical Research, Finnish Meteorological Institute, Helsinki, Finland

Robert F. Pfaff, Jr. and Thomas L. Aggson

NASA Goddard Space Flight Center, Greenbelt, MD

The effects induced by electron density and temperature gradients on double probe electric field measurements are investigated. Gradients with components along the antenna direction produce a potential difference ΔV between the probes, which appears as a spurious effect in the electric field measurements. This effect is usually negligible in dense plasma regimes (such as in the ionosphere) but may account to a few percent or even larger for measurements in tenuous plasmas unless the probes are properly biased. The relationship between gradients and ΔV is controlled by a number of parameters, such as the electron density and temperature (T_e), the saturation photoelectron current density (j_{ph0}), the photoelectron temperature, and the bias current (I_b). ΔV is particularly small when the probes are biased to a few volts positive with respect to the ambient plasma. The largest ΔV occurs if the probes become negative to the plasma potential, and the magnitude of ΔV is directly proportional to T_e. However, a large j_{ph0} usually keeps the probes at a positive potential in tenuous plasmas. If additionally the probes are biased, the gradient effects are insignificant. A small I_b is usually sufficient to keep these effects rather small, whereas the use of a large I_b may sometimes result in large ΔV signals, for instance, during energetic plasma injections.

INTRODUCTION

The double probe technique is a well-established method for determining dc and ac electric fields in space plasmas. This technique has been reviewed by *Maynard* [1997] for measurements in dense plasmas (e.g., ionospheres) and *Pedersen* [1997] for measurements in tenuous plasmas (e.g., magnetospheres). Figure 1*a* shows a typical configuration of the double probe antenna, where two probes, usually spherical or cylindrical in shape, are separated by a baseline **L** and are located at the tips of two opposing

Measurement Techniques in Space Plasmas: Fields
Geophysical Monograph 103

booms. The electric potential in the ambient plasma near these two probes is V_{01} and V_{02}, while the probe potential is V_1 and V_2, respectively, which are usually different from the ambient plasma potentials. Thus, $V_1 = V_{01} + \Delta V_1$ and $V_2 = V_{02} + \Delta V_2$. When the plasma conditions are similar at both probes, that is, $\Delta V_1 \approx \Delta V_2$, the measured potential difference $V_1 - V_2$ is approximately equal to $V_{02} - V_{01}$, which is, on the other hand, the electric field component along the instrument baseline, $\mathbf{E \cdot L} = V_{02} - V_{01}$.

In this paper, we investigate cases where the plasma conditions are not exactly the same at the two probe locations. Figure 1*b* presents the case of a plasma density gradient, where the electron density is n_e at probe 1 and $n_e + \Delta n_e$ at probe 2, while the electron temperature T_e is the same at both probes. Figure 1*c* presents the case for a plasma temperature gradient: the electron temperature is T_e at probe 1 and $T_e + \Delta T_e$ at probe 2, while the electron density n_e is the same at both probes. The problem of plasma gradients was initially analysed by *Laakso et al.* [1995], presenting both a numerical and an analytical treatment of the problem. Those authors demonstrated that the magnitude of the error signal strongly depends on the values of n_e, T_e, j_{ph0}, T_{ph}, and I_b, where j_{ph0} is the saturation photoelectron current density from the probe, T_{ph} is the photoelectron temperature, and I_b is the bias current.

This study continues the analysis of plasma density and temperature gradient errors, but now considers a large range of plasma conditions. We also examine in detail how the magnitude of the plasma gradient effect is affected by the values of the instrument parameters j_{ph0} and I_b.

For reference, for the numerical calculations in this paper, the instrument baseline is selected to be 100 meters and the probe radius is 4 cm; the gradient errors are, however, essentially independent of these dimensional parameters [*Laakso et al.*, 1995].

CHARGING OF PROBES

When a conducting probe is immersed in a plasma, it immediately acquires a potential for which the sum of all currents to the probe is zero. This potential is called the probe potential. We simplify the situation by considering only four current components: the ambient electron current (I_e), the ambient ion current (I_i), the photoelectron current (I_{ph}), and the bias current (I_b). The probe potential is obtained by solving the current balance equation $I_e + I_b - I_i - I_{ph} = 0$, where the explicit expressions for the currents will be given below.

Ambient Electron and Ion Collection

In this paper we examine only cases where the probe (4 cm in radius) is small with respect to the Debye length, and the effect of space charge can then be neglected [e.g., *Whipple*, 1981]. The space charge effect on the double probe measurements has been recently investigated by *Diebold et al.* [1994].

Let us assume that the probes are spheres (for the treatment of other geometries, see *Laakso et al.* [1995]), that the velocity distribution of the plasma is Maxwellian, that there is only one species of ions, and that the electron temperature is equal to the ion temperature. For convenience, the plasma potential is chosen to be zero. The ambient electron current collected by the probe is given by [*Whipple*, 1981]

$$I_e = I_{e0} \exp\left(\frac{V}{V_e}\right), \quad V \leq 0 \qquad (1a)$$

$$I_e = I_{e0}\left(1 + \frac{V}{V_e}\right), \quad V \geq 0 \qquad (1b)$$

where $I_{e0} = S\, e\, n_e \sqrt{eV_e/2\pi m_e}$ is the electron current at plasma potential, V is the probe potential (with respect to the ambient plasma), $V_e = T_e/e$, e is the electron charge, S is the surface area of the probe, and m_e is the electron mass.

Similarly, the ambient ion collection is described by the following equations: $I_i = I_{i0}\left[1 - V/V_e\right]$, for $V \leq 0$, and $I_i = I_{i0}\exp(-V/V_e)$, for $V \geq 0$, where $I_{i0} = I_{e0}/M$, $M = \sqrt{m_i/m_e}$, and m_i is the ion mass; we can assume that $n_e \approx n_i$.

Photoelectron and Bias Currents

Under the influence of solar illumination, a surface immersed in a space plasma emits photoelectrons, which can significantly affect the value of the probe potential. When the body is at a negative potential with respect to the plasma, all the photoelectrons emitted from its surface can escape, resulting in a saturation photoelectron current, I_{ph0},

$$I_{ph} = I_{ph0}, \quad V \leq 0 \qquad (2a)$$

where $I_{ph0} \approx (S/4)\, j_{ph0}$ for a sphere. The value of j_{ph0} is influenced by a number of factors, like the surface properties of the probe, solar activity and atmospheric density [*Brace et al.*, 1988].

When the probe assumes a positive potential relative to the plasma, a flux of low-energy photoelectrons returns to the probe. The escaping current depends on the probe potential, the energy distribution of photoelectrons, and the shape of the probe. If the velocity distribution of photoelectrons is Maxwellian and the size of the probe is smaller than the Debye length, the photoelectron current for a sphere can be written as [*Grard*, 1973]

$$I_{ph} = I_{ph0}\left(1 + \frac{V}{V_{ph}}\right)\exp\left(-\frac{V}{V_{ph}}\right), \quad V \geq 0 \qquad (2b)$$

where $V_{ph} = T_{ph}/e$.

It has recently been found that the velocity distribution function of photoelectrons is effectively bi-Maxwellian [*Laakso and Pedersen*, 1994; *Pedersen*, 1995]. A major population appears at ~2.5 eV as a result of solar Lyman α radiation (rather than at ~1.5 eV usually measured in the laboratory, see e.g., *Grard* [1973]), and a minor population exists near ~5 eV due to 55–110 nm solar radiation; the latter population becomes significant when the surface potential exceeds +10 volts, i.e., in the very tenuous plasma of the magnetotail [*Laakso and Pedersen*, 1995; *Pedersen*, 1995]. This case will not be considered in this paper.

Large j_{ph0} leads to very positive surface potentials in tenuous plasmas. Then a spurious current fluctuation ΔI_e (for instance, due to Δn_e or ΔT_e) can generate a large fluctuation ΔV in the probe potential, decreasing the quality of double probe measurements. With a high impedance current source, an artificial electron flux, called a bias current, is forced from the spacecraft to the probes. Then the probes assume a new potential where the probe impedance is much smaller, producing more accurate double probe measurements.

The bias current effectively limits the magnitude of the positive probe potential in tenuous plasmas. For instance, the maximum positive potential occurs when there is no ambient plasma near the probe: for a nonbiased probe, this potential is very large. Driving a bias current into the probes, the maximum probe potential is obtained by solving the equation $(1+x)e^{-x} = I_b/I_{ph0}$, where $x = V/V_{ph}$. If $I_b/I_{ph0} = 0.5$, the solution is $x = 1.7$ which equals to $V \sim 2.6$ volts if $V_{ph} = 1.5$ volts.

MAGNITUDES OF PLASMA GRADIENT EFFECTS

Density Gradient Effect

This section deals with plasma density gradient effects in electric field measurements. Such a situation is presented by Figure 1b, where T_e is equal at both probes but the plasma density is n_e at probe 1 and $n_e + \delta n_e$ at probe 2. This will cause a gradient error given by [Laakso et al., 1995]

$$\delta E_n = -C_n \frac{V_e}{L} \frac{\delta n_e}{n_e} \qquad (3)$$

which is directly proportional to the plasma temperature and the relative plasma density variation. The factor C_n is

$$C_n = \frac{M \exp\left[\dfrac{V}{V_e}\right] + \dfrac{V}{V_e} - 1}{M \exp\left[\dfrac{V}{V_e}\right] + 1}, \quad V \leq 0 \qquad (4a)$$

$$C_n = \frac{1 + \dfrac{V}{V_e}}{1 + \dfrac{I_{ph0}}{I_{e0}} \dfrac{V}{V_{ph}} \dfrac{V_e}{V_{ph}} \exp\left[-\dfrac{V}{V_{ph}}\right]}, \quad V \geq 0 \qquad (4b)$$

The expressions of C_n for other probe geometries are presented by Laakso et al. [1995].

Let us assume 1% density gradients along the instrument baseline, that is, $\Delta n_e/n_e = 0.01$ (e.g., n_e is 100 cm^{-3} at probe

1 and 101 cm^{-3} at probe 2). The magnitude of the resulting gradient effects are shown in Plate 1; the color scale on the top indicates the magnitude of the induced effect in mV/m. Equations (1–2) assume that the probe size (radius a) is small compared to the Debye length (λ_D) and therefore Plate 1 shows results only when $\lambda_D > 3a$. The horizontal axis shows the plasma density range from 10^{-2} to 10^5 in cm^{-3} at one probe (and 1% higher at another probe). The plasma temperature, shown on the vertical axis, ranges from 0.1 eV to 1 keV. The photoelectron parameters are: $j_{ph0} = 4$ nA cm^{-2} (i.e., $I_{ph0} \approx 200$ nA) and $T_{ph} = 1.5$ eV. The top panel is for a nonbiased antenna and the bottom panel is for a biased antenna, where the bias current is a half of the saturation photoelectron current (i.e., $I_b = -0.5 \cdot I_{ph0}$).

Plate 1 shows that the gradient effects are very small in tenuous plasmas if the probes are biased, whereas for the nonbiased probes, the effects can be relatively large in the same plasma regime. The largest effects appear when the probes are exactly at or somewhat negative to the plasma potential. The zero potential appears as a clear separator in Plate 1.

When applying the results indicated in Plate 1, one should remember that in the Earth's plasma environment the instruments normally encounter only a small part of the plasma regimes presented by this figure; i.e., either cases where the plasma temperature is high and the plasma density is low, or opposite cases, where the plasma temperature is low and the plasma density is high. The latter case is typical for ionospheric measurements where clearly both biased and non-biased probes work quite well. Although very large errors can occur if plasma gradients appear in a plasma of high temperature (e.g., $T_e \sim 1$ keV) and modest density(e.g., $n_e \sim 10$ cm^{-3}), such conditions are rare, even in extreme situations of energetic substorms. On the other hand, gradients larger than 1% in less severe conditions may induce errors that could be a few percent or larger of the ambient electric field. In all cases, equations (3–4) can be used to estimate possible magnitudes of the plasma density gradient errors in the data.

Temperature Gradient Effect

This section deals with temperature plasma gradient effects, that is, n_e is equal at both probes, and the plasma temperature is T_e at probe 1 and $T_e + \delta T_e$ at probe 2. This situation is shown by Figure 1c. The magnitude of the temperature gradient effect is [Laakso et al., 1995]

$$\delta E_T = -\frac{1}{2} C_T \frac{\delta V_e}{L} \qquad (5)$$

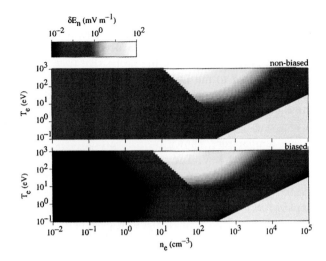

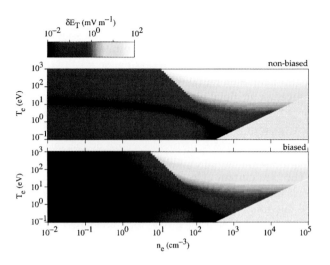

Plate 1. Magnitude of the gradient effect induced by 1% n_e gradients between 10^{-2} and 10^5 cm^{-3}, shown on the horizontal axis; T_e, shown on the vertical axis, is between 0.1 eV and 1 keV. The upper panel is for a nonbiased probe and the lower panel for a biased probe. The instrument parameters are $j_{ph0} = 4$ nA cm^{-2}, $T_{ph} = 1.5$ eV, $r = 4$ cm, $L = 100$ m, and $I_b = 100$ nA.

Plate 2. Magnitude of the gradient effect induced by 1% T_e gradients between 0.1 eV and 1 keV, shown on the vertical axis; n_e, shown on the horizontal axis, is between 10^{-2} and 10^5 cm^{-3}. The upper panel is for a nonbiased probe and the lower panel is for a biased probe. The instrument parameters are as in Plate 1.

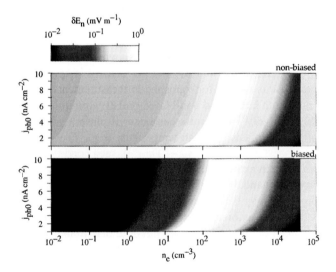

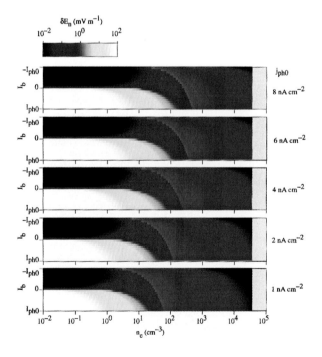

Plate 3. Effect of j_{ph0} on the magnitude of 1% n_e gradient effects in a 10 eV plasma. The horizontal and vertical axes show n_e and j_{ph0}, respectively. The upper panel is for a nonbiased probe and the bottom panel is for a biased probe.

Plate 4. Effect of I_b on the magnitude of density gradient effects in a 10 eV plasma. The panels from top to bottom represent five different values of j_{ph0}. The horizontal and vertical axes show the ambient electron density and the bias current, respectively.

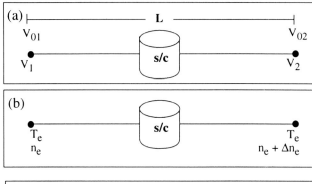

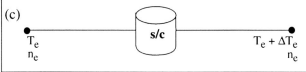

Figure 1. (a) Configuration of a double probe experiment, composed of two spherical probes at the tips of two opposing booms. (b) Plasma density gradient along the instrument baseline; the plasma temperature is constant. (c) Plasma temperature gradient along the instrument baseline; the plasma density is constant.

variation and a factor C_T

$$C_T = C_n - 2\frac{V}{V_e}, \quad V \leq 0 \tag{6a}$$

$$C_T = C_n \frac{V_e - V}{V_e + V}, \quad V \geq 0 \tag{6b}$$

Plate 2 presents numerical results for electric field signals induced by 1% plasma temperature gradients; otherwise the plasma and instrument parameters are the same as in Plate 1. As in Plate 1, results are shown only when $\lambda_D > 3a$. Again the top panel is for a nonbiased antenna and the bottom panel is for a biased antenna, where the bias current is a half of the saturation photoelectron current. The temperature gradient effects are very small in tenuous plasmas when the probes are biased, whereas for the nonbiased probes the errors are relatively large. The largest errors appear when the probes are near or negative to the plasma potential. Similar to Plate 1, the zero potential appears as a distinct separator in Plate 2. The temperature gradient error shows an anomalous behavior at $V \sim V_e$ where the error disappears, as is also indicated by expression (6b). This is, however, a very localized phenomenon and cannot be experimentally utilized.

Influence of Saturation Photoemission Current

The above analysis has demonstrated how plasma gradient effects in different plasma regimes may produce deviations in electric field double probe measurements. So far we have assumed that the saturation photoelectron current density j_{ph0} is 4 nA cm^{-2}, which is not strictly a valid as-

sumption in many cases. For instance, the solar photon flux has a strong 11-year temporal variation, which causes a similar variation in j_{ph0} [see e.g., *Brace et al.*, 1988]. Or, as also shown by Brace et al., the atmospheric contamination on the probe surface can decrease j_{ph0} quite drastically. For instance, j_{ph0} increased by a factor of two when the Pioneer Venus satellite moved outside the dense atmosphere of Venus. As a conclusion, both contamination and changes in solar EUV radiation can easily change the saturation photoelectron current by a factor of two. Therefore we now consider density gradient effects for different saturation photoelectron current densities.

Plate 3 presents results for 1% plasma density gradients in a 10 eV plasma (cf. Figure 1b). The horizontal axis indicates the ambient electron density ranging from 10^{-2} to 10^5 cm^{-3}, and the vertical axis shows j_{ph0} from 1 to 10 nA cm^{-2}. The top panel is for a nonbiased antenna and the bottom panel is for a biased antenna. The bias current I_b is selected as a half of the saturation photoelectron current I_{ph0}. Since I_{ph0} increases from about 50 nA to 500 nA along the vertical axis, I_b increases simultaneously from 25 nA to 250 nA in the bottom panel (while $I_b = 0$ in the top panel).

In tenuous plasmas, the density gradient effect is quite insignificant for the biased probes. The error increases only when the probe potential becomes near the plasma potential. Therefore probes with low photoemission current ($j_{ph0} < 3$ nA cm^{-2}) can more often work in situations where the probe potential is negative and then sensitive to the plasma density variations. On the other hand, a double probe antenna with high photoemission current (such as $j_{ph0} = 6$ nA cm^{-2} or more) can measure electric fields relatively accurately even in rather energetic plasma where gradients with various scale lengths may occur. It is a fortune that in space, the photoemission varies so that small j_{ph0} values appear in the ionosphere where the plasma temperature is low and large j_{ph0} values appear in the magnetosphere where the plasma temperature can be quite high. Then the gradient effects usually remain small.

Influence of Bias Current

In Plates 1–3, the bias current was selected as $I_b = I_{ph0}/2$. Now this requirement is relaxed, and we investigate how the density gradient effect varies if the bias current is changed. Plate 4 shows results for 1% plasma density gradients in a 10 eV plasma (cf. Figure 1b), when the plasma density ranges from 10^{-2} to 10^5 in cm^{-3}. The panels from top to bottom represent five different j_{ph0}: 8, 6, 4, 2, and 1 nA cm^{-2} (which approximately correspond to $I_{ph0} \approx 400$, 300, 200, 100, and 50 nA, respectively, for a probe of 4 cm radius). In each panel, the horizontal axis shows the ambient electron density from 10^{-2} to 10^5 cm^{-3}, and the vertical axis shows the bias current that ranges from $-I_{ph0}$ to I_{ph0}; the density gradient is chosen 1%. According to Plate 4, in tenuous plasmas, the magnitude of the bias current is not so important, as long as it does not change sign. In dense plasmas, where the electron and ion collection determine the probe potential and where the probe potential is negative, the use of the bias current does not reduce the gradient effects significantly. However, in such cases the plasma temperature is usually less than 10 eV which is used in Plate 4, and then gradient effects are also weaker than those shown in Plate 4.

Figure 2 shows an example of the effect of the bias current on the magnitude of the 1% density gradient occurs in a plasma with $n_e = 10^{-2}$–10^5 cm^{-3} and $T_e = 10$ eV. The bias current is assumed to have values from $-I_{ph0}$ to I_{ph0} with a step of 20 nA ($I_{ph0} = 200$ nA). When I_b equals to $-I_{ph0}$, the situation corresponds to a body in shadow; in such a case, no plasma gradient errors occur, because the probe potential is independent of the electron density. The plateau in Figure 2 varies with T_e and $\Delta n_e/n_e$ (see (3)). This figure clearly shows that in tenuous plasmas it is important that the probes are biased but the magnitude of the bias current does not actually matter very much. Already a small bias current can effectively improve the accuracy of the double probe experiment. In fact, large bias currents should be avoided, because otherwise the probes may become negative if the ambient electron flux suddenly increases, for instance, during substorms (for instance, notice how the line with $I_b = -180$ nA behaves as a function of plasma density).

SUMMARY

In this paper, the effects induced by electron density and temperature gradients on double probe electric field measurements have been investigated analytically with numerical solutions (for details of the analytical expressions, see *Laakso et al.* [1995]). It has been demonstrated how gradients in plasma density and temperature with components along the antenna direction can produce a potential difference ΔV between the probes which appears as a spurious effect in the electric field measurements. It has been shown that such effects are negligible (<< 1 mV m^{-1}) in high density, low temperature plasmas (such as the Earth's ionosphere). In tenuous plasmas, such as in the Earth's magnetosphere, plasma gradient effects have been shown to be greatly mitigated if the probes are biased. Significant errors can in such cases occur when the probes are at or negative to the plasma potential, and in such cases, the magnitude of the error is directly proportional to the electron temperature.

We conclude that the plasma gradient effects on double probe electric field measurements are not significant in the Earth's ionosphere, and should be rarely important in the Earth's magnetosphere, provided the probes are properly biased.

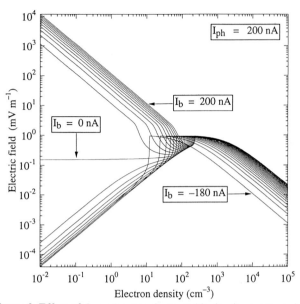

Figure 2. Effect of I_b on the magnitude of 1% density gradients in a 10 eV plasma. The represented lines are from −200 nA to +180 nA with a step of 20 nA, and I_{ph0} is 200 nA. Note that all positive bias currents fall in the dark border when the density exceeds 100 cm^{-3}.

REFERENCES

Brace, L. H., W. R. Hoegy, and R. F. Theis, Solar EUV measurements at Venus based on photoelectron emission from the Pioneer Venus Langmuir probe, *J. Geophys. Res.*, *93*, 7282–7296, 1988.

Diebold, D., N. Hershkowitz, J. DeKock, T. Intrator, and S.–G. Lee, Space charge enhanced, plasma gradient induced error in satellite electric field measurements, *J. Geophys. Res.*, *99*, 449–458, 1994.

Grard, R. J. L., Properties of the satellite photoelectron sheath derived from photoemission laboratory measurements, *J. Geophys. Res.*, *78*, 2885–2906, 1973.

Laakso, H. and A. Pedersen, Satellite photoemission characteristics, in *Materials in a Space Environment*, edited by H. T. D. Guyenne, pp. 361–365, ESA SP–368, ESTEC, Noordwijk, 1994.

Laakso, H., T. Aggson, and R. Pfaff, Plasma gradient effects on double probe measurements in the magnetosphere, *Ann. Geophys.*, *13*, 130–146, 1995.

Maynard, N. C., Electric field measurements in moderate to high density space plasmas with passive double probes, this monograph, 1997.

Pedersen, A., F. Mozer, and G. Gustafsson, Solar wind and magnetosphere plasma diagnostics by spacecraft electrostatic potential measurements, *Ann. Geophys.*, *13*, 118–129, 1995.

Pedersen, A., Electric field measurements in a tenuous plasma with spherical double probes, this monograph, 1997.

Whipple, E. C., Potentials of surfaces in space, *Rep. Prog. Phys.*, *44*, 1197-1250, 1981.

Thomas L. Aggson and Robert F. Pfaff, Code 696, NASA Goddard Space Flight Center, Greenbelt, MD 20771

Harri Laakso, Geophysical Research, Finnish Meteorological Institute, P.O. Box 503, 00101 Helsinki, Finland

Ionospheric Electric Fields From Stratospheric Balloon-Borne Probes

R. H. Holzworth

Geophysics Program, University of Washington, Seattle, WA

E. A. Bering, III

Physics Dept., University of Houston, Houston, TX

The balloon-borne, double Langmuir probe technique for measuring vector electric fields in the stratosphere has been used successfully since the late 1960s. The cost effectiveness, flexibility and relatively high time resolution of the technique has resulted in stratospheric electric field measurements being used to address a wide variety of ionospheric and magnetospheric problems. Indeed, the technique continues to hold certain advantages over other measurements for some types of problems and remains an important tool in the 1990s where it is being used by a growing number of groups. Experience gained from a large number of flights, using modern electronics and telemetry is summarized here to describe the technique, provide examples of its use, and to discuss its advantages and limitations.

INTRODUCTION

Unlike magnetic field measurements, electric field measurements at the surface of the earth are greatly perturbed by local tropospheric sources and generally cannot be used to study ionospheric phenomena such as plasma currents and dynamics. However, since the atmospheric conductivity is approximately exponentially increasing with altitude, the stratospheric electric field is closely related to the large scale average ionospheric electric field overhead [c.f. Mozer and Serlin, 1969; Mozer, 1971; Holzworth, 1977]. For an ionospheric electric field with 300 km horizontal scale size, the attenuation at balloon altitudes will be only a few percent, while a source field of 10 km scale size will be attenuated almost completely. Indeed, today this well tested technique of balloon-borne, vector electric field measurements in the stratosphere is

an important tool for studies of large scale ionospheric and magnetospheric plasma phenomena. This paper will provide a brief background to the technique, provide recent evidence of the utility of these measurements to diagnosing present questions of ionospheric and magnetospheric dynamics, and provide a discussion of the strengths and possible weaknesses of the technique in various circumstances.

VECTOR ELECTRIC FIELD MEASUREMENTS IN THE STRATOSPHERE

The vector electric field in the stratosphere can be simply and effectively measured using the double Langmuir probe technique operated in a very high impedance mode (10^{14} Ohms). Because the local relaxation time in the stratosphere is of the order of one second, field mills cannot be used for quasistatic electric field measurements at time constants of a few seconds or longer. The double probe technique, as used in the stratosphere, is explained in detail in many publications [c.f. Kellogg and Weed,

Measurement Techniques in Space Plasmas: Fields
Geophysical Monograph 103

1968; Mozer and Serlin, 1969; Holzworth, 1977] This section provides a brief overview of the method. In today's instruments very low leakage FET (field effect transistor) operational amplifiers (Op Amps) with leakage current < 70 femtoamps are used as the preamplifiers, along with guard ring layout techniques. The basic design of the experiment is to arrange three orthogonal pair of conducting probes mounted on insulating (fiberglass) booms. The probes make direct electrical contact with the charge carriers in the air and need to be 15 cm in diameter or larger to be able to collect enough charge to drive the electronics. The probes must be far enough from the central payload body to minimize the field perturbation caused by the gondola structure (typically 1.5 m boom length to each probe, for a 3 m tip-to-tip probe arrangement using a 0.3m linear dimension gondola). For studies requiring frequency response below about 100 kHz it is sufficient to place the preamplifiers directly into the central payload (as opposed to inside the probes themselves, say). The ion collecting probes are either thin spun aluminum spheres or coated styrofoam spheres. The external probe surfaces as well as the ground plates and other external payload surfaces are coated with aqueous carbon solution such as Aquadag. This helps make the work function uniform over the probe and, perhaps more importantly, decreases the photoemission currents since the work function of the carbon solution is larger than for Al [cf. see Byrne et al., 1990]. The high impedance FET Op Amp is connected to the probe in a unity gain voltage follower circuit which provides an impedance transformation from the high impedance atmospheric source to an easy to use, low impedance voltage.

The wire connecting the probes (spheres) to the input of the electrometer (Op Amp) is the center conductor of a Teflon (high impedance) coaxial cable (such as RG188) with the shield driven by the unity gain Op Amp output. Driving the coaxial shield at the same voltage as the input also means that there is no voltage drop between inner and outer conductor in the wire, which effectively eliminates the otherwise large load capacitance of the coaxial wire at these low frequencies. Electrostatic cleanliness is required for all hardware and handling must be done very carefully.

The electric field in the direction of the boom is found by subtracting (differencing) the voltages of two opposite probes. The signal ground is a star point ground connected to a set of isolated ground planes which surround the central payload. If the spheres are nearly identical the difference voltage will be near zero except for the voltage difference caused by slightly different floating voltages of the separated isolated conductors. Byrne et al.

[1990] showed that the probes typically float at 0.1 to 0.2 V from the ground level. This voltage is the result of both the ambient electric fields (what you want to measure) and offsets due to sheaths and other background effects. The entire payload is rotated a few times per minute to eliminate these dc offsets in the horizontal probes caused by residual work function differences, non-zero Op Amp offsets, or other effects that are constant over the period of rotation. The measured electric field is just the maximum voltage difference divided by the boom length (and any electronic gain) in the direction the booms are oriented at the time of the maximum. A constant horizontal electric field appears as a sine wave (at the rotation rate) in the output voltage. Sensitivities of 1 mV/m for the horizontal electric field component and 15 mV/m for the (nonrotated) vertical component are typical for the dc field. Wideband VLF sensitivity of 10^{-5} V/m has been typically achieved with these 3 m tip-to-tip antennas. The payloads are often lowered below the balloon by a few balloon radii (typically 10 to 15m radius) to minimize distortion fields associated with possible induced charge on the balloon caused by the ambient field.

Self contained balloon payloads include power systems, telemetry, data systems and auxiliary measurements. Careful design of the power system is needed to minimize electromagnetic interference (EMI) which particularly affects the wave data (ac measurements). Telemetry systems vary widely from high speed line of sight systems to satellite relayed ones. For telemetry links that use either low altitude, or geostationary orbit satellites the data rates are often limited and require extensive on-board data processing and compression [see Powell, 1983; Holzworth et al., 1993; or Hu, 1994].

Data from stratospheric balloon electric field instruments have been applied to magnetospheric and ionospheric problems for decades. Theoretical calculations [Mozer, 1971; Dejnakarintra and Park, 1974] clearly demonstrated that ionospheric electric fields with spatial scale size greater than 100 km map to 30 km altitude in the atmosphere with little attenuation. Mozer [1973] showed that during active auroral conditions the electric field components, measured at the balloon, were those expected to drive the ionospheric Hall currents. Furthermore, it has been clearly demonstrated that the large scale, two cell, ionospheric convection pattern at auroral latitudes is seen by the balloons [Mozer and Lucht, 1974; Holzworth et al., 1977]. So the horizontal stratospheric electric field at middle and high latitudes is an average of the ionospheric field components with large scale size. Structure on the size of auroral arcs (0.1 to 10 km, say) is usually attenuated well above balloon altitudes. Therefore, with a single point measurement (at the balloon) of the two horizontal electric field components one can monitor the large scale ionospheric convective motions overhead. Furthermore, from these measurements, and with the help of an ionospheric conductivity measurement or model, one can make dynamic estimates of the large scale ionospheric heating (due to fields driving Pedersen currents) with 1 minute time resolution.

Several experimental studies have shown [Weimer et al., 1985; Mozer, 1971; Holzworth et al., 1981] that equipotential mapping from the magnetic equator to the auroral zone is a good approximation for slow field variations (> 10 sec) and large scale sizes (>100 km at the ionosphere). Thus, stratospheric electric field measurements represent average ionospheric fields which, in turn can be mapped with confidence to the equator to study magnetospheric electric fields.

QUASISTATIC IONOSPHERIC AND MAGNETOSPHERIC ELECTRIC FIELDS

Following the introduction of the double probe technique in the 1960s, hundreds of papers have been written using double probe, balloon-borne, electric field data. In one of the early major efforts [Holzworth et al., 1977] a world wide balloon experiment was conducted in which multiple balloons were launched simultaneously from 6 locations spaced globally in the auroral zone. The inferred ionospheric electric fields were equipotentially mapped to the equator and used to calculate the radial diffusion of radiation belt particles [Holzworth et al., 1979]. Such global, simultaneous magnetospheric electric field data cannot be obtained by any present or planned satellite experiment.

In more recent balloon flights Bering et al [1991] found ionospheric electric field signatures associated with SAID (Sub-Auroral Ion Drift) events [cf. Spiro et al., 1979] and studied ULF electric wave phenomena.

High Latitude Impulsive Events

The study of short duration impulsive phenomena has emerged in the last decade as a very important subfield of ULF research. Observationally, these events are seen by high latitude dayside ground magnetometers and stratospheric balloon borne electrometers as large amplitude (several 10's of nT and/or several 10's of mV/m) impulsive (5 - 15 min duration) perturbations. These events are usually interpreted as signatures of transient process(es) occurring at the magnetopause such as flux transfer

events (FTE's) [c.f. Lanzerotti et al., 1986; Sibeck, 1993 or Konik et al., 1994]

Electric field data from the 1985-86 South Pole Balloon campaign [Bering et al., 1987 and 1990] have been used extensively for studies of high latitude impulsive events. Figure 1 presents balloon-borne electric field data from January 7, 1986 [Bering et al., 1988]. As discussed in this paper, three candidate events are shown, at 1302, 1448 and 1551 UT. The event at 1302 UT was also used in the work reported by Lanzerotti et al. [1991] and Konik et al. [1994]. The first important conclusion from this work was that TCV's and MIE's are spatially extended enough to be observed by the balloon technique. A model study of the 1551 UT event found that it was necessary to include contributions from both a coaxial 2-d monopole or "twisting" current system [Lee, 1986] and a 2-d dipole or "towing" current system [Southwood, 1985, 1987]. It can be seen in the bottom panel of this figure that these events are also associated with clear perturbations in the vertical component of the electric field. The simultaneous presence of this unique signature and of appropriate perturbations in the horizontal components of $\vec{E}$ has enabled us to select a set of events with a lower amplitude threshold than other workers have used. The top two panels of the figure show that these particular events were temporally associated with high energy electron precipitation. This association may or may not be coincidental in this case. However, the simultaneous occurrence of >40 keV electron precipitation and TCV's is a relatively rare phenomenon.

TECHNIQUE ADVANTAGES AND POTENTIAL LIMITATIONS

The balloon-borne double probe technique for measuring ionospheric and magnetospheric electric fields has been used for decades and shown to work well for the detection of large scale average ionospheric electric fields with scale sizes >100 km and time scales longer than a few seconds. The technique has an accuracy of 1 mV/m for ionospheric electric field components perpendicular to $\vec{B}$. This 1 mV/m level is also the threshold sensitivity and thus the technique is not limited to some higher minimum electric field strength, as is the case for some radars. This balloon-borne technique is flexible in application and can be conducted nearly anywhere in the world with relatively short notice and with low cost. The simplest form of the balloon-borne vector electric field payload is that which is used for short duration balloon flights. Such payloads are typically about 15 kg and have a marginal cost of about $5,000 to $10,000 (the cost to

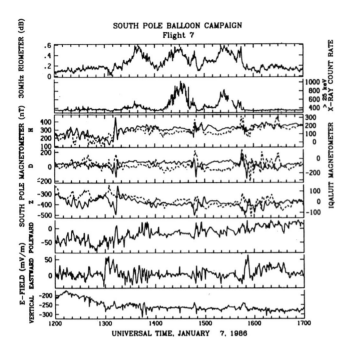

Figure 1. Balloon electric field and South Pole magnetometer and riometer data on January 7, 1986 showing high latitude impulsive events which have been linked to magnetopause activity. (Magnetometer data provided by L. Lanzerotti, C. Maclennan and A. Wolfe of AT&T Bell Laboratories).

make one more during fabrication of a new series of payloads). The technique has now been applied to long duration flights of many months duration [see Holzworth et al., 1993]. For these cases, while the payload costs are higher (about an order of magnitude), the total data set collected is orders of magnitude larger. In the ELBBO (Extended Life Balloon-Borne Observatories) experiment the use of 5 payloads resulted in over 410 payload-days of vector electric field collected. Also, during ELBBO the PPB (Polar Patrol Balloon) experiment was flown from Antarctica [Bering and Benbrook, 1995 and Bering et al., 1995] making the total balloons aloft as many as 6. Thus the total data set obtained during the ELBBO/PPB experiment time was 483 payload-days. This data set duration rivals the length of data collected with some small satellite projects, and at a fraction of the cost.

From the point of view of magnetospheric physics, the average ionospheric field can be mapped along magnetic field lines with high confidence for large scale (>100 km) slowly varying (<10 sec) fields. Multiple, simultaneous balloon flights can then provide a "snapshot" of the instantaneous average magnetospheric electric field. Such data could not be provided by satellites, or even multiple satellites. Even a dozen satellites at the magnetic equator

on auroral field lines could not obtain the large scale average, instantaneous (1 minute resolution, say) vector electric field. The satellite data would have to be averaged during the orbital motion over thousands of km to perform the average; which would take hours for an orbit at this altitude. The time resolution obtainable by the balloon electric field technique is also provided by radars, but only bistatic ones such as the Super-DARN [see Greenwald et al., 1995]. Monostatic radars, either HF or incoherent scatter type, must sample different parts of the ionosphere (different look directions) and assume uniformity to deduce vector fields. Of course radars of all types together only cover a small part of the global ionosphere whereas balloons can in principle be located anywhere for a focused campaign. Therefore, today, even decades after their first introduction, balloon-borne electric field instruments provide an important type of ionospheric and magnetospheric data which is difficult to obtain or unavailable from other techniques.

The balloon-borne electric field technique is limited by several factors. The payload itself disturbs the environment (as for any probe experiment), there are other sources of electric fields (such as thunderstorms) which may interfere with detection of the high altitude sources of interest here, and the scale size of the ionospheric field is not known independently. The balloon payload can be charged up as the balloon moves through the atmosphere (frictional charging) resulting in electric fields at the probes which saturate the electronics. This is usually the reason that we obtain little or no useful data on ascent or descent (since the electrometers are set for sensitivity at altitude, and the frictional charging of the payload can be very large.) The conducting surfaces, including the probes and payload ground plane, will emit a photoelectron current which can be a significant perturbation above 25 km in the daytime [see Byrne et al., 1990; or Hu, 1994]. If this photoelectron current is balanced (same for each probe), the differential voltage measurement (giving the electric field) will be unaffected. However photoemission from payload surfaces such as the solar panels can cause an error signal from a charged payload which may vary with the rotation, and therefore be difficult to detect. This can be minimized by extending the booms to longer distances (to increase the electric field signal) and by coating all external conducting surfaces with a high work function material such as Aquadag.

Other sources of electric field such as thunderstorms, while very interesting to study in their own right, will, nevertheless, limit the ability to study ionospheric electrodynamics from the balloon. In past experience, at northern auroral latitudes in the summer, this has resulted in a "loss" of about 10 percent of the data to ionospheric studies [see Holzworth, 1981]. Additionally, at mid latitudes, a recently discovered source of horizontal electric field in fair weather [Holzworth, 1989; Hu, 1994] could also mask the ionospheric signal for payloads below 30 km altitude in the daytime.

Determination of the ionospheric and magnetospheric electric field assumes an ionospheric electric field scale size greater than about 100 km [see Dejnakarintra, 1974]. However, from a single point measurement in the stratosphere, the scale size cannot be determined uniquely. In general the conductivity profile (from the balloon to the ionosphere) is such that ionospheric electric fields with scale sizes on the order of, say, 10 km, are severely attenuated at the altitude of the balloon (30 km, say)[see Dejnakarintra, 1974]. However, it is possible that small scale (10 km) features of great intensity (100 mV/m) could be seen by the balloon-borne probes and therefore be confused with a large scale field. Such times can often be determined with global scale imagery (such as DMSP, DE, Viking or other satellite images) and usually appear as short lived events. In principle this scale height ambiguity can be eliminated by using multiple payloads on one balloon. The payloads would have to be vertically separated by a large fraction of the conductivity scale height (several km). Then it would be possible to directly measure the attenuation factor as calculated by Mozer and Serlin [1969] and others, and thus estimate the source field scale size. Such a dual payload experiment has been attempted once in the past but was met with technical difficulties of the reel-down mechanism (F. Mozer, personal communication, 1976). The advantages and limitations of the balloon-borne electric field technique for ionospheric electric fields are summarized in Table 1.

CONCLUSIONS

The technique for measuring large scale ionospheric electric fields using balloon-borne double probes in the stratosphere has been well tested and is actively in use today. Modern implementation of the technique uses up to date electronics, telemetry and power systems but the fundamentals are basicly the same as introduced in the late 1960's and well tested in the 1970's. Ionospheric and magnetospheric phenomena from auroral convection to radiation belt dynamics can be studied with this technique. Several active balloon electric field programs are under way today including midlatitude and polar region experiments. New advances have been made in under-

Table 1.

Advantages and Limitations of Balloon Electric Field Technique

---

Advantages:
1. 10 sec, large scale averages of $\vec{E}$ perpendicular to $\vec{B}$
2. 1 mV/m absolute accuracy and sensitivity level
3. No amplitude threshold (as some radars)
4. Remote distribution can be done relatively inexpensively
5. Long heritage of testing the technique

Possible Limitations:
1. Payload perturbations (photoemission, charged surfaces)
2. Troposphere, stratosphere source perturbations
3. Source scale size is unknown so mapping can be inexact

---

standing the technique in the area of ac or transient response and of the vertical electric field component at the balloon in relationship to ionospheric fields. Even with today's large coverage radars such as Super-DARN, there is a niche for balloon electric field measurements in a campaign or long duration mode at locations not covered by the radars, or for average to quiet magnetospheric studies and for global coverage. The technique has some limitations which are overcome by careful implementation and by auxiliary data sets.

Acknowledgements. This work was supported by NASA grants NAGW-4147 and NAG5-668 and by NSF grants ATM8920428 and ATM9402764 at the University of Washington. Design and implementation support by John Chin is acknowledged and greatly appreciated. At the University of Houston, this research was supported by National Science Foundation grants DPP 8415203, DPP 8614092, DPP-8917464, DPP 9019567 and OPP 9318569, and by two grants from the Texas Advanced Research Program. We thank L. Lanzerotti, C. Maclennan and A. Wolfe of AT&T Bell Laboratories for the magnetometer data in Figure 1.

REFERENCES

Bering, E. A., III, and J. R. Benbrook, Conjugate ionospheric electric field measurements, Ann. Geophys., 5A(6), 485-502, 1987.

Bering, E. A., III, and J. R. Benbrook, Intense 2.3 Hz electric field pulsations in the stratosphere at high auroral latitude, J. Geophys. Res., 100, 7791-7806, 1995.

Bering, E. A., III, et al., Impulsive electric and magnetic field perturbations observed over South Pole: Flux transfer events?, Geophys. Res. Lett., 15, 1545-1548, 1988.

Bering, E. A., III, et al., Solar wind properties observed during high-altitude impulsive perturbation events, Geophys. Res. Lett., 17, 583, 1990.

Bering, E. A., III, et al., The intense magnetic storm of December 19, 1980: Observations at L=4, J. Geophys. Res., 96, 5597-5617, 1991.

Bering, E. A., III, et al., Balloon measurements above the South Pole: Study of ionospheric transmission of ULF waves, J. Geophys. Res., 100, 7807-7820, 1995.

Byrne, G. J., et al., Solar radiation (190 - 230 nm) in the stratosphere: implications for photoelectric emissions from instrumentation at balloon altitudes. J. Geophys. Res. 95, 5557 - 5566, 1990.

Dejnakarintra, M., "A Theoretical Study of Electrical Coupling between the Troposphere, Ionosphere, and Magnetosphere," Technical Report 3454-3, Radiosci. Lab., Stanford Electron. Lab., Stanford University, 1974.

Dejnakarintra, M. and C. G. Park, "Lightning induced electric fields in the ionosphere," J. Geophys. Res., vol. 79, p. 1903, 1974.

Greenwald, R. A. et al., DARN/SuperDARN. A global view of the dynamics of high-latitude convection. Space Science Reviews, 71 (1-4), 761-96, 1995.

Holzworth, R. H., Large Scale DC Electric Fields in the Earth's Environment, Ph. D. Dissertation, Physics Department, University of California, Berkeley, 1977.

Holzworth, R. H., High Latitude Stratospheric Electrical Measurements in Fair and Foul Weather under Various Solar Conditions, J. A. T. P., 43, 1115-1125, 1981.

Holzworth, R. H., A new source of horizontal electric fields in the middle latitude stratosphere, J. Geophys. Res. 94, 12795, 1989.

Holzworth, R. H., and F. S. Mozer, Direct evaluation of the radial diffusion coefficient near L=6 due to electric field fluctuations, J. Geophys. Res., 84, 2545, 1979.

Holzworth, R. H. et al., The Large-scale Ionospheric Electric Field: its Variation with Geomagnetic Activity and relation to Terrestrial Kilometric Radiation, J. Geophys, Res., 82, p. 2735-2742, 1977.

Holzworth, R. H., et al., "Global Ionospheric Electric Field Measurements in April 1978," J. Geophys. Res., vol. 86, no. A8, pp. 6859-6868, 1981.

Holzworth, R. H., et al., "ELBBO: Extended Life Balloon Borne Observatories," URSI Radioscientist, vol. 4, pp. 33 - 37, 1993.

Hu, Hua, "Global and local electrical phenomena in the stratosphere," PhD thesis, Univ. of Washington, Geophysics Program, Seattle, Washington, 1994.

Kellogg, P. J. and M. Weed, "Balloon measurements of ionospheric electric fields," Proceedings of Fourth International Conference on the Universal Aspects of Atmospheric Electricity, Tokyo, 1968.

Konik, R. M., et al., Cusp latitude magnetic impulse events, 2. Interplanetary magnetic field and solar wind conditions, J. Geophys. Res., 99, 14831-14853, 1994.

Lanzerotti, L. J., et al., Possible evidence of flux transfer events in the polar ionosphere, Geophys. Res. Lett., 13, 1089-1092, 1986.

Lanzerotti, L. J., et al., Cusp latitude magnetic impulse events. 1. Occurrence Statistics, J. Geophys. Res., 96, 14009-14022, 1991.

Lee, L. C., Magnetic flux transfer at the Earth's magnetopause, in Solar Wind - Magnetosphere Coupling, edited by Y. Kamide, and J. Slavin, Terra Scientific Pub., Tokyo, 1986.

Mozer, F. S., "Balloon measurements of vertical and horizontal atmospheric electric field," PAGEOPH, vol. 84, p. 32, 1971.

Mozer, F. S., "Analysis of techniques for measuring d.c. and a.c. electric field in the magnetosphere," Space Science Reviews, vol. 14, p. 272, 1973.

Mozer, F. S. and P. Lucht, "The Average Auroral Zone Electric Field," J. Geophys. Res., vol. 79, pp. 1001-1006, 1974.

Mozer, F. S. and R. Serlin, "Magnetospheric electric field measurements with balloons," J. Geophys. Res., vol. 74, p. 4739, 1969.

Powell, S. P., An on-board microprocessor system for processing electric field signals on superpressure balloons, Masters Thesis, Electrical Engineering, Cornell University, p. 111, 1983.

Sibeck, D. G., Transient magnetic field signatures at high latitudes, J. Geophys. Res., 98, 243-256, 1993.

Southwood, D. J., Theoretical aspects of ionospheric-magnetospheric-solar wind coupling,Adv. Space Res., 5, 7-1985.

Southwood, D. J., The ionospheric signature of flux transfer events, J. Geophys. Res., 92, 3207-3213, 1987.

Spiro, R. W., et al., Rapid subauroral ion drifts observed by Atmospheric Explorer C, J. Geophys. Res. Lett., 6, 657, 1979.

Weimer, D. R., et al., "Auroral zone electric fields from DE 1 and 2 at magnetic conjunctions," J. Geophys. Res., vol. 90, pp. 7479-7494, 1985.

E. A. Bering, III, Dept. of Physics, Science and Research Bldg. 1, Rm. 532 University of Houston, 4800 Calhoun Blvd, Houston, TX 77204-5506 (email: bering@space.phys.uh.edu)

R. H. Holzworth, Space Science Division, Geophysics Program, Univ. of Washington, Box 351650, Seattle, WA 98195-1650. (email: bobholz@geophys.washington.edu)

Scalar Magnetometers for Space Applications

Fritz Primdahl

Department of Automation, Technical University of Denmark, Lyngby, Denmark

A survey of existing instrumentation and developments is presented emphasizing instrumentation for in-flight calibration of vector magnetometers on magnetic mapping missions. Proton free or forced precession magnetometers are at the focus as calibration references, because the proton gyromagnetic ratio is a basic atomic constant for the SI units of magnetic field and electric current. The classical proton free precession, the Overhauser forced oscillation and a new field cycling Overhauser magnetometer are presented. Alkali metal vapor magnetometers, although not absolute in the same sense as the classical proton magnetometer, offer stability and resolution well suited for the calibration purposes. Recent developments are discussed. The metastable Helium magnetometer also offers quasi-absolute scalar measurements, and the use of semiconductor tuned lasers replacing an RF-excited Helium lamp holds great promise for improved accuracy and reduced power consumption.

INTRODUCTION

Absolute determination onboard satellites of the near Earth's vector magnetic field is particularly challenging, because progress in modelling and separating fields originating in the Earth's liquid core and solid crust, fields from sources in the magnetosphere and solar wind exterior to the satellite orbit, and the identification of field aligned currents crossing the orbit has reached the point where instrument performance determines the limits, rather than the power of modern mathematical tools [*Langel, 1991*]. This is one reason for reexamining the availability of high resolution absolute scalar magnetometers.

Science calls for separate mathematical models of the Earth's magnetic vector field from the different classes of sources mentioned above. To achieve this we need to know the vector components of the field with sub-nanoTesla absolute accuracy simultaneously in a large number of densely and evenly distributed points on a closed surface

around the Earth. Global coverage in the shortest possible time span is the key factor for successful modelling, and the only way to do this is to fly magnetometers onboard one or several simultaneous polar orbiting satellites.

Attempts to derive vector models from scalar data have demonstrated that certain terms in the spherical harmonic analysis of little influence on the magnitude of the field are ill defined, and that the vector field derived from such models may contain large errors. This is termed the Backus ambiguity [*Stern and Bredekamp, 1975*], and it emphasizes the need for vector data in order to get vector models.

Apart from the fact that vector magnetometers need an external arc sec absolute orientation reference, such as a star camera, then all high resolution vector instruments are based on compensation of the external field by some tri-axial arrangement of coils with currents controlled by a zero field indicator. The geometrical dimensions of the coils enter the determination of the field components, and so do the zero errors of the field transducers. Over time and over a temperature range, coils and zero offsets will change, and for satellite instruments this presents a problem.

Vector magnetometers thus need calibration against an absolute magnetic field standard, and a scalar magnetometer can supply such a standard, if a sufficiently large data set of

Measurement Techniques in Space Plasmas: Fields
Geophysical Monograph 103

simultaneous scalar and vector measurements is available, evenly distributed over all directions and magnitudes. This was demonstrated for the Magsat data set. The key to the intercalibration between the scalar and the vector magnetometer is to have as many different directions and magnitudes of the external field as possible giving good coverage of the measurement ranges. In this respect the Earth's field changes fairly slowly along the orbit, so that the scalar data rate needs not be higher than one sample per second. Even one sample per several minutes will do the job [*Langel, private communication, 1994*].

The scalar magnetometer sensor should preferably be omni-directional, i.e. it should give reliable and accurate readings for all directions of the external field relative to the sensor. The Earth's magnetic field is everywhere close to the meridian plane, and onboard a gravity gradient stabilized satellite with attitude control it is always possible to keep a single exclusion line of the scalar sensor away from the magnetic field vector. However, this may impose unnecessary restrictions on the satellite dynamics leading to frequent attitude control actions, which in turn may cause magnetic perturbations on the sensors from torquer coil currents and possibly induce other perturbations that will complicate the satellite dynamics and the attitude restoration.

Magnetic mapping missions require the field sensors to be removed some distance from the satellite in order to avoid the local magnetic fields. Also, the vector magnetometers based on compensation of the external field will have some external magnetic perturbation, requiring a minimum distance to the scalar sensor. A long boom for supporting the suitably separated scalar and vector field sensors is necessary to obtain the accuracy. Deployment on a boom puts severe restrictions on the number of cables and on the total deployed weight of the sensors.

In the following, the performance of some likely candidate scalar instruments will be investigated with due regards to the demands imposed by the satellite environment as outlined above. Factors like weight, power, number of cables needed for the sensors, and the overall complexity of the system will carry great weight, as will the anticipated measurement accuracy.

THE PROTON FREE PRECESSION SCALAR MAGNETOMETER

The principle of the simplest and first of the Earth's field quantum magnetometers rests on the fact that the proton has an angular momentum or spin, and at the same time a magnetic moment [*Block, 1946; Stuart, 1972*]. A classical mechanical description views the proton as a small mechanical gyroscope with the rotation axis made up of a bar magnet (see Figure 1). The angular momentum (spin) L and the magnetic dipole moment μ are atomic constants, they are parallel, and related by the gyromagnetic ratio $\gamma_p = \mu/L$. If the proton is placed in a magnetic field, then its magnetic moment will feel a mechanical torque $\mathbf{T} = \mu \times \mathbf{B}$, and the spin will cause it to precess about the magnetic field B with the angular velocity $2\pi f = \gamma_p B$ [*Ness, 1970*].

The Earth's magnetic field tends to align the protons in a liquid sample such as water, kerosene, hexane etc., but thermal agitation at room temperature counteracts the alignment, so that the effective magnetization of the protons is too small to be detected. However, a strong dc polarization field applied for a certain time to the sample will enhance the magnetization, and make a detection of the precessing proton magnetic moments possible. The polarization field should be applied at approximately right angles to the Earth's field, it must be much stronger, of the order of 10 mT or more, and it must be switched off in a time much shorter than a precession period in the Earth's field.

Varian Associates flew in 1961 a proton magnetometer on the Vanguard III satellite. The sensor consisted of a liquid

Proton Precession Magnetometer

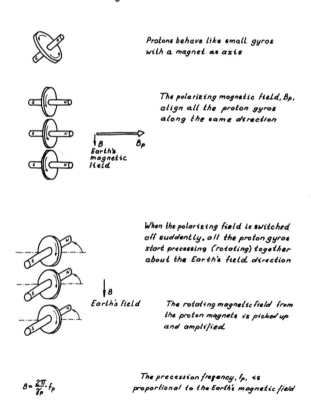

Protons behave like small gyros with a magnet as axis

The polarizing magnetic field, B_p, align all the proton gyros along the same direction

$\uparrow B$ Earth's magnetic field $\longrightarrow B_p$

When the polarizing field is switched off suddenly, all the proton gyros start precessing (rotating) together about the Earth's field direction

$\uparrow B$ Earth's field

The rotating magnetic field from the proton magnets is picked up and amplified

$B = \frac{2\pi}{\gamma_p} \cdot f_p$

The precession frequency, f_p, is proportional to the Earth's magnetic field

Figure 1. Macroscopic classical view of measuring the Earth's magnetic field by nuclear magnetic resonance.

proton rich sample surrounded by a single solenoid used for both polarization and precession signal pickup as shown in Figure 2. The solenoid sensor has a null line coincident with the coil axis along which the magnetometer cannot measure the magnetic field, because the precession signal disappears. The sensor is connected to the front end of the electronics via a single twisted shielded pair carrying first the polarization current, and then connecting the microVolt precession signal to the amplifier. After amplification the exponentially decaying signal (Figure 3) is analysed for the dominant frequency carrying the information about the scalar magnitude of the magnetic field.

During polarization the solenoid acts as a strong magnetic dipole disturbing the surroundings, and the solenoid picks up external noise in the near Earth's field band of the amplifier from 700 to 2,500 Hz. The measurement rate is one sample per second with 0.5 second for polarizing and 0.5 second for the frequency analysis. Using phase-locked-loop techniques and fast counters resulted in resolutions down to 0.1 nT; but modern fast digital signal analysis may certainly improve the sampling rate and the resolution.

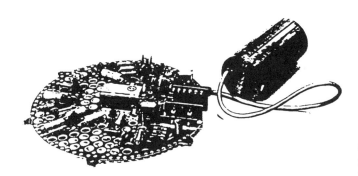

X-4942 Proton Magnetometer

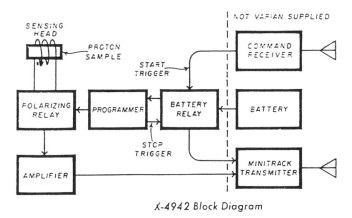

Figure 2. Varian Associates' magnetometer on Vanguard III.

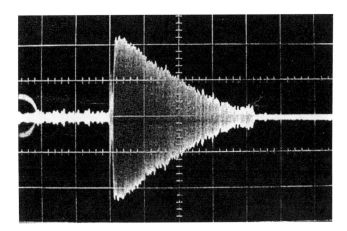

Figure 3. Proton Precession Signal. Horizontal div. in seconds and vertical div. in microVolts. NASA Official Photo.

An omnidirectional noise immune sensor with low external polarization disturbance was described by Serson [1962]. It consists of a toroidal doughnut shaped vessel with the proton liquid sample, and a coil wound through the toroid [Acker, 1971] with evenly distributed windings along the circumference (Figure 4). The sensor is omnidirectional because some parts of the coil will always have the axis perpendicular to the field. The sensor is also insensitive to homogeneous external noise fields in the band pass, and the toroidal geometry cancels the dipole term of the magnetic field from the polarization current. The toroid sensor still only needs a cable with one shielded twisted pair to operate, and it weighs about 1 kg. The polarization current dissipates 50 W to 100 W in the coil. However, the average power can be reduced by reducing the sample rate to once per minute, which is adequate for in-orbit calibration purposes. Modern digital signal processing may reduce the demands on the signal-to-noise ratio thereby allowing the size of the sensor and thus the polarizing power to be reduced. Development work on the classical omnidirectional proton free precession magnetometer is ongoing [Pierce et al., 1995].

The gyromagnetic ratio for a free proton in vacuum is $267,522,128.81$ rad T^{-1} s^{-1}. However, the protons in a liquid sample are not free, they are affected by the molecule they belong to, and by the diamagnetic (or paramagnetic) dipole shielding effect of all the other molecules in the sample. The gyromagnetic ratio for water in a spherical sample is:

$$\gamma_p' = 267,515,255.81 \text{ rad } T^{-1} \text{ } s^{-1} \quad \text{(protons in water)}$$

[Cohen and Taylor, 1987]. In the Earth's field of about 50,000 nT, this effect from the water molecule amounts to a fixed shift of 1.28 nT. For samples like short cylinders or

Figure 4. Serson omnidirectional toroid proton magnetometer sensor. Cut-away made for post vibration inspection.

toroids an extra correction of the order of 3 ppm or less [*Belorizky et al., 1990; 1991*] corresponding to less than 0.15 nT in a field of 50,000 nT is introduced by the nonspherical shape of the vessel. This shape or demagnetization shift is expected to change slightly with the external field direction, because the demagnetizing factor of a nonspherical sample will change with direction. Other proton rich liquids will have fixed shifts different from that of water, and such a magnetometer will have to be calibrated against a water-based standard instrument in order to get absolute accuracy at the 0.1 nT level. Other sources of error exist. The frequency standard, against which the precession frequency is determined, must be stable and calibrated at the 1 ppm level to get 0.1 nT accuracy. The most prominent error source is magnetic impurities in the sensor or in its immediate surroundings adding a small vector field fixed in the sensor coordinate system. This results in sensor heading errors, which should be carefully checked and mapped for a high precision instrument.

The protons always precess in an inertial frame, and any rotation of the signal coil will shift the apparent frequency and introduce a bias in the field determination. If the platform rotates with the angular velocity Ω, then the correction will be within $\pm\Omega/\gamma_p$' and depend on the angle between B and Ω [*Hartmann, 1972; Ness, 1970; Alexandrov and Primdahl, 1993*]. A platform rotation rate of 15° per second will give shifts of up to ±1 nT. This can be corrected by monitoring the carrier attitude; but large irregular changes will make corrections to sub-nanoTesla levels difficult.

THE OVERHAUSER EFFECT PROTON PRECESSION MAGNETOMETER

Continuous enhancement of the proton polarization is possible if some free unpaired electrons (i.e. electrons whose spins are uncompensated) are introduced into the proton rich liquid. By RF-ESR (Electron Spin Resonance) the electrons can be excited or polarized to saturation in the Earth's field (or in a nuclear or externally applied large field), and the various couplings between the polarized electrons and the protons result in the creation of a large magnetization of the protons antiparallel to the Earth's field [*Hartmann, 1972*]. This is the dynamic nuclear polarization (DNP). The theoretical enhancement of the proton polarization is about 330 times their natural polarization in the same field.

The RF resonance frequency of a free electron is about 28 GHz T^{-1}, and so in the Earth's field the RF must be tuned to follow the variations to maintain the polarization. The ESR line width determines the tuning accuracy and the RF power needed to saturate the electrons.

As was the case for the protons, the molecule or the nucleus hosting the free electron has some influence on the resonance frequency, and this is used in the DNP Overhauser proton precession magnetometers built by GEM Systems Inc. in Canada [*Hrvoic, 1990*] and by LETI in France [*Duret et al., 1994; 1995; Kernevez and Glenat, 1991; Kernevez et al., 1992*].

In a nitroxide free radical dissolved in a proton rich liquid, the free radical unpaired electron dwells in the local magnetic field of the nitrogen nucleus of effectively about 2.1 mT, so that its RF resonant frequency is not 1.4 MHz (as in the Earth's field of 50,000 nT), but about 60 MHz. This is the zero field splitting of the free radical Tempone used in these magnetometers. Because the electron polarization takes place in a much larger field, then the theoretical enhancement of the proton polarization exceeds a thousand times against the 330 times obtainable in the Earth's field [*Hrvoic, 1990*].

Figure 5 from Kernevez et al. [*1992*] shows the perdeuteriated Tempone ^{15}N radical dissolved in the proton rich liquid, and its energy levels versus the applied external magnetic field. The frequency separation 'A' at zero external field indicates the local magnetic field of the nitrogen nucleus seen by the free radical electron.

Following DNP the protons can be reoriented into exponentially decaying free precession by a short dc current 90° deflection impulse in a solenoid surrounding a cylindrical proton sample. Creation of continuous oscillation is possible by applying a weak excitation magnetic field rotating in the plane of precession. This field will slightly deflect the protons from being antiparallel to the Earth's field, and they will precess with a frequency close to the excitation frequency, as long as it is within the line width of the proton resonance. The proton spectral line width can be within 2 nT, and inaccuracies can be limited to a small fraction of a nanoTesla [*Hrvoic, 1990*]. However, this

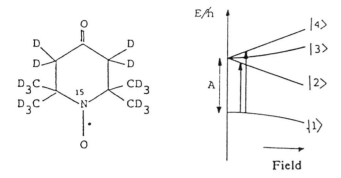

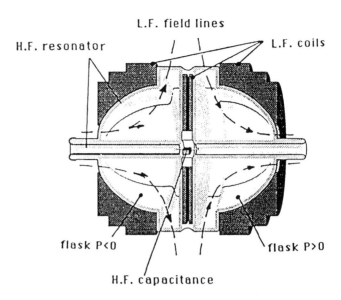

Figure 5. Left: Perdeuteriated Tempone ^{15}N radical (the free electron is indicated by the dot). Right: Energy levels versus the externally applied field. "A" indicates the zero field splitting.

inaccuracy will add to the inaccuracies previously discussed for the dc polarized free precession proton magnetometer, which are also valid for the dc impulse deflected proton free precession of the DNP Overhauser magnetometer.

As for the simple dc polarized free precession magnetometer the single solenoid DNP sensor has a null line along the coil axis. The sensor needs a coaxial cable for the 60 Mhz RF excitation and at least one twisted shielded pair for sustaining continuous oscillation. The ESR line for the nitroxide is fairly broad, from about 20 μT and up to 100 μT, and as the saturation RF power depends on (at least) the square of the ESR line width, then the nitroxide free radical demands a fair amount of excitation power.

The GEM Systems magnetometers have noise levels of 0.01 nT to 0.05 nT for one to two readings per second for an oscillating magnetometer, and 0.02 nT for a dc 90° pulsed magnetometer measuring once per second. The proton oscillator sensor needs 5 W of continuous RF power, where a pulsed magnetometer accurate to 0.1 nT needs only 1 W sec (Joule) per reading [*Hrvoic, 1990*]. The Danish Ørsted Geomagnetic Research satellite project uses a GEM Overhauser dc pulsed magnetometer as a secondary standard for calibration of the satellite magnetometers. The unit was tested against the Danish Brorfelde magnetic observatory standard and showed an overall standard deviation of 0.25 nT, which is the claimed absolute accuracy of the magnetic observatory [*Laursen and Højlev, 1994*].

OMNIDIRECTIONAL CONTINUOUSLY OSCILLATING OVERHAUSER MAGNETOMETER

A silvered flask (Figure 6, from LETI's brochure) with the free radical dissolved in Methanol (MeOH) or Dimethoxyethane (DME) constitutes LETI's basic sensor for the Danish Ørsted geomagnetic research satellite [*Duret et*

Figure 6. Schematic cut view of LETI's omnidirectional and continuously oscillating Overhauser proton magnetometer sensor. The two half cells have opposite proton polarisations.

al.,1994, Perret and Llorens, 1995]. The perdeutoriated Tempone free radical makes the ESR line narrower and reduces the RF power, and they obtain an amplification factor of about 1500. The LF coil gives a highly inhomogeneous field for the excitation of the proton oscillation, so whatever the direction of the Earth's field there will always be some part of the liquid where the excitation field is perpendicular to the external field.

The proton oscillator is made of two cells in a differential coupling to the input amplifier (Figure 7). The proton oscillation excitation signal is applied in parallel to the two coils giving only a common mode signal to the input amplifier, whereas the proton magnetizations are antiparallel in the two flasks, and they supply the differential inputs to the amplifier. Homogeneous external noise signals cancel each other in the oppositely wound coils.

The two different solvents in the two flasks (DME and MeOH) give different shifts of the ESR spectra, each having a positive and a negative nuclear polarization peak (see Figure 8). The shifts induced by the solvents cause the negative peak of the DME to fall at the same RF frequency as the positive peak of the MeOH, so that one common RF will excite positive polarization in one and negative polarization in the other in accordance with the needs of the oscillator circuit. The separation between the positive and the negative peak in the ESR spectrum for a particular solvent is proportional to the external field. For very low fields the two peaks move closer together and tend to cancel each other resulting in a poorer signal-to-noise ratio. The

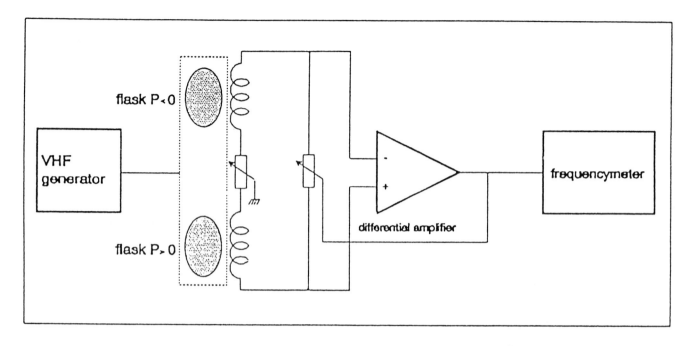

Figure 7. LETI's Overhauser magnetic resonance oscillator.

positive/negative peak structure is a peculiarity of the ^{15}N nitroxides, and it is associated with the zero field splitting.

The magnetometer operates in the Earth's field from 16,000 nT to 54,000 nT along the Ørsted orbit. The sample rate is one per second with an upper 3 dB frequency response of 0.2 Hz and an absolute accuracy of 0.5 nT. The input amplifier is nonmagnetic and placed at the sensor, the RF power is 1 W including the transmission line losses, total power is less than 3 W and the total mass is 2.9 kg including 1.4 kg for the sensor. The sensor anisotropy (heading error) is less than 0.5 nT. The cables supporting the sensor are one coaxial cable for the RF, two twisted shielded pairs for the detector signals, and a twisted pair for temperature measurement. [*Perret and Llorens, 1995*].

THE DC AND RF DOUBLY POLARIZED
OVERHAUSER MAGNETOMETER

A free radical Trityl produced by Nycomed Innovation, Malmø, Sweden is used as an alternative to Tempone. It does not have the zero field splitting of Tempone, and it has only a single ESR resonance line and not the double positive/negative structure of Tempone resulting in increased noise at low fields.

The free radical has a very narrow ESR line, about 2.5 μT compared to more than 20 μT for the nitroxide (see Figure 9, [*Ardenkjær-Larsen, private communication*]). The RF saturation power is proportional to (at least) the square of

the line width, and so the power needed for saturation is milliWatts compared to Watts for the nitroxide free radical. Figure 10 [*Ardenkjær-Larsen, private communication*] compare the saturation levels against RF power for the Nycomed Trityl and for the ^{15}N perdeuteriated Tempone. This means that Trityl can be fully saturated at a modest RF power level, and that the free radical concentration can be optimized for maximum polarization coupling between the electrons and the protons, because there is room for some ESR line broadening and for the corresponding modest increase in power.

Dynamic nuclear polarization can either take place in the Earth's field, which requires the RF to track the field (1.4 Mhz in 50,000 nT), or in an additionally applied homogeneous dc field of the order of 10 mT giving an extra polarization enhancement proportional to the ratio between the applied dc field and the Earth's field [*Ardenkjær-Larsen, 1994*].

A simple cylindrical sensor with a single null axis can be made very small. A laboratory instrument built at the Technical University of Denmark, with a 7 millilitre sample gives a signal-to-noise ratio of 55 dB Volts per root Hz in 50,000 nT. The dc polarization power decreases linearly with volume, and so only an average of 600 mW for the 10 mT dc polarization and 80 mW for the 260 MHz RF polarization is needed. With 200 ms for polarization and 150 ms for readout the instrument takes about 3 samples per second. A digital signal processor determines the the time constant and the proton free precession frequency of the exponentially decaying spin signal. About 1000 points of the signal can be digitized in 100 ms and subsequently real time analysed using the powerful modern algorithms available for signal retrieval.

The principle seems promising for the development of a low power free precession proton magnetometer with high sensitivity, high sampling rate, and with an absolute accuracy equalling that of the classical proton magnetometer.

ALKALI VAPOR OPTICALLY PUMPED MAGNETOMETERS

The alkali metal atoms have an electron with unpaired spin and thus with an uncanceled magnetic moment. A vessel with alkali metal gas of suitable concentration (avoiding inter-atomar collision effects and wall effects) will absorb photons with energies matching the transitions of the electron and atom quantum spin system. In doing so, the electrons can be brought into a state of net magnetic moment anti-parallel to an externally applied magnetic field, and the electron magnetic moments can be made to precess

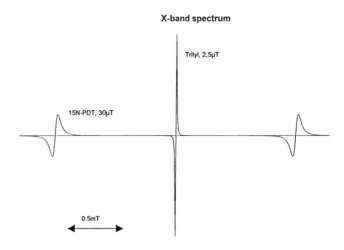

Figure 9. The DNR spectra for Trityl and for Tempone (PDT).

about the B-vector with the Larmor frequency, which for Rubidium, Cesium and Potassium falls in the RF range from about 50 kHz to 460 kHz in the Earth's field [*Hartmann, 1972; Alexandrov and Bonch-Bruevich, 1992*].

The Larmor frequency is related to the external magnetic field B_o by the gyromagnetic ratio, which for Rb^{85} is 4.667 Hz nT^{-1}, for Rb^{87} it is 6.996 Hz nT^{-1}, for Cs^{135} it is 3.499 Hz nT^{-1}, and for K^{39} it is about 7 Hz nT^{-1}. The gyromagnetic ratio is about a factor of 100 larger than the proton gyromagnetic ratio, and a magnetometer based on alkali metal atoms is thus proportionally less sensitive to the platform rotation [*Alexandrov and Primdahl, 1993*].

Figure 11a [*Ness, 1970*] shows the practical implementation of an optically pumped magnetometer. The cell with the alkali atoms in the gas phase, maintained at a suitable temperature and pressure, is illuminated with circularly polarized light from a similar cell with light emission stimulated by a magnetic VHF field from the oscillator. An interference filter attenuates an unwanted line (D_2) at 780.0 nm (for Rb^{85}) letting the optical pumping line (D_1) at 794.7 nm through to the gas cell [*Farthing and Folz, 1967*]. The light transmitted through the gas cell is focussed by the lens onto the photocell, and the output from the amplifier is a measure of the light absorption or the opacity of the cell. The external magnetic field B_o must be parallel to (or have a certain component along) the optical axis, this system thus has a null zone for B_o-fields close to perpendicular to the optical axis.

When first sending light to the cell the photons are absorbed, and the electrons lifted up to the excited energy state. The cell becomes opaque until all the upper energy level states are occupied by electrons, and the states are saturated.

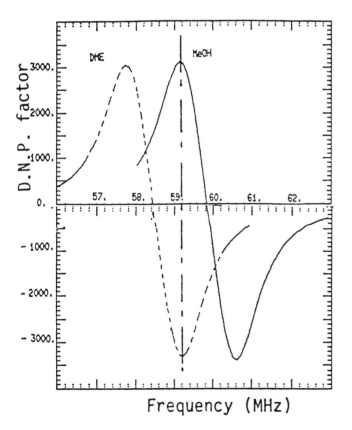

Figure 8. DNP spectrum for the two solvents in LETI's sensor.

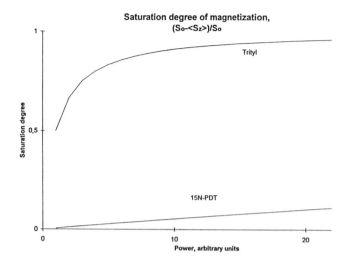

Figure 10. Polarization saturation degree versus ESP RF power for Trityl and for Tempone.

Then there are no more electrons to absorb the light, and the cell becomes transparent again.

When an RF magnetic field at the Larmor frequency f_o is applied to the cell perpendicularly to the optical axis (and to the external B_o-field), then the polarized electrons are induced to precess with the frequency f_o about the B_o-field, whereby they are taken out of the saturated state and randomized. This leaves room for re-exciting electrons into the upper energy level again, and for absorbing photons of the proper energy from the lamp. The cell opacity increases again, and the transmission coefficient is permanently reduced by about 10% to 20%.

The cell transmission coefficient close to the Larmor frequency f_o changes with the applied frequency of the RF B-field as shown graphically to the right in Figure 11a. Standard feedback techniques allow the RF oscillator to be locked to the center of the resonance, and the center frequency is proportional to B_o via the gyromagnetic ratio for the particular gas.

This magnetometer measures B_o-fields with directions less than about 80° from the positive or negative direction of the optical axis.

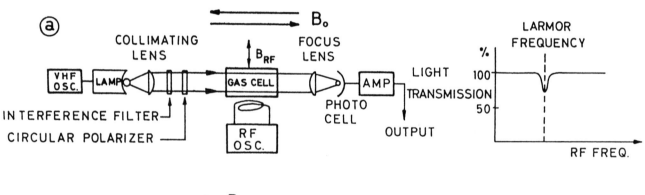

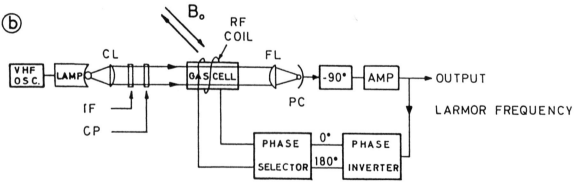

Figure 11. Optically pumped Alkali vapor magnetometers. (a): Single cell Mz-mode. (b): Single cell oscillating Mx-mode. [*Ness, 1970*].

The system is termed the Mz-mode, because the light absorption along the external magnetic field (the z-axis) is observed. The resonating RF B-field is applied perpendicularly to the B_o-field and the optical path. The Mz-mode can not be made self oscillating, it relies on locking the RF to the spectral line by feedback techniques, and it has a null zone perpendicular to the optical axis.

Figure 11b [Ness, 1970] shows the principle of an optically pumped alkali magnetometer, which is self-oscillating at (or near) the Larmor frequency. The B_o-field at the gas cell is directed 45° or 135° from the light beam propagation direction, depending on the sense of circular polarization of the light or on the phasing (0° or 180°) of the feedback RF B-field. The pumping light propagation direction must have a certain component along the B_o-field for pumping the electrons into the upper energy level, where the spins (and magnetic moments) have a component antiparallel to the B_o-field. The RF B-field resonating at the Larmor frequency is here applied along the optical path, and the component of this RF B-field perpendicular to the external static B_o-field sets up a precession of the electron spins (and magnetic moments) about the B_o-field.

When the electron spins are oriented most parallel to the optical path (or antiparallel, depending on the sense of circular polarization and the feedback phasing), then the cell absorption is minimum, and light is passing through to the photocell. Half an f_o precession period later, the electron spins are closest to antiparallel (parallel) to the optical path, and absorption can take place dimming the light through the cell.

The light intensity seen by the photocell is thus modulated at the Larmor frequency f_o, and with proper phasing the signal can be fed back to the gas cell as the RF B-field. The oscillation frequency contains the information on the scalar magnitude of the B_o-field.

This is termed the Mx-mode, because light absorption across the B_o-field is used. As explained, this system can self-oscillate, and it has null zones along and perpendicular to the optical axis.

The Rubidium and Cesium electron spin system has not one but 6 different spectral lines spaced so closely that they can not be resolved with the described technique. The lines have different intensities, which vary with the direction of the external field relative to the optical axis, the pumping light intensity and with the magnitude of the magnetic field.

The resultant oscillator frequency is a weighed average over the lines, and it may change 182 nT for Rb[85], 82 nT for Rb[87] and 6 nT for Cs[133] [Ness, 1970; Alexandrov and Bonch-Bruevich, 1992].

Unlike Cesium and Rubidium, Potassium has a completely resolved spectrum throughout the whole range of the Earth's magnetic field, and a very high resolution and high absolute accuracy laboratory magnetometer has been developed based on a mixture of K[41] and K[39]. Sensitivities of one picoTesla Hz[-½] and no systematic error exceeding 10 pT in the range from 100 nT to 100,000 nT was obtained [Alexandrov and Bonch-Bruevich, 1992].

Another feature of the single cell self-oscillator is the rather limited solid angle in which the external B_o-field can be measured with acceptable accuracy. Figure 12 [J.-M. Gibon, Varian Associates, private communication, 1972] shows a cut of the solid angle, which is rotation symmetric about the sensor axis. The measurement zone extends only from ±15° to ±80° from the axis.

A dual cell system (see Figure 13) was developed by Varian Associates [Ruddock, 1961; Morris and Langan, 1961] that makes the measurement solid angle zones symmetric about the equator plane (see Figure 14, [Gibon, private communication, 1972]), and, equally important, makes the spectral line system more symmetric, whereby the static error and the heading variations are dramatically reduced. The dual cell system uses one RF-excited lamp to drive two antiparallel single cell systems. The photocell of one system drives the RF B-field on the gas cell of the other system, so that the RF B-field phase of each cell always matches the direction of the external field. The dual cell system still has two ±15° dead zones around the poles and a ring shaped dead zone ±10° on each side of the equator plane (Figure 14).

Using two dual cell systems at an angle of 55° reduces the null zones to two small opposite regions of approximately 20° by 10° in size [Ness, 1970]. Figure 15 from Farthing and Folz [1967] shows a system of two dual cell Rb[85] magnetometers flown onboard OGO-II launched in 1965.

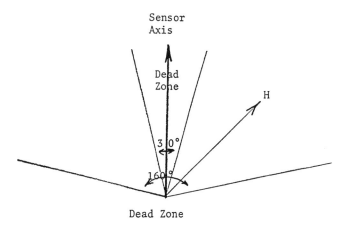

Figure 12. Mx-mode single cell self-oscillator dead zones rotationally symmetric about the optical axis, Varian Associates.

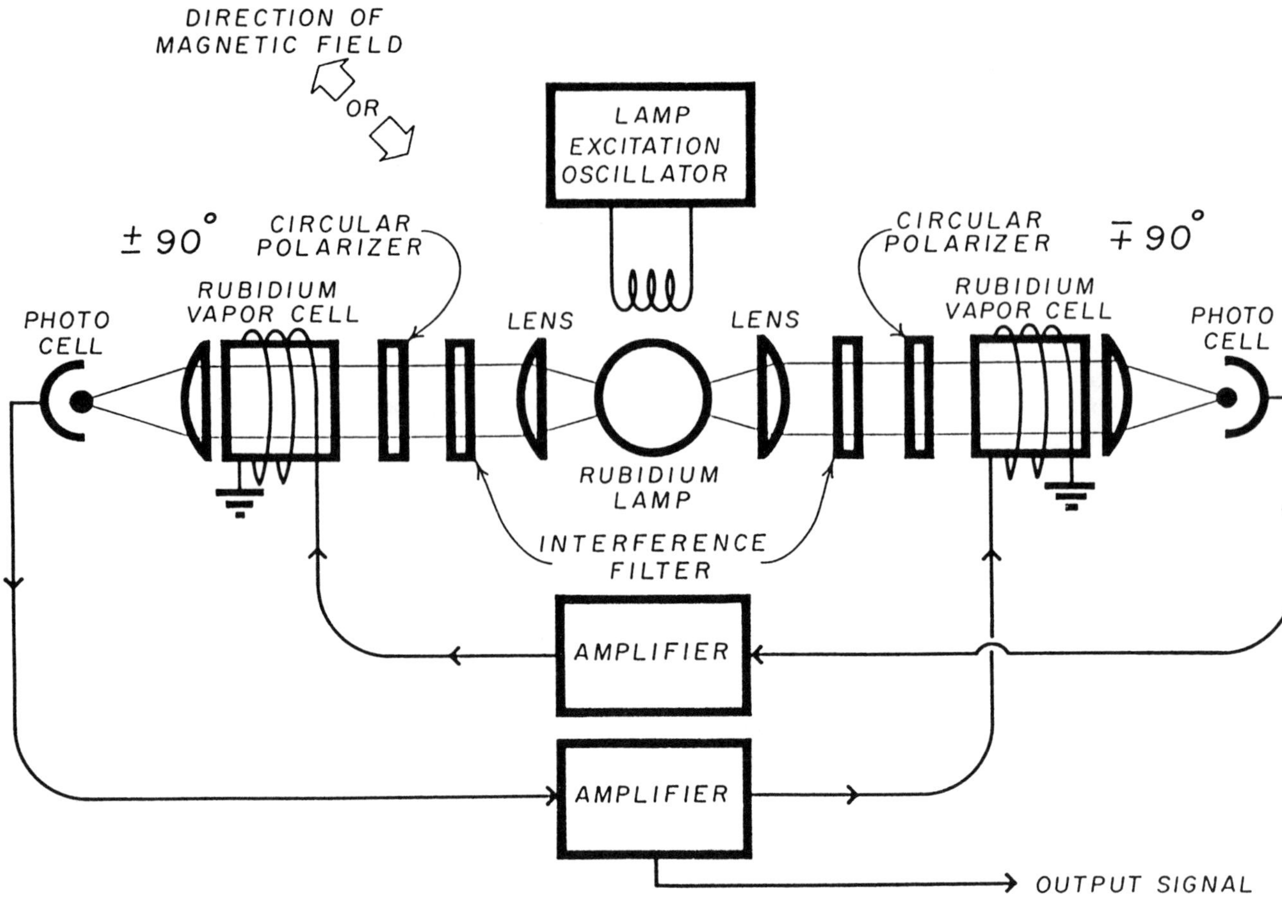

DIRECTION OF
MAGNETIC FIELD

LAMP
EXCITATION
OSCILLATOR

± 90° CIRCULAR POLARIZER

RUBIDIUM VAPOR CELL

PHOTO CELL

LENS

LENS

CIRCULAR POLARIZER ∓ 90°

RUBIDIUM VAPOR CELL

PHOTO CELL

RUBIDIUM LAMP

INTERFERENCE FILTER

AMPLIFIER

AMPLIFIER

OUTPUT SIGNAL

Figure 13. Dual cell Mx-mode self-oscillating sensor. Varian.

The twin dual cell system measured fields in the range from 15,000 nT to 64,000 nT with errors from ±0.5 nT to ±1.5 nT. For complete elimination of the null zones a third dual cell sensor could be added; but this would make an already complex system even more complicated.

The Magsat scalar magnetometer was a Cesium[133] twin dual cell system with ≈55° between the axes in order to reduce the null zones to two small regions oriented transverse to the satellite polar orbit. Since the Earth's magnetic field is generally in or near the orbit plane, then no data was lost because of the null zones [*Mobley et al., 1980*]. The magnetometer provided 8 samples per second, and it was accurate to about 1.5 nT [*Langel et al.,1982*].

The magnetometer sensor weight was 1.9 kg [*Mobley et al., 1980*], and the power for excitation of the Cs lamps about 6 W. Lamp and cell heating is typically 1 W each [*Farthing and Folz, 1967*]. Heaters are needed for tight temperature control of the gas cells for maintaining the optimum cell vapor pressure, for Cs the temperature range is 25°C to 30°C, and for Rb it is about 10°C higher.

Misalignment of the RF feedback coil with the optical axis will cause quite large frequency shifts, in particular when the field direction is close to the null zones. The winding of the coil and positioning of the optics must be tightly controlled [*Farthing and Folz, 1967*].

The cables needed to support a twin dual cell system is typically a coax cable for the excitation of the lamps, a number of twisted pairs for the individually controlled heaters and the thermistors, and at least 4 twisted shielded pairs or coaxial cables for the photocells and the RF feedback coils [*Varian Associates, Palo Alto, CA, V-4969 Data Sheet, 1972*].

The number of cables may be reduced by placing the lamp inside the electronics box and guiding the light to the absorption cells via an optical fibre. Farr and Otten [*1974*] describe a wide range single cell laboratory magnetometer sensor using this principle. Besides the optical fibre only 3 coaxial cables are needed for the lamp heater and temperature control thermistor, for the transverse RF B-field and for the solar cell light detector.

THE METASTABLE HE⁴ OPTICALLY PUMPED MAGNETOMETER

Metastable He⁴ is produced from ordinary He⁴ in a glass cell by a weak 27 MHz electrodeless RF glow discharge. Figure 16 [*Smith et al., 1993; from McGregor, 1987*] shows the spectrum for optical pumping of metastable He⁴, where one of the triplet lines (D_0) is well separated from the two others (D_1, D_2) and fully resolved by an Mz-mode system, as shown in Figure 11a and described above in the section on the alkali magnetometers. A Helium lamp excited by 27 MHz RF power sends infrared light of 1083.0 nm to the metastable He⁴ cell via a circular polarizer (Figure 17, from Smith et al., [*1991*]). The optical pumping quickly saturates the D_0-line, and the cell becomes transparent. An RF B-field at the Larmor frequency applied to the cell perpendicularly to the optical path and to the external B_0-field redistributes the electrons, and the infrared sensor detects a decrease in the cell transmittivity. The sweep oscillator makes the RF generator scan a range of frequencies, and the detector and integrator lock the center of the RF to the Larmor frequency, which is a measure of the external field B_0. He⁴ has the largest gyromagnetic ratio, 28 Hz nT⁻¹, of all the optically pumped magnetometers.

The sweep frequency is 200-400 Hz, and the electron Larmor precession line width is larger than 1 kHz (~4 kHz or 149 nT, [*Slocum and McGregor, 1974*]), so despite being an Mz-mode instrument, the system is relatively fast responding.

The cell and the lamp are insensitive to temperature, because He is a gas in the relevant range, and no heating for

Figure 15. Double dual cell system with one linear dead zone. Magnetometer flown on OGO-II [*Farthing and Folz, 1967*].

temperature control is needed. This means that the instrument is operative, as soon as it is switched on.

In lock, the output of the infrared detector contains a large signal at the second harmonic of the frequency modulation sweep frequency. This is used to indicate lock. Slightly off the center frequency the light detector signal also contains a first harmonic component proportional to the frequency deviation, and this is used to control the feedback loop [*Smith, 1985*]. Unlike a self-oscillating system, this system will need provisions for initial acquisition of lock by letting the RF oscillator slowly search the entire measurement range until a strong second harmonic of the sweep oscillator frequency indicates proper lock. The modulation sweep range is ±1,400 Hz corresponding to ±50 nT [*Smith, 1985*].

The instrument has a dead zone for magnetic field directions perpendicular to the optical axis, but it has no null lines for the directions along the optical path, as the alkali Mx-mode magnetometers have. Figure 18 [*Smith et al., 1991*] shows the directional response of a single sensor. Out to ±60° from the optical axis is the signal-to-noise ratio sufficient to maintain an accuracy of 1 nT [*Smith, 1985; Smith et al. 1991*]. Combining two crossed sensors, as shown in Figure 18 top, reduces the dead zones to a single null line, which is acceptable for an Earth field mapping mission [*Smith, 1985*].

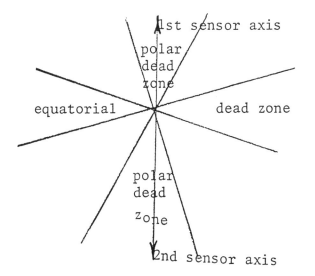

Figure 14. Dual cell self-oscillator dead zones. Rotational symmetry about the optical axis. Varian Associates.

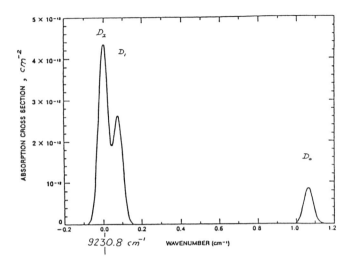

Figure 16. Spectrum of optically pumped metastable Helium. The D_0-line is fully resolved [*McGregor, 1987*].

Because of the fully resolved D_0-line the instrument can be considered absolute in the sense that the frequency-to-field conversion depends on fundamental constants.

The sensor has no heading error, but a small frequency shift depending on the intensity of the pumping light does exist [*Alexandrov and Bunch-Bruevich, 1992; Smith, 1985*]. This light shift may be reduced by reducing the light intensity; but as this decreases the sensitivity and increases the null zones, it does not seem a way out. The overall sensitivity is about 10 times less than that of the alkali magnetometers; so a fairly large light intensity is needed. The light shift proves to be the dominant error, but with great care in the construction it can be reduced to ±0.5 nT. The alignment between the optics and the RF excitation coil has to be accurately established and maintained.

Figure 19 shows an absolute comparison between a station proton free precession magnetometer and the Jet Propulsion Laboratory scalar Helium magnetometer [*Smith, 1985*]. A carefully mapped gradient of 1.5 nT exists between the sites of the two sensors, and the scatter between the instruments is only ±0.5 nT including the errors from both instruments.

The sampling rate may be as high as 2 to 20 samples per second at 1 nT accuracy [*Smith et al., 1991; Smith, 1985*], and the response error to a Gauss-curve shaped bump on the field, 5 nT in amplitude and 1 sec in duration, is within 0.05 nT [*Smith, 1985*].

A two-cell sensor weighs about 1.2 to 1.6 kg and consumes 0.9 to 1.2 Watts. The cabling needed for boom operation amounts to 2-4 coaxial cables for the lamp and cell operation and one for the RF excitation. Two shielded twisted pairs and one twisted pair are necessary for the infrared detectors and the thermistor.

A vector field measuring version of the He[4] magnetometer is included onboard the Ulysses mission [*Balogh et al., 1992*].

Using laser excitation replacing the He[4] lamp has been investigated by McGregor [*1987*]. It opens for the possibility of getting only a single strong line at 1083.0 nm for exciting the D_0 spectral line, and not the mixture of lines from the He[4] lamp; but it does not, unfortunately, alleviate the light shift problem [*Alexandrov and Bunch-Bruevich, 1992*]. Smith et al. [*1993*] suggest a further reduction in size and a simplification of the He sensor by using optical fibers to guide the light from a lamp or a laser in the electronics box to the sensor and also to detect the light output from the sensor via a fiber with an infrared sensor in the electronics.

DISCUSSION AND CONCLUSION

The simplest of the highly accurate scalar magnetometers is the dc polarized free precession proton magnetometer. Using the toroid geometry it is omnidirectional, it needs only a single twisted pair to operate, and the weight of a practical sensor (about 1 kg) is comparable to the weight of other scalar sensors. A basic drawback shared with all proton magnetometers is the small gyromagnetic ratio of about 0.04 Hz nT[-1], which may be a concern if the platform rotation is not monitored. Other concerns are the polarization power (about 50 Joules per sample) and the external perturbation field from the polarization current. Reducing the sample rate to one per minute, then the average power is comparable to the operation power of other sensors, and with a suitable distance to other magnetometer sensors, the external field from the polarization will not be seen. Reexamination and optimization of the simple classical dc polarized proton free precession magnetometer is ongoing.

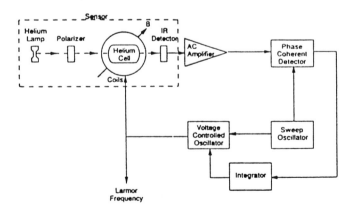

Figure 17. Bloch diagram of the Mz-mode scalar Helium magnetometer (SHM) [*Smith et al., 1991*].

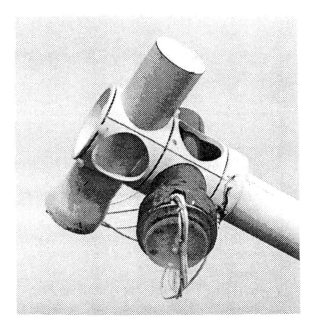

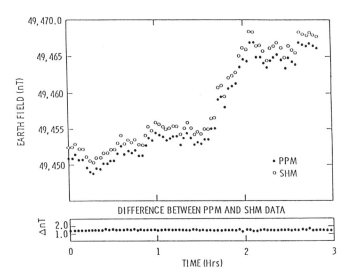

Figure 19. SHM versus proton precession magnetometer tracking test. **JPL.** [*Slavin, 1984*].

The nitroxide free radical based Overhauser effect proton magnetometers consume less power, if used in the pulsed mode and at a rate of one sample per second. The cost is the RF polarization, which demands a loop gap resonator around the sensor, and the sensor cannot simply be made omnidirectional. A magnetometer of this type was tested against an observatory standard and found absolute accurate to within 0.25 nT. The nitroxide based system can be made into a self-oscillator resonating at the proton precession frequency, and omnidirectional systems are available. The oscillator electronics will, however, introduce the danger of a small frequency shift, leading to systematic errors, and the costs are in weight and in a fairly large number of cables for supporting the system.

Free radicals with much narrower line widths than that of the nitroxide exist, and a combination of the dc and RF polarization techniques has demonstrated that this leads to reduced sensor size needing only milliWatts for operating at 3 samples per seconds. Further development of this sensor is ongoing.

The alkali vapor optically pumped magnetometers can not be termed absolute because the self-oscillator resonance frequency depends on tight temperature control, on the relative orientation between the optical path and the magnetic field, and on the fact that the spectral lines are not resolved leading to a compromise in the oscillator frequency. Two crossed dual-cell units are needed to get a system with a null zone along one line only, and the weight, power demands, and support cabling is not competitive. The advantages are the fast response and continuous operation, and the fact that the gyromagnetic ratios of 3 to 7 Hz nT^{-1}

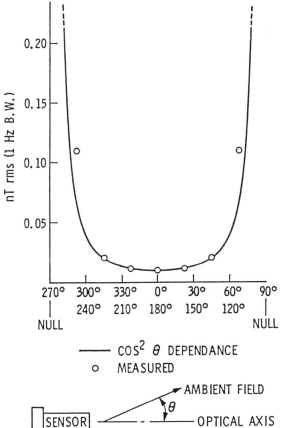

Figure 18. Top: Photograph of two orthogonal SHM sensors. Bottom: SHM signal noise versus angle between the external magnetic field and the optical axis. **JPL.** [*Smith et al., 1991*].

makes the magnetometer readings virtually independent of the platform rotation. The best accuracy is obtained with Cesium, and it is about 1.5 nT.

Using Potassium the spectral lines can be resolved, and a laboratory model is reported to have greatly improved accuracy. Development work on this instrument is also ongoing.

Of the optically pumped scalar magnetometers the metastable He^4 instrument is the most stable and accurate. It has the largest gyromagnetic ratio of all the atomic magnetometers, 28 Hz nT^{-1}, and the spectral lines are completely resolved. However, the resonance frequency is affected by the pumping light shift at the ± 0.5 nT level, and it takes two crossed cells to build a sensor with only one null line. Direct comparison with a station free precession proton magnetometer shows 0.5 nT random deviation between the instruments. Promising new developments with a semiconductor laser light source and the use of optical fibres is in progress.

In conclusion, several scalar magnetometers are available with adequate accuracy and with successful flight records, and a number of interesting new developments are in progress.

Acknowledgements. My institute colleagues Ib Laursen, G.B. Jensen and Jan Henrik Ardenkjær-Larsen are thanked for discussions on the quantum mechanics behind the working principles.

REFERENCES

Acker, F. E., Calculation of the Signal Voltage in a Toroidal Proton Precession Magnetometer Sensor, *IEEE Trans. Geosci. Electronics, GE-9*, 98-103, 1971.

Alexandrov, E. B. and V. A. Bonch-Bruevich, Optically Pumped Magnetometers After Three Decades, *Optical Engineering, 31*, 711-717, 1992.

Alexandrov, E. B. and F. Primdahl, On Gyro-Errors of the Proton Magnetometer, *Meas. Sci. Technol., 4*, 737-739, 1993.

Ardenkjær-Larsen, J. H., Overhauser Magnetometer, *Danish Patent Application*, PTC/DK 94/00480, 21. Dec., 1994.

Balogh, A., T. J. Beek, R. J. Forsyth, P. C. Hedgecock, R. J. Marquedant, E. J. Smith, D. J. Southwood and B. T. Tsurutani, The Magnetic Field Investigation on the Ulysses Mission: Instrumentation and Scientific Results, *Astronomy & Astrophysics, Suppl. Ser., 92*, 221-236, 1992.

Belorizky, E., W. Gorescki, M. Jeannin, P. H. Fries, P. Maldivi and E. Gout, Sample-Shape Dependence of the Inhomogeneous NMR Line Broadening and Line Shift in Diamagnetic Liquids, Chemical Phys. Lett., 175, 579-584, 1990.

Belorizky, E., P. H. Fries, W. Gorecki and M. Jeannin, Demagnetizing Field Effect on High Resolution NMR Spectra in Solutions with Paramagnetic Impurities, *J. Phys. II, 1*, 527-541, 1991.

Block, F., Nuclear Induction, *Phys. Rev., 70*, 460-485, 1946.

Cohen, E. R. and B. N. Taylor, The 1986 Adjustment of the Fundamental Physical Constants, *Rev. Modern Phys., 59*, 1121-1148, 1987.

Duret, D., N. Kernevez and T. Thomas, Functional Description of the LETI OVH Magnetometer for the Oersted Mission, *Rep.*, 8 pp., LETI, Grenoble, France, 21. Feb., 1994.

Duret, D., J. Bonzom, M. Brochier, M. Frances, J.-M. Leger, R. Ordru, C. Salvi, T. Thomas and A. Perret, Overhauser Magnetometer for the Danish Oersted Satellite, INTERMAG 95, *IEEE Trans. Mag., MAG-31*, 3197-3199, 1995.

Farr, W. and E.-W. Otten, A Rb-Magnetometer for a Wide Range and High Sensitivity, *Appl. Phys., 3*, 367-378, Springer-Verlag, 1974.

Farthing, W. H. and W. C. Folz, Rubidium Vapor Magnetometer for Near Earth Orbiting Spacecraft, *Rev. Sci. Instrum., 38*, 1023-1030, 1967.

Hartmann, F., Resonance Magnetometers, *IEEE Trans. Mag., MAG-8*, 66-75, 1972.

Hrvoic, I., Proton Magnetometers for Measurement of the Earth's Magnetic Field, GEM Systems Inc., *Proc. Int. Workshop on Geomagn. Observatory Data Acquisition and Processing, Nurmijärvi Geophysical Observatory, Finland, May 1989, Sect. 5.8*, 103-109, Finnish Meteorological Institute, 1990.

Kernevez, N. and H. Glenat, Description of a High-Sensitivity CW scalar DNP-NMR Magnetometer, *IEEE Trans. Mag., MAG-27*, 5402-5404, 1991.

Kernevez, N., D. Duret, M. Moussavi and J.-M. Leger, Weak Field NMR and ESR Spectrometers and Magnetometers, INTERMAG 92, *IEEE Trans. Mag., MAG-28*, 3054-3059, 1992.

Langel, R. A., G. Ousley, J. Berbert, J. Murphy and M. Settle, The Magsat Mission, *Geophysical Research Lett., 9*, 243-245, 1982.

Langel, R. A., Mission Objectives and Scientific Rationale for the Magnetometer Mission, *Proceedings of the Workshop on Solid-Earth Mission ARISTOTELES, Anacapri, Italy, 23-24 September 1991, ESA SP-329*, 17-26, December 1991.

Laursen, I. and A. Højlev, Test of Overhauser GEM-19 at Brorfelde, *Rep. Oersted Technical Note TN 119, Department of Electrophysics*, 6 pp., Technical University of Denmark, March, 1994.

McGregor, D. D., High-Sensitivity Helium Resonance Magnetometers, *Rev. Sci. Instrum., 58*, 1067-1076, 1987.

Mobley, F. F., L. D. Eckard, G. H. Fountain and G. W. Ousley, MAGSAT - A New Satellite to Survey the Earth's Magnetic Field, *IEEE Trans. Mag., MAG-16*, 758-760, 1980.

Morris, R. and L. Langan, Varian Associates' Space Magnetometers 1956-1961, *Geophysical Technical Memorandum No.8*, 5 pp., Quantum Electronics Division, Varian Associates, Palo Alto, California, 1961.

Ness, N. F., Magnetometers for Space Research, *Space Sci. Rev., 11*, 459-554, 1970.

Perret, A. and J. C. Llorens, Ørsted Project Overhauser Magnetometer Interface Specification, *Rep. OMIS*, 27 pp., CNES, Toulouse, France, Rev.1, 2. February, 1995.

Pierce, D., R. C. Snare and F. Primdahl, A Small Proton Precession Magnetometer for Space Measurements, Poster, AGU Chapman

Conference on Measurement Techniques for Space Plasmas, Santa Fe, New Mexico, 3.-7. April, 1995.

Ruddock, K. A., Optically Pumped Rubidium Vapor Magnetometer for Space Experiments, *Space Research II*, 692-700, 1961.

Serson, P.H., A Simple Proton Precession Magnetometer, *Dominion Observatory Report*, 13 pp., Dominion Observatory, 1 Observatory Crescent, Ottawa, Canada, May 1962.

Slavin, J. A., JPL Scalar Helium Magnetometer, *Rep. Technical Note*, 8 pp., Jet Propulsion Laboratory, 1984.

Slocum, R. E. and D. D. McGregor, Measurement of the Geomagnetic Field Using Parametric Resonance in Optically Pumped He⁴, *IEEE Trans. Mag., MAG-10*, 532-535, 1974.

Smith, E. J., R. J. Marquedant, R. Langel, M. Acuna, ARISTOTELES Magnetometer system, *Proceedings of the Workshop on Solid-Earth Mission ARISTOTELES, Anacapri, Italy, 23-24 September 1991, ESA SP-329*, 83-89, Dec. 1991.

Smith, E. J., R. J. Marquedant, B. Wilson and S. Forouhar, The Helium Magnetometer: Potential Reductions in Physical Requirements Without a Loss in Performance, *Small Instruments for Space Physics, B.T. Tsurutani (Ed.), Proceedings of the Small Instruments Workshop held in Pasadena, California, 29-23 March, 1993*, pp. 2-2 to 2-9, Space Phys. Div., NASA, November 1993.

Stern, D. P., and J. H. Bredekamp, Error Enhancement in Geomagnetic Models Derived from Scalar Data, *J. Geophys. Res., 80*, 1776-1782, 1975.

Stuart, W. F., Earth's Field Magnetometry, *Rep. Prog. Phys., 35*, 803-881, 1972.

Fritz Primdahl, Department of Automation, Building 327, Technical University of Denmark, DK-2800 Lyngby, Denmark.

A History of Vector Magnetometry in Space

Robert C. Snare

Institute of Geophysics and Planetary Physics, University of California, Los Angeles

The first vector magnetic measurements in space were made with fluxgate sensors. In order to achieve higher accuracy a variety of vector magnetometers using alkali vapor cells were developed. These were abandoned as the ring core sensor was developed with improved offset and low noise performance. The improved ring core fluxgate magnetometer is used almost exclusively for vector measurements, with the exception of the vector Helium magnetometer which is used on some deep space missions. Over the years a variety of innovative sensor mounting configurations were designed to overcome the lack of spacecraft resources such as mass, power and telemetry bandwidth. The advent of modern integrated circuits such as the low power amplifier, the analog-to-digital converter and the microprocessor has enhanced data recovery from these instrument systems.

INTRODUCTION

To make the precise vector magnetic measurements in space required by scientific investigations one needs a magnetometer system, each of whose elements pushes the state of the art in linearity, low noise, stability and accuracy. These elements include the vector sensor and its associated circuits, an analog-to-digital converter (ADC) and a high fidelity data system. The sensor should be highly linear and have scale factors that are stable with temperature. The sensor should truly make vector measurements as the angular response should be equal to B cos f where f is the angle between the field vector and the sensor axis. The noise level should be low. It should maintain this low noise while providing a wide dynamic range with low, stable zero levels or offsets. The electronics that operates the sensor should be simple and reliable with no limited life components. The data system should have proper anti-aliasing filters for the data sampling rates used. The ADC should be monotonic, linear, and have high resolution. The data processing system should have the speed to process the sampled data quickly with a variety of algorithms. Although we will not discuss

Measurement Techniques in Space Plasmas: Fields
Geophysical Monograph 103

magnetic cleanliness in detail herein it is assumed that a magnetically clean spacecraft is available with a boom of suitable length for sensor mounting. The spacecraft should provide time and attitude data in the telemetry that will support high time resolution magnetic field measurement data reduction.

From the earliest spacecraft to those of today there has been constant improvement in the quality of magnetic measurements made in space. This improvement has been the result of the availability of greater spacecraft resources such as power, mass and telemetry bandwidth. An even greater impact is the result of advances in the electronics industry such as radiation hardened, low power analog and digital integrated circuits and the advent of microprocessor systems. A variety of innovative techniques have been employed with sensor and data systems to overcome the lack of abundant spacecraft resources. These factors affected not only magnetometers but all instruments and spacecraft systems.

Vector Sensors

The majority of vector measurements in space have been made with fluxgate sensors. The first fluxgate sensors were developed by Aschenbrenner and Goubau [1936]. This was followed by rapid development during 1940's and 1950's.

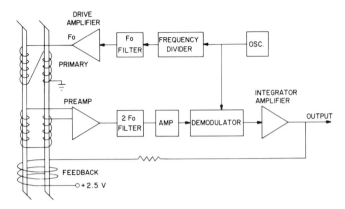

Figure 1. Block diagram of fluxgate magnetometer [Snare and Benjamin, 1966].

Fluxgate magnetometer systems were installed in aircraft to search for submarines, to aid in making navigation charts and for geophysical exploration, Rumbaugh and Alldredge [1949].

The alternating drive current in the fluxgate sensor drive winding drives the permeable core material alternately deep into saturation. (See Figure 1). Because of the non-linear coupling due to core saturation the induced voltage in the sense winding is rich in harmonics. The amplitude of the even harmonics is proportional to that component of the ambient magnetic field aligned with the sense winding. Normally the second harmonic is filtered and synchronously detected to produce a voltage proportional to the field. The output voltage is fed back through a scaling resistor to the feedback winding on the sensor. With proper design the result is a highly linear, stable vector measuring instrument.

There are a number of magnetic core configurations that can be used with the fluxgate magnetometer. The fluxgate sensor that was used in the early years of space exploration was the parallel core design both in the Vacquier and the Förster configurations [Primdahl 1979]. These configurations are shown in Figure 2. The parallel core design requires high drive power. They can be troubled by high offsets that have a tendency to vary with time.

Cylindrical core magnetometers have been constructed by Schonstedt [1959] and Primdahl [1970]. The Schonstedt sensor consists of a ceramic cylinder with thin permalloy tape helically wound around the cylinder, thus the name HELIFLUX sensor. The drive winding is applied toroidally through the tube with a solenoid feedback coil around the tube. The HELIFLUX sensor exhibited low noise and stable offsets. The Schonstedt HELIFLUX sensor has been used on a number of space flights. In his cylindrical sensor Primdahl used a tube of ferrite material for the core with toroidal and solenoidal windings for the drive and feedback as shown in

Figure 3. These sensors were used for sounding rocket flights and ground observatories.

Aschenbrenner and Goubau [1936] constructed a magnetometer core in 1928 using a circular bundle of soft iron florist wire. This configuration was in effect the first ring core sensor [Primdahl, 1979]. The ring core design was revived by Geyger [1962]. Refinements were made at Naval Ordnance Laboratory, White Oak. The modern ring core uses several layers of thin (0.0254 mm) permalloy wound around the edge of a non-magnetic stainless steel ring. The excitation winding is toroidally wound on the ring. Sense and feedback windings are wound on a bobbin on the outside of the ring and define the sense axis of the sensor as shown in Figure 4. Both noise and offset performance were enhanced with the development of 6-81.3 Mo-Permalloy (6% Mo, 81.3 Ni, 12.7% Fe) low-magnetostriction tape material by Gordon et al. [1968].

The first ring core fluxgate magnetometers used in space were in the Lunar Surface Magnetometer package left on the surface of the moon by the astronauts of Apollo 16 in April 1972 [Dyal and Gordon 1973]. There was a rapid transition in the use of the ring core during the 1970's. Today nearly all scientific vector measurements in space are made with ring core magnetometers. This is because of the low mass and simplicity of the circuitry.

Acuna [1981] reported nonlinear fluxgate response when MAGSAT sensors were exposed to large, >5,000 nT, uncompensated transverse fields. Primdahl et al. [1992] further investigated this phenomena and found that a fluxgate

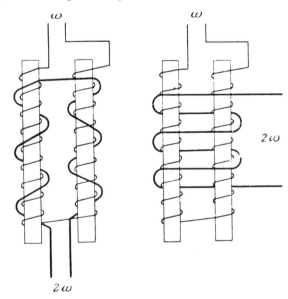

Figure 2. Parallel core magnetometers can use two configurations of windings, the Vacquier configuration shown on the left and Foster configuration on the right [Dolingov et al., 1960].

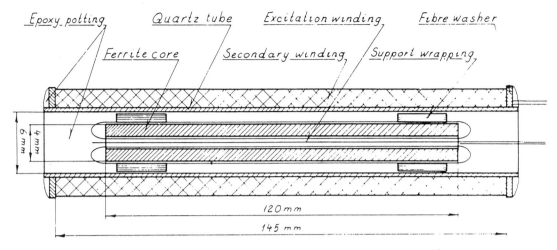

Epoxy potting   Quartz tube   Excitation winding   Fibre washer
Ferrite core   Secondary winding   Support wrapping

6 mm
4 mm

120 mm
145 mm

Figure 3. Tabular ferrile core fluxgate sensor from [Primdahl 1970].

on a spinning vehicle produced signals with magnitude ~10^{-4} of the applied field at the third harmonic of the spin frequency as a result of this nonlinearity. The large transverse field must be in the plane of the ring core to see the nonlinearity. The parallel core designs such as Vacquier and Foster do not exhibit this problem.

To compensate for this problem Primdahl and Jensen [1982] constructed a three axis fluxgate sensor with the three feedback windings on the surface of a sphere. The magnetic cores with their drive windings resided inside the sphere. Thus the ring cores were not exposed to uncompensated transverse fields. Other investigators have used cubical configurations to achieve similar results.

A variety of optically pumped alkali vapor scalar magnetometers have been used in space. Casium 133, Rubidium 85 and Rubidium 87 has been employed in sensors [Ness, 1970]. The alkali vapor magnetometers utilize transitions between the Zeeman sublevels in the ground state of the atom. The energy separation for the Zeeman splitting in proportional to the magnetic field. In the Rubidium self oscillating magnetometer a light from a Rubidium lamp is passed through a filter that only passes the 7948 Angstrom Rubidium light. The light passes through a circular polarizer and then through a cell with Rubidium vapor. The light is collected with a photoelectric cell. It is then amplified, phase shifted and then applied to a feedback coil around the Rubidium vapor cell see Figure 5. The circuit will oscillate at the Larmor frequency. The frequency for Rb 87 is approximately 6.99 Hz nT^{-1} which for a field of 50,000 nT would be ~350 KHz. The oscillation depends upon the presence of a magnetic field and ceases when the field vector is within "12° of parallel or "7° of normal to the optical axis of the sensor. There are offsets in the frequency as the field vector is reversed. This is because the circuit oscillates with

a broad unsymmetrical resonance that shifts with field direction [Ruddock, 1961]. A design that minimizes this problem is one that has a single Rubidium lamp illuminating twin absorption cells, as shown in Figure 6. The Rubidium magnetometer does not qualify to be classified as an absolute instrument with sensitivities derived directly from atomic constants. They have variable offsets that require careful calibration [Allen, 1968].

If one places a scalar magnetometer in a Helmholtz coil and collects data with first no current applied then with current applied and again with the current reversed the angle of the ambient field with respect to the coil axis can be calculated see Shapiro et al. [1960]. By using two sets of deflection coils one can measure the total field and two angles, thus defining the field vector. On a spinning spacecraft only one set of deflection coils is required if the field is steady [Heppner, et al., 1963].

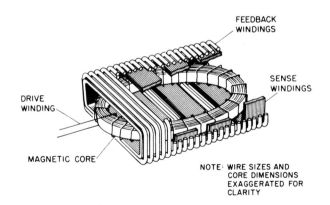

FEEDBACK WINDINGS

SENSE WINDINGS

DRIVE WINDING

MAGNETIC CORE

NOTE: WIRE SIZES AND CORE DIMENSIONS EXAGGERATED FOR CLARITY

Figure 4. Simplified view of the ring core sensor developed jointly by the Naval Ordnance Laboratory and NASA Ames Research Center, [Dyal and Gordon, 1973].

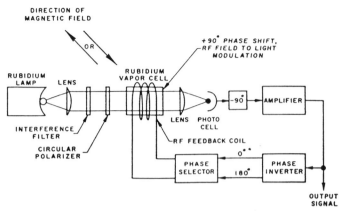

Figure 5. Block diagram of self oscillating single cell rubidium vapor magnetometer [Ruddock 1961].

As previously mentioned both the single cell and the twin cell Rubidium magnetometers have null zones. Explorer 10 used two twin cell magnetometers with optical axes aligned 55° to one another as shown in Figure 7. This arrangement reduced the null area to 500 square degrees solid angle.

The Rubidium vector magnetometer does not provide a continuous readout of the vector. It takes a finite time to apply the deflection fields. Therefore, the data are time aliased if there are temporal variation. Another problem is that in practice the Rubidium lamp operates only from 30° to 50° C with the optimum performance within a few degrees of 40° C. Vector alkali vapor magnetometers were used on several spacecraft in the 1960's and early 1970's. The circuitry of the alkali vapor vector magnetometer is more complicated than that of the fluxgate magnetometer. After the fluxgate magnetometer's performance was enhanced in the 1970's with the introduction of the ring core sensor the vector alkali magnetometer was no longer seen on spacecraft.

Another optically pumped sensor is the Helium magnetometer, a Helium cell can be made to resonate at the Larmor frequency in a manner similar to the alkali vapor magnetometer to form a scalar sensor. The details of the Zeeman splitting is somewhat different from Rubidium. For details see Keyser et al., [1961].

The low field Helium vector magnetometer does not resonate the Helium cell at the Larmor frequency but uses the cell as a null detector. As shown in Figure 8 circularly polarized light with 1.08 mm wavelength is passed through the Helium cell. The ambient magnetic field affects the pumping efficiency of the metastable Helium population. The Helmholtz coils around the cell are driven to create a rotating magnetic field alternately in two planes. The light received by the detector is the vector sum of the rotating

sweep field and the ambient field. The detector signal is then amplified and phase detected using the sweep field as a reference. This signal is then applied to feedback coils in each axis. The feedback keeps the Helium cell in zero field. The current required to null the system is then proportional to the ambient field in each axis. The maximum dynamic range of the Helium vector magnetometer is limited to a few hundred nT. The sensitivity, stability and offsets are the result of careful design and must be accurately calibrated for verification of the performance of the instrument [Slocum and Reilly, 1963 and Smith et al., 1975]. The vector Helium magnetometer has been modified to be a combination vector-scalar magnetometer. This will be first flown on the Cassini spacecraft. The scalar mode will be used at close approach to Saturn where the scalar Helium magnetometer will make absolute measurements accurate to 1 nT [Kellock et al., 1996].

Search coils have occasionally been used on spacecraft as the primary magnetometers. On spinning spacecraft a coil transverse to the spin axis can measure two components of the static field in the spin plane. The search coil magnetometer uses a long thin core of highly permeable alloy containing Nickel and Iron. Both crystaline and amorphous metals have been used. A winding of many turns of fine wire is added to the center of the magnetic core. The sensitivity of the search coil is a direct function of the number of turns of wire. The low frequency noise of the coil is the thermal noise of the resistance of the wire. Therefore, there are tradeoffs between the number of turns, wire size, the desired resonance and the mass of the coil. The amplifier connected to the coil must have low noise referred to the input. The standard design practice is to match the coil thermal noise to the amplifier input noise.

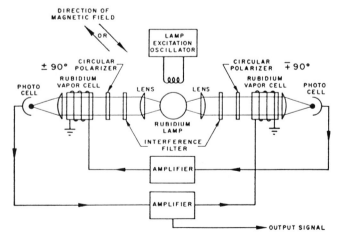

Figure 6. Block diagram of twin cell rubidium vapor magnetometer [Ruddock 1961].

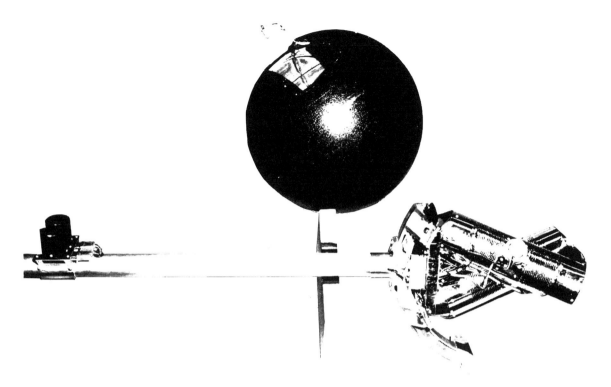

Figure 7. Prototype of EGO rubidium vapor magnetometer showing crossed twin cells with spherical thermal cover removed. For vector measurements bias coils were mounted on the inside surface of the spherical thermal cover [Heppner 1963].

Magnetometer Designs

During the late 1940's and 1950's several scalar magnetometers were flown by the United States on V-2 and Aerobee sounding rockets at White Sands Proving Ground, New Mexico, [Maple et al., 1950, Heppner et al., 1958]. These flights achieved maximum altitudes greater than 150 km.

The first magnetometer was carried into earth orbit May 1958 aboard the Soviet artificial earth satellite Sputnik 3. The three axis fluxgate sensor was mounted in an assembly which could be rotated in two axes by servo motors, [Figure 9]. The objective was to align the primary sensor with the ambient field. This was achieved by driving the mechanism until the transverse magnetometer sensors outputs were nulled. The primary sensor field magnitude and shaft angles were telemetered to earth. The spacecraft was not stabilized and had no attitude reference. It had a rotation period of approximately 136 sec. Therefore, the magnetometer angles defined the spacecraft attitude and provided no angle information for the magnetic field, [Dolginov et al. 1960]. Thus, the instrument as designed functioned as a scalar magnetometer. The magnetometer with electromechanical

orientation unit is similar to those used in aircraft as reported by Rumbaugh and Alldredge [1949]. It appears to be rather heavy but the Soviet spacecraft did not have the same mass restrictions as those faced by other countries.

Pioneer 1 and 5 and Explorer 6 spacecraft each carried a single search coil magnetometer [Judge et al., 1960]. The search coil was chosen because of its low mass and simplicity. These were mounted normal to the spacecraft spin axis and measured the static field perpendicular to the spin axis. Additionally Pioneer 5 and Explorer 6 had sun aspect detectors that provided a spin reference for the spacecraft. The magnetic vector direction in the spin plane could then be determined. Pioneer 5 also employed a fluxgate sensor mounted parallel to the spin axis but the fluxgate did not provide data during the flight.

The first fully vector magnetic measurements in space were made aboard Lunar 1 and 2. These spacecraft used 3 separate single axis fluxgate magnetometers. The sensors were mounted orthogonally. Each sensor had its separate electronics each with unique drive and second harmonic frequencies [Dolginov et al., 1961].

The Pioneer 6, 7 and 8 were small spacecraft with a launch mass of 66 Kg. An unusual approach was used to

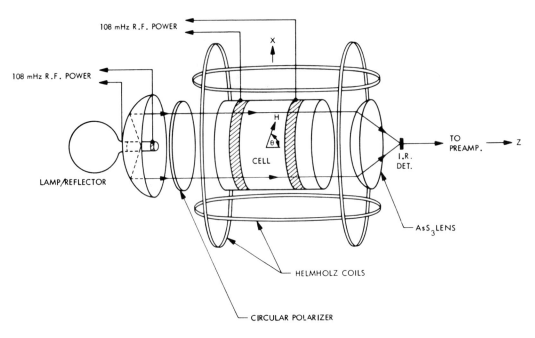

Figure 8. Functional Sketch of vector helium magnetometer sensor. Absorption of circularly polarized light from the He lamp by the He cell is modulated by vector sum of ambient field H and rotating field generated by coils; resulting variations are monitored by infrared detector. For clarity, only two of the three coils are depicted [Smith et al., 1975].

achieve vector measurements [Ness et al., 1966]. In order to conserve instrument mass a single fluxgate sensor was mounted 2m from the spacecraft spin axis. The sense axis of the fluxgate was tilted away from the spin axis by the angle 54° 45' as shown in Figure 10. The magnetometer was digitally sampled each 120° of spacecraft rotation. Thus, any three samples form an orthogonal measurement of the field. The spacecraft spin period was one second with a magnetometer bandwidth of 5 Hz which is 3 times the Nyquist sampling frequency of 1.5 Hz. This technique works in a quiet magnetic field, but not with an active field [Fredericks et al., 1962]. Many of the most scientifically interesting magnetic fields are the most active.

The first Applications Technology Satellite (ATS-1) had a two axis magnetometer. Each sensor was mounted 45° to the spacecraft spin axis, (see Figure 11). The output of the two magnetometers were passed through sum and difference amplifiers to yield field components parallel and normal to the spacecraft spin axis [Barry and Snare, 1966]. The design provided redundancy as data could still be retrieved if one of the magnetometers failed.

The astronauts of Apollos 12, 15 and 16 placed magnetometers on the surface of the moon. The sensor configuration was rather unique. A single axis sensor was mounted at the end of an orthogonal set of arms 1 meter in length as shown in Figure 12. The three sensors measured

the vector magnetic field on the Lunar surface. An articulating mechanism could reverse the pointing of each sensor by 180°. This feature enabled one to calculate the sensor zero levels. The mechanism could also place each sensor parallel to each X, Y, Z coordinate. This feature provided data for the calculation of gradients between the sensors in all three directions [Dyal et al., 1970]. The Lunar Surface Magnetometer for Apollo 16 used the new ring core sensors developed by Naval Ordnance Laboratory, White Oak [Dyal and Gordon, 1973].

The Pioneer Venus spacecraft had three sensors. One sensor was parallel to the spacecraft spin axis one was normal to the spin and the third was at a point 2/3 of the length of the 5m boom and tilted at a 45° angle in the radial direction. At low frequencies in quiet fields, calculation of the vector field from the sine wave amplitude and phase and the steady component parallel to the spin axis could be provided by either the inboard tilted sensor on the two outboard sensors. The difference between these two yielded a measure of the spacecraft field. At low spacecraft telemetry rates only the two outboard sensors were sampled and at lowest data rates when the sampling rate was near or below the spacecraft spin period, the data were despun on board using a Walsh transform. At high data rates the measurements from all these sensors were combined to provide instantaneous vector measurements of the field and

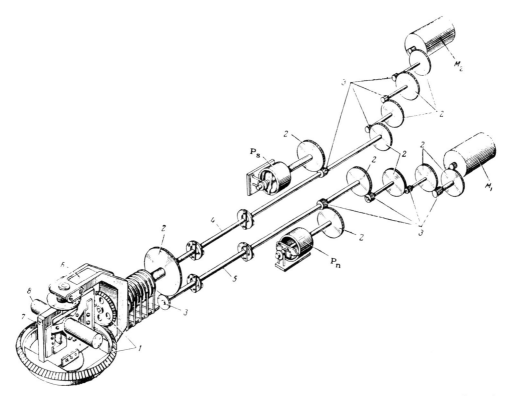

Figure 9. Diagram of magnetometer orientation unit for Sputnik 3. The motors and drive gears orient the main sensor 8 parallel to the earth's magnetic field [Dolginov, 1960].

were despun on the ground. This configuration also provided some redundancy if a sensor should fail [Russell et al., 1980].

The most accurate vector measurements made in space to date were those of the MAGSAT spacecraft [Acuna et al., 1978]. The project provided data for improved modeling of the time varying magnetic field generated within the core of the earth and to map variations in the strength and vector characteristics of the crustal magnetization [Langel et al., 1982]. The fluxgate sensor was designed and calibrated with great care in order to provide accurate vector data. A Cesium vapor scalar magnetometer provided absolute magnetic field data for calibrating the fluxgate in space. An elaborate system of attitude measurement devices provided spacecraft and magnetic sensor attitude data.

Another project with objectives similar to that of MAGSAT is the Danish spacecraft OERSTED to be launched soon. It will carry the compact, spherical three axis fluxgate magnetometer which was previously described [Primdahl and Jensen, 1982]. The fluxgate magnetometer is mounted on an optical bench that also carries a nonmagnetic star imager camera. The camera provides data for determining the magnetometer attitude in inertial space. For absolute field measurements the spacecraft has an Overhauser, continuous wave, proton magnetometer [Duret et al., 1995 and Primdahl, 1997].

Offset Determinations

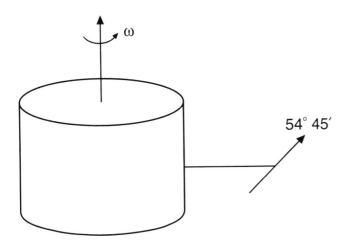

Figure 10. Diagram of tilted sensor on a spinning spacecraft as used by Pioneer 6, 7 and 8.

Accurate measurement of zero levels or offsets, as seen by vector magnetometers has long been a problem. The

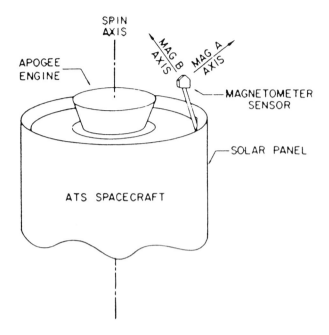

Figure 11. The two sensors for ATS-1 were each mounted 45⁰ to the spin axis [Barry and Snare, 1966].

sensors and sensor electronics can have finite offsets and variations in these offsets with time and temperature. The spacecraft can also have fields that add to the offsets as measured by the magnetometer. Depending on their source these spacecraft fields may be static or also vary with time.

On spinning spacecraft the sensors mounted normal to the spacecraft spin axis are easy to evaluate by averaging the data over several spin periods. This yields the sum of the sensor offset and the rotating spacecraft field. This technique could be extended to the third axis by mechanisms that rotate the sensors such that the non spinning sensor is placed in the spin plane for offset calibration. This technique, however, does not help with determining fields that do not rotate with the sensor. The flipping mechanisms were initially driven by thermal devices such as wax pellet actuators and later by bi-metallic springs. Another technique called electronic flipping is used. Switches reverse the polarity of the sensor in such a manner that the offsets can be measured [Behannon et al., 1977].

Ness et al. [1971] proposed using two triaxial magnetometers on one boom to determine the spacecraft field by measuring its gradient. This was followed by Neubauer [1975] studying the use of as many as four magnetometers on one boon and determining the capabilities and errors of such a system. Spacecraft fields are complex, rarely exhibiting a simple dipole structure that most of the calculations rely on. Moreover, the zero levels of sensors are

important contributors to the differences between sensors. Therefore, defining the static spacecraft magnetic field by two or more magnetometer is seldom feasible. However, a second magnetometer closer to the spacecraft is useful in identifying and calibrating spacecraft dynamic magnetic fields and provides redundancy. Today nearly all spacecraft of any size carry two magnetometers. Often the outboard magnetometer is for low field measurements and the inboard magnetometer for higher fields. On some deep space missions the magnetometers have been mixed types such as the fluxgates and Helium sensors of Ulysses and Cassini [Balogh et al., 1992; Southwood et al., 1992].

Scales and Ranging

The problem of resolving small changes in the presence of large fields is a common problem for magnetometers. However, the dynamic range of the telemetry system often constrained early measurements. The first spacecraft used frequency modulation (FM) telemetry. The voltage to be measured was passed to a voltage controlled oscillator (VCO). This circuit transformed voltage to frequency. Multiple VCOs each with its unique frequency band were mixed together. This composite spectrum was then sent to earth by a high frequency radio carrier. At the earth the frequencies were separated by bandpass filters and discriminated to reconstruct the original signals. The accuracy and resolution of such a system approached 1% if all elements of the system were calibrated and functioning properly. In practice FM telemetry systems often degraded to 3% measurements.

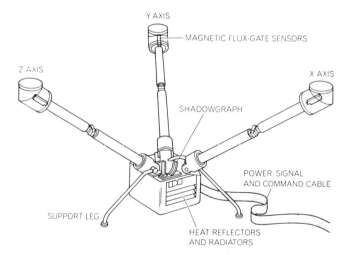

Figure 12. Lunar surface magnetometer with a single axis fluxgate at the end of each 1 meter arm [Dyal and Parkin, 1971].

The introduction of the analog-to-digital converter (ADC) and pulse code modulation somewhat improved telemetry resolution. The first ADC's had only 6 to 8 bits with resolution not much better than the FM telemetry. Presently 16 bit ADC's are routine with advanced ADC's of 20 to 24 bit ADC's in the development laboratory.

Before advent of modern high precision analog to digital converters, field offset systems were developed to extend the dynamic range of magnetometers and maintain their resolution. Such systems are needed around magnetized planets where the inverse cube dependence of the magnetic field on radial distance causes the field strength to vary by several orders of magnitude. In many senses this technique is simply creating an instrument with ADC. One problem with this technique is that the offset fields add noise and must be very well calibrated as they add step function noise when they switch. The system that was used on Sputnik 3 is shown in Figure 13. The magnetometer had a dynamic range of " 2400 nT. This was telemetered to the ground by two separate VCO's, one for the positive voltage and another for the negative voltage [Dolginov et al., 1960]. The offset system passed current from a relay controlled resistor network through a winding on the sensor. The offset incremented in steps of 3000 nT and had a total range of 64,000 nT.

A similar system using solid state components was used on a number of spacecraft including ATS-1, ATS-6 and OGO-5. The system is shown in Figure 14. When the voltage from the basic magnetometer exceeds its dynamic range of "10 volts the level detector causes the counter to increment up or down and apply steps of offset current to a winding on the sensor. This system is actually a digital-to-analog circuit built with discrete parts.

Another technique that has been widely used is to automatically switch the magnetometer dynamic range when

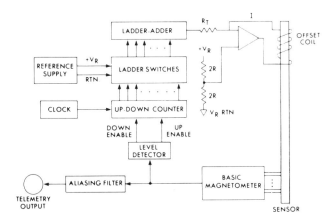

Figure 14. Solid state field offset systems used for ATS-1, OGO-5 and ATS-6 [McPherron et al., 1975].

the measured field exceeds the present range. This technique varies the absolute resolution of the magnetometer but maintains the relative or percent resolution. This technique has been used on a great number of spacecraft and is presently planned for the Cassini magnetometers. The switching ranges for Voyager 1 and 2 are shown in Figure 15. The magnetometer automatically selects the proper range to keep the signal in the center of the magnetometer dynamic range. If the average field remains in the area of the guard bands for several seconds the magnetometer will switch up or down as required. The range switching can be overridden by ground command [Behannon et al., 1977].

Figure 16 depicts the dynamic range and the resolution, i.e. least-significant-bit (LSB), of the telemetered data for several spacecraft. Voyager 1 and 2 had magnetometers with 12 bit ADCs. The main low field magnetometer ranges and resolution are depicted in the figure for each of the 8 gains available for the magnetometer.

Both the ISEE 1 and 2 and the Galileo magnetometer used 12 bit ADCs that were specially manufactured with each step trimmed to an accuracy of better than 3 the value of the LSB. The data was sampled at a high rate and averaged to produce output data that had an accuracy of 15 bits. The magnetometer for GGS-Polar used an ADC with full 16 bit resolution. With the use of 16 bit ADC's and modern techniques the need for multiple range switching is reduced. One can see the effect that 15 and 16 bit ADC's has on extending the dynamic range without sacrificing the high resolution of these magnetometers.

Data Processing

The telemetry bandwidth from spacecraft is generally determined by the power available on the spacecraft, because

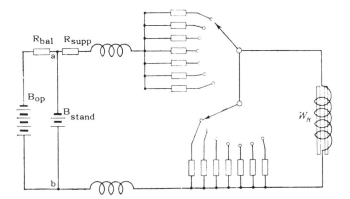

Figure 13. Diagram of field offsetting circuits for Sputnik 3 magnetometer [Dolginov, 1959].

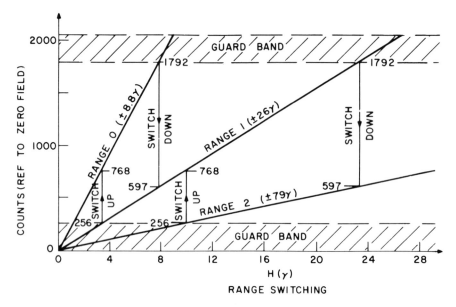

Figure 15. Illustration of a portion of the total magnetometer range switching strategy for the Voyager 1 and 2 magnetometers [Behannon et al., 1977].

scientists can make good use of every available bit and tend to request the greatest possible bandwidth. The available telemetry is divided between various scientific instruments and spacecraft engineering systems. The telemetry allotted to magnetometers is often insufficient to transmit all of the information desired. This has resulted in a variety of data processing techniques to partially analyze the data on board the spacecraft and telemeter the results. This is particularly important on spinning spacecraft whenever the telemetry rate is insufficient to resolve the spin tone in the spin plane associated with the rotation of the spacecraft in the ambient magnetic field. We recall that the Nyquist criterion states that at least two points must be measured on any sine wave per wave period in order to determine its amplitude and phase. If this is not achieved, then the spin tone will appear to be at an incorrect frequency. This effect is called aliasing because a signal appears to occur at a frequency other than its true frequency, i.e. it assumes an alias. Of course all frequencies not just the D.C. field are affected.

The first such systems were executed with analog circuitry and were simple sum and difference amplifiers as previously mentioned on ATS-1. The Apollo 15 Subsatellite Magnetometer had one fluxgate sensor parallel to the spin axis and one transverse to the spin axis. The parallel magnetometer was sampled directly. The transverse signal was rectified and filtered before sampling and the result represented the transverse field magnitude. The time between the zero crossing of the transverse signal and a sun

pulse provided phase information relative to the spacecraft sun line [Coleman et al., 1972].

Explorer 33 was the first spacecraft to carry a magnetometer with a spin demodulator. The two transverse sensor outputs were multiplied by sine and cosine signals derived from the spacecraft sun pulse [Sonett 1966; Sonett et al. 1968]. These functions were implemented with analog circuits.

The Pioneer Venus Orbiter had a 12 sec spin rate and because of the varying distance from earth the spacecraft telemetry bit rate varied from 4096 to 8 bits per second. At the lowest bit rate the magnetometer could only send one vector each 21 or 64 spins depending on which telemetry format was being used. Thus it was necessary to despin the measurements before transmittal. Power and mass for the magnetometer were not enough to fully implement a sine-cosine demodulator. The approach used was to multiply the data by a Walsh transform. The Walsh transform is basically the multiplication of the spinning data by two square waves one in phase and one in quadrature to the spacecraft spin.

The spinning data were digitally sampled and sent to averaging registers. The data were clocked in during the first half spacecraft rotation. During the second half spacecraft rotation the data was inverted. This process is repeated in parallel but with inverted signals from the one-quarter to three-quarter mark of the rotation. This functions as a full wave sun synchronous demodulator and a low pass filter. All of the spin demodulator functions were constructed using

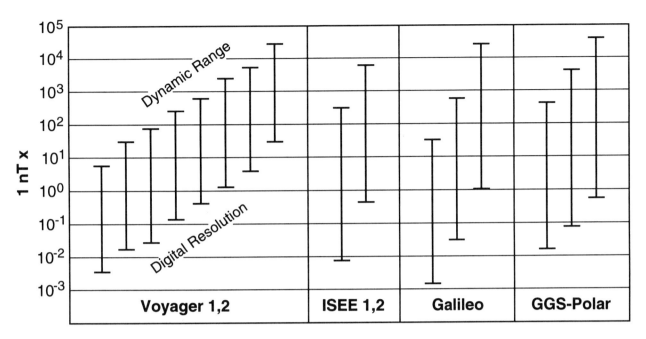

Figure 16. Dynamic Range and Resolution for Several Spacecraft.

hard wired CMOS digital circuits [Russell et al., 1980]. The Pioneer Venus magnetometer also created "floating point" words rather than "fixed point" to make maximum use of the telemetry system.

The ADC and digital processing allowed one to recover greater accuracy than that achievable with analog circuits. The advent of the microprocessor gave the instrument designer a menu of computational functions. Often the data are sampled at a high rate and averaged to create a low pass filter. The digital averaging produces a precise filter algorithm. Recursive filters have been used when computational power is inadequate to handle a large averaging chore. However, recursive filters are asymmetric in time and hence introduce phase shifts.

Another technique is to monitor the data for rapid changes. When a particular shock or similar phenomena is detected, high rate data can be stored in a large solid state memory. The stored high rate data is then returned to Earth slowly at a lower data rate. More sophisticated functions such as the Fast Fourier Transform (FFT) have been programmed to enable the study of the physics of the fine-scale structure of shock waves, directional discontinuities and boundary structures [Lepping et al., 1995; Reidler et al., 1986]. An example of the data processing flow using two 80C86 Processors is that for the GGS-POLAR magnetometer shown in Figure 17. The prime data is sampled at 500 samples per second, filtered with a recursive filter and decimated to 100 samples per second. The data is again filtered and decimated to 10 vectors per second which is the output rate to the spacecraft telemetry. At 10 vectors per second and a spin period of 6 seconds the transverse data did not require despinning [Russell et al., 1995]. However, there is adequate processing capability to despin the data on board and provide for easy quick look interpretation of the data.

CONCLUSION

The continued development of booster rockets with the addition of orbit injection stages has enhanced the capability to launch larger heavier scientific spacecraft. Thus the power available for instruments has increased and the enhanced telemetry transmitter power has increased the data recovery rate several orders of magnitude over that of early satellites.

Most modern spacecraft have the capability of deploying long booms such that instruments sensitive to spacecraft emissions can achieve a quieter environment. Magnetometers and plasma wave antennae are usually found on such booms.

The rapid development of radiation hardened, miniature, semiconductors and other electronic devices has enabled the instrument designer to build more sophisticated electronics and data systems. The result of these developments is a richer return of scientific data from spacecraft. Much of this

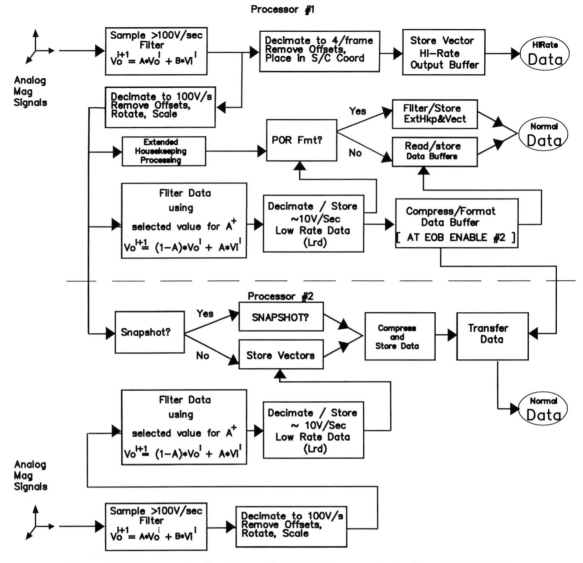

Figure 17. Data Processing flow diagram for GGS-Polar magnetometer [Russell et al., 1959].

progress has been driven by military spending. With the present reduction in both military and scientific funding one might question the direction that scientific instrumentation will take in the future.

Acknowledgments. I would like to thank Paul Coleman, Margaret Kivelson, Robert McPherron and Christopher Russell for giving me the opportunity to make a career of magnetometery. The constructive comments of the reviewers have contributed by pointing out oversights and misstatements made by the writer. This work was supported by the National Aeronautics and Space Administration contract NAS 5-30373 and Jet Propulsion Laboratory contract 959483.

REFERENCES

Acuna, M. H., C. S. Scearce, J. B. Seek and J. Scheifele, The MAGSAT Magnetometer - A Precision Fluxgate for the Measurement of the Magnetic Field, NASA Technical Memorandum 79656, September 1978.

Acuna, M. H., MAGSAT - Vector Magnetometer Absolute Sensor Alignment Determination, NASA technical Memorandum 79648, September 1981, NASA Goddard Space Flight Center, Greenbelt, Maryland 20771.

Acuna, M. H. and N. F. Ness, The Pioneer XI Hiah Field Fluxgate Magnetometer, *Space Sci. Instrum.*, pp177-188, 1975.

Allen, J. H., Long Term Stability of Self-Oscillating Rubidium Magnetometers, *J. Geomag. Geoelect.*, 20, pp197-204, 1968.

Aschenbrenner, H. and G. Goubau, Eine Anordnung Registrieung rascher magnetischer Storungen, *Hochfreg Tech. Elektroakust.* 47, pp178-181, 1936.

Balogh, A., T. S. Beek, R. S. Forsyth, P. C. Hedgecock, R. S. Marquedant, E. J. Smith, D. J. Southwood and B. T. Tsurtani, The Magnetic Field Investigation on the Ulysses Mission: Instrumentation and Preliminary Results, *Astronomy and Astrophys.*, 92, pp221-236, 1992.

Barry, J. D. and R. C. Snare, A Fluxgate Magnetometer for the Application Technology Satellite, *IEEE Trans. Nucl. Sci.*, NS-13, No.6, pp326-331, 1966.

Behannon, K. W., M. H. Acuna, L. F. Burlaga, R. P. Lepping, N. F. Ness and F. M. Neubauer, Magnetic Field Experiment for Voyagers 1 and 2, *Space Sci. Rev.*, 21, pp235-257, 1977.

Coleman, Jr., P. S., G. Schubert, C. T. Russell and L. R. Sharp, The Particles and Fields Subsatellite Magnetometer Experiment, *Apollo 15 Preliminary Science Report*, pp22-1 - 22-9, NASA-SP289, 1972.

Dolginov, S. Sh., L. N. Zhuzgov and V. A.. Selyutin, Magneto-meters in the Third Soviet Earth Satellite, *Artificial Earth Satellites*, 4, pp358-396, 1960.

Dolginov, S. Sh., E. G. Yeroshenko, L. N. Zhugov and N.V. Pushkov, Investigation of the Magnetic Field of the Moon, *Geomagnetism and Aeronomy*, 1, pp21-29, 1960.

Duret, D., J. Bonzom, M. Brochier, M. Frances, J. M. Leger, R. Odru, C. Salvi, and T. Thomas, Overhauser Magnetometer for the Danish Oersted Satellite, *IEEE Trans. Magnetics*, Vol. 31, No.6, pp3197-3199, November 1995.

Dyal, P., C. W. Parkin and C. P. Sonett, Lunar Surface Magnetometer, *IEEE Trans. Geosci. Electronics*, GE-8, pp203-215, 1970.

Dyal, P. and C. W. Parkin, The Magnetism of the Moon, *Scientific American*, 224, No. 2, pp62-73, Aug. 1971.

Dyal, P. and D. I. Gordon, Lunar Surface Magnetometers, *IEEE Trans. Magn.*, MAG-9, pp226-231, 1973.

Fredericks, R. W., E. W. Greenstedt and C. P. Sonett, Magneto-dynamically Induced Ambiguity in the Data from Tilted Spinning Fluxgate Magnetometers: Possible Application to IMP-1, *J. Geophys. Res.*, 72, pp367-382, 1967.

Geyger, W. A., The Ring Core Magnetometer - A New Type of Second-Harmonic Flux-Gate, *AIEE Trans. Communications and Electronics*, 81, pp65-73, 1962.

Gordon, D. I., R. H. Lundsten, R. A. Chiarado and H. H. Helms Jr., A Fluxgate Sensor of High Stability for Low Field Magnetometry, *IEEE Trans. Magnetics*, MAG-4, pp397-401, 1968.

Heppner, J. P., J. D. Stolarik and L. H. Meredith, The Earth's Magnetic Field above WSPG, New Mexico, From Rockef Measurements, *J. Geophys. Res.*, 63, pp277-289, 1958.

Heppner, J. P., N. F. Ness, C. S. Scearce and T. L. Skillman, Explorer 10 Magnetic Field Measurements, *J. Geophys. Res.*, 68, pp1-46, 1963.

Heppner, J. P., The World Magnetic Survey, *Space Sci. Rev.*, 2, pp315-354, 1963.

Judge, D., L. McLoud and A. R. Sims, The Pioneer 1, Explorer VI and Pioneer V High-Sensivity Transistorized Search Coil Magnetometer, *IEEE Trans. on Space Electronics and Telemetry*, SET6, pp114-121, 1960.

Kellock, S., P. Austin, A. Balough, B. Gerlach, R. Marquedant, G. Musmann, E. Smith, D. Southwood and S. Szalai, Cassini Dual Magnetometer Instrument, Proceedings: The International Society for Optical Engineering, Vol. 2803, pp146-152, August 1996.

Keyser, A. R., J. A. Rue and L. D. Schearer, A Metastable Helium Magnetometer for Observing Small Geomagnetic Fluctuations, *J. Geophys. Res.*, 66, pp4163-4169, 1961.

Langel, R., G. Ousley, J. Berbert, J. Murphy and M. Settle, The MAGSAT Mission, *Geophys. Res. Ltr.*, Vol. 9, No.14, pp243-245, April 1982.

Lepping, R. P., M. H. Acuna, L. F. Burlaga, W. M. Farrell, J. A. Slavin, K. H. Schatten, F. Mariani, N. F. Ness, F. M. Nebauer, Y. C. Whang, J. B. Byrnes, R. S. Kennon, P. V. Panetta, J. S. Scheifele and E. M. Warley, The Wind Magnetic Field Investigation, *Space Sci. Rev.*, 71, pp207-229, 1995.

Maple, E., W. A. Bowen and S. F. Singer, Measurement of the Earths Magnetic Field at High Altitudes at Wite Sands, New Mexico, *J. Geophys. Res.*, 55, pp115-126, 1950.

McPherron, R. L., P. J. Coleman, Jr. and R. C. Snare, ATS-6 UCLA Fluxgate Magnetometer, *IEEE Trans. Aerospace and Electronic Sys.*, AES-11, pp1110-1116, 1975.

Ness, N. F., C. S. Scearce and S. Cantarano, Preliminary Results from the Pioneer 6 Magnetic Fields Experiment, *J. Geophys. Res.*, 71, pp3305-3313, 1966.

Ness, N. F., Magnetometers for Space Research, *Space Sci. Rev.*, 11, pp459-554, 1970.

Ness, N. F., K. W. Behannon, R. P. Lepping and K. H. Schatten, Use of two Magnetometers for Magnetic Field Measurements on a Spacecraft, *J. Geophys. Res.*, 76, pp3565-3573, 1971.

Neubauer, F. M., Optimization of Multimagnetometer Systems on a Spacecraft, *J. Geophys. Rev.*, 80, pp3235-3240, 1975.

Primdahl, F., A Ferrite Core Fluxgate Magnetometers, *Danish Meterological Institute Geophysical Papers*, R-12, pp1-66, 1970.

Primdahl, F., The Fluxgate Magnetometer, *J. of Phys. E: Scientific Instruments*, 12, pp241-253, 1979.

Primdahl, F. and P. Anker Jensen, Compact Spherical Coil for the Fluxgate Magnetometer, *J. Phys. E: Sci. Instrum.*, Vol. 15, pp221-226, 1982.

Primdahl, F., H. Luhr and E. K. Lauridsen, The Effect of Large Uncompensated Transverse Fields on the Fluxgate Magnetic Sensor Output, Danish Space Research Institute Report 1-92, 1992.

Primdahl, F., Scalar Magnetometers for Space Applications, This Monograph, 1997.

Riedler, W., K. Schwingenschuh, Y. G. Yeroshenko, V. A. Stayshkin and C. T. Russell, Magnetic Field Observations in Comet Halleys Coma, *Nature*, 321, pp288-289, 1986.

Ruddock, K. A., Optically Pumped Rubidium Vapor Magneto-meters for Space Research, *Space Res.*, II, pp692-700, 1961.

Rumbaugh, L. H. and L. R. Alldredge, Airborne Equipment for Geomagnetic Measurements, *Trans. Amer. Geophys. Union*, 30, pp836-848, 1949.

Russell, C. T., R. C. Snare, J. D. Means, D. Pierce, D. Dearborn, M. Larson, G. Barr and G. Le, The GGS/POLAR Magnetic Fields Investigation, *Space Sci. Rev.*, 71, pp.563-582, 1995.

Russell, C. T., R. C. Snare, J. D. Means and R. C. Elphic, Pioneer Venus Orbiter Fluxgate Magnetometer, *IEEE Trans. on Geosci. and Remote Sensing*, GE18, pp32-35, 1980.

Schonstedt, E. O. and H. R. Irons, Airborne Magnetometers for Determining all Magnetic Components, *Trans. AGU.* 34, pp363-378, 1953.

Shapiro, I. R., J. D. Stalarik and J. P. Heppner, The Vector Field Proton Magnetometer for IGY Satellite Ground Stations, *J. Geophys. Res.*, 65, pp913-920, 1960.

Slocum, R. E. and F. N. Reilly, Low Field Helium Magnetometer for Space Applications, IEEE *Trans. on Nuclear Sci.*, NS-10, pp165-171, 1963.

Smith, E. J., B. V. Connov and G. J. Foster, Jr., Measuring the Magnetic Fields of Jupiter, *IEEE Trans. Magnetics*, MAG-11, pp962-980, 1975.

Snare, R. C. and C. R. Benjamin, A Magnetic Field Instrument for the OGO-E spacecraft, *IEEE Trans. Nucl. Sci.*, NS-13, No.6, pp333-340, 1966.

Sonett, C. P., Modulation and Sampling of Hydrodynamic Radiation, *Space Research*, 6, pp280-322, 1966.

Sonett, C. P., P. S. Colburn, R. G. Currie and J. D. Mihalov, The Geomagnetic Tail: Topology, Reconnection and Interaction with the Moon, *Physics. of the Magnetosphere*, D. Reidel Publishing Co., Dordrecht-Holland, pp461-484, 1968.

Southwood P. J., A Balogh and E. J. Smith, Dual Technique Magnetometer Experiment for the Cassini Orbiter Spacecraft, *J. British Interplanetary Soc.*, 45, pp371-374, 1992.

Robert C. Snare, Institute of Geophysics and Planetary Physics, University of California, Los Angeles, CA 90095-1567

Magnetic Field Measurements in Orbit and on Planetary Surfaces Using a Digital Fluxgate Magnetometer

Rustenbach, J[1]., H.U. Auster[2], A. Lichopoj[3], H. Bitterlich[1], K.H. Fornacon[2], O. Hillenmaier[4], R. Krause[1], H.J. Schenk[1]

The first model of a three component fluxgate magnetometer based on a near sensor digitization of the fluxgate signals (digital fluxgate magnetometer) will be presented. High flexibility as well as low power and weight requirements are the main arguments to qualify the digital magnetometer for planetary missions. Tests have shown that the low noise level and long term stability of the reduced electronics and the described algorithm are good enough, so that only the sensor noise and stability limits the accuracy of the magnetometer.

1. INTRODUCTION

Magnetic field measurements, especially in space science, are limited by severe technical requirements. Factors as weight and power consumption are contradictory to the required low noise level, long term stability and a flexible dynamic and frequency range.

The paper presents a technical solution applied in several fluxgate magnetometers developed in the Max-Planck-Institute for extraterrestrical physics. Based on the usual analogue fluxgate electronics with different data processing standards a digital fluxgate electronics was developed to offer a magnetometer for future missions which satisfies all

scientific requirements in spite of limited resources [Auster et al., 1975].

The presented digital fluxgate magnetometer converts the sensor signal directly after the preamplifier. The following signal processing will be done only by software.

Off-line analysis of the near sensor signal was done successfully by Primdahl and colleagues [1994] with the help of a cross correlation between a reference signal and the sensor signal. We have developed hardware in which the near sensor signal is on-line digitized, its information is evaluated and a feedback field is generated. The digital magnetometer is proposed for the Rosetta Lander RoLand.

2. THE RINGCORE SENSOR

The beginning of the fluxgate ringcore sensor development in our institute dates back to the year 1982. The aim was to develop a deep space magnetometer for the Russian Phobos mission.

A special 6-81-Mo permalloy material, recommend by Acuna [1974], was manufactured in the Institute for Material Development in Dresden. We have used fabrication techniques where tailoring of parameters was facilitated for core processing such as like tape isolation and adjustment of width and thickness of the soft magnetic tape, the number of windings, the diameter of the ringcore, the timing and temperature of thermal processes and advanced technologies

[1] Max-Plank Institut für extraterrestrische Physik, Außenstelle Berlin, 12489 Berlin, Germany

[2] TU Braunschweig, Institut für Geophysik und Meteorologie, 38106 Braunschweig, Germany

[3] DLR Berlin, Institut für Planetenerkundung, 12489 Berlin, Germany

[4] Magson GmbH, 12489 Berlin, Germany

Measurement Techniques in Space Plasmas: Fields
Geophysical Monograph 103

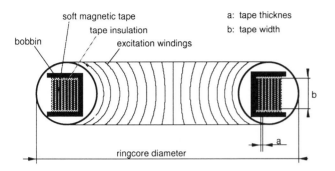

Figure 1. Principle ringcore design

for fixing of the tape and for the bobbin manufacture. The principle ringcore design is shown in Figure 1. In Table 1 ringcores with different parameters are listed.

3. THE ELECTRONICS

3.1. *Magnetometer concepts*

Figure 2 shows the concepts of a usual fluxgate magnetometer and Figure 3 that of the digital fluxgate. The main advantage of the digital fluxgate is, that excitation (the main power consumer), feedback and data acquisition are controlled by the internal DPU.

In the following text two classes of analogue fluxgate electronics and the digital fluxgate are presented.

3.2. *Magnetometer 2..4kg / 2Watt class*

The main parameters of the orbiter magnetometers for the PHOBOS, MARS-96, EQUATOR-S and RELICT missions are given in Table 2. The experiment hardware consists of:
- the triaxial sensor system
- the analogue and drive electronics
- the AD-converter
- the digital processing unit DPU based on a signal processor
- internal memory
- the range and compensation logic
- the spacecraft interfaces
- DC/DC power converter to supply the voltages required by the instrument (this part was delivered by IC London).

As a result of the PHOBOS-FGMM experiment the magnetometer for MARS-96 was designed more flexible in reacting to changing telemetry rates. Instead of a small number of working modes (regimes) with fixed telemetry rates, we have implemented only two data sampling modes (data from one or two sensors) but both with a flexible data rate. To reduce the internal sample rate from

Table 1. Table of available ringcores

ringcore diameter	no. of windings	band width	band thickness	noise at 1Hz
13mm	6	2mm	20mm	5pT/√Hz
18mm	7	2mm	20mm	4pT/√Hz
24mm	7	2mm	12mm	2pT/√Hz

4096vectors/min to the telemetry rate, the data are digitally filtered.

For increasing the effective data rate a version of a noiseless data compression originating from image processing and modified for our purposes is used.

The expected compression effect is 2-5. The noiseless data compression decreases the normally high redundancy of magnetometer data without any data loss. One of the consequences of data compression is that since the real compression effect is unknown there is no direct connection between the telemetry quota and the time resolution (vectors per second). So with a fixed telemetry quota the time resolution has to be controlled in dependence on the current data. A second consequence is that one bit of compressed data contains more information than one bit of uncompressed data. So in the case of noisy data transmission the data loss will be much higher. Upon using the best (maximum) coder one damaged bit may cause the loss of a whole telemetry frame. For this case a simplified (minimum) coder with fixed

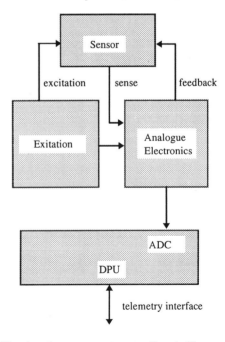

Figure 2. Usual analogue magnetometer (Interball)

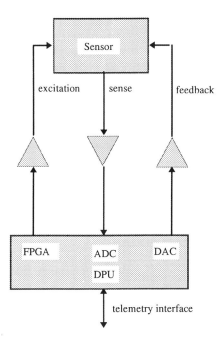

Figure 3. Digital fluxgate magnetometer (ROMAP)

98 balloon missions are given in Table 3. The experiment hardware consists of :
- the triaxial sensor system
- the analogue and drive electronics
- the AD-converter
- only two ranges and no compensation
- commercial DC/DC power converter (27V to ±5V)

The internal controller (exists only in the Auroral Probe Magnetometer) acts as telemetry interface and controls the analogue to digital converter. The data will be converted with a constant primary sampling rate of 64Hz and mean values of the field will be calculated (no digital filters, no data compression). All other tasks will be done by the host computer.

3.4. *Digital fluxgate magnetometer*

Our aim was to decrease weight and power of the electronics without losing efficiency. Starting from the usual analogue electronics which we have used for the MARS-94 magnetometer, we have designed for the Equator-S magnetometer a more compact version by taking away the notch filter. As a further step to miniaturize the device we have substituted the whole analogue electronics by using a MARS-94 DPU. Figure 4 shows how the reduction of the analogue electronics was done step by step.

In the first case, the field independent frequencies (f_o and $3f_o$ with an amplitude of 2V after the preamplifier) are suppressed by notch filters.

In the second case (no filters at all) the requirements for the phase-sensitive detector are increased by the factor V1.

In the third case (the digital fluxgate magnetometer) the $2f_o$ signal is digitized at its second harmonic and mean values are phase-sensitively calculated online.

Figure 5 shows the sensor output signal ($f_o+2f_o+3f_o$) and separately drawn the field dependent signal $2f_o$ (12.5kHz). The data are digitized synchronously with this second harmonic of the excitation frequency in each maximum (M^+) and in each minimum (M^-). The difference $D=M^+-M^-$ is calculated to suppress all offset errors of preamplifier and ADC. Afterwards meanvalues are calculated online up to the open loop bandwidth (for instance up to 10Hz). This simple but fast arithmetic will be done in a separate Field

bit positions may be used (via digital command) so the loss of one bit results in the damage of one vector only.

To increase the accuracy a range adjustment controlled either by command or automatically is implemented. The analogue output voltage will be multiplied by a digital word and converted to the feedback current. MAREMF is designed for ranges between ±16nT and ±512nT, EQUATOR-S for ranges between 512nT and 64000nT.

An additional compensation in combination with different ranges allows special working regimes for high resolution measurements. Furthermore the compensation will be used for inflight calibration modes.

Onboard software is written in assembler and C languages, dependent on timing and memory requirements.

3.3. *Magnetometer 0.5kg class*

The main parameters of the magnetometers for the Interball Tailprobe, the Interball Auroral Probe and the Mars-

Table 2. The main characteristics of orbiter magnetometers for several missions

Parameter	PHOBOS	MARS-96 / RELICT	EQUATOR-S
Dynamic range	±128nT	±32nT ... ±512nT	±256nT ... ±48000nT
Quantization	62pT	1pT ... 16pT	8pT ... 1.4nT
Sampling rate	0.5 Hz	68 Hz	128 Hz
Memory	64Kbit	1Mbit	1Mbit
Redundancy	DPU (cold)	DPU (cold)	complete (cold)

Table 3. The main characteristics of magnetometers for small satellites

Parameter	Interball Tailprobe	Interball Auroral Probe	Mars-98 Balloon
Dynamic range	±128nT	±3200nT	±2000nT
Quantization	62pT	98pT	61pT
Sampling rate	up to 64Hz	2.4Hz / 24Hz	0.1Hz
Operation mode	continuously	continuously	1sec on / 9sec off
Redundancy	no	no	no

Programmable Logic (FPGA) to relieve the fast CPU. The CPU controls the feedback, serves as data logger and is responsible for the communication with the host computer via RS232. A minimum configuration of the hardware is shown in Figure 6. Main advantage of the digital fluxgate is that the DPU controls the excitation (the main power consumer) and the data acquisition. To save power the magnetometer is able to operate in a pulsed working mode. Feedback (DAC) and preamplifier are working continuously, but excitation and ADC are activated only for some periods of the second harmonic. An oversampling factor allows the calculation of mean values with a tolerable aliasing error.

So, deep space missions limited to low power consumption, as well as magnetic survey stations under extreme conditions, for instance, in the polar regions, are special areas for the application of the digital fluxgate magnetometer.

In Figure 7 an example is given for the power management of the Rosetta Lander magnetometer [Auster et al., 1997]. Controller, preamplifier and feedback driver have a background power consumption of 275mW. The controller starts the measurement cycle each 100ms. During the measurement ADC and excitation need additionally 300mW. Mean values from 10 measurements (1s) will be transmitted to the Rosetta Orbiter.

The digital magnetometer is able to change the transfer function by simultaneously changing of feedback current and mean values calculation.

Furthermore the system is completely open to adding other signals like that of a proton magnetometer with the help of a second FPGA and/or to receive DCF/GPS time information. So, magnetic field measurement could be done on a high level of automation.

Typical parameters for a digital fluxgate magnetometer designed to perform measurements on a planetary surface are given in Table 4.

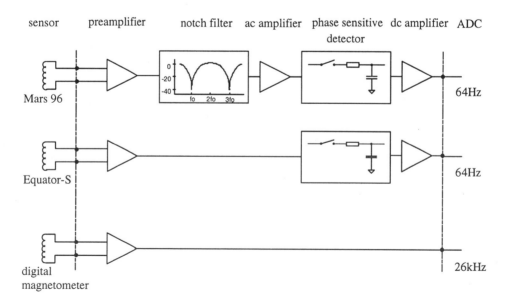

Figure 4. Block diagram of the tested electronics versions

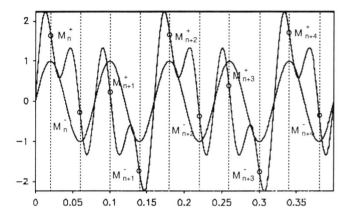

Figure 5. The data acquisition algorithm

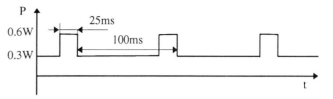

Figure 7. Power Management of the digital magnetometer

An excellent opportunity that we have used for tests, were the IAGA Workshops on Geomagnetic Observatory Instruments (Tihany 90, Chambon 92, Dourbes 94, Niemegk 96). Magnetometers could be tested during a long period by comparison to the observatory standard.

4. MEASUREMENT RESULTS

In all tested electronics versions (analogue and digital) the sensor noise of $5pT/\sqrt{Hz}$ (at 1Hz) limits the accuracy of the magnetometers. Only in the Equator-S Earth field range the feedback current generates an additional noise of $10pT/\sqrt{Hz}$.

The offset temperature dependence of the analogue electronics was 10pT/K in the low field ranges and less than 200pT/K in the Equator-S Earth field range. The influence of the temperature on the ADC of the digital magnetometer, tested between -20°C and 60°C, was not significant (less than 0.1nT total).

To evaluate the long term stability of the digital magnetometer we have compared analogue and digital electronics versions connected in parallel on the same sensor. Figure 8 (top) shows a 24 hours registration in the Adolf Schmidt Observatory of Niemegk and Figure 8 (bottom) the difference between the outputs of analogue and digital electronics versions.

5. CONCLUSION

With the substitution of the analogue electronics we can reduce weight, dimension and power consumption of fluxgate magnetometers. The reliability is improved because the number of components is reduced compared to electronics with an additional analogue part. The accuracy of the magnetometer is limited independently from the electronics versions by the noise level of the fluxgate sensor. Temperature dependence of the analogue electronics, especially of the filter, and the symmetry of the phase sensitive detector are no longer a problem. All tasks of the analogue section will be taken over in the proposed digital magnetometer by software. Furthermore, it becomes possible to change frequency and dynamic range only by means of software commands, without resorting to any hardware changes.

Acknowledgement. The authors wish to thank the DARA for supporting this development.

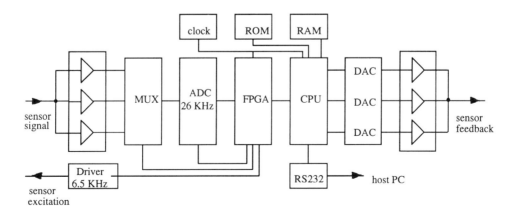

Figure 6. The minimum configuration of the hardware

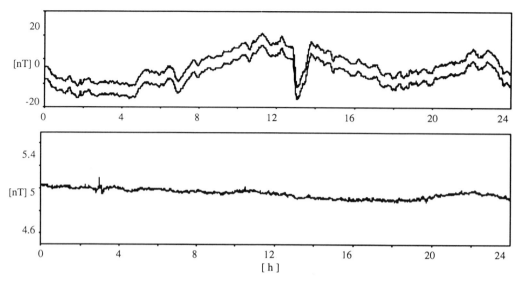

Figure 8. 24 hours parallel registration of digital and analogue electronics (upper and lower curve in the top window) and the difference between both (bottom window)

Table 4. Parameters of a digital fluxgate magnetometer

Parameters	digital magnetometer
mass	
sensor	50g
electronics	150g
size:	
sensor	sphere, $\varnothing$ 4cm
electronics	$150cm^2$ board
frequency range	0...100Hz
dynamic range	±10000nT
quantization	10pT
power	
1 vector/sec	375mW
100 vectors/sec	575mW
temperature range	
sensor	-170...+80 C
electronics	-55...+60 C

REFERENCES

Acuna, M. Fluxgate magnetometer for outer planet exploration, *IEEE Trans. on Magn.*, Vol. 10, Nr. 3, 519-523, 1974.

Auster, H.U., A.Lichopoj, J.Rustenbach, V.Auster, H.Bitterlich, K.H.Fornacon, O.Hillenmaier, R.Krause and H.J.Schenk, Concept and first results of a digital fluxgate magnetometer, *Meas. Sci. Technol.* 6, 477-481, 1995.

Auster, H.U., K.Baumgärtel, A.Bogdanov, H.Feldhofer, K.H.Fornacon, K.H.Glaßmeier, G.Haerendel, E.Kührt, W.Magnes, D.Möhlmann, G.Musmann, C.Russell, J.Rustenbach, K.Sauer, P.A.Schnegg, R.Schrödter, K.Schwingenschuh and R.Wäsch, ROMAP *Experiment Interface Document* for the Rosetta Lander, 1997

Primdahl, F., B.Hernando, J.R.Peterson, and O.V.Nielsen, Digital detection of the flux-date sensor output signal, *Meas. Sci. Technol.*, 5, 359-362, 1994.

H.U.Auster, Institut für Geophysik und Meteorologie der TU Braunschweig, Mendelssohnstraße 3, 38106 Braunschweig, Tel. +49 531 3915241, Fax: +49 531 3915220, E-mail hua@geophys.nat.tu-bs.de

H.Bitterlich, MAGSON GmbH, Rudower Chaussee 6, 12489 Berlin, Tel: +49 30 63923943, Fax: +49 30 63923944

K.H.Fornacon, Institut für Geophysik und Meteorologie der TU Braunschweig, Mendelssohnstraße 3, 38106 Braunschweig, Tel. +49 531 3917237, Fax: +49 531 3915220, E-mail khf@geophys.nat.tu-bs.de

O.Hillenmaier, MAGSON GmbH, Rudower Chaussee 6, 12489 Berlin, Tel: +49 30 63923932, Fax: +49 30 63923944

A.Lichopoj, DLR Institut für Planetenerkundung, Rudower Chaussee 5, 12489 Berlin, Tel: +49 30 67055396, Fax: +49 30 67055303, E-mail: Alexander.Lichopoj@DLR.de

J.Rustenbach, Max-Planck Institut für extraterrestrische Physik, Labor Berlin, Rudower Chaussee 5, 12489 Berlin, Tel: +49 30 63923923, Fax: +49 30 63923929, E-mail: jr@mpe.FTA-Berlin.de

H.J.Schenk, Max-Planck Institut für extraterrestrische Physik, Labor Berlin, Rudower Chaussee 5, 12489 Berlin, Tel: +49 30 63923926, Fax: +49 30 63923929, E-mail: jr@mpe.FTA-Berlin.de

Principles of Space Plasma Wave Instrument Design

Donald A. Gurnett

Department of Physics and Astronomy, The University of Iowa, Iowa City, Iowa

Space plasma waves span the frequency range from somewhat below the ion cyclotron frequency to well above the electron cyclotron frequency and plasma frequency. Because of the large frequency range involved, the design of space plasma wave instrumentation presents many interesting challenges. This chapter discusses the principles of space plasma wave instrument design. The topics covered include: performance requirements, electric antennas, magnetic antennas, and signal processing. Where appropriate, comments are made on the likely direction of future developments.

INTRODUCTION

In this chapter we discuss the principles of space plasma wave instrument design. The study of space plasma waves has its origins in early ground-based studies of very-low-frequency (VLF) radio signals. The earliest record of a VLF radio signal of natural origin is by *Preece* [1894], who describes a number of unusual audio frequency signals that he detected while studying the propagation characteristics of a long-distance transmission line. Some years later, *Barkhausen* [1919] using a rudimentary vacuum-tube amplifier identified a class of whistling VLF signals now known as whistlers. Progress on the understanding of these radio signals was slow. After several years of investigation, *Eckersley* [1935] correctly postulated that whistlers were produced by lightning. However, the propagation path and mechanism that dispersed the signal into a long whistling tone was unknown. It was not until the early 1950s that *Storey* [1953] was able to provide a satisfactory explanation of whistlers. He showed that the dispersion occurred as the signal propagated along the Earth's magnetic field in a plasma mode of propagation now known as the whistler mode.

In addition to whistlers, a number of other signals were discovered during the early ground-based era of VLF research that were clearly not produced by lightning. Among the best known of these are "dawn chorus" [*Storey*, 1953; *Allcock*, 1957] and "auroral hiss" [*Burton and Boardman*, 1933; *Duncan and Ellis*, 1959; *Dowden*, 1959]. These signals eventually came to be known as VLF emissions, since they were believed to be emitted by charged particles in the ionized upper levels of the Earth's atmosphere [*Ellis*, 1957; *Gallet*, 1959]. However, the exact emission mechanism remained largely unknown. For a review of early ground-based observations of whistlers and VLF emissions, see *Helliwell* [1965].

With the launch of the first Earth-orbiting satellites in the late 1950s, radio receivers were soon being carried into orbit to investigate the origin of VLF emissions. The first satellites with receivers specifically designed to study VLF emissions were Alouette 1 (launched Sept. 29, 1962) and Injun 3 (launched Dec. 13, 1962). Alouette 1 carried an electric dipole antenna [*Barrington and Belrose*, 1963], and Injun 3 carried a magnetic loop antenna [*Gurnett and O'Brien*, 1964]. These and other similar instruments soon showed that a wide variety of complex wave phenomena existed in the Earth's magnetosphere, which is a region of magnetized plasma surrounding the Earth. These included not only whistler-mode emissions, but also a variety of new

Measurement Techniques in Space Plasmas: Fields
Geophysical Monograph 103

modes of propagation that had never previously been observed. These waves are now known as space plasma waves. For a review of space plasma waves, see *Shawhan* [1979].

It is now known that plasma waves play a fundamental role in the physics of space plasmas. Virtually all space plasma investigations now include a plasma wave instrument. The purpose of this paper is to discuss the basic principles of plasma wave instrument design. The presentation is organized into four topics. Section II discusses performance requirements, Section III describes electric antennas, Section IV describes magnetic antennas, and Section V describes on-board signal processing.

PERFORMANCE REQUIREMENTS

Before going into a detailed discussion of the design of a plasma wave instrument, it is useful to first discuss the basic performance requirements. Three types of requirements are discussed: (1) the number of field components measured, (2) the frequency range, and (3) the sensitivity and dynamic range.

Number of Field Components Measured

As is well known, two types of waves can propagate in a plasma: electromagnetic and electrostatic [*Stix*, 1962]. Electromagnetic waves have both an electric field and a magnetic field, and electrostatic waves have only an electric field. Since these two types of waves interact quite differently with the plasma, it is important to be able to distinguish electromagnetic waves from electrostatic waves. The best way to do this is to use both an electric antenna and a magnetic antenna. If the wave is detected on both types of antennas, then it is an electromagnetic wave. If the wave is detected only with the electric antenna then it is an electrostatic wave. Thus, the first basic design requirement is to include both electric and magnetic antennas.

Next, one must next decide on the number of components to be measured. Of course, full three-axis measurements are ideal. The main reason for including three axis measurements is to determine the direction of the propagation vector, $\vec{k}$. For electromagnetic waves, magnetic field measurements provide the best method of determining the $\vec{k}$ vector direction. Maxwell's equation, $\vec{\nabla} \cdot \vec{B} = 0$, guarantees that $\vec{k}$ is perpendicular to plane of rotation of the magnetic field. Except at high frequencies (i.e., in the free space regime), there is no comparable relationship for the electric field. To determine the propagation direction of electrostatic waves, three axis electric field measurements are

required. If mass or other restrictions prohibit three-axis measurements, the next best alternative is single-axis measurements. The $\vec{k}$ direction cannot be obtained from two-axis measurements.

Frequency Range

An important consideration in the design of any plasma wave instrument is the frequency range. The frequency range of interest extends from somewhat below the ion cyclotron frequency to somewhat above either the electron plasma frequency or the electron cyclotron frequency, whichever is larger. The electron plasma frequency is given by $f_{pe} = 9 \sqrt{n}$ kHz, where n is the electron density in cm^{-3}, the electron cyclotron frequency is given by $f_{ce} = 28 B$ Hz, where B is the magnetic field strength in nT, and the ion cyclotron frequency is given by $f_{ci} = (m_e/m_i)f_{ce}$, where m_e/m_i is the electron-to-ion mass ratio. In space plasmas, the ion cyclotron frequency is often quite small. For example, in the outer regions of the Earth's magnetosphere, the proton cyclotron frequency is only about 0.1 Hz. In the Earth's magnetosphere, the highest frequency of interest is usually the electron plasma frequency, which extends up to about 5 to 10 MHz in the ionosphere. At Jupiter, where the magnetic field is much stronger, the electron cyclotron frequency can be as high as 40 MHz, well above the electron plasma frequency. As can be seen from the above examples, the frequency range of potential interest for plasma wave measurements is enormous, extending from approximately 0.1 Hz to 10 MHz at Earth, and from approximately 0.01 Hz to 40 MHz at Jupiter. One of the most difficult challenges of modern plasma wave instrument design is to provide satisfactory measurements over such a large frequency range.

When considering the frequency range to be investigated, it is often useful to make a distinction between local and remote measurements. If the spacecraft spends most of its time far from the planet, as is often the case for planetary flybys and eccentric orbiting spacecraft, the local electron cyclotron frequency and plasma frequency are usually much smaller than the corresponding values near the planet. Under these circumstances the lower part of the frequency range is dominated by locally generated plasma waves, and the upper part of the frequency range (above the local electron plasma frequency and cyclotron frequency) is dominated by remotely generated radio emissions. Both parts of the spectrum are important, but require somewhat different techniques. For locally generated plasma waves, both electric and magnetic fields must be measured in order to distinguish electromagnetic waves from electrostatic waves. On the other hand, for remotely generated radio emissions there is little point to measuring both the electric and the magnetic fields, since the waves are known to be electromagnetic. Thus, at frequencies above the local electron plasma frequency and cyclotron frequency, it is usually adequate to measure either the electric field or the magnetic field, but not both.

Sensitivity and Dynamic Range

To be successful, a plasma wave instrument must be able to detect waves over a very large range of intensities. Locally generated plasma waves, such as in planetary bow shocks and in auroral acceleration regions, are often extremely intense. At the opposite extreme, various types of thermally excited waves are extremely weak and difficult to detect. Although weak, some of these waves have proven to be extremely valuable for plasma diagnostics (see, for example, *Meyer-Vernet* [1979], and *Meyer-Vernet et al.* [1997]). To give a rough indication of the dynamic range that must be accommodated, Figures 1 and 2 show representative electric and magnetic field spectrums for various plasma wave phenomena observed in the Earth's magnetosphere. As one can see, the dynamic range requirements are very severe, typically 100 to 120 dB.

ELECTRIC ANTENNAS

Two types of electric antennas have been used for wave electric field measurements: monopoles and dipoles. A

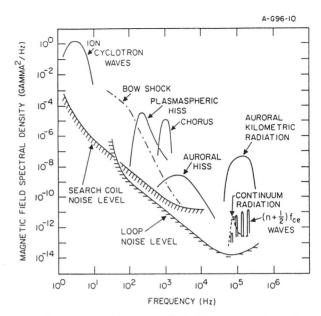

Figure 2. Representative magnetic field spectrums for various plasma wave phenomena observed in the Earth's magnetosphere.

monopole responds to the potential difference between a single antenna element and a ground plane, usually the spacecraft body, whereas a dipole antenna responds to the potential difference between two antenna elements extending outward in opposite direction. The term "monopole," as used above, is actually a misnomer, since the spacecraft body effectively acts as the second element of a dipole. Although monopole antennas have been used on a number of spacecraft [*Scarf et al.*, 1965; *Warwick et al.*, 1977], this type of antenna is very susceptible to spacecraft-generated electrical interference, so dipoles are almost always used for modern plasma wave instruments.

Over the years, two types of electric dipole antennas have evolved: (1) cylindrical dipoles, and (2) spherical double probes. They differ mainly in the geometry of the sensing elements. For a cylindrical dipole, the elements consist of two conducting cylinders extending outward in opposite directions from the spacecraft body, as in the top panel of Figure 3. The radius of the elements, a, is normally much smaller than the tip-to-tip length, L. Typical dimensions are a = 0.1 to 1 cm and L = 100 m. A differential amplifier in the spacecraft body provides a signal, ΔV, that is proportional to the potential difference between the elements. For a spherical double probe, the two elements consist of conducting spheres mounted on the end of booms, as in the bottom panel of Figure 3. The center-to-center separation, L, between the spheres is normally much larger than the radius, r, of the spheres. Typical dimensions are r = 10 cm and L = 100 m.

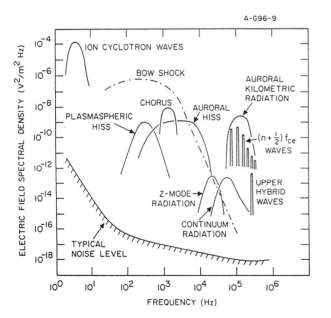

Figure 1. Representative electric field spectrums for various plasma wave phenomena observed in the Earth's magnetosphere.

A-G96-2

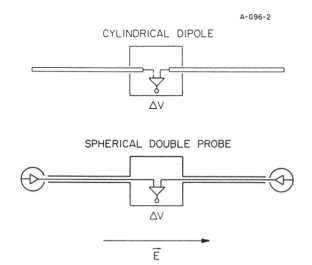

Figure 3. The geometry of a cylindrical electric dipole antenna and a spherical double probe. A differential amplifier in the spacecraft body provides a voltage output, ΔV, that is proportional to the voltage difference between the elements.

Preamplifiers are usually located in the spheres to provide low impedance signals to a differential amplifier in the spacecraft body.

For most types of waves detected in space plasmas, it is usually the case that the wavelengths are much longer than the length of the antenna ($\lambda \gg L$). The response of an electric dipole antenna is then characterized by a quantity called the effective length. The effective length is defined by

$$L_{eff} = \frac{\Delta V}{E} \quad , \tag{1}$$

where E is the electric field component along the axis of the antenna and ΔV is the open circuit potential difference between the elements (i.e., the voltage difference that would exist in the absence of any electrical load). The effect of an electrical load is usually represented by an equivalent circuit of the type shown in Figure 4. The voltage generator, $\Delta V_{in} = E \, L_{eff}$, represents the voltage induced by the electric field, Z_A represents the antenna impedance, and Z_L represents the load impedance imposed by the spacecraft at the base of the antenna. The effective length and antenna impedance depend on a number of factors, including (1) the geometry of the antenna, (2) the frequency and wavelength of the incident wave, and (3) the parameters of the plasma. A quantitative analysis is only possible in certain limiting cases, as discussed below.

High Frequency Limit

If the wave frequency is well above the electron plasma frequency and cyclotron frequency, the effects of the plasma are essentially negligible. The antenna then responds as if it were in free space. In free space the response of an electric dipole antenna is usually analyzed by using the reciprocity theorem [*Jordan*, 1950], which states that the coupling between a transmitter and a receiver is unaffected by interchanging the receiver and the transmitter. Thus, instead of analyzing the receiving properties of the antenna, one can just as well analyze the radiation properties of the antenna.

For a cylindrical dipole that is short compared to the wavelength, it is well known that the radiation properties can be represented by charges, $\pm Q$, that are uniformly distributed along the two antenna elements [*Jordan*, 1950]. Since the centers of the two charge distributions are located at the centers of the elements, it is obvious that the equivalent dipole moment is obtained by separating the charges by a distance of $L/2$. The effective length is then $L_{eff} = L/2$. For a spherical double probe the analysis is even simpler. Since the charges on the two elements are separated by a distance L, it immediately follows that the effective length is $L_{eff} = L$.

Since the fields around an electrically short ($\lambda \ll L$) dipole are almost purely electrostatic, the antenna impedance is almost purely capacitive and can be represented by a capacitive reactance, $Z_A = 1/(i\omega C_A)$, where C_A is the capacitance of the element relative to free space. For a cylindrical dipole, the antenna capacitance is given to a good approximation by

$$C_A = \frac{2\pi\epsilon_0(L/2)}{[\ell n(L/2a) - 1]} \quad , \tag{2}$$

where a is the radius of the element and ϵ_0 is the permittivity of free space [*Jordan*, 1950]. For a spherical

A-G96-18

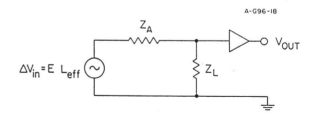

Figure 4. The equivalent circuit of an electric antenna. ΔV_{in} is the voltage induced by the applied electric field, Z_A is the antenna impedance and Z_L is the load impedance at the base of the antenna.

double probe, the antenna capacitance is given to a good approximation by

$$C_A = 4\pi \epsilon_0 r \quad . \tag{3}$$

where r is the radius of the sphere.

To provide maximum power transfer from the source to the load, it is well known that the load impedance must be the complex conjugate of the source impedance, $Z_L = Z_A^*$. This relationship is known as the maximum power transfer theorem [*Jordan*, 1950]. Since the impedance of a short (L ≪ λ) electric dipole antenna is almost purely capacitive, maximum power transfer is achieved when the load impedance is almost purely inductive, with $Z_L = -1/(i\omega C_A)$. In practice it is extremely difficult to achieve this impedance matching condition. Instead, the usual procedure is to maximize the voltage delivered to the preamplifier. This condition is achieved by making the load impedance large compared to the antenna impedance, $Z_L \gg Z_A$. At high frequencies the load impedance can usually be represented by a capacitive reactance, $Z_L = 1/(i\omega C_L)$. The capacitance C_L includes: (1) the capacitance from the antenna element to the spacecraft structure, (2) the capacitance of the cable that connects the antenna element to the preamplifier, and (3) the input capacitance of the preamplifier. When the load impedance is purely capacitive, one can see from the equivalent circuit in Figure 4 that the ratio of the output voltage to the input voltage is given by a simple capacitive divider

$$\frac{V_{out}}{E\,L_{eff}} = \frac{C_A}{C_A + C_L} \quad . \tag{4}$$

Maximum voltage transfer is achieved by making C_A much larger than C_L. This condition ($C_A \gg C_L$) is called the ideal voltmeter condition. As one can see from Equations 2 and 3, the antenna capacitance is proportional to the size of the sensing elements. Since the length, L/2, of a cylindrical dipole element is much larger than the radius, r, of a spherical double probe, the capacitance of a cylindrical dipole is much larger than the capacitance of a spherical double probe. Thus, it is usually much easier to achieve ideal voltmeter operation for a cylindrical dipole than for a spherical double probe.

Low Frequency Limit

If the wave frequency is comparable or less than the electron plasma frequency and cyclotron frequency, then the effects of the plasma can no longer be ignored. When an object is placed in a plasma, it is well known that a plasma sheath forms around the object. If the object is exposed to ultraviolet light, which is usually the case in space, then two types of sheaths can occur: (1) a positive ion sheath, and (2) a photoelectron sheath. A positive ion sheath occurs when the electron current from the plasma exceeds the emitted photoelectron current. Under these conditions the surface charges to a negative potential, thereby repelling some of the electrons and leaving the sheath with a positive charge (hence the term, positive ion sheath). Since ion currents are usually negligible, equilibrium occurs when the electron current reaching the surface is equal to the emitted photoelectron current. A positive ion sheath has a characteristic thickness that is given by the Debye length, $\lambda_D = 6.9\,(T/n)^{1/2}$ cm, where T is the electron temperature in °K, and n is the electron density in cm^{-3}. The Debye length ranges from about 1 cm in the ionosphere to about 10 m in the solar wind.

If the plasma density is so low that the electron current from the plasma cannot compensate for the emitted photoelectron current, then the surface charges to a positive potential. The positive potential inhibits the escape of photoelectrons, thereby forming a negatively charged sheath of photoelectrons around the object. This type of sheath is called a photoelectron sheath. The thickness of a photoelectron sheath depends on the intensity and spectrum of the photoelectrons and is almost completely independent of the plasma parameters. For the typical photoelectron spectrum emitted from a metal surface at 1 A.U. (astronomical unit), the shielding thickness is about 30 cm. For the plasma temperatures that are normally encountered in the Earth's magnetosphere, the transition from a positive ion sheath to a photoelectron sheath usually occurs as the electron density drops below about 10^2 to 10^3 cm^{-3}.

A full treatment of the response of an electric antenna to a wave in a magnetized plasma is quite complicated and is well beyond the scope of this review. A relatively simple model that seems to work well under a wide range of plasma parameters was introduced by *Storey* [1965]. Also, see *Fahleson* [1967] and *Kelley et al.* [1970]. This model assumes that the wave can be described locally by an electrostatic potential, and that the potential is linearly coupled to the antenna through the sheath that surrounds the antenna. If the thickness of the sheath is small compared to the tip-to-tip length of the antenna (i.e., $\lambda_D \ll$ L), then the sheath forms a layer of nearly constant thickness around the antenna. The approximate geometries of the sheaths that form around the antenna are then as shown in Figures 5 and 6. To analyze the response of the antenna, Storey assumed that the potential in the plasma is

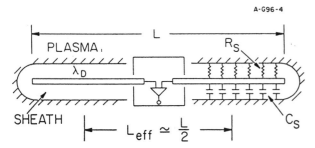

Figure 5. The geometry of the plasma sheath that forms around a cylindrical electric dipole antenna. The thickness of the sheath is given by the Debye length, λ_D. Potentials in the plasma are coupled to the antenna via the sheath resistance, R_S, and the sheath capacity, C_S. The effective length, L_{eff}, of a cylindrical dipole is one-half the tip-to-tip length, L.

coupled to the antenna via a sheath resistance R_S and a sheath capacitance C_S that are uniformly distributed over the surface of the antenna (see the right-hand sides of Figures 5 and 6).

Since ion currents are usually negligible, the primary contribution to the sheath resistance comes from the electrons. If we restrict the analysis to frequencies well below the electron plasma frequency and cyclotron frequency, the sheath resistance can be computed from Langmuir probe theory. Langmuir probe theory deals with the equilibrium voltage-current (V-I) characteristic of an object immersed in a plasma [Langmuir, 1929]. It is well known that the V-I curve for a positive ion sheath has the form shown in Figure 7. For small amplitude signals, the resistance of the sheath is determined by the differential slope, $R_S = \partial V/\partial I$, evaluated at some bias point I_{Bias}. For a Maxwellian plasma, it is easy to show that the resistance of a positive ion sheath is given by

$$R_S = \frac{U_e}{I_p + I_i + I_{Bias}} \quad . \quad (5)$$

where $U_e = \kappa T_e/e$ is the electron temperature expressed in Volts, I_p is the emitted photoelectron current, and I_i is the

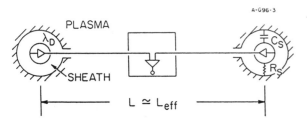

Figure 6. The geometry of the plasma sheath that forms around a spherical double probe. The effective length, L_{eff}, of a spherical double probe is the center-to-center distance between the spheres.

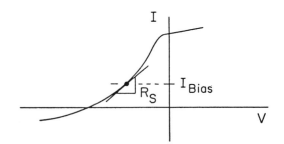

Figure 7. The voltage-current, V-I, characteristics of a conducting object placed in a plasma. The sheath resistance, R_S, is the slope of the V-I curve evaluated at a bias current, I_{Bias}.

incident ion current. For a photoelectron sheath, the sheath resistance is given by

$$R_S = \frac{U_p}{I_e - I_i - I_{Bias}} \quad , \quad (6)$$

where $U_p \approx 1.5$ Volts is the characteristic energy of the photoelectron spectrum (see Cauffman and Gurnett [1972]).

To compute the sheath capacity, C_S, it is usually assumed that the sheath consists of a capacitor, with the antenna element acting as one conductor and the plasma acting as the other conductor. Since electrons are essentially excluded from a positive ion sheath, the interior of the capacitor is assumed to be free space. A similar assumption is used for a photosheath. Ignoring end effects, the capacitance of a cylindrical dipole is then given by

$$C_S = 2\pi\epsilon_0(L/2)/\ell n(\lambda_D/a) \quad , \quad (7)$$

where λ_D is the thickness of the sheath and a is the radius of the element. The above formula assumes that $a \ll \lambda_D$, a condition that is almost always satisfied. For a spherical double probe, the capacitance is given by

$$C_S = 4\pi\epsilon_0 r\left(1 + \frac{r}{\lambda_D}\right) \quad , \quad (8)$$

where λ_D is the thickness of the sheath and r is the radius of the sphere.

Having described the essential characteristics of the sheath that forms around an antenna, we next address the question of the effective length. If an electric field E_0 is applied along the axis of the antenna, then the potential in the plasma is given by $\Phi = -E_0 z$, where z is the distance along the axis of the antenna. If the R-C coupling is uniform along the axis of the antenna, it is easy to see that the potential of an element is simply the average of the potential along the axis of that element, i.e., $V = -E_0(L/4)$. The voltage difference between the two elements is then given by $\Delta V = -E_0(L/2)$, so the effective length is $L_{eff} = L/2$. Since a spherical double probe responds to the potentials at $z = -L/2$ and $z = L/2$, it is easy to see that the voltage difference between the elements is given by $\Delta V = -E_0 L$, so the effective length is $L_{eff} = L$. It is interesting to note that in the low frequency plasma limit, the effective lengths are exactly the same as in the high frequency-free space limit.

Based on the above discussion, it follows that the response of an electric dipole antenna can be represented by the equivalent circuit shown in Figure 8. The antenna impedance Z_A is determined by the sheath resistance R_S and the sheath capacitance C_S, and the load impedance Z_L is determined by the load capacitance C_L and the load resistance R_L. The ratio of the output voltage to the input voltage is then given by a simple impedance division

$$\frac{\Delta V_{out}}{E\,L_{eff}} = \frac{Z_L}{Z_A + Z_L} \quad . \tag{9}$$

Since we want the electric field measurement to be as independent of the plasma parameters as possible it is clear that the antenna should be operated as an ideal voltmeter (i.e., $Z_L \gg Z_A$, so that $\Delta V_{out} = E\,L_{eff}$). At high frequencies the impedance ratio is mainly controlled by the capacitances and is given by

$$\frac{\Delta V_{out}}{E\,L_{eff}} = \frac{C_S}{C_S + C_L} \quad . \tag{10}$$

This is the same result that was obtained in the free space limit, except the antenna capacitance is now replaced by the

sheath capacitance. At low frequencies the impedance ratio is mainly controlled by the resistances and is given by

$$\frac{\Delta V_{out}}{E\,L_{eff}} = \frac{R_L}{R_S + R_L} \quad . \tag{11}$$

To operate as an ideal voltmeter, the sheath capacitance must be much greater than the load capacitance ($C_S \gg C_L$), and the sheath resistance must be much less than the load resistance, i.e., ($R_S \ll R_L$). The transition between the capacitive and resistive regimes is controlled by the sheath capacity and sheath resistance. If the $C_S \gg C_L$ and $R_S \ll R_L$, it is easy to show that this transition occurs at a R-C transition frequency given by $\omega_{RC} = 1/R_S C_S$. Resistive coupling dominates for $\omega < \omega_{RC}$ and capacitive coupling dominates for $\omega > \omega_{RC}$.

Unfortunately, it is not always easy to achieve the conditions required to operate the antenna as an ideal voltmeter (i.e., $C_S \gg C_L$ and $R_S \ll R_L$). Because of its much larger capacitance, a cylindrical dipole again has a significant advantage over a spherical double probe. For example, if $\lambda_D = 1$ m, $L = 100$ m, $a = 1$ cm, and $r = 10$ cm, then the capacitance of a cylindrical dipole is about 1000 pf, whereas the capacitance of a spherical double probe is only about 10 pf. Thus, in the capacitive coupling regime it is much easier to achieve ideal voltmeter operation for a cylindrical dipole than for a spherical double probe. A cylindrical dipole also has another advantage that is related to the relative variability of the sheath resistance and the sheath capacity. Of these two parameters, the sheath resistance is much more variable than the sheath capacitance. This can be seen by comparing Equations 5 and 6 for the sheath resistance with Equations 7 and 8 for the sheath capacity. The sheath resistance is strongly controlled by the electron temperature, U_e, and the electron current, I_e, both of which vary over large ranges. In contrast, the antenna capacitance is controlled by the logarithmic term in Equation 7 and by the $(1 + r/\lambda_D)$ term in Equation 8, neither of which depend strongly on the plasma parameters (usually $r \ll \lambda_D$). Since we want the sheath coupling to be as independent of the plasma parameters as possible, it is clear from the above discussion that the antenna should be operated in the capacitive regime over as large a frequency range as possible. This means that the RC transition frequency, $\omega_{RC} = 1/R_S C_S$, should be made as low as possible, which implies that the capacitance should be as large as possible. Since a cylindrical dipole has a much larger capacitance than a spherical double probe, a cylindrical dipole can be operated in the capacitive regime over a substantially larger frequency range than a spherical double probe.

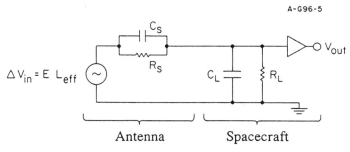

A-G96-5

Figure 8. The equivalent circuit of an antenna immersed in a plasma. The resistance and capacitance of the sheath are represented by R_S and C_S, and the resistance and capacitance of the load are represented by R_L and C_L.

Although ideal voltmeter operation can usually be achieved in the capacitive coupling regime ($\omega > \omega_{RC}$), at low plasma densities it is extremely difficult to achieve ideal voltmeter operation in the resistive coupling regime ($\omega < \omega_{RC}$). In the presence of solar ultraviolet radiation, the sheath is always a photoelectron sheath at low plasma densities (i.e., when $I_p + I_i + I_{bias} > I_e$). For this type of sheath the antenna resistance is inversely proportional to the incident electron current (see Equation 6). Since the incident electron current varies in direct proportion to the electron density, as the electron density decreases the antenna resistance increases, eventually reaching the point where the ideal voltmeter condition, $R_S \ll R_L$, is violated. For typical cylindrical dipole dimensions ($L = 100$ m and $a = 0.1$ cm) and a preamplifier resistance of $R_L \approx 10^9$ to 10^{10} ohm, the ideal voltmeter condition is usually violated at electron densities below about 1 to 10 cm^{-3}. The RC transition frequency at this density is typically a few tens of Hz. For a spherical double probe ($r = 10$ cm), the RC transition frequency under similar conditions is typically several kHz. Thus, in a low density plasma a cylindrical dipole provides sheath independent (i.e., capacitively coupled) measurements over a substantially greater frequency range than is possible with a spherical double probe. To improve the operation of a spherical double probe at low plasma densities, a bias current is often used (see *Mozer* [1969]). As shown by Equation 6, the sheath resistance can be decreased by applying an external bias current I_{Bias}. The bias current must be negative, which means that a current is drawn from the plasma. This current bias technique is commonly used for static electric field measurements, where measurements must be made down to essentially zero frequency.

Short Wavelength Effects

Electrostatic waves can sometimes have very short wavelengths. The theoretical minimum wavelength is given by $\lambda = 2\pi/\lambda_D$. If the wavelength is comparable or shorter than the length of the antenna, then the response deviates considerably from the long wavelength limit discussed in the previous section. The approach to analyzing the short wavelength response is basically the same as in the previous section, except the potential in the plasma is given by $\Phi = -(E_0/k) \sin(kz)$ instead of $\Phi = -E_0 z$. For a cylindrical dipole, the voltage difference between the elements is determined by integrating the plasma potential over the length of the antenna while taking into account the R-C coupling through the sheath. As before, end effects are ignored. Using this simple model, *Fuselier and Gurnett* [1984] have shown that in the ideal voltmeter limit, $Z_A \ll$

Z_L, the response of a cylindrical dipole is given by

$$\frac{\Delta V_{out}}{E_0 L_{eff}} = \frac{\sin^2 x}{x^2} \quad , \qquad (12)$$

where $x = k L_{eff}/2$ is the normalized wave number and $L_{eff} = L/2$. For a spherical double probe, the analysis is even simpler, since the spheres can be regarded as two point probes. The corresponding result for a spherical double probe is given by

$$\frac{\Delta V_{out}}{E_0 L_{eff}} = \frac{\sin x}{x} \quad , \qquad (13)$$

where in this case $L_{eff} = L$.

Plots of the antenna response, $\Delta V/(E_0 L_{eff})$ as a function of the normalized wave number are shown in Figure 9 for a cylindrical dipole and a spherical double probe. In both cases the response has a transition at $x = 1$. For $x \ll 1$, the response asymptotically approaches the long wavelength response. For $x > 1$ the response decreases rapidly with increasing wavenumber, varying as $1/k^2$ for a cylindrical dipole and as $1/k$ for a spherical double probe, with nulls at $x = n\pi$ where $n = 1, 2, 3, \cdots$. These nulls can cause

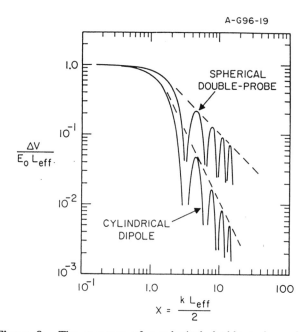

A-G96-19

Figure 9. The response of a spherical double probe and a cylindrical dipole to wavelengths comparable or shorter than the antenna length. The parameter x is the normalized wave number ($x = kL_{eff}/2$), and L_{eff} is the effective length at long wavelengths.

interference patterns in the electric field spectrum [*Temerin*, 1979; *Fuselier and Gurnett*, 1984; *Feng et al.*, 1993]. The interference patterns are particularly noticeable on a spinning spacecraft. As the spacecraft rotates destructive interference occurs whenever the projection of the effective antenna length onto the wave vector corresponds to an integral number of wavelengths, $L_{eff} \cos \phi = n\lambda$. The geometry that leads to a null is illustrated in Figure 10. The shape and spacing of the nulls can be used to give information on the wavelength and mode of propagation of the wave. Since the response of a spherical double probe decreases less rapidly with increasing wavenumber than for a cylindrical dipole, interference effects are usually more clearly defined for a spherical double probe. It is also sometimes advantageous to measure the fractional density fluctuations, $\delta n/n$, at each probe, rather than the electric field. For a review of the techniques used to measure wavelengths in space plasmas, see *LaBelle and Kintner* [1989].

Noise Levels

Two types of noise affect the performance of an electric antenna: preamplifier noise and plasma noise. Since the impedance of an electric antenna tends to be rather large, current noise is often the most important type of noise generated by the preamplifier. This noise flows from the preamplifier into the antenna and has two components: a low-frequency component that varies as $1/f$, and a high-frequency component that is nearly independent of frequency. The spectral density, $I_p^2/\Delta f$, of the current noise is determined by the type of input transistor. Usually FET

A-G91-608-1

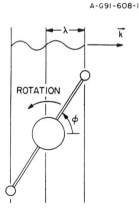

Figure 10. A sketch showing the null patterns produced by a rotating antenna when the wavelength is small compared to the length of the antenna. Nulls are produced when the projection of the effective length onto the $\vec{k}$-vector direction, $L_{eff} \cos \phi$, corresponds to an integral number of wavelengths, $n\lambda$.

transistors provide the lowest noise, although other types of transistors can be used depending on the detailed design of the preamplifier. For a typical FET preamplifier, the noise current in the high frequency range has a spectral density in the range from about 10^{-26} to 10^{-30} amp^2/Hz. The transition to the $1/f$ spectrum usually takes place at a frequency of about 10 to 100 Hz.

There are two types of noise generated by the plasma. The first is thermal (Nyquist) noise generated by the sheath resistance R_S, and the second is shot noise caused by charged particle impacts on the antenna. The thermal noise can be represented by a current source with a spectral density $I_R^2/\Delta f = 4\kappa T_S/R_S$ in parallel with the sheath resistance R_S, where κ is Boltzmann's constant, and T_S is the sheath temperature [*Hancock*, 1961]. For a positive ion sheath the appropriate temperature for the sheath is the plasma electron temperature, since the V-I characteristic of the antenna is mainly determined by the plasma electrons. For a photoelectron sheath the appropriate temperature is the photoelectron temperature (~18,000°K). Since the electron flux is usually much higher than the ion flux, shot noise is mainly due to electron impacts. It is easy to show that the current spectral density due to electron impacts is given by $I_e^2/\Delta f = e^2 \nu_e$, where e is the electronic charge and ν_e is the impact rate [*Hancock*, 1961].

If the antenna is operated as an ideal voltmeter (i.e., $Z_A \ll Z_L$), then all of the above noise currents flow through the sheath. The differential voltage spectral density between the two elements due to these noise currents is then given by

$$\frac{(\Delta V)^2}{\Delta f} = 2 |Z_A|^2 \left(\frac{I_p^2}{\Delta f} + \frac{4\kappa Ts}{R_S} + e^2 \nu_e \right) . \qquad (14)$$

The factor of two in the above equation occurs because the noise powers from the two elements must be added when computing $(\Delta V)^2$. Using the antenna impedance model described earlier, the electric field noise level then becomes

$$\frac{E_N^2}{\Delta f} = \frac{2}{L_{eff}^2} \frac{R_S^2}{1 + \omega^2 R_S^2 C_S^2} \left(\frac{I_p^2}{\Delta f} + \frac{4\kappa Ts}{R_S} + e^2 \nu_e \right) . \qquad (15)$$

From the above equation one can see that the primary factors affecting the electric field noise level are the effective length L_{eff}, the sheath resistance R_S, the sheath capacitance C_S, and the electron impact rate ν_e. Clearly the effective length should be made as large as possible, subject only to the constraint that the antenna should not be longer than the wavelength ($L < \lambda$). To analyze the remaining factors, the resistive and capacitive coupling regimes must be considered separately. In the resistive coupling regime, $\omega R_S C_S \ll 1$, the noise level is minimized by making R_S and ν_e as small as possible. However, this leads to opposing requirements. Both R_S and ν_e depend on the collecting area of the antenna. From Equations 5 and 6 one can see that R_S is minimized by increasing the area of the antenna (i.e., by making I_p and I_e as large as possible). On the other hand, increasing the area of the antenna increases the impact rate, hence the shot noise. Usually the emphasis is on maintaining ideal voltmeter operation (i.e., maximizing the area in order to minimize R_S) even though this increases the shot noise. In the capacitive coupling region, $\omega R_S C_S \gg 1$, the noise level is minimized by making R_S and C_S as large as possible, and ν_e as small as possible. Of these three factors, the most important is the antenna capacitance. Since a cylindrical dipole has a much larger capacitance than a spherical double probe, the noise level of a cylindrical dipole can be made substantially lower than a spherical double probe, often by as much as 20 to 40 dB. Of the remaining two factors, R_S and ν_e, both can be minimized by making the area of the antenna as small as possible. For a cylindrical dipole, this means making the radius as small as possible. Unfortunately, this condition is not consistent with the condition for maintaining ideal voltmeter operation in the resistive coupling regime, so a compromise must be made between these two competing requirements. For a spherical double probe, the area cannot be reduced without significantly reducing the capacitance, which is a critical factor in maintaining ideal voltmeter operation in the capacitive coupling regime.

In closing the discussion of electric antenna noise levels, we note that the above model does not address the noise level near the electron plasma frequency and the electron cyclotron frequency. The impedance of the antenna has resonances near these frequencies that strongly affect the noise level. The analysis is beyond the scope of this review. For a detailed discussion of these resonance effects, see *Meyer-Vernet* [1979], *Kellogg* [1981], *Sentman* [1982], and *Meyer-Vernet and Perche* [1989].

MAGNETIC ANTENNAS

Two types of magnetic antennas are commonly used for wave magnetic field measurements: loops and search coils. Schematic diagrams of each are shown in Figures 11 and 12. In both cases the basic principle of operation is Faraday's law, which states that in a circuit of N turns a voltage

$$V = N \frac{d\Phi_m}{dt} \qquad (16)$$

is induced whenever the magnetic flux $\Phi_m = \int \vec{B} \cdot d\vec{A}$ through that circuit changes. The main difference between a loop and a search coil is that a search coil uses a high permeability core to concentrate the magnetic flux through the circuit, whereas a loop does not. For the dimensions normally employed for a magnetic antenna, 1 meter or less, the plasma has no effect on the response of the antenna. Also, short wavelength effects are usually not important.

For a properly designed antenna system, the noise level is controlled by the first element in the system, which for a magnetic antenna is the resistance noise of the wire in the sensing circuit. This simple consideration has important consequences for the design of a magnetic antenna. Since a loop is simpler, we start by discussing a loop antenna. Because mass is usually a critical factor, the antenna must be designed to make optimum use of the available mass. For a loop of circumference ℓ_c, the resistance is given by $R = \ell_c N/(\sigma s)$, where N is the number of turns, σ is the conductivity, and s is the cross-sectional area of the wire. The corresponding mass of the loop is given by $m = s\ell_c N\rho$, where ρ is the mass density of the material. If the area of the loop is A, it is easy to show using the above equations and Equation 16 (rewritten as $V = NA\omega B$), that the magnetic field noise level is given by

$$\frac{B_N^2}{\Delta f} = \left(\frac{\rho}{\sigma}\right)\left(\frac{\ell_c}{A}\right)^2 \frac{4\kappa T}{m\omega^2} \qquad . \qquad (17)$$

The above equation shows that the noise level is independent of the number of turns. As the number of turns increases the signal voltage increases, but the resistance also increases, which increases the noise power ($V_R^2/\Delta f = 4\kappa TR$) in direct proportion to the signal power. If the total mass of the loop is fixed, the above equation also shows that the lowest noise level is achieved by choosing a material with the smallest possible ratio of mass density to conductivity, ρ/σ, and a shape that has the largest possible ratio of area to circumference, A/ℓ_c. Of the materials that usually can be used, aluminum is best, although copper and silver are only slightly inferior. A circular loop provides the best ratio of area to circumference. For a search coil similar considerations also apply, except that one must take into account the weight of the high-permeability core and the more complicated geometry of the winding around the core. Because of the large number of variables involved, a simple analytical solution does not exist for the optimum distribution of mass between the winding and the core. For most successful search coil designs the mass is divided about equally between the winding and the core.

A-G95-95

LOOP MAGNETIC ANTENNA

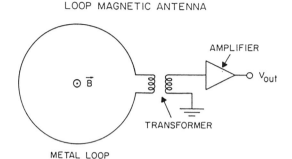

Figure 11. A sketch showing the principle of operation of a magnetic loop antenna. A loop antenna responds to the magnetic field component, B, perpendicular to the plane of the loop.

An important factor in the design of a magnetic antenna is the bandwidth. The bandwidth of a magnetic antenna is controlled by the inductance L and capacitance C of the sensing circuit and its associated electronics. A simple equivalent circuit that describes the frequency response of a magnetic antenna is shown in Figure 13. It is obvious from this simple circuit that a resonance exists at a frequency given by $\omega_{LC} = 1/\sqrt{LC}$. The resistor R_D is used to damp this resonance. Since the output voltage decreases very rapidly above the resonance frequency, ω_{LC} effectively determines the upper frequency limit of the antenna. To maximize the bandwidth, it is obvious that both the inductance and capacitance must be made as small as possible. First, we consider the inductance. Since the inductance varies as N^2, the number of turns must be made

A-G95-94

SEARCH COIL MAGNETIC ANTENNA

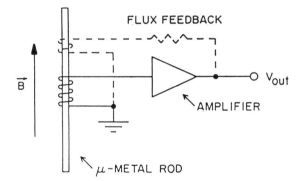

Figure 12. A sketch showing the principle of operation of a search coil magnetic antenna. A search coil responds to the component of the magnetic field, B, parallel to the axis of the μ-metal rod. Flux feedback is sometimes used to maintain a flat frequency response.

A-G96-27

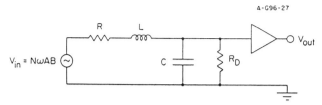

Figure 13. The equivalent circuit of a magnetic antenna. The quantities R, L, and C are the equivalent resistance, inductance and capacitance of the antenna. The resistor R_D is used to damp the resonant response of the antenna.

as small as possible. For a loop antenna, the minimum inductance is achieved by choosing N = 1 (i.e., one turn). Since the resistance of a one-turn loop of typical size and mass (1 meter diameter and 1 kg) is only a few milliohms, this optimization leads to a resistance noise that is much lower than the noise level of a typical transistor preamplifier. To assure that the resistance noise is above the noise level of the preamplifier, a transformer must be included between the loop and the preamplifier (see Figure 11). A typical turn ratio for this transformer is about 1:500. One of the penalties of using a loop is that the transformer introduces a low-frequency cutoff. The cutoff frequency is controlled by the mass and permeability of the transformer core [*Skilling*, 1959]. For a transformer of reasonable weight and size, the low-frequency cutoff is typically about 50 Hz.

For a search coil, the usual approach is to avoid the use of a transformer. This has the advantage of extending the frequency response down to much lower frequencies, the only limitation being the basic $d\Phi_m/dt$ response of the coil. The disadvantage is that to assure that the resistance noise of the wire is above the noise level of the preamplifier, a large number of turns must be used, which reduces the bandwidth. As the number of turns is increased, the cross-sectional area of the wire must be decreased (in order to maintain a fixed mass), which increases the resistance, hence increasing the resistance noise. Eventually a point is reached where the resistance noise of the wire exceeds the noise level of the preamplifier. This condition determines the minimum number of turns, N_{min}, hence the inductance of the coil. Since the upper cutoff frequency, $\omega_{LC} = 1/\sqrt{LC}$, is controlled by the inductance, the noise level of the preamplifier plays a crucial role in determining the bandwidth. To maximize the bandwidth, it is critical that the noise level of the preamplifier be made as low as possible.

Next we consider the capacitance. For a loop, the capacitance is determined mainly by the secondary winding on the transformer. For a search coil, the capacitance is determined mainly by the capacitance of the coil.

Minimizing these capacitances is a complicated process. However, procedures do exist that lead to near optimum solutions. For a discussion of the techniques involved, see *Welsby* [1960]. Once the capacitance has been determined, the frequency response of the antenna (not including the low-frequency cutoff of the transformer) can be represented rather accurately by the equation $V_{out} = \omega G(\omega)BA_{eff}$, where $G(\omega)$ is a normalized frequency response given by

$$G(\omega) = \frac{1}{\sqrt{\left(1 + \dfrac{R}{R_D} - \omega^2 L^2 C^2\right)^2 + \left(\dfrac{\omega L}{R_D} + \omega RC\right)^2}} \quad . \quad (18)$$

and A_{eff} is an effective area that takes into account the geometry of the antenna. Typical plots of the normalized frequency response for a loop and a search coil are shown in Figure 14. For comparable sizes and sensitivities the upper cutoff frequency of a loop antenna is usually considerably higher than the upper cutoff frequency of a search coil. However, as discussed earlier, the loop has a low-frequency cutoff that does not exist for a search coil.

For a loop, the effective area is simply the area of the loop, $A_{eff} = A$. For a search coil, the effective area depends on several factors, the most important of which are the geometry and permeability of the core. Although the core is usually made of rectangular layers of μ-metal (to reduce eddy current losses), for modelling purposes it is usually assumed that the core consists of a long thin ellipsoid of revolution. If the core has a relative permeability μ, it can be shown [*Bozorth*, 1951] that the effective permeability, $\mu_{eff} = B_{inside}/B$, of an ellipsoidal core is given to a good approximation by the equation

$$\frac{1}{\mu_{eff}} = \frac{1}{\mu} + \left(\frac{2a}{L}\right)^2 [\ell n(L/a) - 1] \quad , \quad (19)$$

where a is the radius at the center of the core and L is the tip-to-tip length of the core. Since the permeability μ is dependent on temperature and various other factors, it is desirable that the first term on the right-hand side of the above equation be small compared to the second term, i.e., $\mu \gg (L/2a)^2$. Since the largest μ values that can be achieved are about 10^5, this condition places an upper limit on the length to diameter ratio of the core ($L/a \lesssim 100$). If the above condition is satisfied, the effective permeability of the core is then given to a good approximation by the equation

$$\mu_{eff} = \frac{(L/2a)^2}{\ell n(L/a) - 1} \quad . \quad (20)$$

Since the cross-sectional area at the center of the core is πa^2, the effective area then becomes

$$A_{eff} = \pi (L/2)^2 \frac{1}{\ell n(L/a) - 1} \quad . \quad (21)$$

The above equation shows that the effective area of a search coil is approximately $\pi(L/2)^2$, which is the same as the effective area of a loop with a diameter equal to the length of the core. In practice, the effective area differs somewhat from the above equation due to the finite size of the coil and deviations from an ellipsoidal geometry.

Before finishing the discussion of magnetic antennas, let us return again to the noise level. The earlier analysis of the noise level of a magnetic loop antenna, given by Equation 17, does not take into account the resonant response of the antenna. To correctly take into account the effect of the resonance on the noise level, the real part of the impedance, Z_r, must be computed looking back into the antenna from the terminals of the preamplifier (see Figure 13). The noise voltage spectral density at the terminals of the antenna is then given by $V_N^2/\Delta f = 4\kappa T Z_r$. Proceeding as before, but now computing the noise voltage from the real part of the impedance, it is easy to show that the noise level of a loop antenna is given by

$$\frac{B_N^2}{\Delta f} = \left(\frac{\rho}{\sigma}\right)\left(\frac{\ell_c}{A}\right)\frac{4\kappa T}{m\omega^2}\left[\frac{\left(1 + \dfrac{R}{R_D} - \omega^2 LC\right) + \dfrac{\omega L}{R}\left(\dfrac{\omega L}{R_D} + \omega RC\right)}{\left(1 + \dfrac{R}{R_D} - \omega^2 LC\right)^2 + \left(\dfrac{\omega L}{R_D} + \omega RC\right)^2}\right] \quad . \quad (22)$$

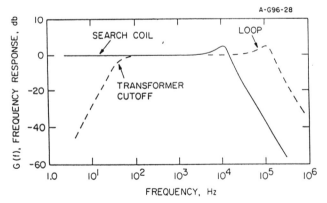

Figure 14. A comparison of the normalized frequency response, G(f), of a loop and a search coil. Loop antennas usually have higher upper frequency cutoffs than search coils. However, loop antennas have a low-frequency cutoff due to the transformer.

A similar result can be derived for a search coil antenna. The term in the rectangular bracket is the ratio of the real part of the impedance to the resistance in the wire, Z_r/R. Usually, the damping resistance, R_D, is small compared to the resistance, R, so we can assume that $R_D \ll R$. At low frequencies it is easy to see that the term in brackets reduces to one, which simply corresponds to the fact that at low frequencies $Z_r = R$. In this limit, Equation 22 agrees with the earlier result given by Equation 17. Proceeding upward in frequency the first evidence of a deviation from $Z_r = R$ occurs at a transition frequency given by $\omega_{RL} = \sqrt{R R_D}/L$. Above this frequency the real part of the impedance increases as ω^2. This frequency dependence is illustrated in Figure 15, which shows a plot of Z_r/R as a function of ω. At even higher frequencies the real part of the impedance eventually reaches a peak at the resonance frequency, ω_{LC}, and then decreases rapidly with increasing frequency, varying asymptotically as $1/\omega^2$.

As can be seen from Figure 15, the noise level of a magnetic antenna is enhanced above the resistance noise of the wire over a wide range of frequencies. Although the enhanced noise level cannot be avoided, the frequency range over which it occurs can be minimized. The low-frequency limit of the enhanced noise level is controlled by the R-L transition frequency, $\omega_{RL} = \sqrt{R R_D}/L$. To minimize the frequency range over which the enhanced noise level occurs, ω_{RL} should be made as large as possible. This means that R_D should be made as large as possible. However, to minimize the peak in the frequency response, the damping resistance should be adjusted to give critical damping. Thus, a compromise must be made between achieving the lowest possible noise level and minimizing the resonant peak in the frequency response.

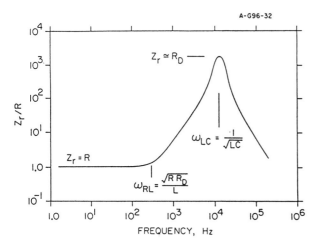

Figure 15. The ratio of the real part of the impedance, Z_r, of a magnetic antenna to the resistance, R, of the wire plotted as a function of frequency.

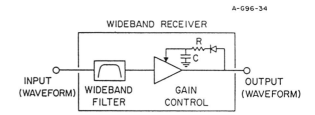

Figure 16. A block diagram of a wideband receiver. Since the entire waveform is transmitted to the ground, this type of receiver provides the highest possible resolution. An automatic gain control is used to reduce the dynamic range of the output waveform.

Usually the damping resistance can be made somewhat larger than the critical damping value without causing an unacceptable peak in the frequency response. For a search coil, flux feedback (see Figure 12) can also be used to control the resonant response of the antenna, thus avoiding the need for the damping resistor.

ON-BOARD SIGNAL PROCESSING

Space plasma waves measurements place great demands on signal processing. The ideal approach, of course, is to transmit waveforms from all the antennas directly to the ground where analyses can be performed with whatever frequency and time resolution are desired. However, one can easily show that for the bandwidths and dynamic ranges involved (1 to 10 MHz and 120 dB) the data rates are much too large. Therefore, a substantial amount of processing must be performed on board the spacecraft. There are two types of on-board processing, analog and digital. Since the techniques involved are quite different, these two types of processing are discussed separately.

Analog Processing

In the early days of space plasma wave research, all on-board signal processing was carried out via analog electronics. Although many different types of analog processing systems have been used, they can be categorized into three main types: (1) wideband receivers, (2) multi-channel spectrum analyzers, and (3) sweep frequency receivers. A block diagram of a wideband receiver is shown in Figure 16. As the name implies, this type of receiver simply transmits all of the signals within a relatively wide range of frequencies. Usually, an automatic gain control is employed to reduce the dynamic range of the signals that must be transmitted to the ground. A block diagram of a multi-channel spectrum analyzer is shown in

Figure 17. This type of spectrum analyzer consists of a bank of continuously active narrowband filters, each followed by an amplifier and some type of diode rectifier or root-mean-square detector. The detector output is usually averaged using a simple RC circuit to provide an output proportional to the average signal strength. A block diagram of a sweep frequency receiver is shown in Figure 18. This type of receiver consists of a single channel that is electronically swept in frequency via a circuit called a frequency converter. A frequency converter uses a nonlinear device to generate a frequency $f_1 = f_0 \pm f$, where f is the frequency of the input signal and f_0 is the frequency of a variable frequency oscillator. By using a fixed-frequency filter at f_1, the frequency of the input signal can be selected by adjusting the frequency of the oscillator. In modern designs, the oscillator is replaced by a frequency synthesizer. A frequency synthesizer generates a frequency that is a fixed integer fraction, M/N, of some basic reference frequency, usually derived from a crystal oscillator.

Each of the above types of receivers has inherent advantages and disadvantages. Wideband receivers come closest to the ideal by transmitting the entire waveform to the ground for analysis. However, wideband receivers inherently involve very high information rates, typically hundreds of kbits/sec. Since such a high information rate usually cannot be provided continuously over long periods of time, wideband receivers are best suited for relatively short bursts of data in regions of special interest. Multi-channel analyzers have inherently good time resolution since each channel is continuously active. However, each channel requires a separate filter and associated electronics, so the required weight and power increase linearly with the number of channels. Although miniaturization and low-power electronics have minimized these limitations, it is seldom possible to afford more than a few tens of channels.

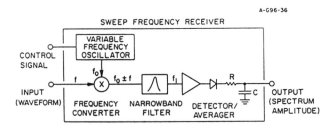

A-G96-36

SWEEP FREQUENCY RECEIVER

Figure 18. A block diagram of a sweep frequency receiver. Since the frequency is continuously variable, this type of receiver provides very good frequency resolution, but relatively poor time resolution.

Thus, although the time resolution of a multi-channel analyzer is very good, the frequency resolution is poor. Sweep frequency receivers suffer from essentially the opposite problem. Since the frequency is variable over an essentially continuous range, the frequency resolution is very good. However, since the receiver must dwell at each frequency for an interval at least equal to the filter response time, $\Delta t = 1/\Delta f$, the time that is required to sweep over the entire frequency range increases linearly with the number of frequency steps. Thus, although the frequency resolution is very good, the time resolution is poor.

Since each of the above receivers has its inherent advantages and disadvantages, an approach that has often been used is to include all three in the same instrument (see for example, *Gurnett et al.* [1995]). Thus, the wideband receiver provides very good frequency and time resolution for short periods of time, the multi-channel analyzer provides continuous spectrums with very good time resolution but relatively poor frequency resolution, and the sweep-frequency receiver provides continuous spectrums with very good frequency resolution but relatively poor time resolution.

Digital Processing

In recent years there has been a strong trend toward the use of on-board digital signal processing. For a plasma wave instrument digital signal processing offers several important advantages, including (1) much greater flexibility, (2) elimination of nonlinear distortion effects that are common in analog circuits, and (3) much smaller phase and amplitude errors.

The main difficulty with on-board digital signal processing is computational speed. Typical computations involve the use of Fourier transforms, digital filters and various types of auto- and cross-correlations. For a discussion of some of the algorithms involved, see *Cunningham* [1992]. Computations of this type are very demanding. For example, a Fourier transform using the

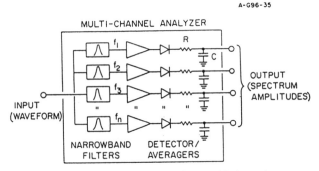

A-G96-35

MULTI-CHANNEL ANALYZER

Figure 17. A block diagram of a multi-channel spectrum analyzer. Since each channel is continuously active, this type of receiver provides very good time resolution, but relatively poor frequency resolution.

Cooley and Tukey [1965] method takes approximately N $\log_2$ N operations to compute a N-point transform. Thus, a 1024-point transform requires about 10^4 operations. To perform such computations in real time for a 1-MHz bandwidth and a dynamic range of 120 dB (20-bit resolution) requires approximately 10^7 20-bit operations per second. Such high computational rates are well beyond the reach of present space-qualified microprocessors. However, rapid advances are being made. A system that performs a 1024-point transform over a 22 kHz bandwidth with 12-bit resolution has already been flown [*Bougeret et al.*, 1995]. One can anticipate that in the not-too-distant future virtually all on-board plasma wave processing will be carried out using digital techniques.

CONCLUSION

In this chapter we have reviewed the principles of space plasma wave instrument design. Among the various topics considered, the most complicated involve electric antennas. Although the basic elements of electric antenna design are known, there are still parameter regimes that are poorly understood. For example, all current theories assume that the Debye length is much smaller than the tip-to-tip length of the antenna, $\lambda_D \ll L$. Although this condition is rarely violated for relatively long, 100 m antennas, significant uncertainties arise when shorter antennas are used, which are more likely to have $\lambda_D \gtrsim L$. Also, the response of an electric antenna to an externally applied electric field is still poorly understood in the vicinity of the electron plasma frequency and the electron cyclotron frequency.

Compared to electric antennas, the principles that affect the design of magnetic antennas are well known. Nevertheless, magnetic antennas suffer from several problems, the most important of which is low sensitivity. The basic sensitivity issue is almost entirely one of size and weight. Whereas electric antennas have typical dimensions of 100 m, for mechanical reasons magnetic antennas seldom have dimensions greater than 1 m. Since both types of antennas draw their energy from a volume of approximately L^3, where L is the characteristic size of the antenna, it is not surprising that for equal energy densities (i.e., E = cB), electric antennas have much better sensitivities than magnetic antennas. The most obvious approach to improving the sensitivity of a magnetic antenna is by cooling the antenna to low temperatures, possibly using super-conducting wires. Techniques of this type deserve further investigation.

The new era of on-board digital signal processing also poses significant challenges. Although the technical capabilities have not yet advanced to the point where the entire bandwidth of interest can be processed in real time, it is likely that such capabilities will some day exist. Since it is unlikely that communication rates will improve significantly, plasma wave investigators will be faced with difficult choices regarding exactly what type of information should be transmitted to the ground. One promising technique that may provide some relief is the rapidly evolving field of data compression. Recently, techniques have been demonstrated for transmitting frequency-time spectrums using compression factors of ten or more with little apparent degradation in the quality of the spectrums. Data compression techniques of this type are likely to play an increasingly important role in the processing of space plasma wave data.

Acknowledgments. The author thanks Ann Persoon, George Hospodarsky, Bill Schintler, and Don Kirchner for their assistance and comments during the preparation of this paper. This research was supported by NASA through contract 958779 with the Jet Propulsion Laboratory.

REFERENCES

Allcock, G., McK., A study of the audio-frequency phenomena known as "dawn chorus," *Australian J. Phys.*, *10*, 286-298, 1957.

Barkhausen, H., Zwei mit Hilfe der neuen Verstärker entdeckte Erscheinungen, *Phys. Z.*, *20*, 401-403, 1919.

Barrington, R. E., and J. S Belrose, Preliminary results from the very-low-frequency receiver aboard Canada's Alouette satellite, *Nature*, *198*, 651-656, 1963.

Bougeret, J.-L., et al., Waves: The radio and plasma wave investigation on the Wind Spacecraft, *Space Sci. Rev.*, *71*, 231-263, 1995.

Bozorth, R. M., *Ferromagnetism*, pp. 845-849, Van Nostrand, N. York, 1951.

Burton, E. T., and E. M. Boardman, Audio-frequency atmospherics, *Proc. IRE*, *21*, 1476-1494, 1933.

Cauffman, D. P., and D. A. Gurnett, Satellite measurements of high latitude electric fields, *Space Sci. Rev.*, *13*, 369-410, 1972.

Cooley, J. W., and J. W. Tukey, An algorithm for the machine computation of Complex Fourier Series, *Mathematics of Computation*, *19*, 297-301, 1965.

Cunningham, E. P., *Digital Filtering: An Introduction*, Houghton-Mifflin, Boston, 1992.

Dowden, R. L., Low frequency (100 kc/s) radio noise from the aurora, *Nature*, *184*, 803, 1959.

Duncan, R. A., and G. R. Ellis, Simultaneous occurrence of subvisual aurorae and radio noise on 4.6 kc/s, *Nature*, *183*, 1618-1619, 1959.

Eckersley, T. L., Musical atmospherics, *Nature*, *135*, 104-105, 1935.

Ellis, G. R., Low-frequency radio emission from aurorae, *J. Atmos, Terrestr. Phys.*, *10*, 302-306, 1957.

Fahleson, U. V., Theory of electric field measurements conducted in the magnetosphere with electric probes, *Space Sci. Rev.*, *7*, 238-262, 1967.

Feng, W., D. A. Gurnett, and I. H. Cairns, Interference patterns in the Spacelab 2 plasma wave data: Lower hybrid waves driven by pickup ions, *J. Geophys. Res.*, *98*, 21,571-21,580, 1993.

Fuselier, S. A., and D. A. Gurnett, Short wavelength ion waves upstream of the Earth's bow shock, *J. Geophys. Res.*, *89*, 91-103, 1984.

Gallet, R. M., The very-low-frequency emissions generated in the Earth's exosphere, *Pro. IRE*, *47*, 211-231, 1959.

Gurnett, D. A., and B. J. O'Brien, High-latitude geophysical studies with satellite Injun 3, 5, Very-low-frequency electromagnetic radiation, *J. Geophys. Res.*, *69*, 65-89, 1964.

Gurnett, D. A., et al., The Polar plasma wave instrument, *Space Sci. Rev., 71*, 597-622, 1995.

Hancock, J. C., *An Introduction to the Principles of Communication Theory*, p. 196, McGraw-Hill, N.Y., 1961.

Helliwell, R. A., Whistlers and related ionospheric phenomena, Stanford Univ. Press, Stanford, C.A., 1965.

Kelley, M. C., F. S. Mozer, and U. V. Fahleson, Measurements of the electric field component of waves in the auroral ionosphere, *Planet. Space Sci.*, *18*, 847-865, 1970.

Kellogg, P. J., Calculation and observation of thermal electrostatic noise in the solar wind, *Plasma Physics*, *23*, 735, 1981.

Jordan, E. C., *Electromagnetic Waves and Radiating Systems*, Prentice-Hall, Englewood Cliffs, N.J., 1950.

LaBelle, J., and P. M. Kintner, The measurement of wavelength in space plasmas, *Rev. Geophys.*, *27*, 495-518, 1989.

Langmuir, I., The interaction of electron and positive ion space charges in cathode sheaths, *Phys. Rev.*, *33*, 954-989, 1929.

Meyer-Vernet, N., On natural noises detected by antennas in plasmas, *J. Geophys. Res.*, *84*, 5373-5377, 1979.

Meyer-Vernet, N., and C. Perche, Tool kit for antennae and thermal noise near the plasma frequency, *J. Geophys. Res.*, *94*, 2405-2415, 1989.

Meyer-Vernet, N., et al., Measuring plasma parameters with thermal noise spectroscopy, this issue, 1997.

Mozer, F. S., Instrumentation for measuring electric fields in space, *Small Rocket Instrumentation Techniques*, pp. 26-34, North-Holland, Amsterdam, 1969.

Preece, W. H., Earth currents, *Nature*, *49* (1276), 554, 1894.

Scarf, F. L., G. M. Crook, and R. W. Fredricks, Preliminary report on detection of electrostatic ion waves in the magnetosphere, *J. Geophys. Res.*, *70*, 3045-3060, 1965.

Sentman, D. D., Thermal fluctuations and the diffuse electrostatic emissions, *J. Geophys. Res.*, *87*, 1455-1472, 1982.

Shawhan, S. D., Magnetospheric plasma waves, in *Solar System Plasma Physics, Vol. III*, edited by L. J. Lanzerotti, C. F. Kennel, and E. N. Parker, pp. 211-270, North-Holland, Amsterdam, 1979.

Skilling, H. H., *Electrical Engineering Circuits*, pp. 336-338, Wiley, N. York, 1959.

Stix, T. H., *The Theory of Plasma Waves*, McGraw-Hill, N.Y., 1962.

Storey, L. R. O., An investigation of whistling atmospherics, *Phil. Trans. Roy. Soc.*, London, A, *246*, 113-141, 1953.

Storey, L. R. O., Antennae electrique dipole pour reception TBF dans l'ionosphere, *L'onde Elec.*, *45*, 1427-1435, 1965.

Temerin, M., Doppler shift effects on double-probe measured electric field power spectra, *J. Geophys. Res.*, *84*, 5929-5934, 1979.

Warwick, J. W., J. B. Pearce, R. G. Peltzer, and A. C. Riddle, Planetary radio astronomy experiment for Voyager missions, *Space Sci. Rev.*, *21*, 309-327, 1977.

Welsby, V. G., *The Theory and Design of Inductance Coils*, pp. 144-151, MacDonald, London, 1960.

_____

Donald A. Gurnett, Department of Physics and Astronomy, University of Iowa, Iowa City, IA 52242.

Plasma Wave Measurements: Skepticism and Plausibility

Paul M. Kintner

School of Electrical Engineering, Cornell University, Ithaca, New York

Plasma waves, at least linear plasma waves, are characterized by four parameters, which may have complex values: one of frequency and three of wave vector. Generally, space plasma wave instruments have only measured wave frequency while assuming wave vector properties, either for instrument design or for interpreting plasma wave data. Exceptions to this general trend, such as instruments that measure wave vector properties, have revealed that in many cases these assumptions are wrong; thus skepticism is appropriate for interpreting plasma wave data, while instrument design is critical for assuring plausible results. In this brief paper we outline the inherent challenges in completely measuring plasma wave properties and examine several case studies where knowledge of wave vector properties has been critical.

INTRODUCTION

The measurement of plasma waves and the correct interpretation of these measurements is problematical in space plasmas for three basic reasons. First, different wave modes can exist at a single frequency. Second, Doppler broadening can shift frequencies or erase distinctive spectral features. Lastly, there is a class of wave modes (electrostatic) that have slow phase velocities and short wavelengths, in some cases shorter than the electric field antenna length of many satellite instruments. The typical plasma wave instrument measures only frequency from a spacecraft reference frame. The consequence of measuring only wave frequency can be summarized in light of these three basic problems as follows:

1. If plasma waves are measured, their interpretation may be ambiguous with multiple choices available.
2. All of the apparent choices may be incorrect.
3. The instrument may have failed to respond to the wave so that items 1 and 2 are the least of our worries.

This last issue depends on the plasma environment and the specific satellite instrument considered; nonetheless, there are

important space plasmas where waves with large electric fields have been missed because the electric field antennas were too long. On the other hand, by either combining information from different sensors (electric, magnetic, and density) or examining spectral and statistical information, many of these problems can and have been resolved. In spite of these successes, nagging doubts remain that some observations have been misinterpreted or simply missed. In this article we consider these issues through several examples and demonstrate some solutions. There is no "magic bullet," and these solutions require careful thought and implementation.

Plasma waves are important to measure in space plasmas for two reasons: they are measures of plasma properties such as density, and they are mediators of energy exchange in collisionless plasmas. In the case where energy only flows from a particle population to a wave mode and where a normal mode that is neither evanescent nor damped exists outside the region of wave generation, such as AKR or chorus, the emissions may be sensed a long distance from their origins, making this class of plasma waves much easier to observe and study; however, when the energy flow is from the plasma waves to the background plasma, the waves should be heavily damped outside their origin. Waves that are heavily damped usually have short wavelengths and slow phase velocities that are resonant with thermal particle populations. This latter wave class is more difficult to ob-

Measurement Techniques in Space Plasmas: Fields
Geophysical Monograph 103

serve because it cannot be detected outside its origin, thereby requiring in situ observations in addition to the issues listed above, which generally apply to this class of wave.

Hence an attitude of skepticism is appropriate for understanding plasma waves in space. This attitude is especially important for designing experiments and interpreting data. In this short paper we briefly review some properties of plasma waves in space and the implications of these properties on instrument design. We next consider examples where data was or could have been misinterpreted and then describe instrumental techniques to resolve ambiguities or at least, find plausible interpretations. Our intent here is to highlight a few illustrative examples of this complex but critically important aspect of space physics. We first address the expected wave properties, then discuss examples of spatial irregularities, lower hybrid waves, and Langmuir waves. Finally, we will conclude with the curious example of apparent double layers, which can be shown to have a spacecraft origin. Our examples are taken from experiments that investigated auroral phenomena primarily because in this region correct interpretation of plasma wave data has been the most critical and difficult.

PLASMA WAVE MODES, PROPERTIES, AND IMPLICATIONS FOR INSTRUMENT DESIGN

In the linear approximation, plasma waves may be described as linear solutions to a dispersion equation, $D(w,k)=0$, where $D(w,k)$ is derived from the Fourier transform of Maxwell's equations and an equation relating $\mathbf{E}$ and $\mathbf{J}$. Analytically the dispersion relation is made tractable through a set of assumptions permitting its solution. These assumptions, such as cold plasma or limited frequency range, have historically led to a "zoo" of plasma waves categorized by the applicable range of assumptions. A more complete technique for solving the dispersion equation and describing plasma waves has been developed by *Rönmark* [1982] and *André* [1985]. They have developed a numerical technique that does not make the assumptions necessary for analytical solutions. Furthermore, their numerical technique is easily adaptable to a graphical presentation.

A graphical example of the solutions to the dispersion equation is shown in Figure 1. In this case each surface corresponds to a solution plotted as a frequency that is a function of normalized perpendicular and parallel wave numbers. The right plot is normalized to O^+ gyroradius and gyrofrequency while the left plot is normalized to H^+ parameters. The magnetic field and plasma density and temperature were chosen to be typical of a quiescent ionosphere at 500-km altitude. Nonetheless, the topological features of this plot are more general and apply to any low beta, two-ion

plasma. We have not shown the dispersion relation solutions normalized to electron gyroradius and gyrofrequency, but they can be seen in *Kintner et al.* [1995]. These plots can be examined for regions where the wave phase velocity is the order of the satellite velocity and where wavelengths are shorter than a typical satellite antenna, several tens of meters to 100 m.

The H^+ gyroradius varies from about 1 m in the F-region ionosphere to about 15 m in the auroral acceleration zone, and the H^+ thermal velocity varies from about 5 km/s to 10 km/s over the same range. Of course, the O^+ gyroradius and thermal velocities are four times larger and smaller, respectively. This implies that most of the electrostatic ion cyclotron waves/ion Bernstein modes, labeled M and L where $k\rho=1$, will be severely Doppler broadened and Doppler broadening will still be significant even for the harmonic modes near the lower hybrid resonance. For $k\rho\geq1$, virtually all of these modes will have wavelengths the order of, or much less than, a 100-m antenna. This problem is even more serious in the parameter range ordered by the electron gyroradius or Debye length, which vary from a few cm in the F-region ionosphere to 10 m in the auroral acceleration zone. Doppler broadening near the electron gyrofrequency and Langmuir frequency is generally not significant at satellite velocities.

Perhaps the clearest lesson that Figure 1 imparts is that only measuring frequency leaves the wave vector interpretation completely ambiguous. At a single frequency the wave number can differ by five orders of magnitude.

EXAMPLES OF SHORT WAVELENGTH AND DOPPLER BROADENING

In this section we present examples of waves with short wavelength that produce Doppler broadening or reduced antenna response or that require a wavelength measurement to interpret the mode correctly. Two of these examples make use of plasma wave interferometers, which need to be explained briefly. Then we discuss measurements of spatial irregularities, lower hybrid waves, and Langmuir waves.

PLASMA WAVE INTERFEROMETERS

The plasma wave interferometer works by measuring a plasma wave property, typically density or electric field, at two or more spatially separated locations. Then the wave phase velocity is inferred by measuring the signal phase shift between the sensors, or in the case of nonlinear structures, by measuring the time of flight. By measuring the phase shift the component of wave vector along the interferometer separation axis is estimated. This technique has been reviewed by *LaBelle and Kintner* [1989].

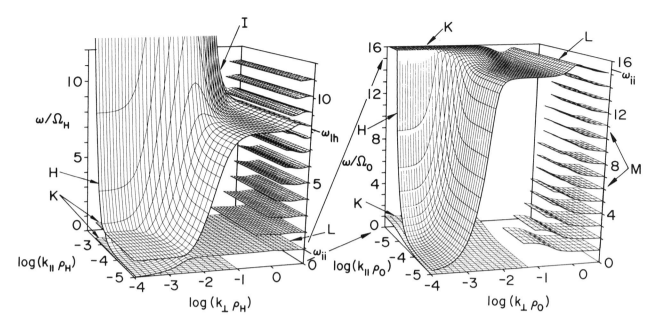

Figure 1. Dispersion surface ((Dcw,**k**) = 0) curves for frequencies up to the lower hybrid frequency assuring a 1% H+, 99% O+ ionosphere at about 500 km altitude. The left-hand plot is normalized to H+ gyrofrequency (Ω_H) and gyroradius (ρ_H), while the right-hand plot is normalized to O+ gyrofrequency (Ω_O) and gyroradius (ρ_O).

Since many wave modes can coexist, the phase shift must be estimated in Fourier space. This is not difficult since the Fourier transform, or more accurately, the discrete Fourier transform, yields a phase for each frequency. The phase at individual frequencies from different sensors can then be compared to estimate phase shift and wavelength; however, there is a problem with this simplistic approach in that comparing two random signals (one at each sensor) by comparing the phase of the two signals at one frequency yields a phase shift and wavelength that are not physical. To distinguish noise signals from physical signals the cross spectrum is used, which is defined below.

We assume two sensors with signals $s_1(t)$ and $s_2(t)$ having the Fourier transforms $S_1(\omega)$ and $S_2(\omega)$. The cross spectrum is then given by,

$$C_{12}(\omega) = \frac{\left\langle S_1(\omega)S_2^*(\omega)\right\rangle}{\left[\left\langle S_1^2(\omega)\right\rangle\left\langle S_2^2(\omega)\right\rangle\right]^{1/2}}$$

where < > implies an ensemble average. This calculation yields a complex number for each frequency. The magnitude of the complex number is called the coherency and the phase of the complex number is **k·d**, which yields the component of the wave vector along the sensor separation axis. The coherency has values between 0 and 1. When the two signals

at each sensor are produced by a single wave, the coherency has a value of one. On the other hand, when the signals at each sensor are uncorrelated (as in noise signals), the coherency has a value of 0; thus by examining the coherency, the noise contribution can be estimated. Unfortunately, multiple uncorrelated waves also yield a low value of coherency; hence this technique is only effective when a single frequency is dominated by a single wave vector.

SPATIAL IRREGULARITIES

Early ionospheric and magnetospheric spacecraft indicated broad regions of waves covering the H+ and O+ gyrofrequencies. In at least some cases these results were interpreted as ion acoustic waves. Figure 2 shows a quantitative interpretation of these waves using a δn interferometer.

The Viking spacecraft carried a δn/n interferometer composed of two δn/n sensors separated by 80 m. Throughout its orbit up to 8000 km altitude the δn/n sensors, as well as the electric field sensors, responded for fluctuations up to a frequency of a few 100 Hz. These fluctuations have maximum amplitudes on auroral field lines but were found virtually throughout the orbit [*Holmgren and Kintner*, 1990]. Figure 2 shows the cross spectrum of this signal produced from the two δn/n sensors separated by 80 m. Below 200 Hz the coherency is large, generally more than 0.5, indicating

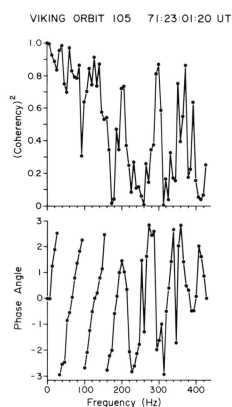

Figure 2. The cross spectrum for spatial irregularities as measured by the Viking interferometer. The upper panel is the coherency while the lower panel is the phase shift. A phase shift of 2π per 80 Hz across the 80-m long interferometer corresponds to a phase velocity of 6.4 km/s.

that both sensors were responding to the same signal. In this same frequency range the phase was a linear function of frequency with a slope of about 2π per 80 Hz. Since 2π of phase shift is produced by one wavelength shift over the interferometer length of 90 m, this corresponds to a phase velocity of 80 m times 80 Hz or 6.4 km/s, the spacecraft velocity; that is, this spectrum is composed of irregularities that are frozen in the plasma reference frame but propagate in the spacecraft reference frame at the spacecraft velocity. The full calculation can be found in *Kintner et al.* [1987]. A similar result for electric field irregularities was demonstrated by *Temerin* [1978], who noted that when multiples of a wavelength matched an electric field antenna length the antenna response has nulls.

Holmgren and Kintner [1990] later used the Viking interferometer to demonstrate that these zero-frequency or spatial irregularities were found everywhere all of the time on Viking. In a spacecraft reference frame these irregularities are Doppler shifted to the same frequencies as ion acoustic, elec-

trostatic and electromagnetic ion cyclotron, and Alfvén waves; this corresponds to the right-hand panel of Figure 1 and the lower third of the left-hand panel. Consequently, this important frequency range is obscured by the irregularities. Simultaneous search coil or magnetometer measurements can help identify the electromagnetic modes through the ratio of E to B, although this is a problematic frequency range for either instrument. On the other hand, no technique has been developed that successfully distinguishes low frequency electrostatic modes in the presence of spatial irregularities, and these modes may have escaped detection because of the widespread existence of irregularities.

LOWER HYBRID WAVES

Lower hybrid waves are generally characterized by surface 1 in Figure 1 and, although Doppler broadening is not an issue, the surface covers 2-3 orders of magnitude in perpendicular wave number. The specific wave number value is essential for determining the importance of lower hybrid waves in producing transversely accelerated ions [*Chang and Coppi*, 1981]. Two techniques exist for determining the wave number of lower hybrid waves: the quadruple technique [*Ergun et al.*, 1991] and the interferometer technique [*LaBelle and Kintner*, 1989]. The quadruple technique yields at each frequency a weighted average of the wave number, and this technique can be employed where multiple wave vectors exist corresponding to a single frequency. On the other hand, the weighted average needs to be interpreted carefully and the longer antennas typically used for this technique can obscure very short wavelength modes. The quadruple technique has been particularly successful in determining wavelengths within auroral hiss.

The interferometer technique has been most successful within lower hybrid cavities or spikelets [*LaBelle et al.*, 1986; *Kintner et al.*, 1992; *Vago et al.*, 1992]. An example of wavelength determined within a lower hybrid cavity or spikelet is shown in Figure 3. The antenna orientation is shown in the lower right-hand panel. Electric measurements are made at each of the four outboard sphere pairs. In this case we will only examine the HF12 and HF43 pairs. Above the antenna configuration panel is a time domain plot with the data to be analyzed highlighted. To the left of each time domain panel is the spectrum of the highlighted data. A typical lower hybrid spectrum with a cutoff at 4 kHz is apparent. The lower left two panels show the coherency and phase shift between the signals at each sensor pair (12 and 43). To interpret the cross spectra we will assume that the wave vector is very close to being perpendicular to the geomagnetic field, i.e. flute mode. Above 5 kHz the signal coherency is typically more than 0.8, which is large. The phase shift above 15 kHz is nearly 0, implying that the ob-

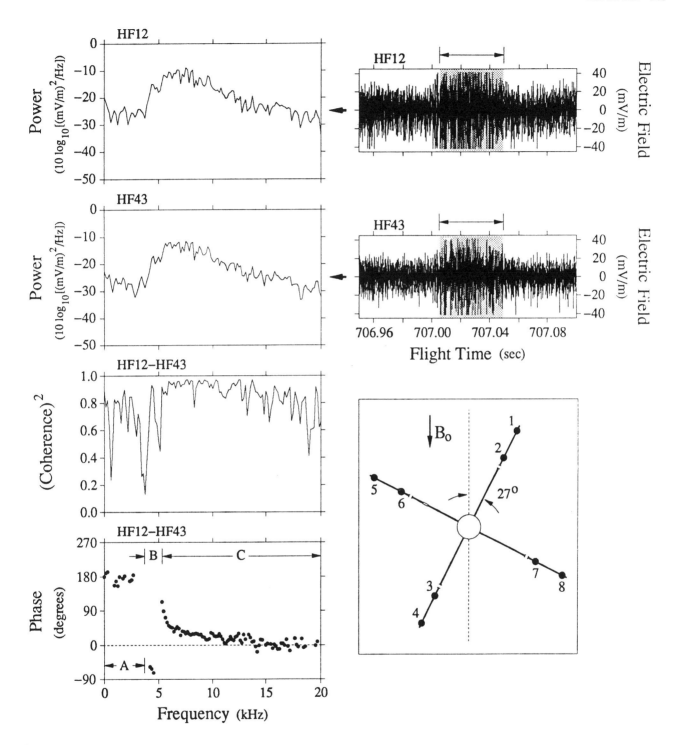

Figure 3. Cross spectra of lower hybrid solitary structures made using the TOPAZ B interferometer. Beginning with the lower right-hand panel and proceeding ccw there is the antenna configuration, the two wave forms, the power spectra corresponding to the two wave forms, the coherency and the phase spectrum.

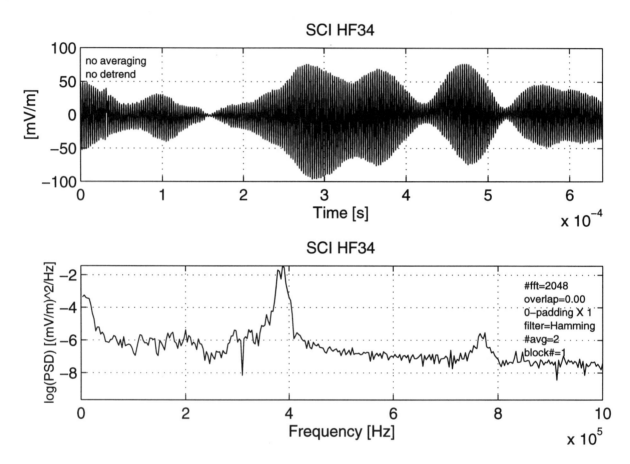

Figure 4. An example of a Langmuir wave snapshot made with the SCIFER sounding rocket. The upper panel shows the wave form while the lower panel shows the power spectrum.

served wavelength is large compared to $\sin(27°)$ x 8 m. In this case the interferometer technique gives a lower bound on the wavelength of roughly 100 m and this response is typical in the presence of EM lower hybrid waves with long wavelengths (1-10 km). Below 15 kHz there is a finite phase shift that increases as the wave frequency decreases until reaching a value of about 90 degrees at 5 kHz. This phase shift corresponds to a wavelength of 360/90 x $\sin(27°)$ x 8 m = 15 m. Below 5 kHz the coherency decreases and is more variable with frequency. Nonetheless, three more phase points are plotted where the coherency exceeds 0.5. These phase shifts exceed 270 degrees, implying a wavelength less than 5 m. Perhaps the most interesting aspect of the phase

spectra occurs below 4 kHz where the phase shift corresponds to 180 degrees. This phase shift implies that the electric field power below 4 kHz was directed radially outward from the payload, that is, the signal below 4 kHz was payload generated.

To summarize, the interferometer technique tells us that this "typical" auroral hiss signal is composed of a range of perpendicular wave numbers varying from $k\rho_{H+} = 0.1$ to $k\rho_{H+} = 10$ with the shortest wavelengths corresponding to the frequencies closest to the lower hybrid frequency. From examining Figure 1, this corresponds to a range spanning the quasi-static whistler mode to the lower hybrid extension on to the H^+ Bernstein mode.

Finally, below the lower hybrid frequency the observed signal is not real. The issue of spacecraft signals that mimic natural signals is mostly ignored in the literature. The interferometer technique demonstrates that this is a risky assumption.

LANGMUIR WAVES

Langmuir waves are arguably the most thoroughly studied (experimentally and theoretically) plasma wave mode in the laboratory and space. They are easily destabilized by electron beams and are clearly shown to exist in many space plasmas, including the solar wind and upstream of planetary shocks (see, for example, *Goldman* [1983] and references therein). Until recently there were few reports of their existence in the magnetosphere, particularly on auroral field lines where auroral electrons should be a plentiful source of free energy. There are four exceptions to this trend, all of which employed "short" antennas and all of which are recent: *McFadden et al.* [1986] with an antenna length of 5.5 m, *Beghin et al.* [1989] with an antenna length of 40 cm, *Kintner et al.* [1995] with an antenna length of 1.2 m, and the example in the next figure. Figure 4 shows an example of Langmuir wave forms observed from the SCIFER sounding rocket using a 30-cm antenna. These wave forms are narrow band and large amplitude. The wave amplitudes in the SCIFER and Freja data sets occasionally exceed 2 V/m$_{p-p}$ and are commonly observed to have amplitudes in excess of 100 mV/m$_{p-p}$.

Somehow, two decades of previous plasma wave measurements failed to detect these large amplitude waves. This paradox can be resolved by recalling that plasma wave antennas often share an antenna with "DC" electric field experiments. To optimize the DC experiments the antennas are made as long as possible, typically tens of meters to in excess of 100 m. Langmuir waves in the auroral zone are expected to have wavelengths of a few meters to a few tens of meters and are likely to be shorter than the typical antenna length. Figure 5 shows the response of double-probe and wire antennas to finite wavelength fields. In both cases the antenna response is greatly reduced at L/λ = 1 and drops sharply for larger ratios, so it appears likely that previous wave instruments with long antennas were not sensitive to these large amplitude signals with shorter wavelengths.

CONCLUDING COMMENT

These previous examples describe observations that are difficult to interpret correctly or may have been missed altogether but nonetheless, are real. To conclude, we would like to present an example of a plasma wave that is not real;

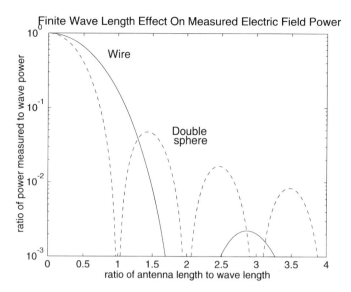

Figure 5. The antenna response to a finite wavelength wave. The solid line is the response of a wire antenna while the dashed line is the response of an equal length double sphere antenna.

instead, it is payload generated. The plasma wave appears to be a double layer that is not strictly a "wave" but a nonlinear structure in the frequency range associated with the ion cyclotron frequencies up to the lower hybrid frequency. Weak double layers were first observed by the electric field instrument on the S3-3 spacecraft within the auroral acceleration region [*Temerin et al.*, 1982]. Later, they were commonly observed by the Viking spacecraft using the δn/n plasma wave interferometer and with the electric field plasma wave instrument [*Boström*, 1988]. They appear as single pulses with time scales of the order of 10-50 ms. In the electric field instrument they appeared either unipolar or bipolar. In the δn/n instrument they appear as simply depletions. The Viking and S3-3 observations of weak double layers are undoubtedly real.

On the other hand, an example of a double layer that undoubtedly is not real is presented in Figure 6. This figure shows a high time resolution plot of three separate differential voltage measurements V21, V34 and V41 made from TOPAZ 3. The numbers refer to spherical sensors spaced along a straight line or boom. The configuration is composed of two outboard sphere pairs, 1-2 and 3-4, and within each sphere pair the spheres are separated by 1 m. The inner spheres, 2 and 3, are separated by 5.5 m with the rocket payload midway between the two spheres. The remaining spheres, 1 and 4, are the outer spheres. The lower right panel of Figure 3 shows the antenna configuration.

The top two panels show a single, roughly 200 mV/m, electric field pulse occurring just after 515.8 s in both the 2-1

TOPAZ III (0–500 Hz)

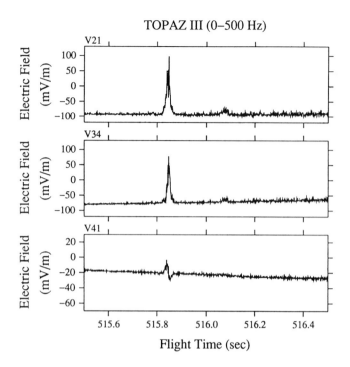

Flight Time (sec)

Figure 6. Evidence of an electric field directed radially outward from the payload, which produces a double layer-like response in a double probe.

and 3-4 sphere pairs. This pulse corresponded to a radially outward single-sided electric field pulse. Note that the two pulses are not exactly simultaneous. About 20 of these pulses were observed during the TOPAZ 3 flight and all occurred within lower hybrid solitary structures. The electric field measured with the outer spheres (V41) is shown in the lower panel where a two-sided electric field pulse is evident. The two-sided pulse can be also be recreated by subtracting the top two panels. We interpret this result as a payload-generated electric field pulse that propagates outward at a velocity faster than the payload velocity but slow enough to arrive at the downstream sphere pair slightly before the upstream sphere pair (the assumed velocity is the order of the ion acoustic velocity). When the pulse is measured by the outer spheres, the radial fields do not exactly cancel out because of the small time differential, and this yields the characteristic double layer signature. Because the double layer signature is produced by an electric field pulse that is radially outward from the payload, we interpret it as a false signature.

In this paper we have outlined the problems associated with "conventional" electric field instruments that measure plasma waves, that is, frequency domain measurements from fast moving platforms with long antennas. This class of measurements is vulnerable to Doppler shifting and broaden-

ing as well as insensitivity to modes with wavelengths the order of or smaller than the antenna length. In addition, wave vector information is difficult to obtain with this kind of instrument, although wavelengths have been determined in some special cases where the relative velocity between the electric field antenna, and the plasma medium is fast compared to the wave phase velocity [*Temerin*, 1978; *Fuselier and Gurnett*, 1984]. Some of these problems can be resolved with multiple sensor instruments or shorter baseline antennas as described in this article. There is no single solution, and the rigorous investigation of plasma waves and wave-particle interactions will require an approach using "conventional" instrumentation as well as interferometers, snapshot receivers, and very short antennas.

Acknowledgements. This research was supported by funding under NASA grants NAG5-691, NAG5-5008, and NAG5-5079 and under ONR grants N00014-92-J-1822 and N00014-95-1-0734.

REFERENCES

André, M., Dispersion surfaces, *J. Plasma Phys., 33*, 1-19, 1985.

Beghin, C., J.L. Rauch, and J.M. Bosqued, Electrostatic plasma waves and HF auroral hiss generated at low altitude, *J. Geophys. Res., 94*, 1359-1378, 1989.

Boström, R., G. Gustafsson, B. Holback, G. Holmgren, H. Koskinen, and P. Kintner, Characteristics of solitary waves and weak double layers in the magnetospheric plasma, *Phys. Rev. Lett., 61*, 82-85, 1988.

Chang, T. and B. Coppi, Lower hybrid acceleration and ion evolution in the suprauroral region, *Geophys. Res. Lett., 8*(12), 1253-1256, 1981.

Ergun, R.E., E, Klementis, C.W. Carlson, J.P. McFadden, and J.H. Clemmons, Wavelength measurement of auroral hiss, *J. Geophys. Res., 96*(A12), 21,299-21,307, 1991.

Fuselier, S.A., and D.A. Gurnett, Short wavelength ion waves upstream of the Earth's bow shock, *J. Geophys. Res., 89*(A1), 91-103, 1984.

Goldman, M.V., Progress and problems in the theory of Type III solar radio emission, *Sol. Phys., 89*, 403, 1983.

Holmgren, G., and P.M. Kintner, Experimental evidence of widespread regions of small-scale plasma irregularities in the magnetosphere, *J. Geophys. Res., 95*(A5), 6015-6023, 1990.

Kintner, P.M., M.C. Kelley, G. Holmgren, H. Koskinen, G. Gustafsson, and J. LaBelle, Detection of spatial density irregularities with the Viking plasma wave interferometer, *Geophys. Res. Lett., 14*, 467-470, 1987.

Kintner, P.M., J.L. Vago, S.W. Chesney, R.L. Arnoldy, K.A. Lynch, C.J. Pollock, and T.E. Moore, Localized lower hybrid acceleration of ionospheric plasma, *Phys. Rev. Lett., 68*(16), 2448-2451, 1992.

Kintner, P.M., J. Bonnell, S. Powell, J.-E. Wahlund and B. Holback, First results from the Freja HF snapshot receiver, *Geophys. Res. Lett., 22*(3), 287-290, 1995.

LaBelle, J. and P.M. Kintner, The measurement of wavelength in space plasmas, *Rev. Geophys., 27*(4), 495-518, 1989.

LaBelle, J., P.M. Kintner, A.W. Yau, and B.A. Whalen, Large amplitude wave packets observed in the ionosphere in association with transverse ion acceleration, *J. Geophys. Res., 91*(A6), 7113-7118, 1986.

McFadden, J. P., C. W. Carlson, and M. H. Boehm, High frequency waves generated by auroral electrons, *J. Geophys. Res., 91*, 12,079, 1986.

Rönmark, K., WHAMP-waves in homogeneous anisotropic multicomponent plasmas, *Kiruna Geophysical Institute report no. 179*, Kiruna, Sweden, 1982.

Temerin, M., The polarization, frequency, and wavelengths of high-latitude turbulence, *J. Geophys. Res., 83*(A6), 2609-2616, 1978.

Temerin, M., K. Cerny, W. Lotko, and F. S. Mozer, Observations of double layers and solitary waves in the auroral plasma, *Phys. Rev. Lett., 48*, 1175, 1982.

Vago, J.L., P.M. Kintner, S.W. Chesney, R.L. Arnoldy, K.A. Lynch, T.E. Moore and C.J. Pollock, Transverse ion acceleration by localized lower hybrid waves in the topside auroral ionosphere, *J. Geophys. Res., 97*(A11), 16,935-16,957, 1992.

Paul M. Kintner, School of Electrical Engineering, 302 Rhodes Hall, Cornell University, Ithaca, NY 14853.

Wave Measurements Using Electrostatic Probes: Accuracy Evaluation by Means of a Multiprobe Technique

A. I. Eriksson and R. Boström

Swedish Institute of Space Physics, Uppsala Division

Ideally, current-biased electrostatic probes provide measurements of voltage differences in a plasma, and voltage-biased (Langmuir) probes give the plasma density and temperature. However, real wave measurements may deviate significantly from the ideal, and a direct interpretation of the data in terms of ideal probe response may lead to large errors. By using a model for the probe-spacecraft-plasma interaction, these errors may be estimated and to some extent compensated for. In particular, the comparison of simultaneous signals from probes in voltage (electric field) mode and probes in Langmuir (current) mode using such a model is very useful for assesing the accuracy of the measurements. We apply this technique to linear and nonlinear waves observed by the Viking satellite in a tenuous plasma, showing that although the probe operations are non-ideal, it is possible to quantify errors and derive information on the real fields in the plasma.

1. INTRODUCTION

Spherical probes on long booms mounted on satellites and sounding rockets are widely used for the study of plasma wave phenomena in space. Probes are either used in voltage (electric field) mode or in Langmuir (current) mode. In voltage mode, the probe is floating freely or is fed with a bias current I_B. Ideally, the observed voltage U_{PQ} between two probes P and Q gives the real voltage Φ_{PQ} in the plasma between those two points. In Langmuir mode, a certain bias potential V_B relative to the spacecraft body is applied to the probe, and the resulting current I_P is measured. In the ideal case, and if the plasma phenomena under study are isothermal, the relative variation in probe current is equal to the relative variations in plasma density n_P at the location of the

probe. Denoting variations by δ and background values by index zero, the ideal measurements of low-frequency voltage and plasma density variations are described by

$$U_{PQ} = \Phi_{PQ} \tag{1}$$

and

$$\frac{\delta I_P}{I_{P0}} = \frac{\delta n_P}{n_{P0}}, \tag{2}$$

respectively.

However, real probe performance may differ from ideal. Potentially important features not included in the equations above include (1) displacement currents (capacitive coupling), (2) the influence of plasma inhomogeneities on voltage mode signals [*Fahleson*, 1967; *Laakso et al.*, 1995], (3) the influence of electric fields on the signal from Langmuir mode probes [*Kelley et al.*, 1975], and (4) rectification and other types of energy transfer in frequency space due to nonlinearities in the probe sheath [*Boehm et al.*, 1994]. If any of these is operating, a naive interpretation of the measured signals may lead to errors in the deduced wave fields (normally

Measurement Techniques in Space Plasmas: Fields
Geophysical Monograph 103

electric field for voltage mode probes and density variation for Langmuir mode probes).

Comparison of wave measurements by voltage and Langmuir probes has mainly been used for deriving additional information on the waves [e. g. *Kelley and Mozer*, 1972], assuming ideal probe operations. By using of a model of the probe-satellite-plasma electrical system, we present a different approach, where the comparison is used to derive information on the real probe performance. The level of sophisticaion of the model limits the types of error sources that can be included: here, we will use a semi-empirical model which allows modelling of all effects listed above. We will show some examples of how the use of this technique gives us possibility to sort out spurious effects in data from the low frequency wave instrument on the Viking satellite.

2. MODELLING WAVE MEASUREMENTS

In this report, we briefly describe the model of the probe-spacecraft-plasma system we will use to study the wave measurements. More details can be found elsewhere [*Eriksson and Boström*, 1995].

2.1 *Probe Theory*

As discussed in the review by *Laframboise and Sonmor* [1993], finding current-voltage relations for sheaths around objects in magnetoplasmas is still a subject of theoretical investigations. To free ourselves from such problems, we adopt a quasi-empirical approach. Langmuir probe sweeps, where the current I_P to the plasma from a probe P is measured as a function of the probe bias voltage V_B, are parametrized by fitting of a function

$$I_P(V_P) = I_e(V_P) + I_{ph}(V_P), \qquad (3)$$

where $V_P = V_{sat} + V_B$ and V_{sat} are the potentials of the probe and the spacecraft relative to the plasma, and

$$I_e(V) = \begin{cases} I_{e0}\left(1 + V/T^*\right), & V > 0 \\[2mm] I_{e0}\exp(V/T^*), & V < 0, \end{cases} \qquad (4)$$

$$I_{e0} = 4\pi r_P^2 e\alpha n \sqrt{\frac{eT^*}{2\pi m_e}} \qquad (5)$$

$$I_{ph}(V) = \begin{cases} -I_{ph,0}\exp(-V/T_{ph}), & V > 0 \\[2mm] -I_{ph,0}, & V < 0 \end{cases} \qquad (6)$$

Here, I_e and I_{ph} are interpreted as the currents due to collection of plasma electrons and emission of photoelectrons, respectively. Currents due to ion collection are negligible for the plasmas we study in this report. The probe radius is r_P, e is the elementary charge, n the plasma number density, and m_e the electron mass. The photoelectron current has empirically been found to be well described by the two parameters $I_{ph,0}$ and T_{ph} in equation (6), with empirical values approximately consistent with the laboratory results of *Grard* [1973]. The form of relation (4) is inspired by the theory for orbital motion limited current collection (OML) in an unmagnetized plasma, where the electron current follows equation (4) with $\alpha = 1$ and T^* being the electron temperature T_e in the plasma. Here, we only use the empirical fact that it is possible to fit probe sweeps to the expressions above, acknowledging the possibility of $\alpha \neq 1$ and $T^* \neq T_e$, thereby freeing ourselves from the constraints of any specific probe theory.

To describe probe interactions with non-stationary phenomena such as waves, a displacement current

$$C_P \frac{dV}{dt} \qquad (7)$$

is added to the expression (3). In the tenuous plasmas to be considered here, the shielding effects in the plasma are small, and the probe capacitance to the plasma C_P can be taken to be the vacuum capacitance of a sphere to infinity, $C_P = 4\pi\epsilon_0 r_P$.

Just as the probes, the satellite body is electrically coupled to the plasma. We treat the satellite as a probe, assuming relations of the type (3) – (7) to hold for the satellite body as well, with a scale factor correcting for the different size. Describing the ratio of typical linear dimensions of the satellite and a probe by D, the material current (3) should scale by area, i. e., as D^2, while the displacement current (7) should scale as D. For Viking, we use $D = 15$.

2.2 *Probe-Satellite-Plasma System*

In order to model wave measurements, we study the complete circuit of the probes and the instrument electronics, the sheaths around probes and spacecraft body, and the ambient plasma with varying density and electric field. Figure 1 shows an example of a circuit including one probe (P2) in Langmuir mode and two probes (P3 and P4) in voltage mode. The extension to any number of probes is straightforward. The shaded areas in the Figure symbolizes the sheaths around probes (circles) and satellite (ellipse) as described by (3) – (7). The only parameter of the instrument electronics we have included is the input capacitance for voltage mode probes, $C_E = 15$ pF, since all other input impedances are neg-

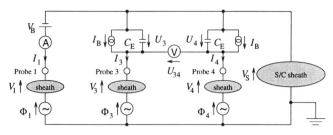

Figure 1. Circuit describing a satellite with one probe in Langmuir mode (P2) and two in voltage mode (P3 and P4).

ligible compared to realistic probe sheath impedances [*Eriksson and Boström*, 1995].

The real fields in the plasma, which are the input to our model, are specified by the voltage in the plasma between the locations of probe number P and the spacecraft body, $\Phi_P(t)$, and by the density at each probe, $n_P(t)$, and at the satellite, $n_{\text{sat}}(t)$. The quantities accesible to measurement are I_2, U_3, and U_4, and the actually sampled signals are $\delta I_2/I_{20}$ and $U_{34} = U_3 - U_4$. By writing down the circuit equations (Kirchoff's laws) for the circuit in Figure 1, we find that the voltages V_P and V_{sat} over the probe and satellite sheaths are given by [*Eriksson and Boström*, 1995]

$$C_{\text{E}}\frac{dV_{\text{S}}}{dt} - (C_{\text{E}} + C_P)\frac{dV_P}{dt} - (1 + \frac{\delta n_P}{n_0})\frac{V_P}{R_{\text{e}}} =$$
$$= (1 + \frac{\delta n_P}{n_0})\frac{T^*}{R_{\text{e}}} - I_{\text{ph}}(V_P) + C_{\text{E}}\frac{d\Phi_P}{dt} + I_{\text{B}}$$
$$(P = 3, 4) \qquad (8)$$

$$([D + 1]C_P + 2C_{\text{E}})\frac{dV_{\text{S}}}{dt} - C_{\text{E}}\frac{dV_3}{dt} - C_{\text{E}}\frac{dV_4}{dt} =$$
$$= D^2 I_{\text{ph}}(V_{\text{S}}) - (D^2\left[1 + \frac{\delta n_{\text{S}}}{n_0}\right] + 1 + \frac{\delta n_1}{n_0})\frac{V_{\text{S}}}{R_{\text{e}}} +$$
$$-D^2\frac{T^*}{R_{\text{e}}}(1 + \frac{\delta n_{\text{S}}}{n_0}) - \frac{T^* + V_{\text{B}} - \Phi_1}{R_{\text{e}}}(1 + \frac{\delta n_1}{n_0}) +$$
$$+C_P\frac{d\Phi_1}{dt} + C_{\text{E}}\frac{d\Phi_3}{dt} + C_{\text{E}}\frac{d\Phi_4}{dt} + 2I_{\text{B}} \qquad (9)$$

for the case of positive probe and satellite potentials, which is appropriate for applications to a tenuous plasma. After solution of these equations, the measured quantities may be found from

$$U_{34} = \Phi_3 - \Phi_4 + V_3 - V_4, \qquad (10)$$

and

$$\frac{\delta I_2}{I} = \frac{\delta n_2}{n_0} +$$
$$- \frac{1 + \frac{\delta n_2}{n_0} + R_{\text{e}}C_P\frac{d}{dt}}{V_{20} + T^*}(\Phi_2 - V_{\text{S}}). \qquad (11)$$

where $R_{\text{e}} = T^*/I_{e0}$ denotes the unperturbed probe sheath resistance in the collection of plasma electrons, which from (5) is

$$R_{\text{e}} = T^*/I_{e0} = \frac{1}{4\pi r_{\text{P}}^2 e\alpha n_0}\sqrt{2\pi m_e T^*/e} \qquad (12)$$

and which approaches the real sheath resistance for probes at high potential.

This mathematical model for how the measured quantities depend on the actual plasma parameters enables us to simulate wave measurements. Due to the nonlinear behaviour of the sheaths, the equations are in general not accessible to analytical solution, necessitating the use of numerical or approximative methods.

3. VIKING APPLICATIONS

The Viking satellite [*Hultqvist*, 1990] was launched in 1986 into a polar orbit. Most measurements are from the altitude region 5,000 – 13,000 km in the northern auroral magnetosphere. The low-frequency wave instrument (V4L) sampled the waveform of two signals (voltage difference or probe current) from a set of four spherical probes (radius 5 cm) mounted on 40 m wire booms in the spin plane (spin rate 3 rpm). Two probes (numbered 3 and 4) were always kept in voltage mode, while the other two (1 and 2) could be used in Langmuir or voltage mode. The sampling frequency was 428 or 856 samples/s, covering the proton cyclotron and sometimes also the lower hybrid frequency in the plasma. In addition to these wave (band pass filtered) signals, the DC (low pass filtered) probe current to the Langmuir probes was sampled at 107 samples/s. The bias voltage was swept at regular intervals (normally five minutes), providing the probe current-voltage characteristic.

We now apply the model described above to two examples of Viking wave observations. First, we use an approximative method to study a Viking observation of small-amplitude electrostatic ion cyclotron waves (EICs). The second example is the detection of nonlinear solitary electrostatic waves (SWs), where we solve the full equations (8) – (9) numerically. Both examples are taken from the same Viking orbit, within 15 minutes from each other. Data are from July 31, 1986, taken around magnetic local time 03.00, invariant latitude 81 degrees, and altitude 9,000 km. The plasma is very tenuous, around 1 cm^{-3}, which is a rather extreme situation for probe measurements.

3.1 *Linear Waves: Electrostatic Ion Cyclotron Waves*

Electrostatic ion (proton) cyclotron waves (EICs) were frequently encountered by Viking [*André et al.*, 1987].

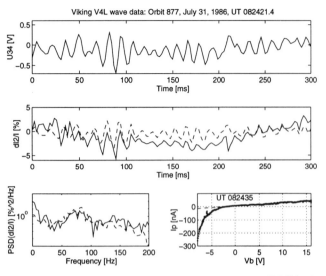

Figure 2. Electrostatic proton cyclotron waves (EICs) observed by Viking. (a) Signal observed by voltage probes. (b) Actual (solid) and simulated (dashed) signal oberved by Langmuir probe. (c) Power spectra of signals in (b). (d) Nearby Langmuir probe sweep.

An example is shown in Figure 2, showing 300 ms of Viking data in a region where the proton cyclotron frequency was 65 Hz. Two probes (3 and 4) were in voltage mode, providing the signal $U_{34} = U_3 - U_4$ seen in Figure 2a. Probe 2 was in Langmuir mode. The relative variation of the current from this probe, $\delta I_2/I_{20}$, is plotted as the solid curves in Figure 2b (time series) and 2c (spectrum). The signature of a wave with frequency just above the proton gyrofrequency is clearly seen in both signals. The low density in the region of measurement is witnessed by the probe bias sweep in Figure 2d. If interpreted using unmagnetized OML probe theory [*Mott-Smith and Langmuir*, 1926], the density estimate from the sweep is below 1 cm^{-3}.

In this application, we may neglect the variation in sheath voltages V_3, V_4, and V_{sat}. This amounts to a linearization of the equations above for small amplitude waves, and an assumption of a dominantly resistive coupling of the voltage mode probes and the satellite to the plasma. Both these assumptions are reasonable in this application [*Eriksson and Boström*, 1995]. Equations (10) and (11) condenses to

$$U_{34} = \Phi_3 - \Phi_4 \qquad (13)$$

and

$$\frac{\delta I_2}{I_{20}} = \frac{\delta n_2}{n_0} - \frac{1}{V_{20} + T^*}\left[\Phi_2 + R_{\text{e}}\,C_{\text{P}}\frac{d\Phi_2}{dt}\right]. \qquad (14)$$

Thus, the voltage measurement is taken to be ideal,

while the estimate of density fluctuations using the probe current will be contaminated by the voltage variations. However, we are now in a position where we can use the measurement of the electric field U_{34} to estimate this contamination of the probe current fluctuations, assuming that the wave electric field is dominantly perpendicular. The satellite spin axis was nearly perpendicular to the magnetic field (85 degrees) at the time of the observations in Figure 2, implying that all probes see essentially the same component of the perpendicular electric field. With knowledge of the angles to the magnetic field of all booms, we can therefore use U_{34} to reconstruct the perpendicular electric field component in the spin plane, and thus calculate the voltage variation Φ_2 at the density probe. By putting $\delta n/n = 0$ in equation (14), we may calculate the part of the probe current fluctuation which is not caused by an actual density variation.

The dashed lines in Figures 2b and 2c show the result of such a procedure, using the vacuum value of 6 pF for C_{P}, $R_{\text{e}} = 710$ MΩ from the sweep, and $I_{20} = 25$ nA from the instantaneous DC probe current (not shown). The good agreement between the observed current fluctuation (solid) and what is calculated from the voltage fluctuations is obvious, in phase as well as amplitude. Hence, the observed Langmuir probe current variations near the ion cyclotron frequency should not be interpreted as due to real density fluctuations in the wave. The discrepancy seen between the two curves in Figure 2 at very low frequencies may indicate the presence of real density fluctuations, but may also be due to variations in the satellite potential, which were neglected in deriving equation (14).

That a good reconstruction is possible certainly shows that a naive interpretation of $\delta I/I$ in terms of density fluctuation leads to erroneous conclusions in this very tenuous plasma. However, it also shows that the model we have for the measurements works. In particular, equation (13) must be approximately correct, so we also have a qualitative verification of the voltage measurement.

3.2 *Nonlinear Waves: Solitary Electrostatic Waves*

Solitary waves (SWs), some of which show the characteristics of weak double layers, are often observed on Viking in the same regions as EICs [*Boström et al.*, 1988, *Mälkki et al.*, 1993]. Figure 3 shows an example of such structures, observed little more than a minute after the waves in Figure 2. A theoretical treatment of the measurement of SWs poses an interesting challenge, since a linearized treatment is no longer possible. As is seen

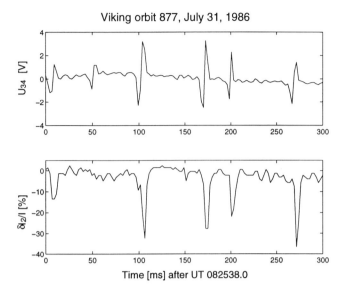

Viking orbit 877, July 31, 1986

Time [ms] after UT 082538.0

Figure 3. Solitary waves (SWs) observed by voltage and Langmuir probes on Viking.

after which the actually measured quantities were calculated from equations (10) and (11). This procedure was repeated with different values of pulse parameters (amplitude in density and voltage, speed, width) until a good fit was obtained.

As is seen in Figure 4, it was possible to obtain a good fit, with speed 20 km/s, amplitudes 4 V and 50 % in voltage and density depletion, respectively, and width 50 m. The number of fitted parameters is rather large, around one fourth of the number of data points in the plot, and the actual significance of the fit may not be obvious. We certainly do not intend to present the fitted values as exact results, but having found the applicabil-

in Figure 3, the probe current may fluctuate by several tens of per cent, and the voltage variation can be several volts. A study of SW measurements [*Eriksson et al.*, 1997] therefore requires the numerical integration of the complete set of equations (8) – (9).

Figure 4 shows a simulation of the measurement of the SW seen around 0.1 s in Figure 3. The SW is modelled as a density depleted pulse of negative potential propagating upwards along the magnetic field, which are the typical characteristics of SWs [*Mälkki et al.*, 1993]. The SW is assumed to be isothermal, which is reasonable in a collission-free plasma. The forms of the SW potential and density depletion were taken to be described by a Gaussian function of the field-aligned coordinate, while the perpendicular dimensions are assumed infinite. The localization in space of the booms is seen in Figure 4d. The SW first passes probe 4, reach the satellite (at $t = 0$), probe 2, and finally passes probe 3. Several values for the amplitude and width of the Gaussian, as well as for the speed, were tested in order to get a good fit between the observed U_{34} and $\delta I/I$ (circles in Figures 4c and 4d) and model (solid curve). Knowing the position of all probes with respect to the satellite and the magnetic field, the potential perturbation Φ_P at all probes is calculated and used as input to the system of equations (8) – (9). Several parameters describing the probe characteristic appears in these equations, and were obtained from the probe bias sweep in Figure 2d: $\alpha n = 0.5$ cm^{-3}, $T^* = 0.2$ eV, $I_{ph,0} = 530$ nA, and $T_{ph} = 1.8$ eV. The equations (8) – (9) were then integrated numerically,

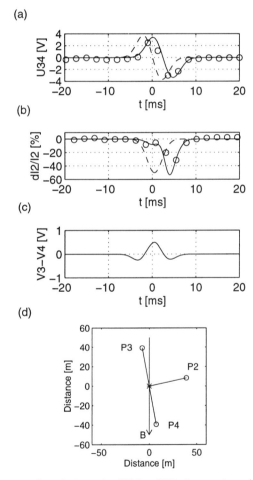

Figure 4. Simulation of a Viking SW observation. (a) Signal to voltage probes: measured (circles) compared to simulated response of actual (solid) and ideal (dashed) probes. (b) Signal to Langmuir probe: measured (circles) compared to simulated response of actual (solid) and ideal (dashed) probe. (c) Simulated error in voltage measurement. (d) Configuration of satellite in space.

ity of the model to several cases we trust in the general results. In particular, it was impossible to obtain anything near a fit if the amplitude of either the density perturbation or the potential was put to zero [*Eriksson et al.*, 1997]. This is particularly important since it shows that even though the probe current to a probe in a tenuous plasma is strongly influenced by voltage variations, as we witnessed in the study of EICs above, the careful analysis of two signals from probes in different bias modes can still reveal an underlying density fluctuation.

The dashed curves in Figures 4a and 4b show what should have been observed by an ideal instrument, following equations (1) and (2). The lag between the curves is the group delay time of the filters in the instrument, present in the data and included in the simulation result (solid curve). Figure 4c shows a quantitative evaluation of the error in the voltage measurement, as can be seen by comparison to equation (10).

4. CONCLUSION

Our main point in this paper can be stated as follows: although voltage as well as current measurements by use of electrostatic probes can be "contaminated" in such a way that they must not be directly interpreted as measurements of plasma density variations and electric fields, respectively, the simultaneous use of probes in voltage and Langmuir mode makes it possible to identify such problems. Moreover, a detailed analysis combining the two signals in a reasonable model of the probe-spacecraft-plasma system makes it possible to quantify the errors and reveal information on the plasma parameters that cannot directly be acquired from any of the signals on their own.

In both cases studied above, a significant fraction of the signals to the current probes could be explained as effects of the electric field, so a direct interpretation in terms of density fluctuations was fallacious. On the other hand, the electric field (voltage) measurement was seen to be accurate. We emphasize that this is in not a general result, but a particular effect of the tenuous plasma we have chosen to study. In other plasma conditions, the situation may be reversed. The message we want to convey is not that electric field measurements are good and density measurements bad – what works and what does not work depends on the plasma conditions. The important point is that *combined* analysis of *both* measurements can tell what works and what does not.

Observations of strong lower hybrid waves and lower hybrid cavities by Freja have been investigated with

methods similar to those applied above [*Eriksson and Boström*, 1995]. The wave instrument on Freja [*Holback et al.*, 1994] used six spherical probes in the spin plane, making it possible to measure two orthogonal electric field components by symmetrical double probe configurations and at the same time observe the probe current on two other probes. Multiprobe measurements will also be possible on the Phoenix/Cluster satellites, where the EFW instrument [*Gustafsson et al.*, 1997] can sample four signals from four probes in voltage or current mode. As the interest in detailed and accurate wave measurements only seems to grow with time, we suggest that the use of multiple probe measurements combined with measurement simulations may be of fundamental importance for future missions.

REFERENCES

André, M., H. Koskinen, G. Gustafsson, and R. Lundin, Ion waves and upgoing ion beams observed by the Viking satellite, *Geophys. Res. Lett.*, *14*, 463, 1987.

Boehm, M. H., C. W. Carlson, J. P. McFadden, J. H. Clemmons, R. E. Ergun, and F. S. Mozer, Wave rectification in plasma sheaths surrounding electric field antennas, *J. Geophys. Res.*, *99*, 21361–21374, 1994.

Boström, R., G. Gustafsson, B. Holback, G. Holmgren, H. Koskinen, and P. Kintner, Characteristics of solitary waves and weak double layers in the magnetospheric plasma, *Phys. Rev. Lett.*, *61*, 82–85, 1988.

Eriksson, A. I., and R. Boström, Measurements of plasma density fluctuations and electric wave fields using spherical electrostatic probes, IRF Scientific Report 220, Swedish Institute of Space Physics, Uppsala, Sweden, 1995.

Eriksson, A. I., A. M. Mälkki, P. O. Dovner, R. Boström, and G. Holmgren, A statistical study of auroral solitary waves and weak double layers. 2. Measurement accuracy and ambient plasma density, *J. Geophys. Res.*, in press, 1997.

Fahleson, U., Theory of electric field measurements conducted in the magnetosphere with electric probes, *Space Sci. Rev.*, *7*, 238–262, 1967.

Grard, R. J. L., Properties of satellite photoelectron sheath derived from photoemission laboratory studies, *J. Geophys. Res.*, *78*, 2885–2906, 1973.

Gustafsson, G., R. Boström, B. Holback, G. Holmgren, A. Lundgren, K. Stasiewicz, L. Åhlén, F. S. Mozer, D. Pankow, P. Harvey, P. Berg, R. Ulrich, A. Pedersen, R. Schmidt, A. Butler, A. W. C. Fransen, D. Klinge, M. Thomsen, C.-G. Fälthammar, P.-A. Lindqvist, S. Christenson, J. Holtet, B. Lybekk, T. A. Sten, P. Tanskanen, K. Lappalainen, and J. Wygant, The electric field and wave experiment for the Cluster mission, *Space. Sci. Rev.*, *79*, 137–156, 1997.

Holback, B., S.-E. Jansson, L. Åhlén, G. Lundgren, L. Lyngdahl, S. Powell, and A. Meyer, The Freja wave and plasma density experiment, *Space Sci. Rev.*, *70*, 577–592, 1994.

Hultqvist, B., The Swedish satellite project Viking, *J. Geophys. Res.*, *95*, 5749–5752, 1990.

Kelley, M. C., and F. S. Mozer, A technique for making dispersion relation measurements of electrostatic waves, *J. Geophys. Res.*, *77*, 6900–6903, 1972.

Kelley, M. C., C. W. Carlson, and F. S. Mozer, Application of electric field and fast Langmuir probes for the in situ observation of electrostatic waves and irregularities, in *Proceedings of the NRL Symposium on the Effect of the Ionosphere on Space Systems and Communications*, U. S. Government Printing Office, Washington, D. C. pp. 453–459, 1975.

Laakso, H., T. L. Aggson, and R. Pfaff Jr., Plasma gradient effects on double probe measurements in the magnetosphere, *Ann. Geophysicae*, *13*, 130–146, 1995.

Laframboise, J. G., and L. J. Sonmor, Current collection by probes and electrodes in space magnetoplasmas: A review, *J. Geophys. Res.*, *98*, 337–357, 1993.

Mälkki, A., A. I. Eriksson, P.-O. Dovner, R. Boström, B. Holback, G. Holmgren, and H. E. J. Koskinen, A statistical survey of auroral solitary waves and weak double layers: 1. Occurrence and net voltage, *J. Geophys. Res.*, *98*, 15521–15530, 1993.

Mott-Smith, H. M., and I. Langmuir, The theory of collectors in gaseous discharges, *Phys. Rev.*, *28*, 727–763, 1926.

A. I. Eriksson and R. Boström, Swedish Institute of Space Physics, Uppsala Division, S-755 91 Uppsala, Sweden. (e-mail: aie@irfu.se; rb@irfu.se)

Mutual-Impedance Techniques for Space Plasma Measurements

L. R. O. Storey

Quartier Luchène, 84160 Cucuron, France

A review is given of measurement techniques for space plasmas based on the mutual impedance of a pair of dipole antennas, which preferably should be of the double-sphere variety. Besides electron density and temperature, capability of measuring the following plasma properties in space has been demonstrated to different degrees: thermal electron drift velocity parallel and perpendicular to the magnetic field; lower hybrid resonance frequency; plasma resistivity at ELF. More research, both theoretical and experimental, is needed in order to fully realize the potential of these techniques.

1. INTRODUCTION

In the late 1960s, the author and some of his colleagues undertook a research program to develop new techniques for space plasma measurements. Two types of measurement were targeted: those of parameters, such as the lower hybrid frequency, for which no means of measurement existed at the time; and those of parameters such as electron density and temperature that could be measured already by other means, but not accurately enough for all purposes. The techniques concerned were based on the coupling between pairs of electric dipole antennas.

The development of these new techniques drew upon much previous work by other people. On the theoretical side, the main input came from the linear kinetic theory of plasma waves, as applied to radiation from antennas in warm plasmas. On the experimental side, it came from the very different field of solid-earth geophysics, where the standard four-electrode method for measuring the resistivity of the soil revealed the importance of using the mutual impedance (MI) as the measure of electric coupling.

In attempts to model the coupling between two antennas in isotropic or anisotropic plasmas, the simple cold-plasma

theory was used initially [*Storey et al.*, 1969]. Warm-plasma wave theory was first applied to plasmas with no magnetic field [*Rooy et al.*, 1972], and only later to magnetoplasmas.

The earliest experiments in space were made on rockets, the first of which carried an antenna system of the kind developed for the French FR-1 satellite [*Storey*, 1965]. This experiment was designed to measure the lower hybrid frequency versus height in the ionosphere, which it succeeded in doing [*Béghin*, 1971]. The second rocket carried two MI experiments, one an improved version of the first [*Béghin and Debrie*, 1972], while the other, using a similar but smaller antenna system, was designed to measure electron density and temperature [*Chassériaux et al.*, 1972]; both were successful. Many more rocket experiments have been made since then, and some of them are discussed in the present paper.

The proven aptitude of the MI method for measuring electron density and temperature led to its use on satellites, initially on the ESA satellite GEOS-1 [*Décréau et al.*, 1978a, 1978b], then on GEOS-2, on the Soviet satellite Aureol-3 also known as Arcad-3 [*Béghin et al.*, 1982, 1983], and on the Swedish satellite Viking [*Bahnsen et al.*, 1988].

During the 1970s, the space experiments were supported by laboratory experiments in a space simulation chamber. The possibility of testing the predictions of the theory and the performance of the instruments in laboratory plasmas helped very much in the design of the flight models.

The present paper reviews this work, which is hard to do in an orderly way because many of the instruments were able to measure more than one plasma parameter. Rather than

Measurement Techniques in Space Plasmas: Fields
Geophysical Monograph 103

classify the subject matter in terms of the measuring instruments or of the parameters measured, we have preferred to do so according to the various plasma resonances around which the measurements were made. As a function of frequency, the MI tends to have maxima at or near the plasma resonances, so that is where the received signals are strongest with respect to the noise. The MI maxima are often also referred to as "resonances" in the literature, so here they will be called *MI resonances* to distinguish them from the plasma resonances. Each of the instruments described below was designed to work with just one of the plasma resonances, in most cases.

The paper is structured as follows. Section 2 presents the basic principles, firstly the theory of electric antennas in warm plasmas, and secondly the experimental MI technique, the latter briefly since it is described in several of the papers cited above. Then the various plasma resonances are considered in turn, beginning in section 3 with the measurements that can be made around the plasma resonance in an isotropic plasma. The resonances of a magnetoplasma are treated in sections 4 through 7, which deal successively with the upper hybrid, lower hybrid, lower oblique, and upper oblique resonances. Section 8 concludes the paper, pointing out the needs for further research. We use the SI system of units.

2. BASIC PRINCIPLES

2.1. Antennas in Warm Plasmas

When an antenna in a plasma is excited electrically, it radiates both electromagnetic (EM) and electrostatic (ES) waves, and these can be received on a second antenna elsewhere in the plasma. If the distance between the transmitting and receiving antennas is sufficiently large, the ES waves, which are prone to collisionless damping, may die out before they reach the receiving antenna. Then the coupling between the two antennas is due essentially to the EM waves, and it can be studied well enough with cold-plasma theory which neglects the thermal motion of the charged particles. However, when the two antennas are close to each other, the ES waves contribute to the coupling and warm-plasma (kinetic) theory must be used in any attempt to study it quantitatively. The difficult part is the calculation of the fields radiated by the transmitting antenna: granted some reasonable assumptions, the signals they induce in the receiving antenna are relatively easy to calculate.

The kinetic theory of EM and ES waves propagating in warm, uniform magnetoplasmas under linear conditions was developed in the 1950s and '60s by *Bernstein* [1958] and others; for a modern review of the subject, see *Stix* [1992]. Its most important result is the expression for the conductivity tensor $\sigma(k, \omega)$, or equivalently for the dielectric tensor

$K(k, \omega)$, either of which specifies the response of the plasma to an applied electric field of wavenumber k and angular frequency ω. This function stands for the plasma both in the dispersion equations for freely propagating waves, and in the equations that govern the fields created by sources such as antennas.

A broad theoretical treatment of electromagnetic processes in dispersive media, based on the dielectric tensor, has been published by *Melrose and McPhedran* [1991]. However, their discussion of warm plasmas covers the isotropic case only. Radiation phenomena in plasmas, including radiation from antennas in warm magnetoplasmas, are surveyed in a recent book by *Ohnuma* [1994].

In the plasma near to a transmitting antenna, the electric field is well represented by the so-called quasi-static approximation. This approximation is obtained by setting the free-space permeability μ_0 to zero in the basic equations, which makes the speed of light infinite. The electric field can then be written simply as the gradient of a scalar potential, $E = -\nabla\phi$, and it is found by solving Poisson's equation instead of the two Maxwell curl equations. The quasi-static approximation will be used implicitly from here onwards.

In this approximation, the simplest conceivable model of a transmitting antenna is not a point dipole but a point monopole, i.e., an alternating point charge. Of course, a time-varying point charge is a non-physical concept because charge has to be conserved, and therefore, strictly speaking, the model should include a thin wire carrying charge to and from the point concerned. In the quasi-static approximation, however, the current in the wire does not radiate, so the presence of the wire can be neglected and the electric potential and field computed solely from the charge at the end of it.

Alternatively, this simple antenna can be regarded as a point source of current rather than of charge, which is more convenient for present purposes. With a source of current $I = I_0 \exp(-i\omega_0 t)$ at the origin of the co-ordinates, the potential at the point r in the plasma at the time t is

$$\phi(r, t) = I_0 \zeta(r, \omega_0) \exp(-i\omega_0 t) \qquad (1)$$

where $\zeta(r, \omega)$ is a scalar Green's function with the dimensions of an impedance. This function is obtained by inverting its spatial Fourier transform, which is given in terms of the dielectric tensor by

$$\zeta(k, \omega) = -[i\omega\epsilon_0 k \cdot K(k, \omega) \cdot k]^{-1} \qquad (2)$$

The constant ϵ_0 is the permittivity of free space, while the definitions of the spatial and temporal Fourier transforms are those adopted by *Melrose and McPhedran* [1991]. These two basic equations were used in most of the theoretical calculations of mutual impedance cited below.

2.2. Mutual-Impedance Techniques

Figure 1 illustrates the mutual-impedance (MI) technique as used for space plasma measurements. The sensor consists of two double-sphere antennas, the electrodes of which are small compared with the distances between them. The figure shows these four electrodes lying at the corners of a square, which is a common layout but not the only one in use. Of the two antennas, one is excited by a signal source that serves as the transmitter, while the other is connected to a receiver, and electric signals are transmitted from the one to the other through the plasma. The transmitter measures the current I that it supplies to its antenna. The receiver has an input impedance high enough that the potential difference $\Delta\phi$ between the two electrodes of its antenna is effectively measured on open circuit. These data are used to compute the MI, defined as $Z \equiv \Delta\phi/I$. Plasma parameters are determined by interpreting measurements of this quantity, generally made as a function of frequency around one of the resonance frequencies of the plasma.

The MI technique is a transposition to space of the standard four-electrode method used in geophysical prospecting for measuring the resistivity of the earth. The rationale for this transposition has been given by *Storey et al.* [1969].

3. PLASMA RESONANCE

A plasma in thermal equilibrium is electrically isotropic in the absence of a magnetic field, and even in the presence of a field it is approximately so whenever the plasma frequency is much larger than the electron gyrofrequency. In near-Earth space, this condition is roughly satisfied at times in the low-latitude F region or in the remote plasmasphere; it is satisfied much better and more commonly in some planetary ionospheres such as those of Venus and Mars.

In an isotropic plasma, the mutual impedance of a sensor such the one shown in figure 1 exhibits a maximum at the electron plasma frequency, provided that the sensor is large enough (see below). This maximum is due to excitation of the

plasma resonance, and a measurement of its frequency yields the electron density.

The electron temperature can be obtained from the curve of MI versus frequency. Figure 8 of *Rooy et al.* [1972] gives examples of its theoretical form, computed from equation (1) on the assumption that the four electrodes can be approximated as points. The different curves in these figures correspond to different values of the parameter $L \equiv l/\lambda_D$, where l is the length of the side of the square in figure 1 of the present paper and λ_D is the Debye length. The smaller the overall dimensions of the sensor system, the more severely the MI resonance at the plasma frequency is damped by collisionless processes. To get a strong, sharp MI resonance, the sensor should be made much larger than a Debye length ($L \gg 1$). Then, by finding the theoretical curve that best fits the measured curve of $Z(\omega)$ we can determine L, and hence λ_D, and hence the electron temperature. Further examples of these curves have been given by *Béghin* [1995].

The first such measurements in space were made from a rocket in the equatorial ionosphere [*Chassériaux et al.,* 1972]. The electrodes of the sensor were metal spheres 3 cm in diameter, at the corners of a square 17 cm on the side. When analyzing the data, corrections had to be made for the effects of the Earth's magnetic field. The results showed that the electron temperature in the lower E region was only a few hundred degrees, in agreement with geophysical theory but contradicting some previous measurements made with Langmuir probes.

The magnetospheric satellites GEOS-1 and GEOS-2 both carried MI probes to measure electron density and temperature, using a sensor with the electrodes laid out more or less in line, as shown in figure 1 of *Décréau et al.* [1978b]. The graphs of figure 2 of *Décréau et al.* [1978a] compare some theoretical and experimental curves of $|Z(\omega)|$; the good agreement is evidence that the measurements are correct.

4. UPPER HYBRID RESONANCE

In a magnetoplasma like that of the Earth's ionosphere, the curve of MI versus frequency often has a higher maximum near the upper hybrid frequency than near the plasma frequency [*Chassériaux et al.,* 1972]. New modes of wave propagation, such as the Bernstein modes, complicate the response curves of the probes. Hence the experimentally measured curves are harder to interpret than they are in the case of an isotropic plasma.

Figure 1 of *Debrie and Thiel* [1981] shows results for $|Z(\omega)|$ from a Franco-Soviet rocket experiment performed in 1974 in the polar ionosphere, using an MI probe with its four electrodes in line. The payload was not stabilized, but the data in the figure are for selected cases where the line of the elec-

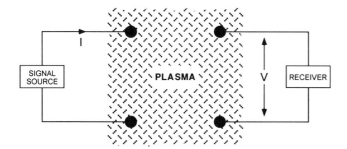

Figure 1. Illustrating the principles of the mutual-impedance techniques.

trodes happened to be perpendicular to the magnetic field. Experiment and theory agree reasonably well at frequencies above the upper hybrid resonance frequency f_{UH}, but there are significant disagreements in the values at the resonance itself and at some lower frequencies.

Another rocket experiment with an MI probe was performed in 1979 as part of the West-German program Porcupine of research on auroral physics during the International Magnetospheric Study. On this occasion the sensor had a rectangular layout with its plane perpendicular to the payload spin axis, which was stabilized parallel to the magnetic field. The electronics included a cross-spectrum analyzer, so that the probe could be operated in a passive mode as well as the usual active mode. Evidence was found for thermal disequilibrium of the plasma, in the form of a small ($\sim$ 1%) population of suprathermal electrons with an anisotropic distribution function, their perpendicular temperature being about four times their parallel temperature [*Pottelette et al.,* 1981; *Pottelette and Illiano,* 1982].

5. LOWER HYBRID RESONANCE

The main interest of being able to measure the lower hybrid frequency f_{LH} is that it is related to the ionic composition of the plasma. A measurement of f_{LH} by an MI probe could be used to check the composition indicated by a mass spectrometer, and might in itself be sufficient information about the composition for some purposes.

As was mentioned in section 1, the lower hybrid resonance was observed in the first rocket experiment with an MI probe. The sensor comprised four metal spheres, each 8 cm in diameter, at the corners of a square of side 2.5 m. The variation of f_{LH} with altitude plainly revealed the transition from molecular ions in the E region to atomic oxygen in the F region [*Béghin,* 1971].

The second such experiment, using a similar sensor, was designed to improve the observation of this MI resonance. Figure 13 of *Béghin and Debrie* [1972] shows one set of measurements of the modulus of the MI as a function of frequency, taken at a moment when the plane of the sensor was perpendicular to the magnetic field.

Besides its relationship with the ionic composition, another interesting feature of the MI resonance at the lower hybrid frequency is its sensitivity to cross-field motion of the plasma. This effect was predicted theoretically before being observed in the first of the two experiments mentioned above. In principle, it could be used to measure the cross-field drift velocity of the plasma with respect to the sensor, and hence the perpendicular electric field; how this might be done has been discussed by *Pottelette* [1972], using cold-plasma theory.

6. LOWER OBLIQUE RESONANCE

The lower oblique resonance exists in a band extending upwards from the lower hybrid frequency to the plasma frequency or the electron gyrofrequency, whichever is the lesser. In this frequency range, the potential ϕ created by a point source of alternating current, as given by equation (1), has the following peculiarity: at a fixed distance r ($\equiv |r|$) from the source, the variation of ϕ versus the angle between the radius vector r and the magnetic field exhibits a maximum at a certain angle depending on the frequency and on the plasma parameters; in cold-plasma theory, and for a collisionless plasma, the potential is infinite at this angle [*Kuehl,* 1962]. At a fixed frequency the loci of all these maxima, for different distances and different azimuthal positions around the field, form a pair of cones, the so-called resonance cones, with their vertices at the source and their axes parallel to the field, which is why this type of resonance is also known as a cone resonance.

Fisher and Gould [1970] were the first to study kinetic effects on MI resonance cones. They showed theoretically that in a warm plasma the potential stays finite on the cone while subsidiary maxima appear on one side of the main maximum; they also demonstrated these phenomena in laboratory plasma experiments.

In experiments on resonance cones, the sensor commonly consists of a pair of monopoles rather than a pair of dipoles. Signals are transmitted from one small electrode, generally spherical, and are received on another. The advantage of this arrangement is that only one propagation path exists, at just one angle to the magnetic field. The disadvantage is that actually the sensor has a third electrode, namely the wall of the plasma chamber in the laboratory or the spacecraft body in space, which is shared by the transmitting and receiving antennas. Hence, the impedance of the sheath between the plasma and the chamber wall or spacecraft body is added to the mutual impedance measured between the two monopoles. Experimenters customarily ignore this source of error.

Most space experiments have been made with sensors of this type. Examples are the rocket experiments of *Gonfalone* [1974], of *Michel et al.* [1975], and of *Thiemann et al.* [1988]. An exception is the experiment of *Koons et al.* [1974] on the polar orbiting satellite OV1-20S, which used short dipole antennas for both transmission and reception. All of these experiments were made at fixed frequency, and used the spin of the spacecraft to vary the direction of the propagation path. The data from them were analyzed to yield electron densities and temperatures.

In my opinion, however, the feature of MI resonance cones that is of most interest for space plasma measurements is their sensitivity to drift motion of the thermal electrons parallel to the magnetic field; they are sensitive to cross-field electron drift as well, but this drift motion can be measured in other ways. A double-monopole probe can readily detect either of these types of drift because in both cases the MI becomes non-reciprocal, meaning that it changes when the direction of transmission is reversed. The non-reciprocity due to parallel electron drift motion was predicted theoretically by *Storey and Pottelette* [1971], and has been observed in laboratory plasmas [*Illiano and Pottelette,* 1979; *Boswell and Thiel,* 1981].

In view of the interest in measuring the thermal electron contribution to field-aligned currents, various attempts have been made to observe this non-reciprocity effect in space, but so far unsuccessfully. Either no such effect was observed [*Koons et al.,* 1974], or the small one expected was masked by a much larger one that did not behave as the theory predicts [*Michel et al.,* 1975; *Storey and Thiel,* 1984]. For over a decade this perturbation stayed unexplained, but recently it has been identified as being due to the $E \times B$ electron drift associated with the permanent electric field in the presheath around the spacecraft body [*Rohde et al.,* 1993]; the inferred fields (~ 0.5 V.m^{-1}) are surprisingly large, however.

7. UPPER OBLIQUE RESONANCE

The upper oblique MI resonance exists in a band extending downwards from the upper hybrid frequency to the plasma frequency or the electron gyrofrequency, whichever is the greater. Qualitatively, its properties resemble those of the lower oblique MI resonance, but quantitatively they are more promising for the measurement of thermal electron drifts, for the following reasons: for a given separation between the source and the receiver, the MI resonance peak is sharper; and, for a given ratio of drift velocity to the electron thermal velocity, the shift of the peak is greater. Figures 2 and 4 of *Pierre et al.* [1992] illustrate these points; theoretical predictions are compared with the results from a laboratory plasma experiment in which the upper oblique resonance was excited in the presence of a field-aligned current. The plasma frequency was higher than the electron gyrofrequency, and the different curves refer to different emitted frequencies, extending from (a) 10% to (e) 25% above the plasma frequency. The asymmetries in the curves with respect to the 90° direction, in particular the displacement of the single peak in curve (a) from this direction, are attributable to field-aligned drifts of the thermal electrons. In space, such an experiment should be capable of measuring the velocity of the drift motion. This is for the future.

8. CONCLUSION

From the work reviewed above, it is fair to conclude that mutual-impedance techniques offer a promising means for making space plasma measurements, but that their potential is not yet fully realized. Further research is needed in order to attain the objectives announced at the outset, namely to improve the accuracy of measurement for the quantities that are already accessible, and to develop methods for measuring quantities that are not fully accessible at the present time. In the latter category, the use of an MI technique for measuring plasma resistivity at ELF is described in a companion paper [*Storey and Cairó,* this volume]. Suggestions as to what more research is needed are made in a longer version of this paper, available from the author on request.

Acknowledgments. My work and that of my colleagues was supported in France by the Centre National de le Recherche Scientifique and by the Centre National d'Etudes Spatiales. I thank the many other authors who told me about their work, and the reviewers for their helpful comments.

REFERENCES

Bahnsen, A., M. Jespersen, E. Ungstrup, R. Pottelette, M. Malingre, P. M. E. Décréau, M. Hamelin, H. de Feraudy, S. Perraut, and B. M. Pedersen, First VIKING results: high frequency waves, *Physica Scripta, 37,* 469–474, 1988.

Béghin, C., Excitation de la résonance hybride basse (LHR) par sonde quadripolaire a bord d'une fusée, in *Space Research XI,* edited by K. Y. A. Kondryatev, M. J. Rycroft, and C. Sagan, pp. 1071–1078, Akademie-Verlag, Berlin, 1971.

Béghin, C., Series expansion of electrostatic potential radiated by a point source in isotropic Maxwellian plasma, *Radio Sci., 30,* 307–322, 1995.

Béghin, C., J. J. Berthelier, R. Debrie, Yu. I. Galperin, V. A. Gladyshev, N. I. Massevitch, and D. Roux, High resolution thermal plasma measurements aboard the AUREOL 3 spacecraft, *Adv. Space Res., 2,* 61–66, 1983.

Béghin, C., and R. Debrie, Characteristics of the electric field far from and close to a radiating antenna around the lower hybrid resonance in the ionospheric plasma, *J. Plasma Phys., 8,* 287–310, 1972.

Béghin, C., J.-F. Karczewski, B. Poirier, R. Debrie, and N. Massevitch, The ARCAD-3 ISOPROBE experiment for high time resolution thermal plasma measurements, *Ann. Geophys., 38,* 615–629, 1982.

Bernstein, I. B., Waves in a plasma in a magnetic field, *Phys. Rev., 109,* 10–21, 1958.

Boswell, R. W., and J. Thiel, Electron drift velocity deduced from lower oblique (whistler) resonance-cone measurements, *Phys. Fluids, 24,* 2120–2122, 1981.

Chassériaux, J. M., R. Debrie, and C. Renard, Electron density and temperature measurements in the lower ionosphere as deduced from the warm plasma theory of the h.f. quadrupole probe, *J. Plasma Phys., 8,* 231–253, 1972.

Debrie, R., and J. Thiel, Perpendicular observation of resonances in a Maxwellian magnetoplasma, *Phys. Lett., 84A,* 477–480, 1981.

Décréau, P. M. E., C. Béghin, and M. Parrot, Electron density and temperature as measured by the mutual impedance experiment on board GEOS 1, *Space Sci. Rev., 22,* 581–594, 1978a.

Décréau, P. M. E., J. Etcheto, K. Knott, A. Pedersen , G. L. Wrenn, and D. T. Young, Multi-experiment determination of plasma density and temperature, *Space Sci. Rev, 22,* 633–645, 1978b.

Fisher, R. K., and R. W. Gould, Resonance cone structure in warm anisotropic plasma, *Phys. Lett., 31A,* 235–236, 1970.

Gonfalone, A., Oblique resonances in the ionosphere, *Radio Sci., 9,* 1159–1163, 1974.

Illiano, J.-M., and R. Pottelette, Measurement of the collective motion of the electrons deduced from the shift of the lower oblique resonance frequency, *Phys. Lett., 70A,* 315–316, 1979.

Koons, H. C., D. C. Pridmore-Brown, and D. A. McPherson, Oblique resonances excited in the near field of a satellite-borne electric dipole antenna, *Radio Sci., 9,* 541–545, 1974.

Kuehl, H. H., Electromagnetic radiation from an electric dipole in a cold anisotropic plasma, *Phys. Fluids, 5,* 1095–1103, 1962.

Melrose, D. B., and R. C. McPhedran, *Electromagnetic Processes in Dispersive Media,* 407 pp., Cambridge University Press, Cambridge, UK, 1991.

Michel, E., C. Béghin, A. Gonfalone, and I. F. Ivanov, Mesures de densité et de température électroniques sur fusée dans l'ionosphère polaire par l'étude du cône de résonance, *Ann. Géophys., 31,* 463–471, 1975.

Ohnuma, T., *Radiation Phenomena in Plasmas,* 316 pp., World Scientific, Singapore, 1994.

Pierre, T., V. Rohde, and A. Piel, Experimental and numerical study of the plasma drift effect on upper-hybrid resonance cones, *Phys. Fluids B, 4,* 2661–2664, 1992.

Pottelette, R., Possibilités de mesurer la vitesse de circulation du plasma magnétosphérique à l'aide d'une sonde quadripolaire utilisée au voisinage de la fréquence hybride basse, *Ann. Géophys., 28,* 257–286, 1972.

Pottelette, R., M. Hamelin, J. M. Illiano, and B. Lembège, Interpretation of the fine structure of electrostatic waves excited in space, *Phys. Fluids, 24,* 1517–1526, 1981.

Pottelette, and J. M. Illiano, Observation of weak HF electrostatic turbulence in the auroral ionosphere, *J. Geophys. Res., 87,* 5151–5158, 1982.

Rohde, V., A. Piel, H. Thiemann, and K.-I. Oyama, In situ diagnostics of ionospheric plasma with the resonance cone technique, *J. Geophys. Res., 98,* 19,163–19,172, 1993.

Rooy, B., M. R. Feix, and L. R. O. Storey, Théorie de la sonde quadripolaire en plasma chaud isotrope, *Plasma Phys., 14,* 275–300, 1972.

Stix, T. H., *Waves in Plasmas,* 566 pp., American Institute of Physics, New York, 1992.

Storey, L. R. O., Antenne électrique dipôle pour réception TBF dans l'ionosphère, *L'Onde Electrique, 45,* 1427–1435, 1965.

Storey, L. R. O., M. P. Aubry, and P. Meyer, A quadripole probe for the study of ionospheric plasma resonances, in *Plasma Waves in Space and in the Laboratory,* Vol. 1, edited by J. O. Thomas and B. J. Landmark, pp. 302–332, Edinburgh University Press, Edinburgh, 1969.

Storey, L. R. O., and L. Cairó, Measurement of plasma resistivity at ELF, this volume.

Storey, L. R. O., and R. Pottelette, Possibilités d'utiliser la sonde quadripolaire pour la mesure de courants électriques dans la magnétosphère, *C. R. Acad. Sci. Paris (Sér. B), 273,* 101–104, 1971.

Storey, L. R. O., and J. Thiel, An attempt to measure the field-aligned drift velocity of thermal electrons in the auroral ionosphere, *J. Geophys. Res., 89,* 969–975, 1984.

Thiemann, H., A. Piel, and S. P. Gupta, In-situ measurements of plasma parameters in the equatorial ionosphere by the resonance cone technique, *Adv. Space Res., 8,* (8)147-(8)150, 1988.

L. R. O. Storey, Quartier Luchène, 84160 Cucuron, France. (e-mail: storey@nssdca.gsfc.nasa.gov)

Multiple-Baseline Spaced Receivers

R. F. Pfaff, Jr. and P. A. Marionni[1]

NASA/Goddard Space Flight Center, Greenbelt, Maryland (USA)

Multiple-baseline, spaced receivers provide a powerful means to determine the wavelength and phase velocity of electrostatic waves. The effectiveness of the technique is demonstrated using waves detected by probes on sounding rockets. Compared to single baseline measurements, advantages of using multiple-baseline spaced receivers include: (1) the ability to provide independent determinations of phase velocity and wavelength; (2) the ability to determine the properties of short-lived or transient wave events with greater confidence; (3) the ability to resolve $2n\pi$ ambiguities particularly for non-spinning components; and (4) the ability to more easily identify spatial resonance effects. Drawbacks of such multi-baseline probes include potentially more complicated deployment systems and the need for additional electronics, memory, and telemetry.

INTRODUCTION

This paper discusses spaced-receiver techniques to determine the wavelength and phase velocity of plasma waves detected by probes on spacecraft. Although the determination of a measured spectrum as a function of frequency is relatively straightforward, the interpretation of the same spectrum in terms of wavenumber is highly ambiguous, requiring knowledge of the spatial distribution of wavevectors and their phase velocities [e.g., Fredricks and Coroniti, 1976]. Even for cases in which the velocity of the probe (i.e., the spacecraft velocity) may be safely assumed to be either much greater than, or much less than, the phase velocity of the waves, such a conversion still depends on the successful determination of the direction of the wavevector, **k**, to convert frequency to wavenumber. Since *in-situ* probes may encounter oscillations comprised of a broad spectrum of wavenumbers and phase velocities, the identification of the 2-d and 3-d rest-frame spectrum or wave distribution may be considerably complex.

A powerful technique to determine both the wavelength and the phase velocity of waves detected in the spacecraft frame is one which measures the time delay of signals detected by two or more sets of wave detectors physically separated on the spacecraft. Such "spaced receivers" (sometimes referred to as "interferometers") were first applied to space measurements on a sounding rocket by Bahnsen et al. [1978] who utilized spaced electric field double probes to study plasma instabilities in the auroral electrojet. The technique has also been used with density probes (including $\delta n/n$ measurements) on both sounding rockets [e.g., Kintner et al., 1984] and satellites [e.g., Boström et. al., 1989; Holmgren and Kintner 1989], and has been used in laboratory plasmas [e.g., Harker and Ilic, 1974; Beall et al., 1982]. Besides providing a mechanism to convert frequency to wavenumber, the spaced-receiver technique addresses other aspects of wave physics, such as wave dispersion and the degree of "coherence" that the waves maintain between two detectors. Thus, the technique can be used to study turbulence and other non-linear interactions in space plasmas [e.g., Pecseli et al., 1989; Vago et al., 1992]. The use of the technique for space plasmas has been reviewed by LaBelle and Kintner [1989].

Whereas it should be intuitively obvious that multiple baselines are superior to a single baseline measurement, this paper demonstrates how multiple-baseline receivers may be used to obtain a more significant determination of the wave properties. It also discusses limitations of the technique.

---

[1] Deceased

Measurement Techniques in Space Plasmas: Fields
Geophysical Monograph 103
Published in 1998 by the American Geophysical Union

THE SPACED RECEIVER TECHNIQUE

We begin by reviewing the principles of the spaced receiver technique. Although we concentrate here on electric field wave measurements, the same general approach may be used for spaced plasma density and electric potential measurements. We then illustrate the main features of multiple-baseline spaced receivers using data gathered on sounding rockets in the Earth's ionosphere.

Relation of Measured Phase to Wavelength and Phase Velocity. Let us consider the "double-double" electric field probes shown in Figure 1 in which the baseline vector, **b**, represents the distance between the mid-points of two opposing outer pairs of double probe detectors. The measured phase shift, $\theta(\omega)_{meas}$, provides the component of **k** along **b** as a function of wave frequency, ω:

$$\theta(\omega)_{meas} = \mathbf{k} \cdot \mathbf{b} - 2n\pi \qquad (1)$$

where n is an integer. It should be clear from (1) that the resolution in the phase measurement increases with increasing separation distance. However, in cases where this distance becomes greater than the wavelength, a $2n\pi$ ambiguity results, as discussed later.

The phase velocity component between the two detectors, V_b, is simply the baseline divided by the time lag, which, as a function of frequency, is $\theta(\omega)/\omega$. Thus, we have:

$$V_b = \frac{\omega_{meas} |\mathbf{b}|}{\theta(\omega)_{meas}} \qquad (2)$$

This is only one component of the phase velocity. Clearly, three-axis measurements, or spin-plane and spin-axis data, are needed to fully determine this vector. Alternatively, the direction of **k**, which is parallel to the phase velocity, may be determined from the wave electric field, δE, since $\delta E \parallel \mathbf{k}$ for electrostatic waves.

The analysis of measured phase differences can be carried out using any pair combination of double probes, provided the distance between the two mid-points of each double probe is not zero. The longest baseline in Figure 1 is created by the two double probe pairs, δE_{12} and δE_{34}, and provides the most sensitive phase measurements, although the short probes gather the least sensitive signal-to-noise electric field measurements (ignoring short wavelength effects). On the other hand, the interleaved double probe pair, δE_{13} and δE_{24}, provides a shorter separation baseline and thus less sensitive phase measurements, but its longer boomlengths increase the signal-to-noise ratio. Utilizing the two baselines together provides a powerful tool for the analysis of electrostatic waves. This is the multi-baseline aspect of the technique that is the focus of this paper.

Remarks on Signal Processing. There are a variety of choices to determine the time lag between measurements

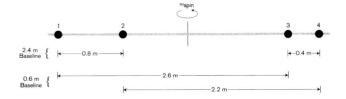

Figure 1. Spaced receiver geometry for a single axis electric field double probe using four spherical detectors.

separated in space, depending on the type of data and problem being investigated. These include an analysis in the time domain of characteristic waveform "features" encountered by the detectors, as used to identify the speed and direction of solitary density structures on the Viking satellite [Boström et al., 1989], and cross correlation techniques, as used by Bahnsen et al. [1978] in an analysis of coherent-like waveforms in the auroral ionosphere. By first passing the data through frequency filters, the cross correlation is computed as a function of frequency. This is the basis of the cross spectral technique which we use here.

The cross spectral technique determines the phase and coherency as a function of frequency through the complex multiplication of the Fourier transforms of the two waveforms detected at each receiver location. The phase information is obtained by the argument, and the coherency by the square of the modulus, of this product. (See Bendat and Piersol, 1971, for definitions.) To obtain reliable phase and coherency measurements, ensemble average factors are needed, which typically vary from 4 to 16, and may be carried out in either the time or frequency domain.

Coherency is an indicator of how well two signals are correlated. It is undefined if the ensemble average is 1 and is increasingly better defined as the number of ensembles increases. Thus, the number of ensemble averages should be as large as possible without compromising stationarity requirements. This is particularly difficult if the data are short-lived in the spacecraft frame or if the wave characteristics change quickly with time. Furthermore, for spinning receivers, the baseline component along **k** is constantly changing, and thus the number of ensemble averages must be kept small so that the baseline may be considered a constant for a given computation. Relatively slow vehicle spin rates aid this analysis. By calculating phase with several *simultaneous* pairs of spaced receiver data along the same axis, the number of ensemble averages required for a given statistical confidence level may be reduced, and therefore, higher time resolution and/or higher fidelity may be achieved. This is a chief advantage of multiple-baseline spaced receiver measurements.

Examples of multiple-baseline, spaced receiver analysis. We now illustrate the spaced-receiver technique using measurements of (nearly) one-dimensional two-stream waves detected with the electric field probes in Figure 1 on a rocket in the equatorial electrojet [Pfaff et al., 1997].

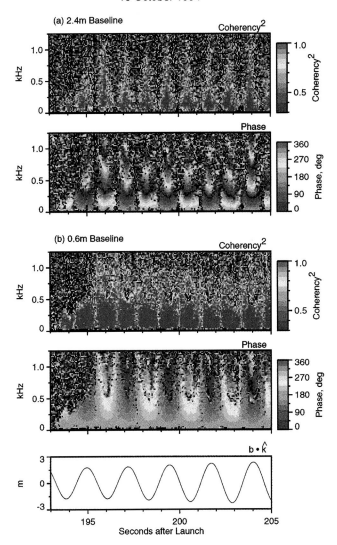

NASA 21.113 -- Alcântara, Brazil
15 October 1994

(a) 2.4m Baseline

(b) 0.6m Baseline

Plate 1. Phase and coherency *vs.* frequency calculated with 2.4m (upper panels) and 0.6m (lower panels) baselines. The lowest panel shows the component of the 3.0m double probe along the magnetic zonal direction [Pfaff et al., 1997].

Plate 1 shows phase and coherency measurements as a function of frequency, measured by parallel double probes separated by 2.4m (upper) and 0.6m (lower) baselines. The lowest panel shows the component of the 3.0m double probe, **d**, in the direction of **k**, measured with on-board attitude data. Notice that the phase measurements change sign twice-per-spin, and that they display crescent patterns which result from the sinusoïdally-changing angle between **k** and **b**. As discussed later, different crescent shapes appear in the 0.6m baseline coherency data which result from spatial resonances between the finite length double

probes and the wavelengths. The low signal/noise for the short double probes, which occur when they are at large angles to **k**, accounts for the more pronounced twice-per-spin modulation in the 2.4m baseline coherency data.

As discussed in Pfaff et al. [1997], the direction of **k** is determined by the measured spin plane phase data and also by the peak in the δ**E** data, since the payload spin axis was reasonably well aligned with the magnetic field direction and since **k** is essentially perpendicular to this direction. The phase velocity could thus be found as a function of both **k** and altitude within the electrojet layer.

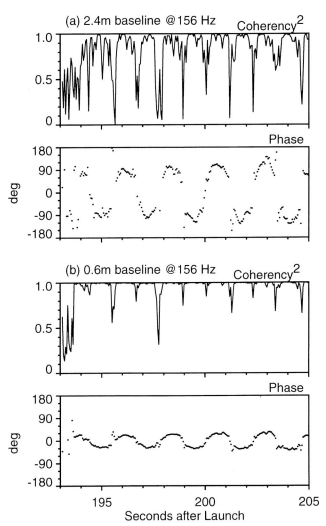

NASA 21.113 -- Alcântara, Brazil
15 October 1994

(a) 2.4m baseline @ 156 Hz

(b) 0.6m baseline @ 156 Hz

Figure 2. Phase and coherency at 156 Hz for the data shown in Plate 1.

In Figure 2, the phase and coherency are plotted for the 156 Hz frequency bin, which corresponds closely to the peak in the power spectrum (not shown). In this figure, the phases are only plotted for coherency values above 0.5. The phases vary with the spin frequency as the component of the baseline varies sinusoidally with respect to the direction of **k**. Notice immediately that the phases from the two independent spaced-receivers scale with respect to their baselines, indicating that each measures the same wavelength and phase velocity values given by (1) and (2).

The coherency in Figure 2 also shows this same spin modulation, but now the longer baseline data gathered by the much shorter double probes, show the deeper dips in the coherency. The corresponding coherency minima for the shorter baseline data, gathered by the longer double probes, do not appear as low and do not occupy as large a fraction of the spin period. Small non-linear departures in the sine waves result from the fact that the effective boomlength changes within the data period required to compute the cross spectrum and ensemble averaging. This distortion varies as a function of the angle between **b** and **k**.

We now plot the phase *vs.* frequency in Figure 3 for the cross spectrum centered at 199.5s when the receivers were aligned with **k**. For the frequencies from 0-400 Hz, a linear phase relation between $\theta(\omega)$ and ω is shown in both sets of data, demonstrating that these waves are dispersionless (at least for this portion of the spectrum). Furthermore, for this interval, the slope of the two curves both provide the same phase velocity of the waves. Above 400 Hz, the phase linearity extends to 1200 Hz in the longer baseline data, yet for the short baseline data, the measured phases show departures from linearity due to spatial effects of the electric field detector (discussed below). This is both a drawback and a feature of spaced receivers that use electric field double probes. If these spatial resonances are accounted for, the phase measurements for the 0.6m baseline can be extended to higher frequencies as well.

SPATIAL RESONANCE EFFECTS

Both the length of the double probe detector as well as the length of the separation baseline are critical design factors for multiple-baseline receivers that utilize electric field detectors. Resonances occur in the data when either of these distances are exactly equal to an integral number of wavelengths. Their identification provides a powerful tool for readily ascertaining the wavelength and wave direction. These effects are most apparent in data gathered from spinning probes for which the wave characteristics are generally stationary over a spin period. These same effects are also present in data gathered by probes on non-spinning spacecraft, although they are more difficult to discern, particularly if only one baseline is available. We now discuss how both the double probe length and the baseline affect the analysis of spaced-receiver phase measurements.

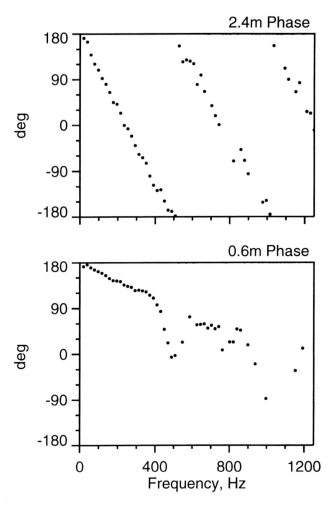

Figure 3. Phase measurements for the two baselines shown in Figure 2 for the data at 199.5s where **b** ‖ **k**.

Double probe resonances. As shown by previous researchers [e.g., Temerin, 1978], when the number of wavelengths is an integral of the electric field double probe component along **k**, the measured potential difference is zero. When properly identified, such nulls provide an immediate measure of the wavelength corresponding to the observed frequency. Similarly, when this condition is met by the double probes used in the spaced receiver analysis, the coherency becomes very low (as no potential difference is measured) and a valid phase measurement is not possible. Identifying such nulls in the cross spectra serve the same purpose as for the electric field power spectral analysis, but are usually more distinct in the coherency data, as these nulls result from a lack of coherency between two signals.

In Plate 1, the spatial resonance nulls created by the finite electric field boomlength are readily apparent as crescents in the 0.6m baseline coherency data. The corresponding phases for this baseline also respond (e.g., the small darkish spots near 500 Hz near the crescent minima) although they are more difficult to discern. The effect in the phase data is more distinct in the lower panel of Figure 3, although their appearance as ordered phase shifts (decreases, in this case) is not well-understood. Similar crescents do not appear in the 2.4m baseline coherency data in Plate 1 since those double probes were short. In fact, nulls created by resonances with short detectors correspond to much higher frequencies, where, for this case, the natural signal strength of the spectrum decreases significantly.

Baseline resonances. A different type of spatial resonance occurs when the spaced receiver baseline component along **k** corresponds to an integral number of wavelengths. In contrast to the electric field response discussed above, the baseline resonance may occur for any quantity measured by a spaced receiver, including plasma density scalar data. In the 2.4m baseline data in Figure 3, the phases clearly "wrap around" where the integral number of wavelengths equals the baseline, as described by the $2n\pi$ term in (1). This can also be seen in the color phase data in Plate 1 for the 2.4m baseline data, where the phases cycle through 360° as a function of frequency.

Notice further that the wrap-around effect in the 2.4m baseline data in Figure 3 occurs at precisely those frequencies which correspond to the nulls in the 0.6m baseline data due to the electric field resonances discussed above. In other words, for **b** ∥ **k**, the nulls in the coherency (seen here as phase decreases) for the 0.6m baseline data and the wrap-around effect in the 2.4m baseline data both reveal that 500 Hz corresponds to 2.4m and 1000 Hz corresponds to 1.2m. Such a correspondence will always be present when outer and interleaved baseline pairs of electric field spaced receivers are compared. Note that in determining the wavelength corresponding to the coherency dips in the 0.6m baseline data, we use the average length of the 2.2m and 2.6m detectors. Utilizing asymmetric double probes widens the frequencies corresponding to coherency nulls whereas symmetric double probes "sharpen" the nulls. Coning effects discussed by Pfaff et al. [1997] must also be taken into account but do not effect the discussion here.

The above examples show that for waves that persist and are stationary for at least one or more spin cycles, the changing component of the baseline with respect to **k** determines the direction of **k** within the spin plane. The rotating baseline also resolves the $2n\pi$ ambiguity for such cases. This is further illustrated in Figure 4 which presents two stream wave data collected with multiple baselines in the auroral electrojet. In this case, the waves are traveling faster, and the phase measured by the 2.4m spaced receiver clearly shows the wrap around effect, whereas the shorter 0.6m baseline data do not. A simulation of the output of

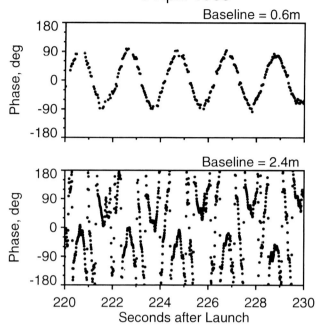

Figure 4. Two-stream phase measurements @300 Hz with 0.6m and 2.4m baseline spaced receivers in the auroral electrojet. The lower panel illustrates the wrap-around effect when the wavelength is comparable to the baseline.

cross spectral analysis of waveforms of 10, 5, and 3 meter wavelengths by electric field double probes spaced 3.9m apart is shown in Figure 5 and demonstrates how the wrap around effect becomes evident only for wavelengths shorter than the baseline [Pfaff, 1986].

Although the data shown here are highly one-dimensional and slowly-varying over several spin periods, plasma waves are often short-lived in the spacecraft frame (i.e., lasting much less than a spin cycle). In this case and for waves detected by non-spinning probe pairs, multiple-baseline receivers along a single axis are essential for determining the wavelength and phase velocity with a high degree of certainty as well as for resolving the $n2\pi$ ambiguity. Orthogonal spaced receivers provide the instantaneous wavenumber and phase velocity vectors without the need to rely on the spacecraft spin to determine these values.

OPTIMUM BASELINE SELECTION

An important factor in the design of multiple-baseline spaced receivers is the selection of baselines. For electric field detectors, symmetrically placed inner spheres provide symmetric double probe pairs which create the sharpest nulls in the coherency data that, in turn, provide spatial

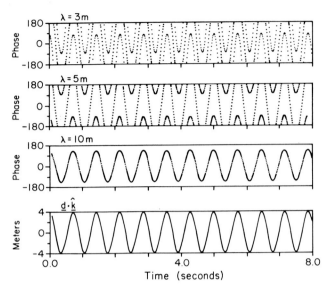

Figure 5. Output of the cross spectral analysis of simulated coherent waveforms of 10m, 5m, and 3m wavelengths by electric field double probes spaced 3.9m apart.

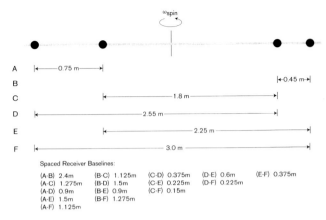

Figure 6. Multiple-baseline receivers using 4 electric field sensors on a single axis, creating 6 unique-length double probes and 15 spaced-receiver pairs with 9 unique baselines.

resonance information. Symmetric probes also provide a stable dynamic balance, important for spacecraft stability.

Irregular spacings provide other advantages, however, notably the increased number of non-zero baselines with which to perform simultaneous phase analysis, reducing the number of ensemble averages needed. For example, if the distances between spheres 1 and 2 and between 3 and 4 were equal in Figure 1, there could be no useful phase difference measurement between the spaced receiver pair, $\delta E14$ and $\delta E23$. For a double probe baseline with four spherical detectors, an example of spacings using minimum redundancy is given in Figure 6. Here, the four spheres provide 6 double probe pairs each with a unique boomlength. These double probes in turn provide 15 spaced receiver pairs, including 9 unique spaced receiver baselines, which is the maximum number allowable for a single baseline using only four spheres. Clearly, adding additional sensors (including the central spacecraft body as a sensor) increases this number considerably.

The number of unique spaced receiver pairs can be exploited further to provide additional spatial filters to determine the coherency scales of a broad spectrum of plasma waves, particularly in cases where the wavelengths of interest are on the order of the various baseline and double probe distances. Such signal analysis lends itself to matched filters that could be constructed to extract spatial attenuation information from weaker signals. The choice of the exact spacings along a given double probe axis is ultimately influenced by the expected spectrum of the waves, for cases where this is known *a priori*.

For studies of plasma turbulence, the spaced receiver analysis determines the coherency lengths over which a given fluctuation retains its identity. Where this is a

fundamental measurement goal, the spacings should be determined in such a way that when the spatial Fourier transforms of the combined spaced receiver baselines are combined, their sidelobes should apodize to the maximum extent. Such a procedure is generally applied in the selection of interferometry baselines in aperture synthesis radio astronomy and is another advantage of the irregular spacings of the spaced receiver pairs.

SUMMARY AND FUTURE OUTLOOK

Compared to single-baseline spaced receivers, multiple-baseline-measurements provide the following advantages:

(1) Independent solutions of wavelength and phase velocity provide increased confidence, particularly for data with a large random component.

(2) Fewer ensemble averages are needed to compute a given phase relation, thus directly increasing the temporal and/or frequency resolution of a measurement, as well as the confidence level. Consequently, the properties of short-lived waves can be determined with greater confidence.

(3) The $2n\pi$ factor may be resolved by simultaneous multiple-baseline measurements, which is particularly important for non-spinning spacecraft and for when the waves do not last for a significant fraction of a spin period.

(4) Spatial resonance patterns in the phase and coherency data can be more easily identified for use as powerful tools to determine wavelength and phase velocity.

There are two main disadvantages of multiple-baseline electric field spaced receivers: First, including additional spheres in-board of the outermost spheres may create the need for a more complex boom deployment mechanism. Such arrangements are commonly included in sounding rocket electric field measurements, and have recently been deployed on the 56m tip-to-tip FAST satellite electric field booms which include a second set of spheres situated 5m in-board from the outermost sensors [Ergun, personal communication, 1995]. Second, additional electronics and telemetry are needed to transmit the information corresponding to all of the wave pairs. Using burst memories and on-board processing, however, this is not a formidable problem in modern spacecraft.

Although the interpretation of wave emissions and structures detected by *in-situ* probes in terms of wavelength and phase velocity is not straightforward, instruments designed to alleviate some of the long-standing ambiguities associated with this problem have enabled recent progress in this field. We have concentrated here on nearly one-dimensional waves in order to illustrate the multiple-baseline technique. A great potential of multiple-baseline detectors, however, lies in their ability to discern, with greater confidence, several wave modes simultaneously, propagating with different velocities. Furthermore, gathering plasma wave data with such multiple-baseline spaced receivers promises to provide a powerful tool with which to investigate in detail the elusive properties of plasma turbulence and other non-linear phenomena that frequently characterize waves in space.

Acknowledgments. We acknowledge useful comments from Dr. H. Freudenreich, NASA/Goddard Space Flight Center.

REFERENCES

Bahnsen, A., et al., Electrostatic waves observed in an unstable polar cap ionosphere, *J. Geophys. Res.,* 83, 5191-5197, 1978.

Beall, J. M., et al., Estimation of wavenumber and frequency spectra using fixed probe pairs, *J. Appl. Phys.,* 53, 3933-3940, 1982.

Bendat, J. S., and A. G. Piersol, *Random Data: Analysis and Measurement Procedures,* New York: Wiley-Interscience, 1971.

Boström, R., et al., Solitary structures in the magnetospheric plasma observed by Viking, *Phys. Scr.,* 39, 782-786, 1989.

Fredricks, R. W., and F. V. Coroniti, Ambiguities in the deduction of rest frame fluctuation spectra from spectra computed in moving frames, *J. Geophys. Res.,* 81, 5591-5595, 1976.

Harker, K. J., and D. B. Ilic, Measurement of plasma wave spectral density from the cross-power density spectrum, *Rev. Sci. Instrum.,* 45, 1315-1324, 1974.

Holmgren, G., and P. M. Kintner, Experimental evidence of widespread regions of small scale plasma irregularities in the magnetosphere, *J. Geophys. Res.,* 95, 6015-6023, 1990.

Kintner, P. M., et al. Interferometric phase measurements, *Geophys. Res. Lett.,* 11, 19-22, 1984.

LaBelle, J., and P. M. Kintner, The measurement of wavelength in space plasmas, *Rev. Geophys.,* 27, 495-518, 1989.

Pecseli, H. L., et al., Low-frequency electrostatic turbulence in the polar cap E region, *J. Geophys. Res.,* 94, 5337-5349, 1989.

Pfaff, R. F., Rocket studies of plasma turbulence in the equatorial and auroral electrojets Ph.D. Thesis, Cornell Univ., Ithaca, NY, 1986.

Pfaff, R. F. Jr., P. A. Marionni, W. E. Swartz, Wavevector observations of the two-stream instability in the daytime equatorial electrojet, *Geophys. Res. Lett.,* 24, 1671, 1997.

Temerin, M., The polarization, frequency, and wavelengths of high-latitude turbulence, *J. Geophys. Res.,* 83, 2609, 1978.

Vago, J. L., et al., Transverse ion acceleration by localized lower hybrid waves in the topside ionosphere, *J. Geophys. Res.,* 97, 16935, 1992.

R. F. Pfaff, Jr., NASA/Goddard Space Flight Center, Greenbelt, MD, 20771 (USA).

The Plasma Frequency Tracker: An Instrument for Probing the Frequency Structure of Narrow-Band MF/HF Electric Fields

E. J. Lund,[1] M. L. Trimpi, E. H. Gewirtz,[2] R. H. Cook,[3] and J. LaBelle

Department of Physics and Astronomy, Dartmouth College, Hanover, New Hampshire

The Plasma Frequency Tracker (PFT) is a rocket-borne instrument designed to make measurements of narrow-band MF/HF electric fields with high resolution in both frequency and time. The PFT is capable of covering a band 140 kHz wide (extendable to over 200 kHz) with a dynamic range of 60 dB. The center frequency of the passband is changed during flight in order to track a characteristic frequency of the ambient plasma such as the plasma frequency, upper hybrid frequency, or electron gyrofrequency. The PFT is designed to use real-time data from another instrument and/or a preprogrammed model to set the passband frequency during flight. Telemetry can be either FM/FM or PCM. We present PFT data in which artificially stimulated upper hybrid waves are observed on the Auroral Turbulence sounding rocket.

INTRODUCTION

The existence of energetic electron beams in aurorae has long been known (see *Evans* [1974] for a review of early work). In the presence of a stationary electron population, such as exists in the ionosphere, an energetic electron beam is unstable and produces Langmuir waves, which are electrostatic oscillations near the plasma frequency [e.g., *Nicholson*, 1983]. These Langmuir waves can become quite intense during active aurorae [*Boehm*, 1987; *Ergun*, 1989]. There are theoretical arguments [*Zakharov*, 1972] and experimental evidence [*Boehm*, 1987; *Ergun*, 1989; *Ergun et al.*, 1991a, b, c] that the most intense Langmuir waves can collapse into highly localized nonlinear waves called Langmuir solitons.

The Plasma Frequency Tracker (PFT) is designed to track the plasma frequency using real-time data from another instrument aboard the rocket. The PFT can obtain detailed information about the frequency and time structure of narrow-band MF/HF electric fields with nearly 100% duty cycle. In the next section we will discuss how the PFT complements other techniques for measuring waves at these frequencies. We will then discuss the design of our instrument, algorithms for tracking the plasma frequency, and results from the maiden flight of this instrument.

MOTIVATION

The problem is how to obtain high-resolution measurements of electric fields at MF/HF frequencies ($f \gtrsim$

---

[1]Now at Space Science Center, University of New Hampshire, Durham

[2]Now at Digital Equipment Corporation, Hudson, Massachusetts

[3]Deceased September 17, 1995

Measurement Techniques in Space Plasmas: Fields
Geophysical Monograph 103

300 kHz). The direct approach of sampling the waveform at a high enough frequency to resolve the Langmuir frequency in the ionosphere (typically 1–5 MHz) requires a far higher data transmission rate than what has normally been available on existing rocket telemetry systems, so we must make some compromise in order to obtain the desired data.

One common compromise is a swept frequency analyzer (SFA), which divides the frequency range of interest into bins and gives the total spectral energy in each bin during each sweep. This approach has the advantage of providing complete frequency coverage. The tradeoff is that for a typical SFA, the frequency resolution is ∼ 10 kHz at best, and the time resolution is limited by the sweep period of ∼ 100 ms. The SFA also gives only the intensity in each frequency band, so no phase or waveform information is preserved. Although the SFA is a useful diagnostic tool, it cannot answer any questions about the fine frequency structure of Langmuir waves.

Another approach is to sample the entire spectrum at high resolution over a brief interval and transmit the data during some later dead time. This "waveform snatcher" provides all of the desired information, including phase information, during the sampled interval. Because the duty cycle is so small, however, the probability that the sort of event reported by *Ergun et al.* [1991c] will be recorded by chance is small. An improved version which selects the snatches having the largest amplitude waves has been constructed for Freja [*Kintner et al.*, 1995]; however, even this version will miss physically interesting periods when the event duration exceeds the length of a snatch or when several events occur during an interval in which only one "snatch" can be telemetered.

A third alternative is to couple a broad-band receiver to a counter in order to measure the amplitude and dominant frequency of observed waves [*Boehm*, 1987]. This technique is effective for providing the amplitude and amplitude modulations when the signal of interest is narrow-banded and dominant within the broader band. These conditions often hold for intense Langmuir turbulence; however, only the envelope of the waveform is preserved. If the signal of interest is weak or broadbanded, this technique may yield inaccurate estimates of both the frequency and the amplitude.

The PFT represents a different approach from all of the above. It exploits the fact that wave modes are concentrated near a few distinct frequencies, especially near and above the plasma frequency. The PFT therefore selects a frequency interval, rather than a time interval.

The result is complete amplitude and relative phase information about waves in a pre-determined band of interest with nearly 100% duty cycle, although this phase can be difficult to compare with the phase of particle modulations and VLF/ELF waves. The tradeoff is that the PFT gives no information about frequencies outside the selected passband. The difficulty which must be overcome in this case is that the passband must be selected in real time based on available data.

DESCRIPTION OF THE INSTRUMENT

General

The PFT consists of a receiver board, a local oscillator board, a digital interface board, an on-board CPU, and an optional automatic gain control (AGC)/PCM encoder board. The latter board is designed for use with a dedicated 4-Mbit/s PCM link. A block diagram of the PFT is shown in Figure 1. The digital interface board and the AGC board are discussed in greater detail in *Gewirtz* [1992]. The concept is similar to the downconverter independently developed by *Haas et al.* [1995], except that our instrument determines its passband frequency from real-time data from another instrument aboard the rocket. We discuss hardware design in this section; frequency selection algorithms will be discussed in the next section.

Two external experiments have been used to tune the PFT. For the Physics of Auroral Zone Electrons (PHAZE) mission (40.003), SFA data were provided to the PFT and the output was telemetered on an FM/FM system. For the Auroral Turbulence mission (40.005), the PFT received data from a Utah State Plasma Frequency Probe (PFP), which actively determines the upper hybrid frequency, a frequency closely related to the plasma frequency [*Baker et al.*, 1985; *Jensen*, 1988]; this mission uses the AGC board to transmit the data on a dedicated PCM link.

Analog Electronics

The PFT receiver is a triple conversion superheterodyne receiver. If the desired passband is centered at frequency f_0, then an incoming signal at a frequency f_{in} is treated as follows: First, the incoming signal is mixed with the first local oscillator $f_1 = f_0 + f_2$, where $f_2 = 10.7$ MHz is the second stage frequency. The first local oscillator is produced by an MC14515-1 phase-locked loop (PLL) which is set by the CPU. The product is passed through an IF strip, which selects the difference frequency $f_0 + f_2 - f_{in}$. The signal is then

Launch Signal (from NASA)

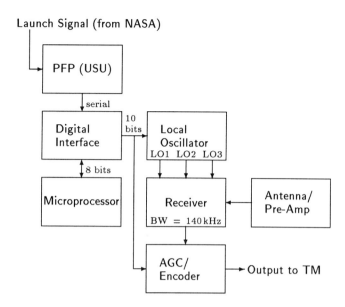

Figure 1. A block diagram of the Plasma Frequency Tracker.

mixed with the second local oscillator f_2 in quadrature. The resulting base band signals are passed through an eight-pole lowpass filter, which selects signals such that $|f_0 - f_{in}| < 0.5\delta f$, where $\delta f = 140$ kHz is the nominal passband width. Finally, the signals are mixed with the third local oscillator $f_3 = f_2/128 = 83.6$ kHz in quadrature and recombined to cancel out the image of the signal. The output thus has frequency f_{in} mapped to $f_{in} - f_0 + f_3$.

Digital Interface and CPU Boards

The on-board CPU is a Q88 made by QSI (Salt Lake City). It has 32K of RAM and a 32K PROM. This CPU was chosen because the manufacturer supplies software which allows easy testing of the PROM program from within Turbo C.

The digital interface board connects the CPU to the outside world. Serial data from an external experiment are converted to parallel form inside a programmable logic device (Altera EP1800), which can easily be modified to accomodate any variation on NASA's PCM format. The data are passed to the CPU, which calculates the frequency to which the PLL should be set and passes this information through the interface board to the local oscillator.

Automatic Gain Control and PCM Encoding

When the PFT output is telemetered via an FM/FM system, no further signal processing is done after the

downconverter output; the PLL frequency is telemetered on a separate channel. The PFT output can also be sent on a dedicated 4-Mbit/s PCM link. In this case, an AGC must be used to fold the PFT's 60 dB dynamic range into 8-bit samples, and the syncwords, set frequency, and gain state changes must be multiplexed with the data. The data stream must then be converted from parallel to serial format and passed through a four-pole Bessel filter before it can be telemetered down.

The design of the AGC board was implemented with programmable VLSI logic (Actel A1020 series), which allows such parameters as the minor frame length and the interval between frequency updates to be varied from flight to flight, and which allows the entire AGC and encoder circuits to fit on a single rocket board. The default setting of the AGC is to pass the lowest seven bits plus a sign bit from a 12-bit ADC to the parallel-to-serial converter. The twenty most recent 12-bit samples (40 μs of data) are stored in a buffer; if both the upper and the lower limits of the high-gain state are exceeded within the buffered interval, the AGC switches to the low-gain state, in which the most significant eight bits from the ADC are sent out, at the first sample in the buffer which overflows the high-gain state. The low-gain state lasts for 1/16 of the interval between frequency settings, after which the AGC repeats the test to determine whether to remain in low-gain mode or switch back to high-gain mode. The set frequency and gain state change times are merged into the data stream while the local oscillator is set to the new frequency.

FREQUENCY SETTING ALGORITHMS

The algorithm by which the local oscillator frequency is calculated clearly plays a key role in the success or failure of the instrument. The difficulty is how to follow a natural emission line without either becoming stuck on an interference line or missing an interesting event at a different frequency.

Our algorithm on the Auroral Turbulence mission was simple, since the upper hybrid and plasma frequencies are simply related. A model magnetic field along the nominal trajectory was stored as a lookup table. The only complication was that we wanted to observe the artificial signature generated by the PFP, so our algorithm increased the passband frequency in steps to cover the entire range between the plasma and upper hybrid frequencies, which are usually separated by more than the 140 kHz bandwidth of the PFT.

Using SFA data to set the frequency, as on PHAZE, is more complicated because of the possibility that interference lines could be misinterpreted as interesting signals. Our algorithm stores a past history of sweeps and looks for constant-frequency signals, which it deems interference lines when they exceed a certain threshold based on the average signal level over the period of time stored. The algorithm will then attempt to track the strongest signal which has not been deemed an interference line. If a non-interference signal exists which is significantly stronger than the current signal, the algorithm jumps to that signal. If no discrete signals are detected, the frequency sweeps over the range of allowed frequencies. This algorithm was tested and tuned with existing auroral zone SFA data.

Both algorithms will detect timeouts if the PFT loses contact with the other instrument; in this case the PFT starts sweeping the passband. The Turbulence algorithm also checks whether the upper hybrid frequency reported by the PFP is greater than the model electron gyrofrequency; if not, the data from the PFP are assumed bad and a sweep mode is initiated. In both cases the PFT resumes normal operation if the other instrument has resumed sending valid data.

RESULTS

PHAZE was launched 27 January 1993 at 1043 UT from Poker Flat Research Range, Alaska but suffered vehicle failure due to a first-stage motor anomaly; the payload returned no scientific data. The Auroral Turbulence rocket was launched 6 March 1994 at 0821 UT from Poker Flat. All systems on the mother payload, including the PFT and PFP, were operational during flight.

The center frequency of the PFT as a function of flight time is shown in Figure 2. Comparison with PFP data from the same interval shows that the PFT tracked the plasma/upper hybrid frequency range quite well during the portion of the flight where the PFP remained locked onto the upper hybrid resonance. Note that the PFT center frequency varied in a sawtooth pattern in an effort to cover the full range between the plasma and upper hybrid frequencies. Due to unfavorable placement of the PFP antenna on the side of the payload, however, the PFP measurement was severely affected by ram-wake effects; in particular, the density in the wake fell below the instrumental threshold well before the ambient density dropped below this limit, and the PFP lost lock when its antenna passed through the wake. As a result of this condition, which was not anticipated before launch, the PFT passband was

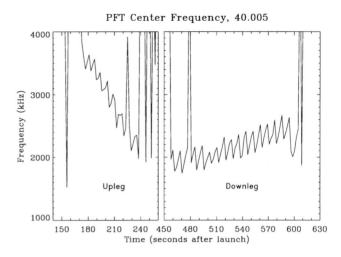

Figure 2. The PFT center frequency as a function of flight time during (a) the upleg and (b) the downleg.

pegged high through the portion of the flight not shown in Figure 2.

An example of the data returned by the PFT is shown in Figure 3a, which shows a spectrogram of 550 ms (0.22 spin) of data centered near 461 s after launch (indicated by an arrow in Figure 2). Note that only those frequencies which were downconverted to 0–250 kHz are shown. The prominent emission line visible in the data is the upper hybrid resonance stimulated by the PFP; Figure 3b, which shows the PFP data from the same interval, confirms this interpretation. The PFP antenna passes through the wake during the interval shown. A second line due to imperfect suppression of the image is visible for part of this interval, as are sidebands discussed below. The marked changes in the background level in Figure 3a are due to application of the AGC described above; the low-gain state is activated at 820 ms and again at 1033 ms due to the signal level.

Figure 4 shows a single spectrum of PFT data which corresponds to the time indicated with an arrow in Figure 3a. The most prominent feature is the peak at 1814 kHz, which is the resonance stimulated by the PFP. This resonance is generally presumed to occur at the local upper hybrid frequency [*Baker et al.*, 1985]. The resonance is about 5 kHz FWHM, implying $\delta f/f \approx 0.003$. Peaks at 1738 kHz and 1906 kHz are images of the main peak; they result from imperfect cancellation of the undesired sideband and the presence of some third harmonic in the quadrature signals, respectively, in the final converter stage of the receiver. The stronger of the image signals is 23 dB weaker than the main peak. The sidebands located about 16 kHz

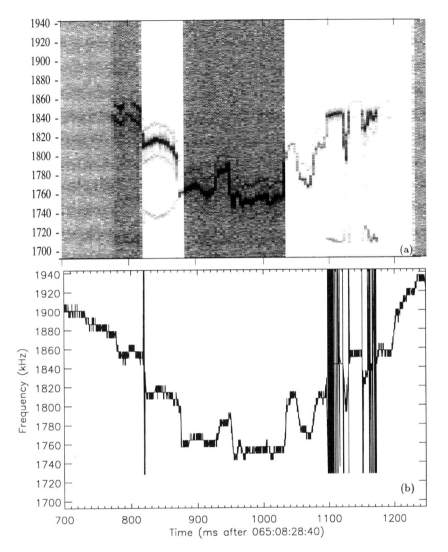

Figure 3. (a) Data from the Plasma Frequency Tracker aboard Auroral Turbulence approximately 460 s after launch. (b) A plot of the upper hybrid frequency as determined by the Plasma Frequency Probe during this period. Comparison shows that this frequency matches the frequency of the most intense line in the top panel within the precision of the measurement.

above and below the PFP line appear throughout the flight, and their separation from the main peak is very nearly constant. The VLF instrument on the same payload recorded auroral hiss at frequencies up to and above 16 kHz over most of the flight, but the hiss generally has most power below 16 kHz, and the frequency of maximum power spectral density varies during the flight. These sidebands most likely due to mixing of the PFP emissions with a stationary signal such as a payload interference line; whether this mixing occurs in the plasma or in one of the instruments has not been determined.

If the resonance frequency is interpreted as the upper hybrid frequency, then its variation with time (see Figure 3b) provides evidence for structure in the wake of the payload, particularly on the gradients. Features such as the electron density peaks at 940 ms and 1050 ms, on either side of the wake, occur throughout the flight. Much of this structure is probably due to advection of plasma by variable electric fields into the wake [*Lund et al.*, 1995].

PFT operation on the Auroral Turbulence sounding rocket underscores the need for a reliable determination of the plasma frequency (or other frequency of interest)

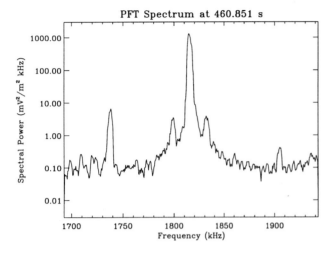

Figure 4.
A single spectrum of PFT data from 460.851 s after launch. The origin of the sidebands at ± 15 kHz from the main peak is unknown but probably due to a payload-generated interference line.

in real time for fully successful operation of the PFT. While a successful determination of the plasma frequency during the affected portions of the flight would have been impossible, its effects could have been reduced by testing for lock loss and resorting to some default behavior such as a slow sweep when lock loss is detected.

SUMMARY

We have shown that the PFT is capable of making wave electric field measurements in a 140 kHz band centered between 0.25 and 5 MHz and that the instrument can tune the passband to track a frequency of interest provided that an algorithm for determining this frequency is properly designed and that the incoming real-time data are correct. The PFT shows great promise as an instrument which can detect fine frequency structure of Langmuir waves in the auroral ionosphere, such as the sum and difference frequencies that would result from modulational instability. We hope that additional flights will detect such features if they exist.

Acknowledgments. The authors would like to thank J. L. Vago for providing SFA data for tests of our algorithm, S. Powell for providing the SFA data for PHAZE, M. D. Jensen for providing the PFP data for Auroral Turbulence, and W. Nerkowski for his advice on telemetry formats.

We acknowledge helpful discussions with M. H. Boehm during planning and R. A. Treumann during data interpretation. We also wish to thank the payload teams on PHAZE and Auroral Turbulence for their help during integration and launch. This work was supported by National Aeronautics and Space Administration grant NAG5-5039, with additional support for EJL from NASA grant NGT-50936.

REFERENCES

Baker, K. D., *et al.*, Absolute electron density measurements in the equatorial ionosphere, *J. Atmos. Terr. Phys.*, *47*, 781, 1985.

Boehm, M. H., *Waves and Static Electric Fields in the Auroral Acceleration Region*, Ph.D. thesis, University of California, Berkeley, 1987.

Ergun, R. E., *Linear and Nonlinear Wave Processes in the Auroral Ionosphere*, Ph.D. thesis, University of California, Berkeley, 1989.

Ergun, R. E., *et al.*, Langmuir wave growth and electron bunching: Results from a wave-particle correlator, *J. Geophys. Res.*, *96*, 225, 1991a.

Ergun, R. E., *et al.*, Observation of electron bunching during Landau growth and damping, *J. Geophys. Res.*, *96*, 11,371, 1991b.

Ergun, R. E., *et al.*, Evidence of a transverse Langmuir modulational instability in a space plasma, *Geophys. Res. Lett.*, *18*, 1177, 1991c.

Evans, D. S., Precipitating electron fluxes formed by a magnetic field aligned potential difference, *J. Geophys. Res.*, *79*, 2853, 1974.

Gewirtz, E. H., *Digital Control of Rocket-Borne Radio Instrumentation*, M. E. thesis, Dartmouth College, 1992.

Haas, D. G., *et al.*, Rocket-borne downconverter system for measuring space plasma turbulence, *Rev. Sci. Instrum.*, *66*, 1056, 1995.

Jensen, M. D., *Investigation of Accuracy Limitations of the Ionospheric Plasma Frequency Probe and Recommendations for a New Instrument*, M. S. thesis, Utah State University, 1988.

Kintner, P. M., *et al.*, First results from the Freja HF Snapshot Receiver, *Geophys. Res. Lett.*, *22*, 287, 1995.

Lund, E. J., *et al.*, Observation of electromagnetic oxygen cyclotron waves in a flickering aurora, *Geophys. Res. Lett.*, *22*, 2465, 1995.

Nicholson, D. R., *Introduction to Plasma Theory*, John Wiley and Sons, New York, 1983.

Zakharov, V. E., Collapse of Langmuir waves, *Zh. Eksp. Teor. Fiz.*, *62*, 1745, 1972, translated from Russian in *Sov. Phys.–JETP*, *35*, 908, 1972.

E. H. Gewirtz, Digital Equipment Corporation, Hudson, MA; (e-mail)

J. LaBelle and M. L. Trimpi, Department of Physics and Astronomy, Dartmouth College, 6127 Wilder Laboratory, Hanover, NH 03755-3528; jlabelle@einstein.dartmouh.edu, mike@einstein.dartmouth.edu

E. J. Lund, Space Science Center, Morse Hall, University of New Hampshire, Durham, NH 03824; Eric.Lund@unh.edu

Phase-Path Measurements in Space Using Receivers With GPS Clocks

H. G. James

Communications Research Centre, Ottawa, Ontario K2H 8S2, Canada

A number of new perspectives on wave processes in space may be obtained using a pair of phase-coherent receivers in close proximity to measure the direction of arrival and other parameters of plasma waves. Waves of either spontaneous or artificial origin are of interest. The particular objective of sensing direction is seen as one of several ways to use a measurement of the total phase difference between signals arriving at two receiver sites with a known baseline. The limitations of phase-path measurement using two conventional radio receivers with Global Positioning System (GPS) clocks are investigated. The inherent precision of GPS time means that GPS-based clocks can support useful phase-difference measurements up to at least High Frequency (3-30 MHz). For instance, when receiver separations of hundreds or thousands of meters are permitted, interferometer modes can be envisaged for synchronous detection of natural electromagnetic waves like auroral kilometric radiation or of artificial waves from ground transmitters. The double-heterodyne receiver concept is found to be more accurate than the direct waveform capture type. A rotating double payload comprising two receivers linked by a nonconducting tether, a "bolas" configuration, is one way to achieve a stable two-receiver direction finder in the ionosphere.

1. INTRODUCTION

Up until recently, in-situ space observations largely have been confined to data gathering in single-point measurement modes. Inherent in single-point observations is the limitation traditionally known as the "space-time" ambiguity. It prevents a data analyst from deciding whether an observed change in any measured quantity is a true temporal change in the rest frame of the space plasma, or rather a spatial variation that produces a temporal variation because of spacecraft orbital motion through the plasma.

A number of multi-satellite studies have been proposed through the years to help sort out space-time ambiguities. The International Sun-Earth Explorers (ISEE) and the Dynamics Explorers (DE) projects were among the first successful attempts. The Russian Interball and the European Cluster are examples of recent programs. In addition, fortuitous rendezvous' of satellites have been sought with particular scientific objectives in mind; these have tended to be specialized investigations of limited scope, and have not been always successful.

What the aforementioned programs, past and present, have in common is the relatively large distance between the satellite pairs. Common measurements of plasma parameters have been made at two different locations. Such measurements may be said to be coordinated but not closely synchronized, and typically the physical separations have been large compared to physical scale sizes of

Measurement Techniques in Space Plasmas: Fields
Geophysical Monograph 103

ionospheric or magnetospheric plasma processes of interest. The result is that physical interpretation remains limited because of the space-time problem.

Simultaneous measurements of waves at two or more sites, meanwhile, continue to be a very desirable objective in space science. Considering the resources and technology currently available for small scientific satellites, independent sensors separated by up to tens of meters can be accommodated on a single spacecraft. Synchronized, multipoint observations of waves with relatively small wavelengths have been made using booms. However, boom technology does not permit the multipoint, coherent investigation of electromagnetic wavelengths that play important roles in plasma dynamics and whose measurements are required.

This communication discusses some implications of the Global Positioning System (GPS) for two-point radio-receiver measurements. The absolute position of spacecraft carrying a GPS receiver can be determined to $\Delta d = \pm 174$ m in three dimensions using the "SPS" code available for civilian use. Equivalently, this says that the Universal Time in a GPS receiver clock can be accurate to $\Delta\tau = \pm 363$ ns. If two scientific spacecraft communicate with exactly the same subset of the GPS spacecraft constellation, then significantly better differential position and time accuracies are theoretically possible. These space-time specifications of GPS suggest that GPS clocks on two separate satellites can be used to the measure the phase-separation of two locations for waves of interest.

This paper is an outgrowth of planning for a proposed orbital tethered experiment called the BIstatic Canadian Experiment on Plasmas in Space (BICEPS) [*BICEPS Team, 1993*]. This concept envisaged two small spacecraft equipped with plasma field and particle detectors, and GPS receivers, operating in a bolas configuration. Figure 1 shows the two BICEPS tethered subpayloads in communication with the same four GPS spacecraft. Over the course of a 6-month mission, the two spacecraft would go through a series of separations from 10 to 1000 m. This would permit them to resolve physical processes in the ionospheric plasma.

In the Section 2 following, assumptions about the position- and time-measuring capabilities of GPS are given. Section 3 then shows what these GPS limitations imply for the two-point measurement of phase-path difference in space. The discussion deals with the case of two-receiver configurations carrying state-of-the-art GPS receivers using the widely known techniques of NATO's "Standard Positioning System". The principle is not confined to BICEPS objectives, but may be applied in other two-point experiments where synchronization is required.

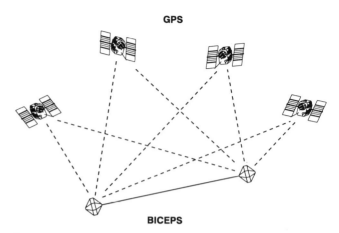

Figure 1. GPS spacecraft communicating location and time information to both ends of a tethered satellite pair, each having GPS receivers, associated clocks and wave receivers.

2. CHARACTERISTICS OF THE GLOBAL POSITIONING SYSTEM

GPS has a constellation of 24 spacecraft which provide time transfer, navigation and surveying services at 1.2 and 1.6 GHz (D-band) for military and civilian users [*Wells et al.*, 1987; *Ackroyd and Lorimer*, 1990; *Leick*, 1990; *Hofmann-Wellenhof et al.*, 1992]. The navigation function is achieved by measuring range to a number of the GPS satellites from the location of a user's GPS receiver. Differential phase measurements of D-band waveforms between at least two user GPS receivers are employed to obtain their precise, relative positions in the surveying type of function. The rather specialized application of GPS to follow, synchronized measurements in two co-orbiting spacecraft, exploits all three capabilities of GPS. As is typical of space-based research in geophysics, it is required to have an accurate knowledge of the range to the subject spacecraft. The bistatic nature of the experiment requires accurate on-board clocks for frequency synthesis and event synchronization. Finally, it is of paramount importance to have a continuous, true knowledge of the separation vector between the two scientific spacecraft.

This discussion about double-satellite observations assumes the location and time accuracies given in Table 1. GPS is essentially a range measuring system. The uncertainty $\Delta\tau$ in the time setting of a GPS receiver clock is the uncertainty Δd of the receiver location divided by the speed of light c.

The objective of this paper is to comment on the potential for useful bistatic (two-point) wave observations using the Standard Positioning System level of navigational

Table 1. GPS accuracies in the Standard Position System, for Coarse Acquisition (C/A) code and for L1 carrier method

Parameter	Absolute	Differential
Location(Δd),m	± 174	± 10 (L1: ± 0.1)*
Time($\Delta \tau$),ns	± 363	± 33 (L1: ± 0.3)*

*using carrier phase difference

accuracy. This is the less accurate of the two coded-pulse schemes of GPS, but the only code available to civilian users. The Coarse Acquisition (C/A) standards for time and three-dimensional single-point location determination given by NATO [NATO, 1991] are in the left column of Table 1.

The values in the right-hand column of Table 1 indicate the considerable improvement that obtains for differential measurements involving two closely orbiting spacecraft, especially if the L1 carrier phase technique is employed. If ionospheric plasma delays are significant, a differential frequency technique using both the L1 and L2 carriers is required for the bracketed numbers in Table 1. The numbers in the right-hand column are theoretical; GPS differential schemes are yet to be flown in orbital applications.

3. TWO-POINT WAVE EXPERIMENTS

3.1 *Phase-Path Separation of two Spacecraft*

In the study of either atmospheric emission or propagation from artificial sources, it is often desired to know the direction of arrival (DOA) of received waves in space. Or, a measure of the refractive index, or wave number, is needed. These are all variants of the requirement for the phase-path separation between two points.

Consider independent wave receivers on two spacecraft having a separation vector **a**. A plane, monochromatic electromagnetic wave impinges on both spacecraft, as shown in Figure 2. The broken line is a wave front. The two-dimensional geometry of the wave vector **k** with respect to **a** makes the phase path separation of the two observing points

$$\phi = \mathbf{a} \cdot \mathbf{k} \qquad (1)$$

The double satellite configuration can be thought of as a steerable beam or interferometer. This can be seen by rewriting (1) as

$$\sin\alpha = \frac{\phi}{ak} = \frac{\phi c}{2\pi a n f} \qquad (2)$$

where the DOA angle α is 90° minus β (the angle (**a**,**k**)) and n is the refractive index. The goal of this Section is to evaluate how GPS receivers limit the accuracy of the measurables on the right side of (2), particularly of ϕ, and thereby set the accuracy of α. In the general case, both n and $\sin\alpha$ in (2) are unknown. Separate information about the azimuth of the **k** vector is needed to arrive at a complete solution. This is true even when $n = 1$.

3.2 *Synchronized super-heterodyne radio receivers*

The case of the double-conversion super-heterodyne receiver is first examined. The electromagnetic wave has a frequency of $f + \delta f$. Reception is possible at any frequency $f \pm \Delta f/2$, where Δf is the constant receiver bandwidth, and f and $\omega = 2\pi f$ denote the central frequency of the band. The waveforms detected by the antennas are $A_2 \sin[(\omega + \delta \omega)t]$ and $A_1 \sin[(\omega + \delta \omega)t + \phi]$. Because of

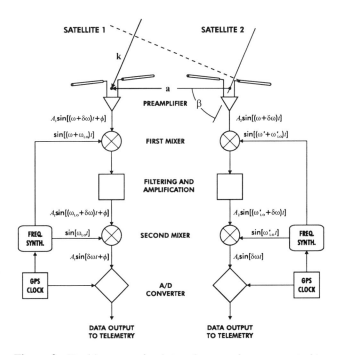

Figure 2. Double conversion heterodyne receivers connected to antennas on two satellites. An electromagnetic wave, with its wave front in broken line, is incident on the satellite pair, such that the wave vector **k** and the separation vector **a** are in the plane of the diagram.

differences in attitudes or in other variables of the two spacecraft, the amplitudes A_1 and A_2 of the signals appearing at the receiving antenna roots are generally not equal. Depending on the application, it may be necessary to retain the distinction between the two signal amplitudes. The present discussion of phase separation does not need to keep track of amplitudes.

Figure 2 also shows the principal signal conditioning functions of the double-conversion receivers. The antenna signal is mixed up to an intermediate frequency $\omega + \omega_{LO}$ where it is filtered and amplified. The second mixer then converts the signal down to a baseband centered at $\Delta f/2$. The analogue-to-digital converters (A/D) at the outputs produce data streams corresponding to the two baseband signals $A_2\sin(\delta\omega t)$ and $A_1\sin(\delta\omega t + \phi)$. Ground processing is used to extract ϕ from these data. It is assumed that prelaunch calibration data permit the output amplitudes to be corrected for differential phase shifts experienced by the signals while passing through the receiver chain.

An estimate of the error in ϕ expressed as a function of all the important variables requires a knowledge of the statistics of the GPS-receiver clock error. A generalized treatment of this subject is well outside the scope of this paper. Discussion is limited to two first-order errors traceable to the clock.

One important GPS-related limitation on the measurement of ϕ enters through the differences of the first mixer frequencies $\omega + \omega_{LO}$ of the two satellites. These frequencies are distinguished by the use of prime signs on the Satellite-2 side of Figure 2. The frequency synthesizer in each satellite is assumed to be driven by the GPS-controlled clock. The band-center frequency ω has an error determined by the clocking rate. Suppose each GPS receiver produces a reference frequency f_r which the wave receiver uses as the master reference. The first mixer frequency $f_{m1} \equiv f + f_{LO} = qf_r$. Typically, the ratio q might be about 10 and $f_r = 10$ MHz. If f_r has an error of Δf_r, the error $\Delta f_{m1} = q\Delta f_r$.

How can Δf_r be estimated? One method is to apply the results of *Conley* [1993] who has shown that GPS vertical position solutions at a given ground location may drift by as much as $\dot d = 100$ m/min. *Conley* found maximum horizontal drifts to be somewhat smaller. It is hypothesized here that $\dot d$ is really a measure of the required GPS-receiver clock differential error. Let us consider the effect over a 1-min interval, a typical length of a BICEPS recording period. In this period, the clock time markers will accumulate a total time error of $\dot d \times 1$ min/c $=$ $100/3\times10^8 = 0.33$ μs. The corresponding error ΔT_r per period of the 10-MHz reference is 0.33 μs/Σ, where Σ is the number of 10-MHz wave periods in 1 min $= 60f_r$.

Now because $f = 1/T$, the relationship between error in frequency Δf and error in period in period ΔT is

$$\Delta f = -f^2\Delta T \qquad (3)$$

So, $\Delta f_r = -f_r^2\Delta T_r = -f_r^2/(3\times10^6\times60\times f_r) = -f_r/(180\times10^6)$, which, for $f_r = 10$ MHz, is about -0.05 Hz.

The error Δf_{m1} in the frequency of the first mixer of Figure 2 is $\Delta f_r(f+f_{LO})/f_r$, while the second mixer frequency error is $\Delta f_r f_{LO}/f_r$. The baseband signal on the A_1 side is $A_1\sin[(\omega + 2\pi\Delta f_m + \delta\omega)t + \phi] = A_1\sin[(\omega + 2\pi\{\Delta f_{m1} - \Delta f_{m2}\} + \delta\omega)t + \phi]$. The GPS-injected phase error $2\pi\{\Delta f_{m1} - \Delta f_{m2}\}t = 2\pi\Delta f_r f t/f_r$ has a magnitude that depends on f. For a wave frequency $f = 1$ MHz and a sampling time $t = 1$ ms, this error is not important.

Errors also enter the calculation of ϕ through the A/D sampling process. The A/D converters in the two receivers record T_s seconds of baseband waveform. The two digitized waveforms have errors of two kinds. First, on account of the above mentioned inexact clock frequency, the waveforms are distorted, either dilated or contracted. The distortion occurs because of the frequency errors $\Delta f_r f/f_r$, in both receivers. If the *Conley*-based estimate of Δf_r is correct, this distortion is negligible in conventional double-conversion receivers for typical ionospheric frequencies of interest at or below 1 MHz. Second, on account of the asynchronism $\Delta\tau$ of the two clocks, the sample periods of length T_s do not start at exactly the same instant. Such a misalignment of their two time axes corresponds to a phase error of $\delta\omega\Delta\tau = 2\pi\delta f\Delta\tau$. This, too, is unimportant for δf of the order of a baseband bandwidth Δf of 50 kHz. If phase errors are unimportant, so too are errors in the wave frequency $f + \delta f$.

Thus, on the basis of the *Conley* results, it would appear that the phase errors in the baseband signals in each satellite, and in the A/D sampling process, resulting from both imprecise differential clocking rates and the absolute-time error $\Delta\tau$, are negligible compared with $\pi/20$. Another way to estimate Δf_r is to apply manufacturers' advertised accuracies of standard frequencies output by their commercial GPS receivers for ground applications. These accuracies are within 1 or 2 orders of magnitude of what was deduced from the *Conley* paper. It is anticipated that future flight observations will confirm whether the Δf_r magnitude used here is realistic.

3.3 Direct-Digitizing Receivers

Recent improvements in A/D technology now permit receivers to directly amplify and digitize rf waveforms at frequencies up to High Frequency with adequate dynamic range for space-science objectives. This can be visualized in Figure 2 simply by removing all of the receiver functions between and including Mixers 1 and 2, and by letting the A/D have an unlimited frequency range of operation.

The waveforms presented to the A/D devices are $A_2\sin(\omega t)$ and $A_1\sin(\omega t + \phi)$, where f and $\omega = 2\pi f$ now stand for the true wave frequency, not for a band center. That is, $f + \delta f$ is replaced by f. Again, an interval T_s seconds of the waveform is captured in the two receivers, starting at a nominal UT time t_0. There appear the same two limitations of the GPS-controlled local clocks as already discussed: the local oscillators inject error into the sampling rates, causing distortions in both waveforms, and the starts of the sampling periods differ by $\Delta\tau$.

Looking first at the sampling-rate inaccuracy, its result will be that, when the sampled wave trains are plotted along the same time axis, one wave train will appear to be a slightly dilated or contracted version of the other. Signal processing might be used to find the time axis multiplier that, applied to the dilated data, contracts them to make their time axis exactly the same as the other. When applied, this relative correction produces two identical waveforms, having a common, unknown, dilation or contraction from their true shape.

As in the double-conversion receiver, this unknown dilation is proportional to the errors in the local clock tick rate. In the direct-digitizing receivers, the phase error will be larger because the frequency f is orders of magnitude larger. Referring again to *Conley* [1992], it was shown that the frequency error in a nominal 10-MHz GPS-derived reference signal is $\Delta f_r \simeq -0.05$ Hz. If the sampling frequency $f_s = mf_r$, then $\Delta f_s = m\Delta f_r$. Realistically, f_s may be as high as 10^8 Hz, giving a error $\Delta f_s = -0.5$ Hz. A sample time T_s has $T_s f_s$ sample periods, a total time error of $T_s f_s \, \Delta f_s / f_s^2$, and hence an equivalent phase error of $2\pi f \, T_s f_s \, \Delta f_s / f_s^2$.

In practical terms, it would be desirable to have $f/f_s = 10^{-2}$. With a sample length $T_s = 1$ ms, $\Delta f_s = -0.5$ Hz, the phase error is $\pi 10^{-5}$. It is again concluded that the distortion on waveforms is negligible. Any associated determination of f with conventional spectral techniques should likewise have negligible error.

It must be appreciated that the present estimates of uncertainty in f are conservative. In these estimates, it is assumed that errors caused by inexact clocking rates at the two satellite locations are noncoherent. A better approach would consider the differential errors in f. This would lead to a lower estimated uncertainty of f in (2). Improved estimates of this sort have not been pursued here simply because the stand-alone frequency synthesis error has been found to be negligible in both receiver types for the scientific measurements under consideration.

This leaves the inaccuracy in ϕ caused by the asynchronous start of the sampling periods at their two locations. On a rf waveform of frequency f, the phase uncertainty is simply $2\pi f \Delta\tau$. The phase itself is determined by standard techniques, for instance, by a calculation of the ϕ lag that maximizes the cross correlation. Given frequencies of interest this error can be significant. In the next Section, this phase error will be used to define the range of f.

3.4 Accuracy of Two-Point Wave Measurements

According to Sections 3.2 and 3.3, the phase error $\Delta\phi$ is significant only in the direct digitizing receiver. A conservative evaluation of the GPS-based interferometer is therefore based on the direct digitizer. A reasonable experimental objective in space science is to measure ϕ to $\pm 5\%$. An important case is the one in which ambiguities of 2π in ϕ are avoided and the "interferometer" of Figure 2 is operated on its main lobe where $\phi < 2\pi$. To stay safely away from ambiguities of 2π in ϕ, let us put $\phi = \pi$, and therefore require that $\Delta\phi \leq \pi/20$. As implied in Section 3.3, $\Delta\phi \leq \pi/20$ becomes $f\Delta\tau \leq 1/40$. This requires that $f \leq 0.76$ MHz for the time error $\Delta\tau = \pm 33$ ns.

Neglecting uncertainty in the refractive index $n(\alpha)$, the differential form of (2) is

$$\frac{d\alpha}{\tan\alpha} = \frac{d\phi}{\phi} - \frac{da}{a} - \frac{df}{f} \qquad (4)$$

which shows that contributions to the uncertainty of the DOA angle α come from uncertainties in ϕ, a and f. The first two terms on the right side depend on $\Delta\tau$, while the third term is negligible according to the preceding argument. Because of their common dependence, it is conceivable that the first two terms in (4) cancel, at least partially. However, without a specific receiver design, it seems better to regard these two errors as noncoherent, in order to obtain a conservative error estimate in this first preliminary discussion.

The $\tan\alpha$ in (4) shows that the best accuracy will be obtained for small α, where $\tan\alpha \simeq \alpha$. The error $\Delta\phi$ is set to $\pi/20$. The separation vector a has errors in magnitude and direction given by the differential location uncertainties of both spacecraft, $\Delta d = \pm 10$ m.

As an example, suppose two co-orbiting receivers are to be used for direction finding on 500-kHz auroral kilometric radiation (AKR). Equation (2) can be used to stipulate the minimal separation a for optimal operation. The condition is the small angle criterion: $\sin\alpha \simeq \alpha < \pi/10$. Putting n

= 1 leads to $a > 300\phi$. The optimal $\phi = \pi$ shows that the baseline should be 1 km or more to have best sensitivity for the measurement of α. In the rotating bolas version of BICEPS, the spin vector is perpendicular to the orbital plane, so it is probable that, over the life of the satellite mission, some auroral-zone passes could have the **a** direction nearly perpendicular to the incident AKR wave vectors, thus providing optimal sensitivity. For the $\Delta\tau = \pm 33$ ns case from Table 1, the percentage uncertainty in a is roughly 10/1000. Then $\Delta\phi/\phi$ dominates the right side of (4) making $d\alpha/\alpha$ of the order of 0.05.

Contrast this with electrostatic waves near the lower hybrid resonance where, say, $n = 100$ and $f = 10$ kHz. It is found that the small-α condition now corresponds to $a > 3000/2\pi$, which is not radically different from the limit found in the AKR case.

3.5 *Related Applications in Wave Observations*

The foregoing treatment of phase path difference between two observation points may be combined with analyses to extract other wave characteristics. First, if the direction of **k** is known in three dimensional space, then (2) can be rearranged to give a measure of n. In the case of DOA measurements on waves from artificial sources, the same approach as above is used. However the coherence of the transmitted waveform presumably is greater, meaning that longer integration times may be employed yielding better accuracy in f and ϕ.

Suppose that the wave receivers in Figure 2 are replaced by particle detectors with correlators. The correlators measure the waveforms left imprinted on the particle fluxes by some wave-particle interaction. Bringing the observed waveforms back to a common time base through the GPS clocks should provide information about where the operative interaction took place.

Receiver and correlator data could be combined in investigations of spatial growth rates of plasma waves in the presence of energetic particle distributions. Suppose that the two satellites are immersed in nonequilibrium electron distributions and that it is desired to test some theory for the spatial growth rate k_i. Assuming convective growth in the space between the two observing spacecraft, the small-signal (linear) instability theory predicts growing amplitudes proportional to $\exp(\mathbf{k}_i.\mathbf{r})$ where **r** is the vector displacement along the group direction. The investigation could start with determining the candidate wave components ω, $\mathbf{k}_r$ as described above, and electron distribution components with velocity **v** obeying some resonance condition $\omega = \mathbf{k}_r.\mathbf{v}$. The accuracy of the resulting scaled $\mathbf{k}_i$ would depend on the characteristic Δd and $\Delta\tau$ uncertainties of the GPS clocks.

4. CONCLUSIONS

Direction finding employing GPS on radiowave emissions observed in space appears to be feasible at frequencies up to at least HF if the L1 differential phase technique is applied. Transionospheric propagation from artificial ground sources can also be studied with this technique.

Assuming that GPS receivers are affordable for small satellites, whereas atomic clock standards are not, the advent of mature, space-qualified GPS receivers should open doors of opportunity in bistatic space-science experiments. GPS receivers have the additional attraction of providing orbital position, thus avoiding traditional ground tracking facilities. Also, the scheme of Figure 2 has the advantages of simplicity and direct access to data over older, more complicated arrangements that were required for this type of measurement [*e.g. Shawhan*, 1979].

REFERENCES

Ackroyd, N. and R. Lorimer, *Global Navigation - A GPS User's Guide*, Lloyd's of London Press, London, U.K., 1990.

BICEPS Team, BICEPS-BIstatic Canadian Experiment on Plasmas in Space, ed. H. G. James, Proposal to Canadian Space Agency, 1993.

Conley, R., GPS performance: what is normal?, *NAVIGATION: J. Institute Navigation 40*, 261-281, 1993.

Hofmann-Wellenhof, B, H. Lichtenegger and J. Collins, *GPS Theory and Practice*, Springer Verlag, Vienna, 1992.

Leick, A., *GPS Satellite Surveying*, Wiley, New York, 1990.

NATO, NAVSTAR Global Positioning System (GPS) system characteristics-preliminary draft, Draft Issue M, Military Agency for Standardization, Standardization Agreement 4294, 1991.

Shawhan, S.D., Description of the ISEE satellite-to-satellite kilometric wavelength interferometer system, *Ann. Télécommun. 35*, 266-272, 1979.

Wells, D.E., N. Beck, D. Delikaraoglu, A. Kleusberg, E.J. Krakiwsky, G. Lachapelle, R. B. Langley, M. Naskiboglu, K.P. Schwarz, J.M. Tranquilla and P. Vaníček, *Guide to GPS Positioning*, Canadian GPS Associates, Fredericton, Canada, 1987.

H.G. James, Communications Research Centre, P.O. Box 11490, Station "H", Ottawa, Ontario K2H 8S2, Canada

A Simulation of the Behavior of a Spherical Probe
Antenna in an AC Field

Robert Manning

DESPA, Observatoire de Paris-Meudon

We have made measurements using a scale model of a typical configuration of a spherical double-probe antenna in an AC electric field. The model was immersed in a partially conductive medium (water) in which a precise uniform AC electric field can be readily applied. This method, which has been called rheographic or rheometric, has often been used to estimate the radiation pattern of a cylindrical antenna in presence of a spacecraft structure of comparable dimensions but we think this is the first time it has been used to study the behavior of a spherical double-probe antenna with guard configuration, about which there has been some concern due to the extension of the spacecraft potential out to the probe vicinity. The results show that the configuration with the guards performs as expected for measurement of AC fields at frequencies below the cut-off frequency of the feedback loop controlling the guard voltage but that the measurements at higher frequencies are significantly degraded.

1. INTRODUCTION

A method of estimating the effective electrical length of an electric antenna in presence of a spacecraft structure of non-negligible dimensions was developed in the 1970's at the Office National d'Etudes et Recherches Aéronautiques (ONERA) in Chatillon, France. This method is only valid for frequencies where the wavelength is long with respect to the antenna and structure dimensions (i.e., electrostatic field structure). The method is described in the next section. DESPA has used it to predict the effective electrical length and the polar diagram orientation of the antennas to be flown on ISEE, Ulysses, Wind, and Cassini [*Rucker, et al.*, 1996] spacecraft. Recently measurements have also been made on a model of the Voyager spacecraft.

Results in the past have shown the method to be very valuable for the evaluation of the effect of spacecraft appendages (for instance magnetometer booms or RTGs) on the behavior of cylindrical electric antennas. We have been interested in the effective receiving efficiency (effective height or length) of the antennas as well as the angular shift in the radiation pattern, which is always dipolar for dimensions which are small with respect to the wavelengths. In this paper we report on a new type of measurement using a model which simulates the conditions of a spherical double-probe configuration with active control of the guard voltage.

1.1 The Spherical Double-Probe Configuration

The spherical double-probe configuration has been developed to circumvent the problems caused by the presence of photoelectrons around sun-lit objects. The technique has been well-described [*Pedersen et al*, 1984].

Figure 1 has been adapted from a figure in [*Gustafsson et al.*, 1993]. It shows the guards and stubs that have been added to the basic double-probe antenna systems (for example on ISEE and Cluster). The guards and stubs are driven to a voltage which is equal to the voltage picked up by the high impedance spherical probe plus a super-

Measurement Techniques in Space Plasmas: Fields
Geophysical Monograph 103

Figure 1. Typical spherical double-probe configuration. Stub and guard configurations vary. This configuration was to be used on CLUSTER I.

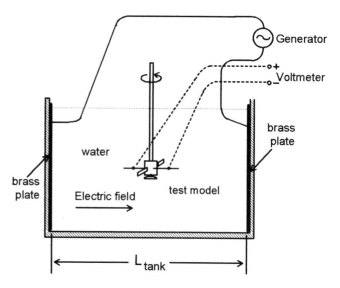

Figure 2. Basic rheographic measurement set-up. The dimensions of the water tank are 200 x 100 x 70H cm. Rotation of the test model gives length and direction of the equivalent dipole.

imposed programmable DC voltage. These superimposed DC voltages are applied to repel the photoelectrons generated by the spacecraft and cable which would "short out" the DC field measurement. This configuration has become accepted as the best method to measure DC electric fields on a spacecraft.

However, the presence of the cable which connects the guards/stubs/spheres to the spacecraft prevents one from considering that the spheres are simple voltage sensors situated at their physical separation. The cable brings the spacecraft ground potential out to the proximity of the spheres and will at least partially short out the measured field. This effect has been discussed briefly in [*Mozer et al.*, 1978] and in [*Pedersen et al.*, 1983].

Some doubt has remained as to the validity of the configuration for the measurement of AC fields. The role of the biasing of the guards and stubs as described in [*Pedersen et al.*, 1983] and [*Gustafsson et al.*, 1993] is to increase the impedance between the cable and the spheres so that the DC measurement is meaningful. This impedance increase will also help very low frequency field measurements but once the impedance becomes dominated by the antenna capacitances the applied DC potential is not expected to make a difference.

2. METHOD

The method consists of immersing a scale model of the spacecraft and antenna system in a large tank of water and applying a low frequency electric field as shown in Figure 2. The drive voltage is applied between the two parallel brass plates completely covering each end of the tank and the high impedance outside the tank insures that the field lines remain straight except in the vicinity of the test model.

The applied field is alternating, in order to avoid electrolysis effects. The operating frequency is usually around 1 kHz. The test models must have very good surface conductivity and thus are generally gold-plated. Insulators are used to isolate the antenna elements from the

spacecraft body and between whatever other elements that need isolation. These insulators must be of good quality and non-porous.

Thin insulated wires are soldered to the antenna elements and the spacecraft body. These are brought out of the tank and the voltages between each are measured using a synchronous detector which filters out all noise not at the drive frequency.

In the general case where it is desired to determine the effective length and orientation of the antenna pattern the model is placed in the middle of the tank using a mounting jig which can rotate the model around the vertical axis. This will give one component of the antenna pattern; the other component is found by re-mounting and turning the model around another axis.

The effective length of the antenna projected into the rotation plane is given by

$$l_{eff} = \frac{V_{out}(max)}{V_{in}} * L * Scale\_Factor$$

where $V_{out}(max)$ is the largest signal measured during the rotation, V_{in} the applied voltage between the ends of the tank, L the length of the tank, and the scale factor is the ratio between the true spacecraft size and that of the model.

The orientation of the dipole is found by determining the angle for which V_{out} is nulled.

In the special case of interest here we are not concerned with the orientation of the antenna pattern. We will be using a symmetrical model with no appendages to shift the orientation of the polar diagram which we thus admit will be exactly aligned with its physical orientation. Therefore we chose to use only a monopole configuration, taking advantage of the symmetry, with no rotation possible. The set-up is shown in Figure 3.

3. RESULTS

Figure 4 shows the dimensions of the model we used for this test. It is intended to be at scale 100:1 for a typical spacecraft mission. The smaller dimensions such as cable diameter and the separators between sections are not scaled properly for practical reasons and the "sphere" was actually modelled with a segment of a cylinder, but these should not affect the qualitative results.

The satellite was modelled with a half-cylinder of 2.4 cm diameter and height. The antenna cable was 50 cm long and 1 mm thick (diameter). The sphere was simulated with a short cylinder, 3 mm diameter and length. Also, only the guards and not the stubs were simulated but neither of these simplifications this should change the qualitative results. The guards were 2x1 and 2x2 cm long (two cases). All of these antenna elements were hollow, allowing passage of a very thin catgut thread with which they were maintained horizontal. Insulation between the elements was assured by small insulating cylinders, 2 mm long and thick.

The first test case was the cable only used as a cylindrical rod antenna, because we knew that this is a case which gives correct results in general. Indeed we measured 26.5

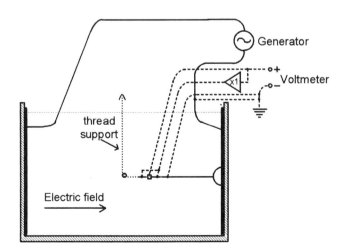

Figure 3. Modified rheographic set-up for spherical probe test. No angular measurements are performed.

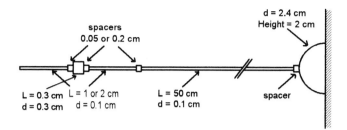

Figure 4. Spherical probe test model.

cm effective length, which is practically identical to the distance to its centerpoint from the ground plane (1.2 + .2 + (50/2) = 26.4).

Then we added the sphere model at the end of the cable and grounded the cable. In this case we found an effective length of 33.5 cm which is 64.7% of the sphere's actual position of 51.75 cm. This would give a loss of 3.8 dB with respect to the theoretical voltage received.

Then we installed the guards. This positioned the sphere at 52.95 cm with the 1 cm guards and 53.95 cm with the 2 cm guards. With the guards driven by an operational amplifier to follow the received voltage on the sphere the effective lengths were found to be 41.7 and 45.0 cm respectively, or 78.8% (-2.1 dB) and 83.4% (-1.6 dB). These results are in general agreement with those reported by [*Mozer et al.,* 1978] and [*Pedersen et al.,* 1983] (60-70%) with the very short (20 cm ≡ .2 cm here) ISEE-1 guards.

The presence of the driven guards helps a bit the AC performance of the antenna but we are concerned about what happens when the receiving frequency is higher than the cut-off frequency of the guard's amplifier. In this case we expect the guard voltages to approach those of the spacecraft ground. So we also simulated this case. This gave respectively 30.2 and 26.5 cm, which are only 57.0% (-4.9 dB) and 49.1% (-6.2 dB) of the physical length.

We also considered the case where the guards' voltage did not go down to zero but were instead floating, as if there were a low-pass filter in series with the amplifier's output. This gives results which are close to those of the normal guard configuration: 41.0 and 44.2 cm, 77.4% (-2.2 dB) and 81.9% (-1.7 dB). However this is probably not a realistic case because cable capacitance would prevent any filter from being effective.

We realized that the rather long simulated insulators (20 cm at full scale) will have an effect of lowering the resulting attenuation. To test this we repeated the 1 cm case for the three options (guards active, grounded, open) with .5 mm spacers (≡ 5 cm which is still long with respect to the real case). The three results with the 2 mm spacers

(-2.1, -4.9, -2.2 dB) became (-2.5, -7.2, -2.6 dB). Thus the degradation was slight for the active and open cases but was severe for the grounded (high frequency) case.

In a final test we took off the outer guards. This did not change things much for the active and open cases but improved the grounded case substantially (-7.2 -> -4.7 dB).

4. CONCLUSION

The spherical double-probe configuration does not measure AC electric fields as accurately as does the simple wire antenna. The presence of guards which are driven to the sphere potential increases the precision at frequencies for which the drive works, but is detrimental above the cut-off frequency of the feedback amplifiers.

The loss of signal due to the presence of the connecting cable is of the order of 4 dB without guards, is reduced to 2 dB with guards, but becomes at least 7 dB above the frequency where the feedback loop is operational. In other words, the effective length is only about 40% of the real length. This is however a frequently observed range since usually the cut-off frequency of the feedback amplifiers is quite low (a few hundreds of Hertz) because of the high cable capacitance. In the recent CLUSTER 1 development the equivalent cut-off frequency for the guards was significantly raised (>100 kHz) by the addition of a capacitor between the output of the preamplifier and the guards themselves (unfortunately a similar action was not taken for the stubs). In addition there will be phase effects in the frequency range around the cut-off frequency. These amplitude and phase uncertainties would be very deleterious for observations such as measurement of planetary and solar radio emissions where good accuracy is important for direction of arrival measurements, and for evaluation of quasi-thermal noise.

Some thought could be given to eliminating the outer guards and stubs and even to have unconnected inner guards in the event of a mission where important science is potentially attainable both at very low and at higher frequencies and it is not feasible to implement both spherical double-probes and cylindrical antennas.

Acknowledgements. We thank Roger Hulin, Jean-Pierre Rivet, et Alain Rapin for their assistance in the fabrication of the test model and the set-up and testing in the water tank.

5. REFERENCES

Gustafsson, G., Boström, R., Holback, B., Holmgren, G., Stasiewicz, K., Aggson, T., Pfaff, R., Block, L.P., Fälthammer, C.-G., Lindqvist, P.-A., Marklund, G., Cattell, C., Mozer, F., Roth, I., Temerin, M., Wygant, J., Décréau, P., Egeland, A., Holtet, J., Thrane, E., Grard, R., Lebreton, J.-P., Pedersen, A., Schmidt, R., Gurnett, D., Harvey, C., Manning, R., Kellogg, P., Kintner, P., Klimov, S., Maynard, N., Singer, H., Smiddy, M., Mursula, K., Tanskanen, P., Roux, A., Woolliscroft, L.J.C., The Spherical Probe Electric Field and Wave Experiment for the Cluster Mission, *ESA SP-1159*, 17, 1993.

Pedersen, A., Cattell, C.A., Fälthammer, C.-G., Formisano, V., Lindqvist, P.-A., Mozer, F., Torbert, R., QuasiStatic Electric Field Measurements with Spherical Double Probes on the GEOS and ISEE Satellites, *Space Science Reviews, 37*, 269, 1984.

Mozer, F.S., Torbert, R.B., Falheson, U.V., Falthammer, C.G., Gonfalone, A., Pedersen, A., Measurements of Quasi-Static and Low-Frequency Electric Fields with Spherical Double Probes on the ISEE-1 Spacecraft, *IEEE Trans. Geoscience Electronics, GE-16*, 258, 1978.

Rucker, H.O., Macher, W., Manning, R., Ladreiter., H.P., Cassini model rheometry, *Radio Science, 31*, 1299, 1996.

Robert Manning, Observatoire de Paris-Meudon, DESPA, 5, place Janssen, 92195, MEUDON CEDEX, France

Electromagnetic Emissions in the Ionosphere - Pulsed Electron Beam System

A.Kiraga and Z.Klos

Space Research Centre, PAS, ul.Bartycka 18A, 00-716 Warsaw, Poland

V.N.Oraevsky, V.C.Dokukin, and S.A. Pulinets

IZMIRAN, Troitsk, Moscow Region, 142092, Russia

Numerous rocket and satellite projects were devoted to study of astrophysical plasma with the aid of active electron beam experiments. The quality and volume of wave data from such experiments did not fulfill original expectations due to complexity of involved processes, technical malfunctions and limited diagnostic. There were several cases when pulsed electron beam had been injected from the APEX satellite into an otherwise unmodified ionospheric plasma. Diversely structured beam induced emissions were registered. In many cases, a very prominent doublet could be convincingly identified as an upper hybrid band. Those spectra with sharp peaks below electron gyroharmonics were successfully resolved in the framework of cold plasma - diluted weakly relativistic beam system. Electromagnetic X, O, Z mode emissions relevant for astrophysical applications were identified. The identification procedure, characteristic spectra and experimental conditions are discussed. Reproducibility of distinct spectral features and their sensitive dependence on the ratio f_n/f_c of ambient plasma f_n and gyro f_c frequencies support the concept of simulation of astrophysical plasmas in active experiments. Controlled creation of such environments along low altitude orbits provides opportunity for evaluation and development of measurement techniques.

1. INTRODUCTION

Active experiments utilizing electron beams in space have been carried out for about 25 years. The main scientific objectives were to try to understand the structure and geophysical processes in the Earth magnetosphere and the elementary interactions in beam - plasma system in application to astrophysical objects. The outcome of these experiments didn't fulfill initial expectations based on the assumption that a practically unbounded magnetospheric plasma provides a simpler environment than a plasma chamber. The main obstacles in space experiments followed from spacecraft neutralization, evolution of an initially dense beam, limited space-time diagnostics, changes of ambient plasma parameters and technical malfunctions. In effect, the experimental conditions were significantly different from anticipated, idealized homogeneous system composed from low temperature plasma pervaded by electron beam with known velocity distribution. Generally

Measurement Techniques in Space Plasmas: Fields
Geophysical Monograph 103

rockets experiments were better instrumented but suffered from background variability and technical malfunctions.

The above listed problems seriously affected wave data particularly in the frequency range close to and above the local electron gyrofrequency f_c, which is important for astrophysical applications. Emissions below f_c were reported most frequently. For ambient plasma frequency $f_n > f_c$, emissions close to f_n were reported [Cartwright et al., 1974; Dechambre et al., 1980] but only in a few cases was an upper hybrid doublet $(f_n, f_u = (f_n^2 + f_c^2)^{1/2})$ resolved and f_n was subsequently used in an attempt to identify diffuse resonances Q_n [Kawashima and Akai, 1985] and D_n [Goerke et al., 1993]. Rich harmonic structure related to f_c was reported [Mourenas and Beghin, 1991] and interpreted as electrostatic Bernstein waves. The electromagnetic character of emissions close to $2f_c$ was claimed [Winglee and Kellog, 1990] but without detailed data about ambient plasma frequency f_n. Emissions in the frequency bands listed above were registered during injections of a pulsed electron beam from the APEX satellite [Kiraga et al. 1995a].

In this paper the interactions inferred from HF spectra are invoked to obtain broad scope of experimental conditions rather than to study in depth the respective physics. Spectra characteristics definitely depend on the ratio f_n/f_c and pitch angle of the electron beam. The overall consistency received in the full experimental range $1.2 < f_n/f_c < 4.1$ supports both the identification procedure and the concept of simulation of astrophysical plasmas in active beam plasma experiments. An experimental arrangements and diagnostics are described in Section 2. The overview of experimental data and identification of (f_n, f_u) doublet are given in Section 3. Section 4 presents the identification procedure of upper hybrid and gyroharmonic X, Z, and O mode emissions in the framework of a cold plasma - diluted, weakly relativistic beam system. Discussion is presented in Section 5.

2. EXPERIMENTAL ARRANGEMENTS

The APEX satellite was launched on December 18, 1991 into an orbit with apogee of 3080km, perigee of 440km and inclination of 82.5^0. The satellite was stabilized along three axes with slowly changing deflections having amplitudes of the order of 20^0. The most detailed desciption of particle accelerators was published by Dokukin [1990]. Two electron guns were located at the aft side of the spacecraft with their axes directed perpendicularly and almost parallel to the satellite vertical axis. The initial divergence of electron beam was close to 4^0. For safety reasons it was planned for the initial phase of operation that time intervals with elec-

tron beam injections would be imbedded into time intervals with Xe^+ plasma injections from an accelerator located in the aft section too. In the synchronized modes, the electron beam formats featured sequences of 1-s periods of electron beam injections with variable rates of $2\mu s$ or $32\mu s$ long pulses and three levels of beam intensity. In a backup mode the electron gun was injecting $2\mu s$ long pulses at a fixed rate of about 45kHz for several minutes. The unstabilized acceleration voltage was of the order of 10keV. Within a pulse the beam intensity did not exceed 0.15A. The HF diagnostics was aimed to analyze signals from three kinds of antennas and from three current sensors of accelerators. The most comprehensive data were expected from two ADC devices with 12MHz sampling rate which were supported by two 64kbyte registers. Synchronous snap shots for various pairs of signals while gun operation in synchronized modes would enable study on a microsecond time scale. However during protecting Xe^+ injections the registered wave activity did not show any structure which could be related to local plasma frequency. No useful signals were received from accelerators as well.

Due to recoverable malfunction of the plasma gun, there were three prolonged intervals, when the electron beam was injected together with neutral Xe leaking at the rate 3×10^{-6} kg s^{-1} . They were preceded by much shorter intervals with e-beam injection only. The electron gun operated in the backup mode. The HF receiver was connected to a dipole antenna of half lengths of 7.5m and operated in passive mode providing one spectrum every 8s. The single spectrum was measured in 1s with an equally spaced mesh of 200 frequencies starting from 100kHz with a step of 50kHz and 3 0kHz bandwidth. The time constant of an amplitude peak detector was set to 0.6ms. The wave data from these experiments are analyzed in following sections.

3. OVERVIEW OF EXPERIMENTAL DATA

The essential features of spectra excited during pulsed beam injections are presented in Figure 1. The largest amplitudes are shown black. Thin solid lines superimposed on the spectra denote the local gyrofrequency of thermal electrons computed from IGRF95 model and its even harmonics. Only in Figure 1b the third harmonic is added. LAT, LON, ALT, MLT denote geographic latitude, longitude, altitude and magnetic local time. The earliest spectrum is at the top of each Figure. The power was supplied to accelerators during intervals pointed by bold horizontal lines on the left margins.

The start of beam injection is marked with the letters a, d and b in Figures 1a, 1b, and 1c respectively. Beam injection was terminated by the power supply switch off. The lower

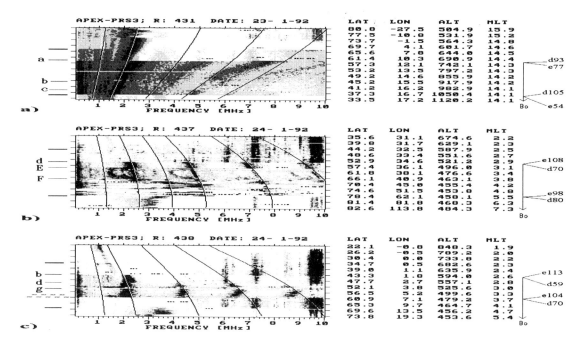

Figure 1. Electron beam induced spectra in afternoon (panel a) and in post midnight ionosphere (panels b and c). The injection UT time intervals are(12:17:51-12:24:20), (00:12:03-00:18:26), and (02:11:01-02:15:20) respectively.

case letters are used to mark the beginning of similar spectral patterns too. The capital letters mark the beginning of intervals in which three or more distinct patterns occurred. Approximate angles between the Earth's magnetic field B_0 and dipole antenna (d) and electron gun (e) are depicted for the start and the end of electron beam injections with values shown by the symbols.

Prior to beam injection in Figure 1a, the most pronounced structure corresponds to resonant reception of waves launched by the satellite electrical systems. Characteristic plasma emissions in the band (f_n, f_u) can not be distinguished due to their small amplitude and mismatch of input network in this band. After injection of an electron beam, the very prominent doublet emerges at frequencies corresponding to the (f_n, f_u) band of the ambient plasma [Kiraga et al., 1995b]. This is illustrated in detail in Figure 2. The relation $f_u = (f_n^2 + f_c^2)^{1/2}$ is well fulfilled for an assignment of doublet peaks to f_n and f_u until lower frequency peak approaches closely to $2f_c$ harmonic. The harmonics of doublet structure are excited (a pattern "a"). Initially only harmonics of the higher frequency doublet member f_u are present. As the lower frequency member f_n approaches $2f_c$ its harmonics also appear. The harmonic structure of f_c solely independent from doublet members is absent. As the f_u comes closer to $2f_c$, the harmonics fade (a pattern "b"). They reappear as the harmonics of lower frequency doublet

member located very close to $2f_c$ at the time when amplitude of f_u peak significantly diminishes (a pattern "c").

The post midnight recordings in Figure 1b are much more complex. Prior to electron beam injection the spectral structure related to the local plasma frequency is hardly visible. Constant frequency bands above 6MHz are due to broadcasting HF transmitters. With the onset of electron beam injection, the reception of these signals is cut off and strong emissions close to even gyroharmonics appears (a pattern "d"). This harmonic structure is subsequently supplemented by a doublet structure below $2f_c$ and faint emissions above $2f_c$ (a pattern "E"). Later on, selectively excited emissions close to $3f_c$ or $4f_c$ dominate in the spectra (a pattern "F"). The assumption that their evolution along the orbit is related to ambient plasma frequency f_n seems to be reasonable. The spectra presented in Figure 1c confirm and complement previous observations. Initially they show the pattern "b" instantaneously with reduction of broadcasting signals. Synchronously with increase of telemetred values of average beam intensity, the pattern "d" appears and the broadcasting signals are cut off. Next, the decoupling from cold plasma gyroharmonics develops (a pattern "g"). Spectra marked with wavy lines are influenced by the degradation of electron gun. Disruption of the telemetred beam intensity results in disappearing of discrete plasma emissions and reappearing of broadcasting signals. The detailed

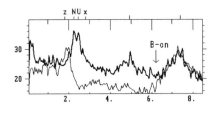

FREQUENCY [f/f$_C$]

Figure 2. Two consecutive spectra showing dramatic change in the vicinity of local plasma frequency due to beam injection. The spectrum given with thin line is characteristic for ambient plasma to the left from the arrow and increases to the right from the arrow due to the onset of beam injection. The next spectrum is given with bold line. Thermal electron even gyroharmonics are marked with long vertical lines. Normalized characteristic cold magneto-plasma frequencies: $n=f_n/f_c$, $u=f_u/f_c$, $z=f_z$ and $x=f_x$ are marked at the top of each spectrum. $f_z=-0.5+[0.25+(f_n/f_c)^2]^{1/2}$ and $f_x=0.5+[0.25+(f_n/f_c)^2]^{1/2}$ are cut off frequencies for Z and X modes. In this and following Figures capital letters U or N point maximum used to calculate characteristic frequencies. Unmarked lines above the upper frame point harmonics of U.

analysis of patterns "E" and "F" reveals electromagnetic nature of gyroharmonic emissions.

4. UNFOLDING OF ELECTROMAGNETIC EMISSIONS

A multitude of processes develop after injection of space limited, pulsed, charge uncompensated electron beam into the magnetosphere due to redistribution of ambient plasma density, evolution of injected beam in phase space and generation of non thermal populations. It is very challenging to assign the details of the spectral structures to definite

sources using only wave data from the single dipole antenna placed in the center of the initial perturbation, with time resolution set for measurements in stable ionosphere. Fortunately, experimental conditions appeared very favorable due to systematic changes and reproducible features in the measured spectra. The data presented in Figures 1a and 2 show that intense emission was recepted at characteristic frequencies close to ambient plasma frequency f_n and that the ratio of f_n/f_c strongly controls the overall spectral structure. Attempts to explain spectra in Figures 1b and 1c in terms of either Q_n or D_n resonances or electrostatic Bernstein waves failed. The break through point in their interpretation was crossed after identification of main peak slightly below $3f_c$ occurring in several spectra in the pattern "F", as a Z mode cyclotron beam emission. The pattern "F" spectra were registered in a polar zone where significant non monotonic changes of ambient f_n occurred along the orbit. The identification is explained in Figure 3. Total of four spectra similar to that presented in Figure 3a were registered. With an assumption that maximum of pronounced emission above $3f_c$ is located at ambient plasma frequency f_n (N mark), the upper hybrid frequency f_u and cut off frequencies f_z and f_x for Z and X electromagnetic modes were calculated. The main maximum is located between Z mode cut off and $3f_c$. An application of this procedure to the spectra presented in Figures 3b-3d ordered according increasing frequency of emission assigned to the (f_n,f_u) band shows clear relation between the calculated value of Z mode cut off frequency and the structure of gyroemission close to $3f_c$. Diminishing and disappearance of $3f_c$ emission is accompanied by appearance of strong emission below $4f_c$ in Figures 3c and 3d which falls in frequency range favoring cyclotron beam O mode generation. The Z mode cyclotron beam emission reappears below $4f_c$

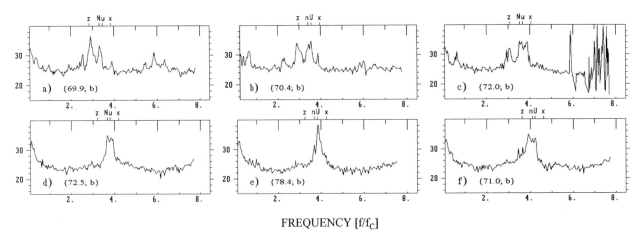

FREQUENCY [f/f$_C$]

Figure 3. Selected spectra illustrating diversity of radiation generation for $3.3<f_n/f_c<4.1$. Amplitude is measured in dB relative to $1\mu V$. A letter and a number in {} give reference to the panel in Figure 1 and the latitude of registration.

in Figure 3f. This is also the case when the greatest value of f_n was encountered.

Electromagnetic origin of gyroharmonic emissions when the band (f_n, f_u) is separated from them is further supported by resolving the pattern "E" spectra. Selected spectra ordered with f_n/f_c are presented in Figures 4a-4f for critical values of f_n/f_c. All identifications were done with assignement of upper hybrid frequency to maximum of selected emission band pointed with "U". For minimum value of f_n/f_c, an intensity of upper hybrid emission is on the background level. Intense, narrow band emission is located between X mode cut off and $2f_c$. Harmonics of this emission are clearly located below gyroharmonics computed for thermal electrons. There exist substantial emission close to f_n. When the X mode cut off $f_x = 1.93f_c$ is close to frequency $F_x = 1.96f_c$ of the maximum below $2f_c$, the low frequency slope becomes very steep (Figure 4b). For further increase of f_x, the maximum diminishes in amplitude and shifts from F_x to $2f_c$ and slightly above $2f_c$ (Figures 4c and 4d). At the same time the harmonics at $2F_x$ are most pronounced so their instrumental origin can be ruled out. The extremely sensitive control of gyroharmonics structure on X mode cut off frequency proves their electromagnetic character and invokes X mode cyclotron maser mechanism. For the data in Figures 4c and 4d efficient maser mechanism operates in the remote region from the satellite what is manifested by the absence of the fundamental F_x and the presence of its second harmonic. Presented characteristics of maser emis-

sions confirm correct identification of upper hybrid emission. For still further increase of f_n/f_c the emission below $2f_c$ reappears in condition favoring O mode cyclotron beam mechanism. There is no gyroharmonic structure but instead the harmonic structure of upper hybrid emission develops what is illustrated in Figures 4e-4h. For f_u sufficiently close to $2f_c$, the O mode beam emission disappears (Figure 4h). It is worthwhile to point out similarity of spectra in Figures 3e and 4h where upper hybrid emission and its harmonics are the only very pronounced spectral structures. For further increase of f_n/f_c, the coupled $(f_n, f_u, 2f_c)$ emissions appear. An example of such a spectrum is presented in Figure 4i.

5. DISCUSSION

It was shown in two preceding sections that divers spectral structures above local gyrofrequency can be consistently interpreted in terms of basic emissions located close to ambient plasma frequency f_n, around ambient upper hybrid frequency f_u, and at frequencies of X, Z, and O cyclotron beam modes. For experimental arrangement characterized by antenna inclination and beam injection close to perpendicular to the Earth magnetic field, the spectral structures are extremely sensitive to the ratio f_n/f_c. The values of f_n were obtained from the spectra. The relativistic mass of 10keV beam electron pulls down its gyrofrequency to the value $0.98f_c$ and locates beam gyroharmonics well below

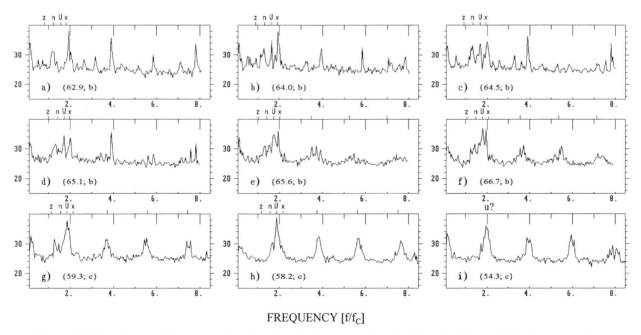

FREQUENCY [f/f_C]

Figure 4. Selected spectra illustrating diversity of radiation generation for $1.2 < f_n/f_c < 2.1$. Amplitude is measured in dB relative to 1μV. A letter and a number in { } give reference to the panel in Figure 1 and the latitude of registration.

thermal plasma gyroharmonics. Excitation of electrostatic waves in a large volume around the satellite is indirectly manifested in the spectra through broadcasting signals cut off. Lack of the Beam Plasma Discharge signatures calls forth an efficient scattering of broadcasting waves on electrostatic waves.

There are many published theoretical treatments of beam plasma interactions [Freud et al., 1983; Kainer et al., 1988; Winglee and Kellog, 1990] for the parameters relevant to APEX data which compare qualitatively well with isolated spectral features, but full understanding of the coexistence and competition between various instabilities requires dedicated simulations which would take into account several unique characteristics of experimental setup. Kinematics considerations show that due to initial divergence of the electron beam, 2-μs beam pulses propagate initially as the separated, approximately 2-wound helixes which start to mix at a distance of several hundred meters into a continuous rotating, annular beam. Its energy spread is significantly less than beam energy [Zbyszynski and Klos, 1995]. So, quite different beam plasma interactions may prevail in the close to the satellite and in the more remote regions. In the continuous beam region, the competition between electromagnetic instabilities with frequencies close to cyclotron harmonics [Sprangle, 1976] and those close to the upper hybrid frequency [Kino and Gerchberg, 1963] may be responsible for gross spectral structures. The beam bunching can excite the harmonics. However, the interaction of a train of beam helixes with the plasma occurs on the time and spatial scales which were not explored neither in laboratory [Chan and Stenzel, 1994] nor in space experiments and challenge the theory as well as diagnostic techniques. The strong electromagnetic radiation is accompanied by intense electrostatic waves evidenced by blocking the broadcasting radiation. The spatial localization of the sources of electrostatic waves is unknown. It may be impossible to unfold from our time averaged spectra the amount of electromagnetic radiation generated due to electrostatic waves as well as the respective mechanisms. The viable mechanisms may involve ion dynamics as in the case of modulational instabilities evolving into strong turbulence [Akimoto et al., 1989; Christiansen et al., 1984; Newman et al., 1994] or occur on shorter time scales [Intrator et al., 1984; Chan and Stenzel, 1994]. The destructive influence of Xe^+ beam on the spectra bolsters assumption that the ion plasma dynamics if significant in pulsed electron beam - plasma interaction should develop in accord with the electron plasma dynamics driven by the electron beam. The diversity and reproducibility of the wave data coupled with presented understanding of underlying physical processes and controlling parameters strongly support the concept of

simulation of astrophysical plasmas in active electron beam - space plasma experiments. An access to such plasmas on a suitably chosen parts of low altitude orbits can be very profitable for evaluation and development of measurement techniques.

REFERENCES

Akimoto K., H. L. Rowland, and K. Papadopoulos, Electromagnetic radiation from strong Langmuir turbulence, *Phys. Fluids,* 31, 2185-2189, 1988.

Cartwright D. G., P. J. Kellog, Observation of radiation from an electron beam artificially injected into the ionosphere, *J.Geophys.Res.,* 79, 1439-1457, 1974.

Chan L. Y., and R. L. Stenzel, Beam scattering and heating at the front of an electron beam injected into a plasma, *Phys. Plasmas,* 1, 2063-2071, 1994.

Christiansen P. J., J. Etcheto, K. Ronmark, and L. Stenflo, Upper hybrid turbulence as a source of nonthermal continuum radiation, *Geophys. Res. Letters,* 11, 139-142, 1984.

Dechambre M., G. A. Gusev, Yu. V. Kushnerevsky, J. Lavergnat, R. Pellat, S. A. Pulinets, V. V. Selegey, and I. A. Zhulin, High-frequency waves during the Araks experiment, *Ann. Geophys.,* 36, (3), 333-339, 1980.

Dokukin V., The scientific instrumentation of the APEX project, *APEX Project. Proceedings of the International Conference on Active Experiments,* Lipetsk, November 1990, (in Russian), pp. 16-29, Nauka, Moscow, 1990.

Freud H. P., H. K. Wong, C. S. Wu, M. J. Xu, An electron cyclotron maser instability for astrophysical plasmas, *Phys. Fluids,* 26, 2263-2270, 1983.

Goerke R. T., P. J. Kellog, and S. J. Monson, Observation of diffuse resonance emissions from low-current electron beam injections in the auroral zone ionosphere, *J. Geophys. Res.,* 98, 5949-5958, 1993.

Intrator T., C. Chan, N. Hershkowitz, and D. Diebold, Nonlinear self-contraction of electron waves, *Phys. Rev. Lett.,* 53, 1233-1235, 1984.

Kainer S., J. D. Gaffey, Jr., C. P. Price, X. W. Hu and G. C. Zhou, Nonlinear wave interactions and evolution of a ring-beam distribution of energetic electrons, *Phys. Fluids,* 31, 2238-2248, 1988.

Kawashima N. and K. Akai, Non linear wave excitation in the magnetosphere by an electron beam emission from the satellite JIKIKEN, *J. Atmos. Terr. Phys.,* 47, 1307-1309, 1985.

Kino G. S., and R. Gerchberg, Transverse field interactions of a beam and plasma, *Phys. Rev. Letters,* 11, 185-187, 1963.

Kiraga A., Z. Klos, V. Oraevsky, V. Dokukin, and S. Pulinets, Observation of fundamental magnetoplasma emissions excited in magnetosphere by modulated electron beam, *Adv. Space Res.,* V15, (12), 21-24, 1995a.

Kiraga A., Z. Klos, V. Oraevsky, S. Pulinets, V. Dokukin, and E. P. Szuszczewicz, Estimation of plasma density from wave data of cold electron plasma, *Adv. Space Res.,* 15, (12), 143-146, 1995b.

Mourenas D., and C.Beghin, Packets of cyclotron waves induced by electron beam injection from the space shuttle, *Radio Science,* 26, 469-491, 1991.

Sprangle P., Excitation of electromagnetic waves from a rotating annular relativistic e-beam, *J. Appl. Phys.,* 47, 2935-2940, 1976.

Winglee R. M., P. J. Kellog, Electron beam injection during active experiments, 1. Electromagnetic wave emissions, *J. Geophys. Res.,* 95, 6167-6190, 1990.

Zbyszynski Z., Z. Klos, Negative body potential as result of modulated electron beam injection into ambient plasma, *Adv. Space Res.,* V15, (12), 29-32, 1995

Dokukin V.C., IZMIRAN, Troitsk, Moscow Region, 142092, Russia; Email: dokukin@charley.izmiran.rssi.ru;

Kiraga A., Space Research Centre, PAS, ul.Bartycka 18A, 00-716 Warsaw, Poland, Email: kiraga@cbk.waw.pl;

Klos Z., Space Research Centre, PAS, ul.Bartycka 18A, 00-716 Warsaw, Poland, Email: director@cbk.waw.pl;

Oraevsky V.N., IZMIRAN, Troitsk, Moscow Region, 142092, Russia

Pulinets S. A., IZMIRAN, Troitsk, Moscow Region, 142092, Russia; Email: pulse@charley.izmiran.rssi.ru;

Radio Remote Sensing of Magnetospheric Plasmas

James L. Green[1], William W. L. Taylor[2], Shing F. Fung[1], Robert F. Benson[1], Wynne Calvert[3], Bodo Reinisch[3], Dennis Gallagher[4], and Patricia Reiff[5]

With recent advances in radio transmitter and receiver design, and modern digital processing techniques it should be possible to perform remote radio sounding of the magnetosphere utilizing methods perfected for ionospheric sounding over the last two decades. Like ionospheric sounding, free-space electromagnetic waves, launched within a low density region will reflect at remote plasma cutoffs. The location and characteristics of the plasma at the remote reflection point can then be derived from measurements of the delay time, frequency, and direction of an echo. A magnetospheric radio sounder, operating at frequencies between 3 kHz to 3 MHz could provide quantitative electron density profiles simultaneously in several different directions on a time scale of minutes or less. The test of this technique will not have to wait long, since the first magnetospheric radio sounder will fly on the Imager for Magnetopause-to-Aurora Global Exploration (IMAGE) mission to be launched in the year 2000. A simulation of radio remote sensing of the magnetosphere from the IMAGE orbit, reported here, was accomplished by using ray tracing calculations combined with specific radio sounder instrument characteristics. The radio sounder technique should provide a truly exciting opportunity to study global magnetospheric dynamics in a way which was never before possible.

1. INTRODUCTION

Like a radar, a radio sounder transmits and receives coded electromagnetic radio pulses. A basic radio sounder measures the time delay between the transmitted pulse and the echo. The time delay measurement is then converted into a distance. In a magnetized plasma the reflection location depends on the wave mode or polarization. There are two wave modes, the ordinary O and the extraordinary X. Reflection of the O mode occurs where the sounder wave frequency equals the plasma frequency (f_p) which is a function of the local electron density (N_e):

$$f_p \approx 9\sqrt{N_e} \qquad (1)$$

(where f_p is expressed in kHz and N_e in cm^{-3}). This condition forms the basis for remotely measuring the plasma density from such reflections since N_e can be directly obtained. An analogous cutoff involving both the local plasma and cyclotron frequencies exists for the X mode [Fung and Green, 1996].

As the sounder frequency is increased, the waves penetrate to greater distances, into regions of larger N_e, yielding echoes with larger delay times. By inverting the resulting echo delay as a function of frequency, N_e as a function of distance from the spacecraft can be determined.

2. BACKGROUND AND HERITAGE

Investigation of the ionosphere using radio sounding techniques dates back more than a half century to the ground-

[1]NASA Godard Space Flight Center, Greenbelt, Maryland
[2]Hughes STX Corporation, Lanham, Maryland
[3]University of Massachusetts, Lowell, Massachusetts
[4]NASA Marshall Space Flight Center, Huntsville, Alabama
[5]Rice University, Houston, Texas

Measurement Techniques in Space Plasmas: Fields
Geophysical Monograph 103

based experiments using selected fixed frequencies by *Breit and Tuve* [1926]. Early experiments indicated the need for swept-frequency sounders which were developed and evolved into a global network of sophisticated instruments [*Brown*, 1959]. The resulting $N_e(h)$ profiles of the bottomside ionosphere provided major input to the goals of the International Geophysical Year [*Berkner*, 1959].

As technology advanced, ionospheric swept-frequency sounders were incorporated into satellites in order to obtain $N_e(h)$ profiles of the topside ionosphere (above the density maximum) which is inaccessible to ground-based sounders. Alouette 1 & 2 and ISIS 1 & 2 satellites provided data critical for the success of one of the most long-lasting international space programs, producing more than 50 satellite-years of swept-frequency ionospheric topside-sounder data [*Jackson*, 1986]. In addition to producing a wealth of information on vertical $N_e(h)$ profiles and orbit-plane N_e images, the Alouette and ISIS satellites also demonstrated that active radio sounders could operate in a manner compatible with other instruments on the same spacecraft. This compatibility was particularly well illustrated with ISIS 2 which included two independent optical instruments and produced the first auroral images from space [*Lui and Anger*, 1973; *Shepherd et al.*, 1976].

The greatest advance in radio sounding techniques has been in bottomside (ground-based) ionospheric sounders over the last two decades [*Hunsucker*, 1991]. In addition to measuring the amplitude and time delay of the returned pulse, as done traditionally by analog ionosondes, advanced ionospheric digital sounders measure the frequency, phase, Doppler spectrum, polarization, and direction of arrival of the echoes. Of relevance to the present work is the Digisonde Portable Sounder (DPS) developed at the University of Massachusetts, Lowell [*Reinisch et al.*, 1992]. An important feature of the DPS is the high degree of software controlled flexibility in mode of operation and measurement format. Large scale ionospheric structures can be imaged by the DPS in the form of sky maps that show the location of a multitude of reflection points [*Reinisch et al.*, 1995].

3. MAGNETOSPHERIC RADIO IMAGING

The design of the Radio Plasma Imager (RPI) on the IMAGE spacecraft is based on the DPS. The IMAGE mission will be launched in early 2000. Analogous to the DPS, the RPI should be able to perform repetitive remote sensing of N_e structures and dynamics in the magnetosphere and plasmasphere. More detailed information on the science objectives to be accomplished by IMAGE and RPI can be found in *Green et al.*, [1996].

A schematic magnetospheric N_e profile is shown in Figure 1. The IMAGE spacecraft is located at its apogee of approximately 7 R_E altitude. At this location, RPI will be

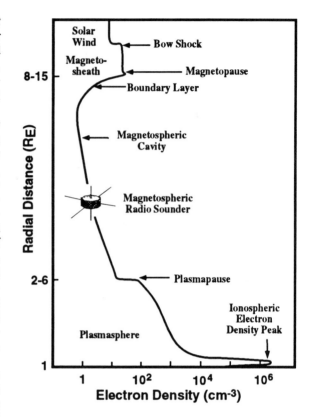

Figure 1. Schematic density profile of the magnetosphere. Under certain conditions the RPI on IMAGE should be able to measure the density profile from the magnetopause to well inside the plasmasphere.

able to remotely measure N_e values from 0.1 to 10^5 cm^{-3} corresponding to f_p (or RPI frequency) values from about 3 kHz to 3 MHz.

4. INSTRUMENT CHARACTERISTICS

The feasibility and characteristics of a radio sounder for remote sensing of magnetospheric plasmas has been studied extensively by *Franklin and Maclean*, [1969], *Green et al.*, [1993] and *Calvert et al.*, [1995; 1997]. The ability of RPI to measure various magnetospheric boundary locations and characteristics depends on RPI's ability to receive detectable echoes and to determine their directions of arrival.

The direction of arrival for each echo can be determined from a 3-axis dipole antenna system. The RPI antenna system will consist of two crossed 500 m tip-to-tip thin wire dipole antennas in the spin plane and a 20 m tip-to-tip thin wire dipole along the IMAGE spin axis. IMAGE will have a 2 minute spin period. Studies indicate that the power amplifiers should have a maximum peak power of 10 W and maximum antenna voltage of 3 kV, and be able to drive the spin plane antennas to transmit right/left-hand circularly (perpendicular to the spin plane) or elliptically polarized signals. The echo arrival angles are calculated from the returning signal voltages and phases on the three orthogonal antennas. The accuracy of the measurement of the echo arrival direction varies directly with the signal-to-noise (S/N) ratio of the echo. For example, for a S/N ratio of 40 dB the accuracy is 1° [*Calvert et al.,* 1995; 1997].

To allow the most efficient use of available spacecraft resources, techniques must be utilized to maximize the S/N of the echo. For magnetospheric sounding, the large echo travel distances, low transmitter power, and small antenna length (relative to the wavelength) will require, in some cases, on-board signal processing in order to produce an adequate S/N ratio. Several techniques are in use in ground-based ionospheric sounders [*Reinisch et al.,* 1992]. In addition to a multitude of pulse operation modes, pulse compression and coherent spectral integration (Fourier Transform) techniques in use by the DPS are essential elements of the RPI instrument.

Pulse compression requires the transmission of a phase coded pulse. The convolution of the echo with the known transmission code then produces an enhancement in the resulting combined signal when the phases of the echo and code match exactly, while the background noise signals would be minimized by its random phases. Spectral integration requires Fourier analysis of multiple echoes. The result is the sum of the echo amplitudes with the same Doppler shift.

The combination of these two techniques produces a S/N gain given by:

$$S/N = S/N_0 \sqrt{mn} \qquad (2)$$

where m is the number of coded pulses, n the number of chips, and N_0 is the original noise before digital integration [*Calvert et al.,* 1995]. A 16 chip pulse compression and a 8 point spectral integration would yield S/N improvement by a factor of 11.3 and hence a 21 dB increase in S/N. Assuming receiver noise plus cosmic noise levels for the RPI instrument, the calculated signal-to-noise ratio before (right axis) and after (left axis) digital processing magnetopause, plasmapause, and plasmasphere echoes are shown in Figure 2 [after *Calvert et al.,* 1995]. The calculations assumed RPI at 6 R_E, the magnetopause at 10 R_E and the plasmapause at the L=4 dipole L shell. The curves indicate a range of possible conditions and assume total reflection, *e.g.,* no O mode echoes are returned at frequencies above f_p at the target location.

The number of sounding frequencies for a given measurement together with the coherent integration time for each frequency determine the total measuring time and hence the time resolution of a single complete measurement of the echoes from a target. The characteristics for the RPI instrument to be flown on IMAGE are summarized in Table 1.

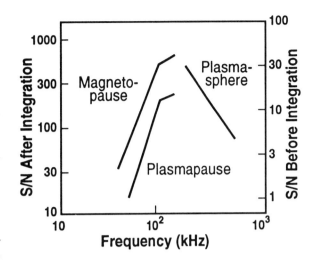

Figure 2. Signal-to-noise ratios for various magnetospheric targets before and after digital integration.

Table 1. RPI instrument characteristics.

Parameter	Nominal	Limits
RF Power	10 W	10 W
Pulse Width	53 msec	3.2 to 125 ms
Receiver Bandwidth	300 Hz	fixed at 300 Hz
Pulse Rate	2 pps	1 to 5 pps
Frequency Range	10 to 100 kHz	3 kHz to 3 MHz
Frequency Steps	5%	≥ 100 Hz
Integration Time	8 seconds	> 2 seconds

Based on these instrument characteristics, Table 2 provides the expected measurement capabilities. While IMAGE is in the magnetospheric N_e cavity the RPI should be able to provide unprecedented global scale magnetospheric observations, detecting the locations and motions of important boundaries and their motions several R_E from the spacecraft.

Table 2. RPI measurement capabilities.

Measurement	Nominal Resolution	Limits
Echo Range	500 km	0.1 to 5 R_E
Angle-of-Arrival	1° at 40 dB S/N	resolution = 2/[S/N]
Doppler	0.125 Hz	75 Hz
Time	8 sec./freq step	4 sec./freq step

5. NATURAL NOISE EMISSIONS

Receiver thermal noise and cosmic noise are incoherent and are usually constant sources of background noise. Dealing with them by RPI will then be a simple matter of sampling integration as already discussed. However, other natural noises, in RPI's frequency range, consist of Type III solar noise bursts and storms, auroral kilometric radiation (AKR), and the non-thermal continuum (escaping and trapped).

Type III solar noise bursts and storms are radiation near the solar wind f_p, generally believed to be excited by outward propagating energetic electrons from the sun. Since the solar wind density decreases outward from the sun, the resulting spectrum is a band of emissions with decreasing frequency with increasing time. Type III storms consist of thousands of Type III bursts produced quasi-continuously, resulting in broadband emissions. The frequency range for Type III emissions extends from above the RPI maximum frequency of 3 MHz to the lowest frequencies that can propagate through the magnetosheath, typically 30 to 100 kHz. Type III bursts are nearly as intense as the most intense AKR, whereas Type III solar storms have typical power fluxes only about an order of magnitude above the cosmic noise background [*Benson and Fainberg*, 1991; *Bougeret et al.*, 1984].

AKR is associated with auroral arcs and originates above auroral regions at about one half to a few R_E altitude. The frequency of the emission is from about 30 kHz to about 700 kHz with peak emission between 100 to 400 kHz, depending on local time and magnetic activity [*Kaiser and Alexander*, 1977]. The maximum power for AKR is many orders of magnitude above the cosmic and receiver noise. Propagation effects [see *Green et al.*, 1977] restrict AKR to higher magnetic latitudes over certain local times. Importantly, both O and X mode AKR are very narrowband, on the order of 1 kHz or less [*Gurnett, et al.*, 1979; *Gurnett and Anderson*, 1981; *Benson et al.*, 1988].

Continuum radiation has two components, trapped and escaping [*Gurnett*, 1975]. The trapped component ranges in frequency from about 30 kHz to the magnetosheath f_p, which is between 30 and 100 kHz. The frequency range of the escaping component varies from the magnetosheath f_p (~30 kHz) to a few hundred kHz. Continuum radiation is believed to be generated primarily in the O mode. The source region appears to be near the low-latitude plasmapause, primarily on the dawn sector. The trapped component is also observed as a broadband emission believed to be the result of frequency diffusion of multiple reflections off the magnetopause [*Kurth et al.*, 1981]. The radiation pattern of the escaping continuum [*Morgan and Gurnett*, 1991] has been found to be from $40°$ to $60°$ about the magnetic equator which will typically not interfere with RPI measurements. There are no reports of trapped continuum radiation, at the same intensity as observed near the equator, at the high latitudes and high altitudes characteristic of IMAGE's orbit.

6. MITIGATION TECHNIQUES

Several special techniques have been developed to eliminate or at least reduce the effects of the above natural noises.

The standard techniques include: signal processing, frequency control, polarization discrimination, and spatial discrimination. These techniques are in use by the DPS and will be necessary for successful operation of RPI on the IMAGE mission.

Frequency avoidance is a technique in which the instrument avoids measurement at a sounder frequency where natural wave emissions are too intense. This technique, commonly used by ground-based ionosondes, could be used in specific regions along an orbit where characteristic natural emissions of sufficiently high intensity are known to exist.

The second frequency control technique is frequency agility. If the RPI receiver determines that, for example, three of the five adjacent frequencies have high signal levels and one is very low, the latter would be chosen for transmission and reception. This technique would be very effective for sounding during periods of narrowband AKR or escaping continuum.

Polarization and spatial discrimination techniques can also be used in RPI operations. In polarization discrimination, differences in polarization between noise and expected signals will be used to increase S/N ratios. For example, AKR is known to be strongly polarized in the X mode. In these cases, RPI signals in the other mode will have a higher S/N ratio. In spatial discrimination, RPI would be able to avoid or reduce the effects of strong emissions in certain locations in the magnetosphere. AKR, for example, is absent or much weaker in the low latitude, dayside [*Green et al.*, 1977] and intense continuum radiation has not been observed over the polar cap.

6. SIMULATIONS

The primary presentation of RPI data will be in the form of plasmagrams, which are the magnetospheric analogs of the ionograms. A plasmagram is a color or gray scale plot of the echo power as a function of frequency and echo delay. Ray tracing calculations have been performed to simulate the return pulses from the RPI instrument on IMAGE located in a model magnetosphere. Detailed descriptions of the ray tracing code used in this simulation can be found in *Green and Donohue* [1988]. The magnetic field model in the simulation is a simple dipole and the plasma density model is a combination of several models (diffusive equilibrium by *Angerami and Thomas* [1964], ionosphere and plasmasphere by *Kimura* [1966], the plasmapause by *Aikyo and Ondoh* [1971] and magnetopause by *Roelof and Sibeck* [1993].

The plasmagram in Figure 3A shows the O mode expected echo amplitudes and delay times as a function of frequency. It is important to note that only receiver noise and cosmic noise have been considered in this simulation. In these calculations, we have included the effects of the RPI antenna length, antenna tuning and matching, and focusing caused by target curvature. The satellite was in the noon meridian, at a latitude of $25°$ and a geocentric distance of 7 R_E. The N_e

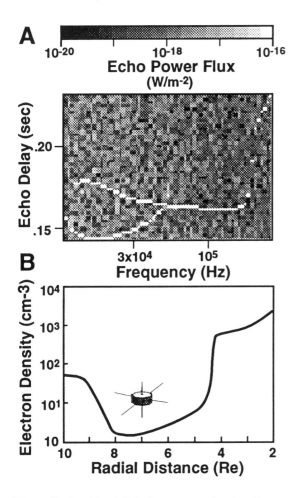

Figure 3. A. Simulated plasmagram showing the expected echo amplitudes and delay times as a function of delay and frequency. **B**. Corresponding radial N_e distribution.

second example is an image created by determining the direction of echoes at a single frequency from a specific magnetospheric feature. Plotting the echo directions could give, for example, an image of a surface wave on the magnetopause. The image plane in this case is perpendicular to the line of sight. Other types of images can also be created [*Reiff et al.*, 1995].

7. SUMMARY AND CONCLUSIONS

Remote plasma structures in the magnetosphere may be observed by probing them with radio waves. The feasibility of the basic concept was clearly demonstrated by the success of the Alouette/ISIS sounders in the late 60's and early 70's. Several recent feasibility studies have also demonstrated that RPI can successfully operate in the Earth's magnetosphere.

The RPI instrument on the IMAGE mission is a swept in frequency adaptive sounder, with on-board signal processing which transmits and receives coded electromagnetic pulses over the frequency range from 3 kHz to 3 MHz. The pulses propagate as free space waves through the magnetosphere and are reflected upon encountering their plasma cutoff frequencies. The RPI will measure the amplitudes, Doppler shift, and direction of arrival as a function of frequency and echo delay. The echo arrival angles will be calculated from the amplitudes and phases of the signals from three orthogonal receiving dipole antennas. Nearly all the sounding techniques utilized on DPS will be used in the RPI.

Situated in the density cavity of the magnetosphere, the RPI should be able to simultaneously determine the location and dynamics of remote boundaries such as the plasmapause and magnetopause. In addition, the RPI should be able to provide N_e profiles in different directions on time scales of a few minutes or less. At specific sounder frequencies, characteristic of remote plasma regions where large scale oscillations are occurring within the range of RPI, images of the magnetospheric structures should be possible.

In summary, the RPI should be able to provide unprecedented global magnetospheric observations.

Acknowledgments. The authors gratefully acknowledge useful discussions with the late Dr. S. D. Shawhan during early stages of magnetospheric sounder development and with the IMI Science Definition Team. The work at Rice University was supported by NASA under grant NAGW 1655, and at Hughes STX under NASA contract NASW-5016.

profile was calculated from the change of echo delay with frequency using the technique developed by *Huang and Reinisch*, [1982], and is shown in Figure 3B. The calculated N_e profile is in excellent agreement with the model N_e profile used in the ray tracing calculations.

Based on the DPS proven technologies and the simulations of RPI in the proposed IMAGE orbit, RPI should provide information on the location of the magnetopause and plasmapause and also their respective N_e values. It should be possible to deduce global scale boundary structures using the directional and range measurements, as well as from a sequence of many plasmagrams along the IMAGE orbit.

RPI data can also be displayed in several types of image formats. For example an image can be created by combining a series of one dimensional N_e profiles into a two dimensional N_e contour image. The image will be formed as a function of time as the spacecraft moves along its orbit and the image plane is then the satellite orbital plane. A

REFERENCES

Aikyo, K., and T. Ondoh, Propagation of nonducted VLF waves in the vicinity of the plasmapause, *J. Radio Res. Labs., 18*, 153, 1971.

Angerami, J., and J. Thomas, The distribution of ions and electrons in the Earth's exosphere, *J. Geophys. Res., 69*, 4537, 1964.

Benson, R. , and J. Fainberg, Maximum power flux of auroral kilometric radiation, *J. Geophys. Res.*, 96, 13749-13762, 1991.

Benson, R., M. Mellott, R. Huff, and D. Gurnett, Ordinary mode auroral kilometric radiation fine structure observed by DE 1, *J. Geophys. Res.*, 93, 7515-7520, 1988.

Berkner, L. V., The international geophysical year, *Proc. IRE*, 47, 133-136, 1959.

Bougeret, J.-L., J. Fainberg, and R. G. Stone, Interplanetary radio storms, 1, Extension of solar active regions throughout the interplanetary medium, *Astro. Astrophys.*, 136, 255-262, 1984.

Breit, G., and M. A. Tuve, A test for the existence of the conducting layer, *Phys. Rev.*, 28, 554-575, 1926.

Brown, J. N., Automatic sweep-frequency ionosphere recorder, Model C-4, *Proc. IRE*, 47, 296-300, 1959.

Calvert, W., R. Benson, D. Carpenter, S. Fung, D. Gallagher, J. Green, D. Haines, P. Reiff, B. Reinisch, M. Smith, and W. Taylor, The feasibility of radio sounding in the magnetosphere, *Radio Science*, 30, 5, 1577-1615, 1995.

Calvert, W., R. Benson, D. Carpenter, S. Fung, D. Gallagher, J. Green, D. Haines, P. Reiff, B. Reinisch, M. Smith and W. Taylor, Reply to R. Greenwald concerning the feasibility of radio sounding, *Radio Sci.*,, 32, 281-284, 1997.

Franklin, C. A., and M. A. Maclean, The design of swept-frequency topside sounders, *Proc. IEEE*, 57, 897-929, 1969.

Fung, S. and J. Green, Global Imaging and Radio Remote Sensing of the Magnetosphere, Radiation Belts: Models and Standards, *AGU Monograph*, 97, 285-290, 1996.

Green, J. L., and D. J. Donohue, Computer techniques and procedures for 3-D ray tracing, *NSSDC Technical Report*, January 1988.

Green, J. L., D. A. Gurnett, and S. D. Shawhan, The angular distribution of auroral kilometric radiation, *J. Geophys. Res.*, 82, 1825, 1977.

Green, J., R. Benson, W. Calvert, S. Fung, P. Reiff, B. Reinisch, and W. W. L. Taylor, A Study of Radio Plasma Imaging for the proposed IMI mission, *NSSDC Technical Publication*, February 1993.

Green, J., S. Fung, and J. Burch, Application of magnetospheric imaging techniques to global substorm dynamics, *Proceedings of the 3rd International Conference on Substorms*, Versailles, ESA, SP-389, 655-661, 1996.

Gurnett, D. A., The Earth as a radio source: The nonthermal continuum, *J. Geophys. Res.*, 80, 2751-2763, 1975.

Gurnett, D. and R. Anderson, The kilometric radio emission spectrum: Relationship to auroral acceleration processes, in *Physics of Auroral Arc Formation, Geophys. Monogr. Ser.*, 25, S.-I. Akasofu and J. Kan, eds., 341-350, AGU, Washington, DC, 1981.

Gurnett, D. A., R. R. Anderson, F. L. Scarf, R. W. Fredricks, and E. J. Smith, Initial results from the ISEE 1 and 2 plasma wave investigation, *Space Sci. Rev.*, 23, 103-122, 1979.

Huang, X., and B. W. Reinisch, Automatic calculation of electron density profiles from digital ionograms. 2. True height inversion of topside ionograms with the profile-fitting method, *Radio Science*, 17, 4, p.837-844, 1982.

Hunsucker, R. D., *Radio Techniques for Probing the Terrestrial Ionosphere*, Vol. 22, Phys. Chem. Space, Springerverlage, Berlin, 1991.

Jackson, J. E., *Alouette-ISIS Program Summary*, NSSDC/WDC-A-R & S 86-09, NASA Goddard Space Flight Center, Greenbelt, MD, 1986.

Kaiser, M. L., and J. K. Alexander, Terrestrial kilometric radiation 3. Average spectral properties, *J. Geophys. Res.*, 96, 17,865-17,878, 1977.

Kimura, I., Effects of ions on whistler mode ray tracing, *Radio Science, 1 (New Series)*, 269, 1966.

Kurth, W. S., D. A. Gurnett, R. R. Anderson, Escaping nonthermal continuum radiation, *J. Geophys. Res.*, 86, 5519-5531, 1981.

Lui, A. T. Y., and C. D. Anger, Uniform belt of diffuse auroral emissions seen by the ISIS 2 scanning photometer, *Planet. Space Sci.*, 21, 799-809, 1973.

Morgan, D. D. and D. A. Gurnett, The source location and beaming of terrestrial continuum radiation, *J. Geophys. Res.*, 96, 9595-9613, 1991.

Reiff, P. H., J. L. Green, S. F. Fung, R. F. Benson, W. Calvert, and W. W. L. Taylor, Radio Sounding of Multiscale Plasmas, To be published in the Physics of Space Plasmas, Cambridge MA, 1995.

Reinisch, B., D. Haines and W. Kuklinski, The new portable Digisonde for vertical and oblique sounding, Proc. AGARD-CP-502, 11-1 to 11-11, 1992.

Reinisch, B. W., T. W. Bullett, J. L. Scali and D. M. Haines, High Latitude Digisonde Measurements and Their Relevance to IRI, Adv. Space Res., Vol. 16, No. 1, pp. (1) 17-(1)26, 1995.

Roelof, E. C., and D. G. Sibeck, Magnetopause shape as a bivariate function of interplanetary magnetic field Bz and solar wind dynamic pressure, *J. Geophys. Res.*, 98, 21421-21450, 1993.

Shepherd, G., J. Whitteker, J. Winningham, J. Hoffman, E. J. Maier, L. H. Brace, J. R. Burrows, and L. L. Cogger, The topside magnetospheric cleft ionosphere observed from the ISIS 2 spacecraft, *J. Geophys. Res.*, 81, 6092-6102, 1976.

Robert F. Benson, Shing F. Fung, and James L. Green,, NASA Goddard Space Flight Center, Greenbelt, MD 20771.

Wynne Calvert and Bodo W. Reinisch, Center for Atmospheric Research, University of Massachusetts Lowell, 600 Suffolk Street, Lowell, MA 01854.

Dennis L. Gallagher, Code ES53, NASA Marshall Space Flight Center, Huntsville, AL 35812.

Patricia H. Reiff, Department of Space Physics and Astronomy, Rice University, Box 1892, Houston, TX 77251-1892.

William W. L. Taylor, Hughes STX Corporation, 4400 Forbes Boulevard, Lanham, MD 20706.

Direct Measurements of AC Plasma Currents in the Outer Magnetosphere

A.A. Petrukovich, S.A. Romanov, and S.I. Klimov

Space Research Institute, Russian Academy of Sciences, Moscow, Russia

Measurements of AC plasma currents may provide valuable information about space plasma oscillations in the ULF/ELF frequency range. Such measurements were performed onboard Prognoz-8 and Prognoz-10 spacecraft with the use of a Faraday cup and a split Langmuir probe. Combining simultaneously measured plasma current, electric and magnetic field data one can identify the wave mode and calculate the wavevector. Direct measurements of currents in space plasmas are significantly affected by the sensor-plasma interaction. We model these effects for several wave modes in the frame of a linear warm plasma approximation. We present two examples of the plasma current measurements conducted in the range 0.1–30 Hz, which prove the validity of proposed method.

1.INTRODUCTION

Identification of the plasma turbulence mode and determination of dispersion characteristics are major tasks of the spacecraft plasma wave experiments. In most previous investigations identification of a wavemode was qualitative and based on the measurements of electric and magnetic field frequency spectra. To calculate the wavelength and thus to determine the wavemode quantitatively one needs to conduct simultaneous measurements of at least five components of electric and magnetic field. However, such set of parameters is rarely measured with sufficient accuracy. This method also cannot be used for the studies of electrostatic waves. Measurements of the plasma current fluctuations provide complementary information about plasma waves and facilitate the determination of the dispersion characteristics. Such measurements were performed with

Measurement Techniques in Space Plasmas: Fields
Geophysical Monograph 103

the use of the split Langmuir probe (SLP) onboard the Prognoz-10 spacecraft [*Romanov et al.*, 1991; *Petrukovich et al.*, 1993] and with the use of the Faraday cup (FC) onboard Prognoz-8 spacecraft [*Büchner and Lehmann*, 1984].

We describe methods of the plasma current data interpretation in Section 2. In Section 3 we apply proposed techniques to the analysis of the ULF/ELF (0.1–25 Hz) plasma turbulence observed by the Prognoz-8,-10 spacecraft near the Earth's supercritical quasi-perpendicular shock are presented in Section 3.

2. DESCRIPTION OF CURRENT SENSORS

2.1. *Experiment description*

Plasma wave experiments onboard Prognoz-8 and -10 spacecraft were optimized for investigations in the frequency range 0.1–50 Hz [*Klimov et al.*, 1986]. The Faraday cup (FC) sensor for plasma flux fluctuation measurements is constructed as a cylinder with the collector on the bottom and four grids. The first grid (near the collector) suppresses the collector's photocurrent. The second and the fourth grids are at zero potential. With the third grid, having the potential V_f, they serve as a selector of particle sorts. If V_f is negative and larger than the electron energy, collector measures ion flux (such probe will be further called ion probe). If the potential is positive and larger than the ion energy then it measures electron flux (electron probe). In the plasma wave experiment onboard Prognoz-8 (1980–1981) one FC sensor was used, designed for the measurements of ion flux fluctuations. It had the V_f potential equal to –120 V and was oriented along the spin axis in the sunward direction.

The split Langmuir probe performs direct measurements of the plasma current fluctuations along its symmetry axis. The plane SLP was used in the rocket experiments [Bering et al., 1973a, 1973b] for the DC current measurements. The spherical SLP splitted in two hemispheric collectors onboard Prognoz-10 had a diameter of about 10 cm and only AC signal was transmitted [Vaisberg et al., 1989]. This has allowed to avoid the influence of the photoelectrons.

2.2. Methods of interpretation of current measurements

The wavevector can be calculated combining plasma current and magnetic field data with the use of Maxwell equations in the low frequency limit [Romanov et al., 1991; Petrukovich et al., 1993]:

$$div\vec{B} = 0, \quad curl\vec{B} = 4\pi/c\vec{J} \tag{1}$$

Ohm's law is more suitable for the analysis of electrostatic waves or while using the Faraday cup data, with separated electron and ion fluxes [Büchner and Lehmann, 1984]:

$$\vec{J}(\omega) = \sigma(\omega, \vec{k})\vec{E}(\omega) \tag{2}$$

Experimental values of the ion or electron conductivity are to be compared with the model conductivity tensor. However, the FC sensor measures flux in the given direction rather than the total current. Also, in the presence of the floating potential the effective collecting area of the sensor is changing. To determine influence of such distortions one needs to know the velocity distribution function of the particles forming the current. The model conductivity tensor and alternative velocity distribution function can be calculated numerically in the frame of the linear model of Maxwellian uniform warm plasma [e.g. Akhiezer et al., 1974], assuming that the dispersion relation is known.

Expressions (1,2) are used with the Fourier harmonics of the signal and only one harmonic (plane wave) should be present on each observed (in the spacecraft frame of reference) frequency to ensure precise enough result. Level of coherency of oscillations can serve as an indicator of a number of harmonics involved. If highly coherent quazimonochromatic structures are seen in the signal then we likely have one harmonic on each frequency. If chaotic signal with flat spectra is detected, we have a mixture of harmonics. Usually, coherency values above 0.8 level are considered as high [e.g. Krasnosel'skikh et al., 1991]. ULF/ELF turbulence observed near the quazi-perpendicular super-critical shock front is a good example for the test of the proposed methods: It contains quazi-monochromatic waves in the range 1–5 Hz and above 5 Hz spectra can be fitted according to a power law [Heppner et al., 1967; Mellott and Greenstadt, 1988; Nozdrachev et al., 1995].

Below we discuss applications of the developed technique to some aspects of the plasma current measurements.

2.3. Floating potential on the SLP

In the rarefied plasma of solar wind and fore-shock the floating potential is usually positive (1–5 V) and smaller than the electron temperature (5–30 eV). In the frequency range of interest (higher than 0.1 Hz) ion part of the total alternative current, carrying the wave, can be neglected in comparison with the electron one. Sensor attracts electrons and the effective area is bigger than the geometrical one. For the monotonic potential the standard expression for the effective area is:

$$S = S_0(1 + 2e\phi/(mv^2)) \tag{3}$$

where S_0 is the geometrical area, ϕ is the floating potential, $mv^2/2$ is the particle's kinetic energy at the infinite distance from the probe, e is the elementary charge. After integration over all alternative velocity distribution function taken far from the probe, one gets:

$$S/S_0 = 1 + (2e\phi/T_e)A \tag{4}$$

where T_e is the electron temperature. A is the complex factor depending from the wavemode, propagation direction, etc.

Values of A for the whistler wave are shown in Figure 1. For the current measurements in the plane perpendicular to the magnetic field vector, A is close to unity for all propagation angles and frequencies and effective

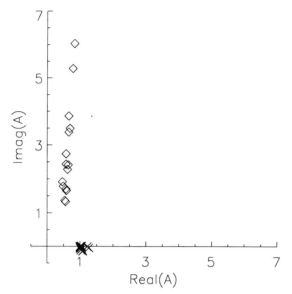

Figure 1. Polar diagram of the A factor for the whistler with frequencies 2–20 Hz and propagation angles with respect to the magnetic field 0–80°. "×" — for the measurement direction perpendicular to the magnetic field vector; "◇" — parallel to the magnetic field vector.

cross-section is determined only by the ratio of the floating potential to electron temperature. For the current measurements in parallel to the magnetic field direction, A depends from the velocity of zero resonance $v = \omega/k_z$ and specific value of the effective cross-section must be computed for each observed harmonic.

2.4. *Model values of the conductivity*

To determine experimentally the conductivity simultaneous measurements of the AC plasma flux (F) and electric field (E) should be performed:

$$F_{i\alpha}/E_j = (C_\alpha/e)\sigma_{ij\alpha} \qquad (5)$$

where C_α allows for the difference between measured flux and real plasma current, $\sigma_{ij\alpha}$ is the conductivity tensor, and α denotes sort of particles. We compute C_α with the use of alternative velocity distribution function. For the ion probe C_i is nearly constant for all whistler harmonics and orientations of the FC and is equal to 0.5.

In the Prognoz-8 wave experiment only one component of the plasma flux was measured and only estimate of the flux to electric field ratio (R_α) was available. Further we will call R_α $(\alpha = e/i)$ electron/ion response function. In the frame of linear cold plasma approximation, [e.g. *Akhiezer et al.,* 1974] the frequency dependence of R_α is different for the magnetized (6) and unmagnetized (7) particles:

$$R_\alpha = C_\alpha(en/m_\alpha)(1/\omega_{c\alpha}), \quad if \ \ \omega \ll \omega_{c\alpha} \qquad (6)$$

$$R_\alpha = C_\alpha(en/m_\alpha)(1/\omega), \quad if \ \ \omega \gg \omega_{c\alpha} \qquad (7)$$

where $\omega_{c\alpha}$ is cyclotron frequency, n is plasma density, m_α is mass of a particle.

In Figure 2 we present the warm plasma estimates of the response functions for the whistler and ion acoustic

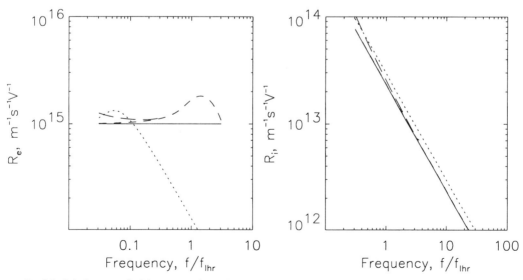

Figure 2. Model electron (left) and ion (right) response functions for cold plasma (solid line), whistler with 0° (long dashes), 70° (long dashes) propagation angles and for ion acoustic wave (dots dashes). Frequency is normalized to the lower hybrid resonance frequency.

waves in the solar wind. Cold plasma approximation is shown by a solid line. Differences between warm and cold plasma approximations are small and steady for the ion response functions. This stability of the ion response function provides the basis for the plasma density measurements, using expression (7). Whistler electron response functions differ from the cold plasma prediction close to the electron cyclotron frequency. Electron response to ion acoustic waves is much smaller than to whistler waves because wavelengths of the ion acoustic oscillations are shorter than whistler ones and are comparable with the electron Larmor radius. Electrons become unmagnetized and slope of their response function changes. Comparison of the model and experimental electron response functions can help in the identification of plasma wave modes in the ULF/ELF frequency range.

3. EXPERIMENTAL DATA

3.1 Example of the Faraday cup measurements

We present the data set collected at the Earth's supercritical quazi-perpendicular bow shock by the Prognoz-8 spacecraft on July 6, 1981. It is described in details by *Nozdrachev et al.*, [1995]. Floating potential on the FC doesn't affect ion measurements as ion energy is very high due to high bulk velocity.

Electrons with the energies higher than the $|eV_f| = 120eV$ are able to reach the collector and introduce interference in the ion current measurements. Number of such electrons can be estimated as $n \exp\left(-eV_f/T_e\right)$. Equating full ion and high energy electron thermal fluxes at the collector one can get rough estimate of the maximum electron temperature at which ion measurements are still possible: $T_{max} = 0.27eV_f$. Upstream (solar wind) electron temperature is usually smaller than this maximum, while behind the shock front it is higher or comparable. Therefore, we can apply our method only in the upstream region.

To get better estimates of the measurement's accuracy one should compare wave mode dependent fluxes instead of thermal ones. In the frequency range of interest the electron response function is constant with respect to frequency and the ion one is decreasing (6,7). At some frequency f_b even small electron interference will become bigger than the main ion signal:

$$R_e(f_b) \exp\left(-eV_f/T_e\right) = R_i; \quad f_b = f_{ci} \exp\left(eV_f/T_e\right) \quad (8)$$

In Figure 3a upstream frequency spectrum of the electric field signal is presented. Monochromatic waves are

forming spectral peak in the frequency range 0.2–2 Hz. Above 2 Hz spectrum is flat. Experimental response function is shown in Figure 3b by solid curve. At the lower frequencies it decreases with the inverse proportionality with respect to frequency. Such behavior is typical for the ion response. At the higher frequencies the electron influence becomes noticeable and experimental curve gradually becomes more horisontal. The model ion response function, computed for the warm plasma with density 8.5 cm^{-3} is shown by dotted line. The sum of model ion response and predicted electron interference ($T_e = 25eV$ as described above is shown by dashed line. The corresponding value of f_b frequency is 6 Hz.

We compute experimental density estimates as point by point ratios of experimental and combined model re-

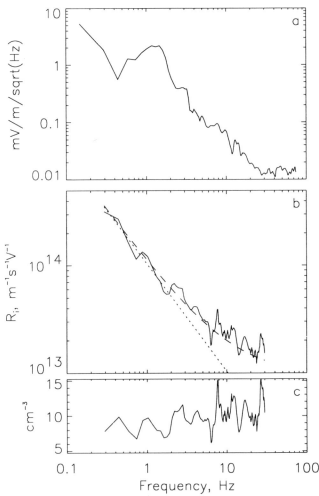

Figure 3. (a) Upstream electric field spectrum. (b) Experimental response function (solid line); ion model response (dots); combined model response (dashes). (c) Plasma density estimates.

sponses (Figure 3c). According to Section 2.2, quantitative result can be obtained only when coherent enough signal is registered and pure ion fluctuations are measured. Analysis of the experimental data supports this conclusion: the difference with the solar wind density measured by IMP-8 (8.5 cm^{-3}) is about 20% in the frequency range with monochromatic waves (0.2–2.0 Hz) and about 50–80% in the range of flat spectrum (higher than 2 Hz).

3.2. *Example of the SLP data*

Described crossing of supercritical quazi-perpendicular Earth's bow shock was registered by the Prognoz-10 spacecraft on October 8, 1985 at 0410 UT and described in details by *Krasnosel'skikh et al.*, [1991] and *Petrukovich et al.*, [1993]. Electron temperature wasn't measured in this crossing and it is impossible to determine influence of the floating potential. However, in our case the sensor axis and magnetic field vector are not parallel and for the average solar wind properties one can expect ratio ($2e\phi/T_e A$) from expression (4) to be less than 0.5. Example of the magnetic field fluctuation spectrum is in Figure 4a. Coherency function of two magnetic field components (Figure 4c) is close to unity for the frequency range 1–8 Hz, while for the frequencies 0.1–1 Hz and 8–12 Hz it is between 0.8 and 0.6. Wavelengths of the oscillations calculated with the help of expression (1) are in Figure 4b. All waves are propagating with the angles higher than 60° with respect to the upstream magnetic field.

Frequencies and wavevectors of waves are determined by the dispersion relations in the Plasma rest Frame of Reference (PFR). Frequencies of waves in the Spacecraft Frame of Reference (SFR) significantly differ from their PFR counterparts due to the Doppler shift imposed by the solar wind bulk flow. In Figure 5 experimentally determined frequencies in the PFR (with extracted Doppler shift, see *Petrukovich et al.*, [1993] for details) are plotted together with the model frequencies for whistler and fast magnetoacoustic waves. Model frequencies are computed using experimental values of the wavevector and dispersion relations for the whistler and fast magnetoacoustic modes. In the intervals 0.1–1 Hz and 8–12 Hz waves likely belong to the magnetoacoustic mode. In the interval 1–8 Hz waves belong to whistler mode. The best coincidence between measured and model frequencies is observed in the frequency range, containing the spectral peaks with monochromatic oscillations (up to 5 Hz). In the frequency region with the flat spectra (higher than 5 Hz) experimental re-

sults are only in a qualitative agreement with the model ones. The proximity of experimental and model results justifies our initial supposition of the small floating potential influence.

4. CONCLUSIONS

Number of the electromagnetic field components measured in some previous plasma wave experiments in the ULF/ELF range doesn't permit quantitative wave mode determination. Measurements of the AC plasma currents can supply additional valuable information.

Sensor-plasma interaction introduces interference in the measured natural signal. These interferences are modelled and extracted with the use of the linear warm plasma approximation. The presence of a sole plane wave on each frequency in the observed signal is proved

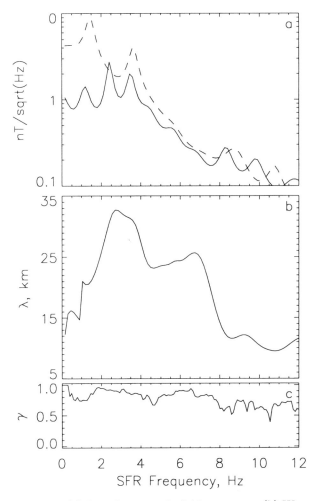

Figure 4. (a) Sample magnetic field spectrum. (b) Wavelengths versus spacecraft frame frequency. (c) Coherency function of two magnetic field components.

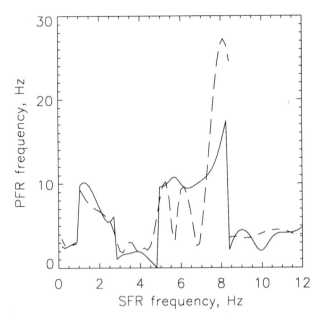

Figure 5. Experimental plasma frame frequency (solid line), model whistler frequency (1–8 Hz, long dashes) and model fast magnetoacoustic frequency (0.1–1 and 8–12 Hz, short dashes) versus spacecraft frame frequency.

to be the critical condition for the quality of such analysis.

Using simultaneous measurements of plasma flux and electric field oscillations it is possible to obtain estimates of the conductivity tensor components. Measurements of the electron flux can help distinguish cyclotron harmonics in the signal and separate short and long wavelength emissions. Measurements of the ion flux, conducted onboard the Prognoz-8 provided the estimate of the plasma density.

Using simultaneous measurements of plasma current by the split Langmuir probe and magnetic field onboard Prognoz-10 spacecraft one obtains the wavevector value.

The proposed method of the electric current data processing and interpretation can be extended to include more sophisticated plasma models and other types of sensors.

Acknowledgments. We are grateful to J.Büchner for useful discussions. We thank the IMP-8 experimenter team for the 1-hour solar wind parameters, available on INTERNET. This work was partially supported by the INTAS-93-2031 grant.

REFERENCES

Akhieser, A.I. (Ed.), Plasma Electrodynamics (in Russian), pp. 190, 205, Nauka, Moscow, 1974.

Bering, E.A., M.C. Kelley, and F.S. Mozer, Split Langmuir probe measurement of current density and electric fields in an aurora, *J. Geophys. Res., 78*, 2201, 1973a.

Bering, E.A., M.C. Kelley, F.S. Mozer, and U.V. Fahleson, Theory and operation of the split Langmuir probe, *Planet. Space. Sci., 21*, 1983, 1973b.

Büchner, J., and H. Lehmann, On the interpretation of ULF fluctuations near the Earth magnetopause, *Adv. Space Res., 4*, 527, 1984.

Heppner, J.P., M. Sugiura, T.L. Skillman, B.G. Ledley, and M. Campbell, OGO-A Magnetic field observations, *J. Geophys. Res., 72*, 5417, 1967.

Klimov, S.I., M.N. Nozdrachev, P. Triska, J. Vojta, A.A. Galeev, Ya.N. Aleksevich, Yu.V. Afanasyev, V.E.Baskakov, Yu.N. Bobkov, R.B. Dunets, A.M. Zhdanov, V.E. Korepanov, S.A. Romanov, S.P. Savin, A.Yu. Sokolov, and V.S. Shmelev, Plasma wave investigation onboard Prognoz-10 satellite (in Russian), *Kosmich. Issl., 24*, 177, 1986.

Krasnosel'skikh, V.V., M.A. Balikhin, H.St.C. Alleyne, S.I. Klimov, A.A. Petrukovich, D.J. Southwood, T. Vinogradova, and L.J.C. Woolliscroft, On the nature of low frequency turbulence in the foot of strong quasi-perpendicular shocks, *Adv. Space Res., 11*, 15, 1991.

Mellott, M.M., and E.W. Greenstadt, Plasma waves in the range of the Lower Hybrid Frequency: ISEE-1 and -2 observations at the Earth's Bow Shock, *J. Geophys. Res., 83*, 9695, 1988.

Nozdrachev, M.N., A.A. Petrukovich, and J. Juchniewicz, ULF/ELF monochromatic oscillations observed by Prognoz -8 and -10 spacecraft during the quasi-perpendicular supercritical shock crossings, *Ann. Geophys., 13*, 573, 1995.

Petrukovich, A.A., S.A. Romanov, and S.I. Klimov, Dispersion characteristics of plasma emissions near the quazi-perpendicular Earth's Bow shock observed by Prognoz-10 spacecraft, in *Proceedings of START conference, ESA WPP-047*, 281, 1993.

Romanov, S.A., S.I. Klimov, and P.A. Mironenko, Experimental derivation of ELF waves dispersion relations and evidence of wave coupling in the Earth's Bow Shock foot region from the results of the Prognoz-10, *Adv. Space Res., 11*, 19, 1991.

Vaisberg, O.L., S.I. Klimov, and V.E. Korepanov, Current density measurements near the shock by the split Langmuir probe (in Russian), *Kosmich. Issl., 27*, 461, 1989.

A.A. Petrukovich, S.A. Romanov and S.I. Klimov, Space Research Institute, 84/32 Profsoyuznaya st, Moscow, 117335, Russia, (e-mail: apetruko@iki.rssi.ru).

Measuring Plasma Parameters With Thermal Noise Spectroscopy

Nicole Meyer-Vernet, Sang Hoang, Karine Issautier, Milan Maksimovic, Robert Manning, and Michel Moncuquet

DESPA/CNRS ura 264, Observatoire de Paris, Meudon, France

Robert G. Stone

NASA/Goddard Space Flight Center, Greenbelt, Maryland

This paper describes the basic principles, the unique features and the limitations of thermal noise spectroscopy as a tool for *in situ* diagnostics in space plasmas. This technique is based on the analysis of the electrostatic field spectrum produced by the quasi-thermal fluctuations of the electrons and ions, which can be measured with a sensitive wave receiver at the terminals of an electric antenna. This method produces routine measurements of the bulk electron density and temperature, and is being extended to measure the ion bulk speed. It has the advantage of being in general relatively immune to spacecraft potential and photoelectron perturbations, since it senses a large plasma volume. We compare this method to other techniques, and give examples of applications in the solar wind as well as in cometary and in magnetized planetary environments.

1. INTRODUCTION

Since particles and electrostatic waves are so closely coupled in a plasma, particle properties can be determined by measuring waves. In a stable plasma, the particle thermal motions produce electrostatic fluctuations which are completely determined by the velocity distributions (and the static magnetic field) [*Rostoker*, 1961]. Hence, this quasi-thermal noise, which can be measured with a sensitive receiver at the terminals of an electric antenna, allows *in situ* plasma measurements.

Except near magnetized planets, the electron gyrofrequency f_g is much smaller than the plasma frequency f_p. Then, the electron thermal motions excite Langmuir waves, so that the quasi-equilibrium spectrum is

Measurement Techniques in Space Plasmas: Fields
Geophysical Monograph 103
Copyright 1998 by the American Geophysical Union

cut-off at f_p, with a peak just above it (see Figure 1). In addition, the electrons passing closer than a Debye length L_D to the antenna induce voltage pulses on it, producing a plateau in the wave spectrum below f_p and a decreasing level above f_p; since L_D is mainly determined by the bulk (core) electrons, so are these parts of the spectrum. In contrast, since the Langmuir wave phase velocity $v_\phi \to \infty$ as $f \to f_p$, the fine shape of the f_p peak is determined by the high-velocity electrons.

Hence when $f_g \ll f_p$, measuring the thermal noise spectrum allows a precise determination of the electron density and bulk temperature (using respectively the cut-off at f_p and the spectrum level and shape around it), whereas the detailed shape of the peak itself reveals the suprathermal electrons (see [*Meyer-Vernet and Perche*, 1989] and references therein). This technique, first introduced in the solar wind [*Meyer-Vernet*, 1979; *Couturier et al.*, 1981; *Kellogg*, 1981], has been applied in a cometary tail [*Meyer-Vernet et al.*, 1986a, b], in the Earth's plasmasphere [*Lund et al.*, 1994], and in the interplanetary medium over a wide range of heliocentric distances and latitudes [*Hoang et al.*, 1992, 1996; *Maksimovic et al.*, 1995; *Issautier et al.*, 1997].

In the environment of magnetized planets the effect of the ambient magnetic field is not negligible and the technique must be modified accordingly; this modification is outlined in Section 5.

2. BASICS

2.1. *Theory*

The voltage power spectrum of the plasma quasi-thermal noise at the terminals of an antenna in a plasma drifting with velocity **V** is

$$V_\omega^2 = \frac{2}{(2\pi)^3} \int d^3k \left| \frac{\mathbf{k} \cdot \mathbf{J}}{k} \right|^2 E^2(\mathbf{k}, \omega - \mathbf{k} \cdot \mathbf{V}) \quad (1)$$

The first term in the integral involves the antenna response to electrostatic waves, which depends on the Fourier transform $\mathbf{J}(\mathbf{k})$ of the current distribution along the antenna. The second term is the autocorrelation function of the electrostatic field fluctuations in the antenna frame. At frequencies $f \gg f_g$, we have

$$E^2(\mathbf{k}, \omega) = 2\pi \frac{\sum_j q_j^2 \int d^3v \quad f_j(\mathbf{v}) \, \delta(\omega - \mathbf{k} \cdot \mathbf{v})}{k^2 \epsilon_0^2 |\epsilon_L(\mathbf{k}, \omega)|^2} \quad (2)$$

$f_j(\mathbf{v})$ being the velocity distribution of the jth species of charge q_j, and $\epsilon_L(\mathbf{k}, \omega)$ the plasma longitudinal dielectric function (see for example [*Sitenko*, 1967]).

For a wire dipole antenna made of two thin filaments each of length L along the **x** axis we have [*Kuehl*, 1966]

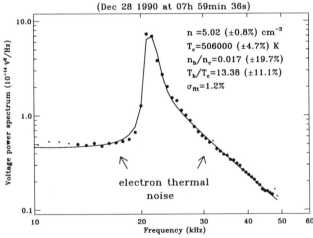

(Dec 28 1990 at 07h 59min 36s)

n =5.02 (±0.8%) cm^{-3}
T_c=506000 (±4.7%) K
n_h/n_c=0.017 (±19.7%)
T_h/T_c=13.38 (±11.1%)
σ_{fit}=1.2%

electron thermal noise

Figure 1. Example of voltage power spectrum measured by URAP on Ulysses in the solar wind (dots) and the deduced plasma parameters. The solid curve is the theoretical electron quasi-thermal noise, with the best-fit parameters indicated. (The statistical uncertainties are obtained from the fitting procedure.)

$$\mathbf{k} \cdot \mathbf{J} = 4 \frac{\sin^2(k_x L/2)}{k_x L} \quad (3)$$

This expression assumes that the current decreases linearly with distance along each antenna arm, or equivalently, that the measured voltage is the difference between the voltages averaged over each arm. This is expected to hold in general if the filament radius is much smaller than the electrostatic wave-lengths, and if $\omega L/c \ll 1$ [*Schiff*, 1971]. The signal at the ports of a receiver is $V_R^2 = V_\omega^2 \times |Z_R/(Z_R + Z_A)|^2$, where the receiver impedance Z_R is mainly due to the antenna base capacitance, and the antenna impedance Z_A is given by

$$Z_A(\omega) = \frac{i}{(2\pi)^3 \omega \epsilon_0} \int d^3k \left| \frac{\mathbf{k} \cdot \mathbf{J}}{k} \right|^2 \frac{1}{\epsilon_L(\mathbf{k}, \omega - \mathbf{k} \cdot \mathbf{V})} \quad (4)$$

2.2. *Measuring the Electron Density and Bulk Temperature, and Estimating the Hot Component*

Eq.(3) shows that a thin wire dipole antenna is mainly sensitive to wave vectors whose projection along its direction is of the order of π/L. This result has important consequences [*Meyer-Vernet and Perche*, 1989]. First, in order to be well adapted to observe Langmuir waves (which satisfy $k < 1/L_D$), the antenna length should exceed a few Debye lengths. Second, if $\pi V/L \ll \omega_p$ (where $V = |\mathbf{V}|$), the Doppler shift is negligible for waves observed near f_p (this statement is also true if V is much smaller than the particle thermal velocities); thus the ion contribution to Eq.(2) is negligible near f_p, since f_p is much larger than the ion characteristic frequencies.

This condition generally holds in the near-ecliptic low-speed interplanetary medium, in cometary environments, and in the outer plasmasphere. A diagnostic of electrons is then obtained from the noise measured around f_p by [*Meyer-Vernet and Perche*, 1989]:
- assuming a model of the velocity distribution,
- calculating the electron quasi-thermal spectrum from Eqs.(1) to (4),
- deducing the parameters of the model by fitting the result to the observations.

Figure 1 shows a typical example of such a fitting, obtained with the 2×35-m wire dipole antenna of the URAP experiment [*Stone et al.*, 1992a] on Ulysses in the in-ecliptic solar wind. The electron distribution is described by a superposition of a cold (c) plus a hot (h) Maxwellian [*Feldman et al.*, 1975], and the fitting yields the electron total density n, cold temperature T_c, and the ratios n_h/n_c, T_h/T_c. The precision is much better for the total density (a few per cent) and core temperature (generally better than 15 %) than for the hot population because an accurate diagnostic of suprathermal particles would require measuring the spectral peak with a very good frequency resolution [*Chateau and Meyer-Vernet*, 1989, 1991].

2.3. *Measuring the Ion Bulk Speed*

When the bulk speed is not negligible, as in the interplanetary medium when $\pi V/L \not\ll \omega_p$, the above method can be used to measure it. Since the proton and electron thermal velocities are respectively much smaller and much larger than V, the bulk velocity has no significant effect on the electron thermal noise, but the proton noise is strongly Doppler-shifted, so that it is no longer negligible, especially for $f < f_p$ [*Meyer-Vernet et al.*, 1986c]. In that case, the proton bulk speed can be deduced by fitting the theoretical electron-plus-proton noise to the observed spectrum [*Issautier et al.*, 1996].

This situation is illustrated in Figure 2 which is based on data obtained with the URAP wire dipole antenna on Ulysses. The fitting yields the proton bulk speed V (and temperature), in addition to the electron parameters n, T_c, n_h/n_c, T_h/T_c [*Issautier et al.*, 1996]. In the present preliminary state of the method, which is not yet fully optimized, the precision on V is generally 10-20%; (in addition, the precision on T_c is better than when the drift velocity is neglected). The method is not well suited to measure the proton temperature T_p since the spectrum is in general sensitive to T_p only at very low frequencies where the shot noise produced by particle impacts and photoemission on the antenna is large.

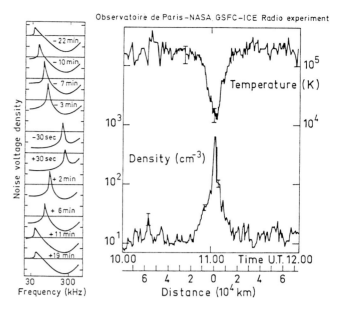

Figure 3. A series of thermal noise spectra recorded by the radio experiment aboard ICE during the crossing of the tail of comet Giacobini-Zinner, and the corresponding profiles of electron density and temperature as a function of the distance from the tail axis at 7800 km from the nucleus. (adapted from [*Meyer-Vernet et al.*, 1986a])

3. EXAMPLES OF APPLICATION

3.1. *Cometary Electrons*

This technique was first used on a large scale with the radio experiment on the spacecraft ISEE-3/ICE when it crossed the tail of comet Giacobini-Zinner (Figure 3). The experiment yielded the profiles of cometary electron density and temperature during the encounter. These results are unique because the ICE electrostatic electron analyzer could not detect adequately the cold cometary electrons in the plasma sheet ($n = 670$ cm^{-3}, $T = 1.3 \times 10^4$ K) as the effects of the spacecraft potential and photoelectrons could not be properly eliminated.

3.2. *Solar Wind*

Plate 1 shows an example of routine plasma measurement with the Ulysses URAP experiment. The upper panel is a radio spectrogram displayed as frequency versus time, i.e., the time evolution of spectra such as shown in Figure 2. The high level near 20 kHz corresponds to quasi-thermal Langmuir waves, close to f_p, which gives the plasma density. The bottom panel shows the density, core electron temperature and bulk speed deduced from the fittings. The structure shown corresponds to Ulysses crossing an interplanetary shock.

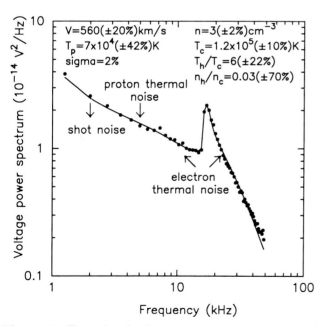

Figure 2. Example of voltage power spectrum measured by URAP on Ulysses in the solar wind when the Doppler-shifted proton contribution is important (dots) and the deduced plasma parameters. The solid curve is the theoretical electron-plus-proton quasi-thermal noise (plus the shot noise, which is only significant at the smallest frequencies). The best-fit parameters are shown.

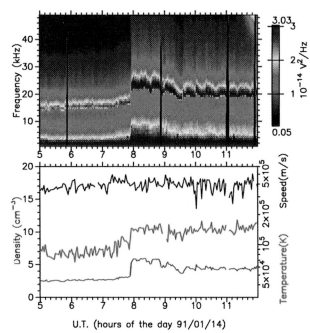

Plate 1. Example of routine plasma measurements obtained from quasi-thermal noise spectroscopy in the interplanetary medium. The upper panel is a radio spectrogram plotted as frequency versus time, with relative intensity indicated by the color bar chart on the right. The bottom panel shows the electron density, core temperature, and bulk speed deduced from the data.

4. WHY AND WHEN DOES IT WORK?

4.1. *Comparison With Other Techniques*

The presence of several instruments measuring electron parameters aboard Ulysses made possible an extensive comparison between them. The electron density was measured by:

- the SWOOPS electron analyzer [*Bame et al.*, 1992],
- thermal noise spectroscopy with the URAP radio receiver [*Stone et al.*, 1992a],
- the URAP relaxation sounder [*Stone et al.*, 1992a].

Maksimovic et al. [1995] compared 12,000 nearly simultaneous measurements (acquired within 1 minute of separation) from SWOOPS and from the thermal noise at several heliocentric distances in the ecliptic; the SWOOPS densities were on average 19% smaller than the thermal noise ones. A comparison was then made with improved SWOOPS results obtained with a vectorial correction of spacecraft charging effects [*Scime et al.*, 1994]; these densities are closer to the thermal noise ones, i.e., only 13% smaller (Figure 4(a)). In contrast, the density measurements from the sounder and from the thermal noise differ by only (6±3)% (Figure 4(b)).

These results suggest that, although the vectorial correction of spacecraft charging effects improves the results of the Ulysses electron analyzer, it still underes-

timates the density by about 10%. The core electron temperature aboard Ulysses was measured by both the electron analyzer and thermal noise spectroscopy. The results are within 17% of each other (Figure 4(c)).

4.2. *Advantages, Drawbacks, and Design*

The above Sections illustrated the main advantages of thermal noise spectroscopy:

- the large volume sensed: since the Langmuir wavelength satisfies $\lambda_L > 2\pi L_D$, the antenna is equivalent to a sensor of surface $S > 2\pi L_D \times L$, making the method in general relatively immune to spacecraft potential and photoelectron perturbations,
- the simplicity of density measurements (in general one has just to locate a frequency on a spectrogram, and this works even in presence of strong radioemissions since electromagnetic waves do not propagate below f_p), and their independence of gain calibration.

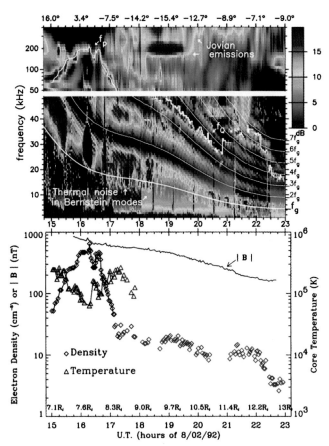

Plate 2. Upper panel: Radio spectrogram measured by URAP on Ulysses in the Io plasma torus and beyond. Lower panel: Corresponding electron density, bulk temperature, and magnetic field |**B**|, deduced from quasi-thermal noise analysis, as a function of time, Jovicentric distance (bottom scale), and magnetic latitude (top scale of upper panel). The color symbols refer to different measurement methods.

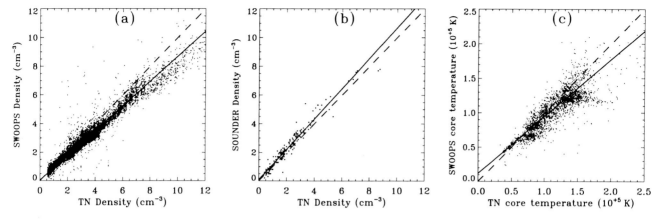

Figure 4. Scatter plot of electron density measurements from the thermal noise (TN) and (a) the electrostatic analyzer SWOOPS on Ulysses or (b) the URAP sounder. In each case, the continuous line corresponds to the line that minimizes the perpendicular dispersion from it and a dotted line of slope 1 is superimposed. (c) Scatter plot of electron temperature measurements from the thermal noise and the electrostatic analyzer, with the same kind of associated lines. (adapted from [*Maksimovic et al.*, 1995]).

In contrast, as already noted, the method is not well-adapted to measure suprathermal particles. In addition, as most wave measurements of bulk plasma parameters, it is perturbed by strong plasma instabilities.

For the method to work, the following conditions have to be met:
- the antenna length must exceed a few L_D in order to detect adequately the Langmuir wave cut-off and peak,
- the antenna must be sufficiently thin in order (i) to minimize the shot noise and (ii) have a radius smaller than L_D (for the simple theory used to hold),
- a good frequency resolution is necessary to resolve the peak,
- a sensitive and well-calibrated receiver is required in order to measure the temperature accurately; this condition is not necessary to obtain the density, which can be measured even with receivers of moderate sensitivity (see [*Gurnett et al.*, 1979; *Lund et al.*, 1994]).

5. MAGNETIZED PLASMA

When the electron gyrofrequency f_g is not negligible compared to f_p, the wave spectrum around f_p is modified by the electron gyration in the magnetic field. In that case, the electron thermal motion excites Bernstein waves, and the observed quasi-thermal noise shows weak bands with well-defined minima at gyroharmonics below the upper-hybrid band [*Meyer-Vernet et al.*, 1993], peaks at the upper-hybrid f_{UH} and f_Q frequencies [*Christiansen et al.*, 1978], and drops in the frequency bands where no Bernstein waves propagate [*Moncuquet et al.*, 1997].

The minima at gyroharmonics allow a simple measurement of the modulus of the magnetic field. Plate 2 shows an application on Ulysses in the Io plasma torus; this determination agrees with the magnetometer results

within a few percent [*Meyer-Vernet et al.*, 1993]. In addition, the quasi-thermal noise levels at maxima yield an estimate of the hot electron temperature [*Sentman*, 1982].

The f_{UH} and f_Q peaks can yield the total electron density [see for example *Birmingham et al.*, 1981; *Hoang et al.*, 1993] (black diamonds in Plate 2). The density can also be deduced from the signal disappearances in the Bernstein wave forbidden bands [*Moncuquet et al.*, 1997] (red symbols); on Ulysses these results were in agreement with the few measurements given by the relaxation sounder [*Stone et al.*, 1992b]. Note however that the accuracy of all these density measurements may be limited by the possibility of confusion between different resonance frequencies, especially in the non-equilibrium case.

The bulk temperature can be obtained either (i) from the thermal noise level at the minima, or (ii) by measuring the Bernstein wave-length (black triangles in Plate 2) [*Meyer-Vernet et al.*, 1993; *Moncuquet et al.*, 1995]. Method (ii) requires a spinning spacecraft; it is based on the fact that the angular pattern of the antenna is a sensitive function of kL when $kL \geq 1$ (see Eq.(3) and [*Meyer-Vernet*, 1994]), so that measuring the thermal noise spin modulation yields k. This technique requires a wire dipole antenna longer than the electron gyroradius r_g, since Bernstein waves have $k \sim 1/r_g$. In addition to the bulk electron temperature, this can also yield an estimate of the density (blue symbols in Plate 2).

6. CONCLUSION AND PERSPECTIVES

Thermal noise spectroscopy is complementary to electrostatic analyzers to measure accurately the density and the bulk electron temperature when spacecraft photoelectron and charging effects cannot be properly eliminated. This method requires a sensitive receiver and a wire dipole antenna; it is routinely used on Ulysses and has been implemented on Wind; it will also be used on Cassini, and is proposed on several future missions.

REFERENCES

Bame, S. J. *et al.*, The Ulysses solar wind plasma experiment, *Astron. Astrophys. Suppl. Ser.*, *92*, 237-265, 1992.

Birmingham, T. J., J. K. Alexander, M. D. Desch, R. F. Hubbard, and B. M. Pedersen, Observations of electron gyroharmonic waves and the structure of Io torus, *J. Geophys. Res.*, *86*, 8497-8507, 1981.

Chateau, Y. F, and N. Meyer-Vernet, Electrostatic noise in Non-Maxwellian plasmas: "Flat-Top" distribution function, *J. Geophys. Res.*, 94, 15,407-15,414, 1989.

Chateau, Y. F, and N. Meyer-Vernet, Electrostatic noise in Non-Maxwellian plasmas: Generic properties and Kappa distributions, *J. Geophys. Res.*, 96, 5825, 1991.

Christiansen, P. J. *et al.*, Geos-I observations of electrostatic waves, and their relationship with plasma parameters, *Space Sci. Rev.*, *22*, 383-400, 1978.

Couturier, P., S. Hoang, N. Meyer-Vernet, and J.-L. Steinberg, Quasi-thermal noise in a stable plasma at rest, *J. Geophys. Res.*, *86*, 11,127-11,138, 1981.

Feldman, W. C., J. R. Asbridge, S. J. Bame, M. D. Montgomery, and S. P. Gary, Solar wind electrons, *J. Geophys. Res.*, *80*, 4181-4196, 1975.

Gurnett, D. A., R. R. Anderson, F. L. Scarf, R. W. Fredericks, and E. J. Smith, Initial results from the ISEE-1 and -2 plasma wave investigation, *Space Sci. Rev.*, *23*, 103-122, 1979.

Hoang, S. *et al.*, Solar wind thermal electrons in the ecliptic plane between 1 and 4 AU: preliminary results from the Ulysses radio receiver, *Geophys. Res. Lett.*, *19*, 1295-1298, 1992.

Hoang, S., N. Meyer-Vernet, M. Moncuquet, A. Lecacheux, and B. M. Pedersen, Electron density and temperature in the Io plasma torus from Ulysses thermal noise measurements, *Planet. Space Sci.*, *41*, 1011-1020, 1993.

Hoang, S., N. Meyer-Vernet, K. Issautier, M. Maksimovic, M. Moncuquet, Latitude dependence of solar wind plasma thermal noise: Ulysses radio observations, *Astron. Astrophys.*, *316*, 430-434, 1996.

Issautier, K., N. Meyer-Vernet, M. Moncuquet, and S. Hoang, A novel method to measure the solar wind speed, *Geophys. Res. Lett.*, *23*, 1649-1652, 1996.

Issautier, K., N. Meyer-Vernet, M. Moncuquet, and S. Hoang, Pole to pole solar wind density from Ulysses radio measurements, *Solar Phys.*, in press, 1997.

Kellogg, P., Calculation and observation of thermal electrostatic noise in solar wind plasma, *Plasma Phys.*, *23*, 735-751, 1981.

Kuehl, H. H., Resistance of a short antenna in a warm plasma, *Radio Sci.*, *1*, 971-976, 1966.

Lund, E. J., J. Labelle, and R. A. Treumann, On quasi-thermal fluctuations near the plasma frequency in the outer plasmasphere: a case study, *J. Geophys. Res.*, *99*, 23,651-23,660, 1994.

Maksimovic, M. *et al.*, The solar wind electron parameters from quasi-thermal noise spectroscopy, and comparison with other measurements on Ulysses, *J. Geophys. Res.*, *100*, 19,881-19,891, 1995.

Meyer-Vernet, N., On natural noises detected by antennas in plasmas, *J. Geophys. Res.*, *84*, 5373-5377, 1979.

Meyer-Vernet, N., On the thermal noise "temperature" in an anisotropic plasma, *Geophys. Res. Lett.*, *21*, 397, 1994.

Meyer-Vernet, N. *et al.*, Plasma diagnosis from quasi-thermal noise and limits on dust flux or mass in comet Giacobini-Zinner, *Science*, *232*, 370-374, 1986a.

Meyer-Vernet, N., P. Couturier, S. Hoang, C. Perche, J.L. Steinberg, Physical parameters for hot and cold electron populations in comet Giacobini-Zinner, *Geophys. Res. Lett.*, *13*, 279-282, 1986b.

Meyer-Vernet, N., P. Couturier, S. Hoang, J.L. Steinberg, and R. D. Zwickl, Ion thermal noise in the solar wind: interpretation of the "excess" electric noise on ISEE 3, *J. Geophys. Res.*, *91*, 3294-3298, 1986c.

Meyer-Vernet, N., and C. Perche, Tool kit for antennae and thermal noise near the plasma frequency, *J. Geophys. Res.*, *94*, 2405-2415, 1989.

Meyer-Vernet, N., S. Hoang, and M. Moncuquet, Bernstein waves in the Io plasma torus: a novel kind of electron temperature sensor, *J. Geophys. Res.*, *98*, 21,163-21,176, 1993.

Moncuquet, M., N. Meyer-Vernet, and S. Hoang, Dispersion of electrostatic waves in the Io plasma torus and derived electron temperature, *J. Geophys. Res.*, *100*, 21,697-21,708, 1995.

Moncuquet, M. N. Meyer-Vernet, S. Hoang, R. J. Forsyth, and P. Canu, Detection of Bernstein wave forbidden bands: a new way to measure the electron density, *J. Geophys. Res.*, *102*, 2373-2379, 1997.

Rostoker, N., Fluctuations of a plasma, *Nucl. Fusion, 1*, 101-120, 1961.

Schiff, M. L., Current distribution on a grid type dipole antenna immersed in a warm isotropic plasma, *Radio Sci.*, *6*, 665-671, 1971.

Scime, E. E., J. L. Phillips, and S. J. Bame, Effects of spacecraft potential on three-dimensional electron measurements in the solar wind, *J. Geophys. Res.*, *99*, 14,769-14,776, 1994.

Sentman, D. D., Thermal fluctuations and the diffuse electrostatic emissions, *J. Geophys. Res.*, *87*, 1455, 1982.

Sitenko, A. G., *Electromagnetic Fluctuations in Plasma*, Academic, San Diego, Calif., 1967.

Stone, R. G. *et al.*, The Unified Radio and Plasma Wave Investigation, *Astron. Astrophys. Suppl. Ser.*, *92*, 291-316, 1992a.

Stone, R. G. *et al.*, Ulysses radio and plasma wave observations in the Jupiter environment, *Science*, *257*, 1524-1531, 1992b.

S. Hoang, K. Issautier, M. Maksimovic, R. Manning, N. Meyer-Vernet, M. Moncuquet, DESPA, Observatoire de Paris, 92195 Meudon Cedex, France. e-mail: meyer@obspm.fr

R. G. Stone, NASA/GSFC, Greenbelt, MD 20771, USA.

Measurement of Plasma Resistivity at ELF

L. R. O. Storey

Quartier Luchène, 84160 Cucuron, France

Laurent Cairó

Université d'Orléans, 45067 Orléans Cédex 2, France

The mutual-impedance (MI) technique can be used to used to measure the resistivity of space plasmas down to frequencies of a few hundred hertz. We outline the physics and technology of an MI probe designed for this purpose, and present the results from a rocket experiment in the auroral ionosphere. Though certain features of the data are hard to interpret, most of the results are consistent with normal resistivity, not modified by plasma turbulence. They suggest, however, that jointly with accurate instruments for electron density and temperature and for ion composition, a probe of this kind would be able to detect anomalous resistivity.

1. INTRODUCTION

The experimental program named Substorm-GEOS formed part of the Swedish national contribution to the International Magnetospheric Study. Its objective was to investigate the physics of magnetospheric substorms, especially auroral phenomena. For this purpose three Black Brant VC rockets (two single-stage, and one with a Nike booster) were launched from the sounding rocket base ESRANGE at Kiruna (67° 53′ N, 21° 04′ E) on the evening of 27 January 1979. Each of the three payloads included a mutual-impedance (MI) probe from the Laboratoire de Physique et Chimie de l'Environnement of the Centre National de la Recherche Scientifique at Orleans, France, the aim of which was to measure the resistivity of the thermal plasma at extremely low frequencies (ELF). The DC electric field experiment by the Royal Institute of Technology, Stockholm, and the AC electric field experiment by the Danish Space Research Institute, Lyngby, shared part of their sensor array with the ELF mutual-impedance probe.

Measurement Techniques in Space Plasmas: Fields
Geophysical Monograph 103

The present paper describes the MI probe and the results it produced on the first flight. On the two subsequent flights its performance was spoilt by various technological mishaps, and discussion of the results is beyond the scope of this paper.

The contents are arranged as follows: section 2 states the objectives of our experiment; section 3 presents the theory of the MI method for measuring plasma resistivity at ELF and shows how to predict the results that would be expected in a stable ionosphere; the technology of the probe is covered in section 4; the experimental results are presented in section 5 and discussed in section 6; section 7 concludes the paper.

2. OBJECTIVES

The ELF mutual-impedance experiment had two main objectives: a technological one, of testing the measuring technique; and a scientific one, of searching for anomalous resistivity. This paper deals primarily with the technique, but some justification for the scientific objective should be offered first.

The main incentive for measuring the resistivity of the ionospheric plasma is to study situations where it is anomalous. For a stable plasma, the resistivity can be calculated if the electron density and the ionic composition are known, along with the various collision frequencies. These frequencies are not always well known, though, and measuring plasma resistivity at ELF is one way of determining them [*Odéro*, 1972]. Nevertheless, this measurement is of greater interest in unstable plasmas, where the apparent electron-ion collision frequency may be enhanced by small-scale fluctuations of the turbulent electric field. Neither the mechanism of the enhancement, nor its possible effects such as particle acceleration or plasma heating, are fully understood at the present time. Hence any measurements of relevant properties of the unstable plasma are welcome for testing the various theories, and ELF resistivity is one such property.

In the auroral ionosphere, three instabilities that can cause anomalous resistivity have been observed remotely by radar, and in some cases directly with rocket-borne instruments as well. They comprise the ion-acoustic instability [*Foster et al.*, 1988; *Rietveld et al.*, 1991], the oscillating two-stream instability [*Papadopoulos and Coffey*, 1974; *Mishin and Schlegel*, 1994], and the Farley-Buneman instability [*Pfaff et al.*, 1984; *Primdahl*, 1986]. The proposal that the anomalous resistivity they may cause could be measured directly by means of an ELF mutual-impedance probe was made by *Storey and Malingre* [1976], and the Substorm-GEOS S23L1 experiment was the first in which it was put to the test.

3. THEORY

As explained in a companion paper [*Storey*, this volume], the basic principles of the mutual-impedance techniques for space plasma measurements are: (1) the warm-plasma theory of waves and antennas in plasmas; and, (2) the four-electrode method for measuring the resistivity of matter in bulk. Transposed to space, the four-electrode method uses a sensor array with the generic form shown in figure 1 of that paper. *Storey et al.* [1969] gave the name *quadripole probe* to an MI probe of this variety, in which the sensor comprises four electrodes making contact with the plasma at well defined points.

Although, in general, warm-plasma theory is required for describing the behavior of an MI probe, under certain conditions the much simpler cold-plasma theory yields a good enough approximation. In the design of our MI probe for measuring plasma resistivity at ELF, we used warm-plasma theory to find these conditions and then took steps to create them. Given that cold-plasma theory applies, the principles specific to the resistivity measurement are easiest to explain in this context, where the relationship between the current density and the electric field is a local one. In cold-plasma theory, the electrodes of the probe can be represented satisfactorily as point contacts with the plasma, but in warm-plasma theory the ratio of the electrode radius to the Debye length turns out to be a key factor. Accordingly the theory is given below in three parts, concerned successively with cold-plasma theory, warm-plasma theory, and finite-radius effects.

In a cold magnetoplasma, under linear conditions, the electric current density vector J is related to the electric field vector E by a conductivity tensor σ; thus $J = \sigma \cdot E$. With the z-axis of the coordinate system parallel to the magnetic field,

$$\sigma \equiv \begin{bmatrix} \sigma_\perp & -\sigma_H & 0 \\ \sigma_H & \sigma_\perp & 0 \\ 0 & 0 & \sigma_\parallel \end{bmatrix} \qquad (1)$$

where $\sigma_\parallel$ is the parallel conductivity, $\sigma_\perp$ the perpendicular (Pedersen) conductivity, and σ_H the Hall conductivity; these quantities all depend on the frequency ω and on the plasma parameters.

In using this theory to estimate the MI of a quadripole probe, two more approximations are frequently made: firstly, the quasi-static approximation, and secondly, the approximation of point electrodes. The quasi-static approximation involves ignoring the magnetic field associated with the current in the plasma, and assuming that the electric field is wholly derivable from a scalar potential. The approximation of point electrodes involves treating the transmitting electrodes as point sources of current, and the receiving electrodes as point antenna elements, none of which perturb the plasma. When both these approximations apply, the mutual impedance Z of the complete probe can be expressed as the sum and difference of contributions Z_{ij} from all the possible pairs of one transmitting and one receiving electrode [*Storey et al.*, 1969]. Hence the basic theoretical problem is to calculate the MI between a pair of point electrodes.

Formally, a point source of alternating current I with the frequency ω, placed at the origin of the coordinates, creates at the point $r \equiv (x, y, z)$ the potential $\phi(r) = I\zeta(r, \omega)$, where $\zeta(r, \omega)$ is the MI between these two points at the given frequency. According to cold-plasma theory [*Odéro*, 1972],

$$\zeta(r, \omega) = [4\pi \rho \, \sigma_m(\omega)]^{-1} \qquad (2)$$

where

$$\rho = (x^2 + y^2 + z^2/a^2)^{1/2} \qquad (3)$$

is a scaled distance, while

$$a = (\sigma_\parallel/\sigma_\perp)^{1/2} \qquad (4)$$

is an anisotropy factor. In the ionosphere this factor can be very large, and then the equipotentials are approximately a set of coaxial cylinders, their common axis being the magnetic field line that passes through the source.

The quantity σ_m in equation (2) is a mean conductivity, defined as below:

$$\sigma_m \equiv \left[(\sigma_\parallel - i\omega\epsilon_0)(\sigma_\perp - i\omega\epsilon_0)\right]^{1/2} \simeq (\sigma_\parallel\sigma_\perp)^{1/2} \quad (5)$$

It is roughly equal to the geometric mean of the parallel and perpendicular conductivities, an approximation that holds good if the plasma is so dense and the frequency so low that the displacement current can be neglected in comparison with the conduction current.

In cold-plasma theory, the mutual impedance Z of a quadripole probe is obtained by setting $Z_{ij} = \zeta(\mathbf{r}_{ij}, \omega)$, where $\mathbf{r}_{ij}$ is the vector distance from the electrode i of the transmitting antenna to the electrode j of the receiving antenna, then summing over the four pairs of one transmitting and one receiving electrode, giving each Z_{ij} its appropriate sign. From equation (2) it then appears that Z is inversely proportional to σ_m, so the resistivity that the instrument measures is σ_m^{-1}; we call this the *mean resistivity*.

Using this expression in conjunction with a representative model of the daytime mid-latitude ionosphere, Odéro [1972] computed the MI of a quadripole probe as a function of altitude at various fixed frequencies in the range 0–2000 Hz. He considered a probe with its electrodes at the four corners of a square, measuring 2 m across a diagonal and oriented with its plane perpendicular to the magnetic field. His results for the modulus of the impedance are given in figure 20 of his paper. As an example, for this probe working at 500 Hz at F-region altitudes, he found that $|Z| \simeq 1$ ohm, which previous experience had shown to be more or less the lower limit of the values that could be measured in practice.

The theoretical study of ELF mutual-impedance probes in warm magnetoplasmas was begun by *Malingre* [1979], who derived an expression for the mutual impedance $\zeta(r, \omega)$ between a pair of point electrodes lying in a plane perpendicular to the magnetic field. He found that ζ can be expressed as the the sum of two terms, the first being the value ζ_c given by cold-plasma theory, while the second, named ζ_w, represents a field of ion acoustic waves excited by the transmitting electrode: $\zeta = \zeta_c + \zeta_w$. Here these two contributions to ζ will be referred to as the *cold-plasma term* and the *warm-plasma term* respectively. Malingre showed that, under conditions typical of the auroral ionosphere, the warm-plasma term would be greater than the cold-plasma term by at least an order of magnitude, and if this were true in practice it would rule out the possibility of measuring plasma resistivity in the way discussed above.

Fortunately the study of finite-radius effects has revealed that the warm-plasma term depends on conditions at the interface between the antenna electrodes and the plasma, and that, by selecting the right conditions, this term can be reduced to an acceptable level. *Malingre* [1984] calculated the potential created in a warm isotropic plasma by a spherical electrode of finite radius, supplied with ELF current from a high-impedance source. A number of simplifications were made, including the adoption of a hydrodynamic (2-fluid) model for the plasma, and the assumption that the electrode was at plasma potential on the average. It was found that as the radius increases from zero, the amplitude of the ion acoustic waves decreases monotonically. Under conditions typical of the Substorm-GEOS experiment the reduction should be substantial, roughly two orders of magnitude.

By reciprocity, a similar reduction should take place at the receiving antenna, on account of the finite radii of its spherical electrodes. The receiving antenna does not respond fully to the fluctuations of plasma potential in the field of the wave, the attenuation being greatest when the average potential of the electrodes is close to that of the plasma [*Fiala and Storey*, 1970]. Laboratory measurements confirming that the response is very weak in this case have been made by *Schott* [1980].

Hence, by biasing all of the electrodes suitably, it should be possible to make the warm-plasma term small compared with the cold-plasma term and thereby create the conditions under which cold-plasma theory applies. When considering the experimental results (section 6), it will be helpful to note that if this theory did not apply, then warm-plasma effects would increase, not decrease, the mutual impedance.

In conclusion, the theory predicts that a quadripole mutual-impedance probe with dimensions of the order of 1 m, and with its electrodes biased close to plasma potential, should be capable of measuring the mean resistivity of the plasma in the Earth's ionosphere.

4. TECHNOLOGY

The rocket-borne instrument for measuring ELF resistivity comprised two main parts: the sensor and the electronics. The factors that influenced its design have been discussed previously by *Storey and Malingre* [1976], and technical details are given below.

The working frequency for the measurements was chosen to be a few hundred hertz, so as to avoid their being unduly sensitive to the ionic composition of the plasma. This meant that the frequency should be as far removed as possible from both the ion gyrofrequencies at several tens of hertz and the lower hybrid frequency at several kilohertz. The actual value was 310 Hz, chosen to avoid harmonics of frequencies used by various other devices on board.

The sensor system involved no less than eight electrodes, four transmitting and four receiving; it is shown in figure 1. The electrodes comprised four cylinders (A,B,C,D) and four spheres (A', B',C',D'), on booms deployed perpendicular to the spin axis of the rocket. The cylinders, 5 cm in diameter by 8 cm long, were the transmitting electrodes, used only by the

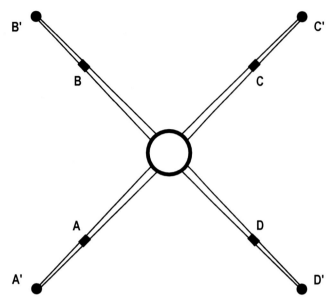

Figure 1. Eight-electrode sensor viewed along the rocket axis.

ELF mutual-impedance experiment. The spherical receiving electrodes, 6 cm in diameter, were on the tips of the booms and were shared with the DC and AC electric field experiments. The distance from each cylinder to its neighboring sphere was 0.8 m, while the distance from the sphere to the spin axis was about 2 m.

Through each of the four transmitting electrodes, a fixed alternating current at 310 Hz was injected into the plasma. Its waveform was square, with a nominal peak amplitude of 1.0 μA; the fundamental component, a sinusoid at 310 Hz, had a peak amplitude of $I = 1.27\mu$A. The square waves applied to the electrodes A and C (on opposite booms) were in phase with each other, but in antiphase with those applied to B and D. Moreover, each electrode was biased with a direct current of 1.2 μA, flowing from the electrode to the plasma and so corresponding to the collection of plasma electrons.

Each of the four receiving electrodes also was biased with a positive direct current from a high-impedance source, but the value of this current was readjusted automatically at regular intervals, in the light of the plasma diagnostic measurements, so as to ensure that the potential of the electrode remained negative with respect to the plasma, yet less so than it would have been in the absence of bias. The automatic bias system formed part of the DC electric field instrument.

Figure 2 shows how the eight electrodes of the sensor were connected to the electronics. The two preamplifiers and one amplifier took the 310 Hz voltages $V_{A'}$, $V_{B'}$, $V_{C'}$, $V_{D'}$ received on the spheres A', B', C', D', and combined them as follows: $V = (V_{A'} - V_{B'}) - (V_{C'} - V_{D'})$. Since the voltage received on each sphere was due mainly to the current injected into

the plasma from the cylinder on the same boom (A, B, C, D respectively), and since these currents had the phase relationships described earlier, it follows that the four voltages were added together constructively. The combined voltage V was amplified by a receiver, not shown in the figure, which had a pass band extending from about 100 Hz to about 500 Hz with a sharp cut-off at the high-frequency end so as to isolate the fundamental component at 310 Hz. The mutual impedance of the complete eight-pole sensor is defined as $Z \equiv V/I$.

The principles of this instrument were those common to all MI probes, but its use of a sensor with eight electrodes instead of the usual four improved the signal-to-noise ratio. The sensor system can be pictured as being made up of two quadripole probes mounted back-to-back, with their outputs combined in such a way that the signals of interest reinforced each other, while much of the noise due to external electric fields was cancelled out; this arrangement was suggested to us by J. M. Chassériaux (personal communication).

5. RESULTS

The first of the three Substorm-GEOS rockets, code-named S23-L1, was launched at 17 h 12 m 32 s UT on January 27, 1979. Its flight path was directed 14° east of geographic north, and peak altitude, 267.5 km above that of the launcher, was reached 260.4 s after launch, at a horizontal range of 40.7 km. The booms for the electric field and ELF resistivity experiments were fully deployed around 70 km altitude.

In the light of their direct measurements of electron density, electron temperature, and electric field, *Marklund et al.* [1981, 1982] have described the ionospheric conditions encountered along the flight path. At about 160 s after launch, the rocket entered an auroral arc. During the next 4 minutes it remained within or above the arc, in a plasma characterized by a rela-

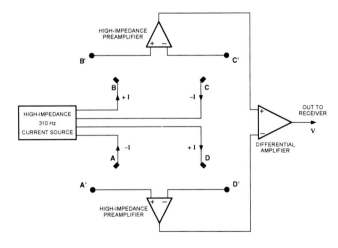

Figure 2. How the sensor was connected to the electronics.

tively low electric field and a relatively high electron density and temperature. Finally, from about 400 s onwards, the ambient auroral activity decreased.

The experimental data from the ELF mutual-impedance probe fell into two groups of very different quality: good data, in which the noise level was at the bottom of the dynamic range of the receiver; bad data, in which the noise was very strong, sometimes saturating the receiver. The occurrence of bad data coincided with an intermittent malfunction of the automatic bias system, resulting in little or no bias current being applied to the receiving spheres. Fortunately only about 40% of the data were spoilt in this way, and the remaining 60% could be analyzed to yield the MI. The complete set of results for $|Z|$ is displayed in figure 3; the good and bad data points are represented by solid dots and open circles respectively.

6. DISCUSSION

It is interesting to compare the experimental values of the MI with those to be expected theoretically, assuming that the plasma resistivity was normal. M. Malingre (personal communication) has estimated how Z should have varied during the flight, using the cold-plasma theory of Odéro [1972]. As data, he took the electron density measured by a Langmuir probe [Marklund et al., 1981], together with a model for the ionic composition. The densities from the Langmuir probe were multiplied by a factor of 4.3 to make them consistent with data from an ionosonde at the Kiruna Geophysical Institute. The results are plotted in figure 3 as the dashed line. Experiment and theory agree fairly well during the middle period of the flight when the rocket was inside the auroral arc, roughly from about 180 s to 400 s flight time, though not so well at the beginning of this period as at the end of it.

Outside the arc, however, where the electron density is lower, the agreement is less satisfactory. In particular, around 100 s flight time on the upleg, before the rocket has entered the arc, the theoretical values exceed the experimental ones by a factor of about 3. Again, in the period 400–420 s when the rocket is leaving the arc on the downleg, the experimental values do not follow the rising theoretical ones, though later they start to do so; note that the last good experimental point is off scale at 35 ohms. We have no explanation yet for these residual discrepancies concerning $|Z|$. However, strong electric fields were observed in the E-region on the upleg of the S23-L1 flight, and on the downleg the observed field was increasing when the measurements ceased at about 150 km altitude; see figure 3 of Marklund et al. [1981]. The effects of these fields were not accounted for in the theoretical calculations of the mutual impedance; possibly, if this were done, the agreement would be improved.

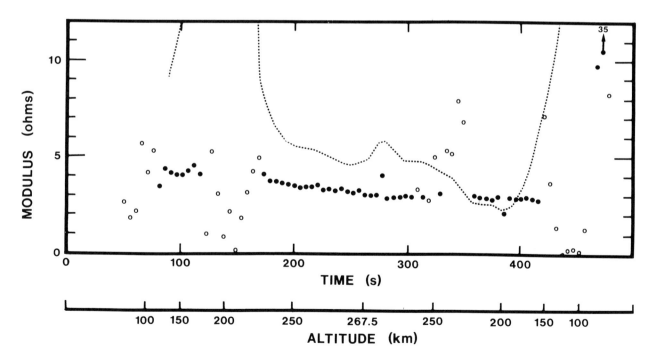

Figure 3. Experimental results for the modulus of the mutual impedance: ● good data; ○ bad data. The dashed curve is theoretical.

7. CONCLUSION

The ELF mutual-impedance experiment described above was the first of its kind, and, as is common in such cases, it revealed many problems that had not been foreseen. None the less, the measurements taken in the middle part of the flight were in fair agreement with theoretical expectations, which suggests that there is nothing basically wrong with the method of measurement. Some of the residual disagreements may have been due to errors of the assumed electron density, electron temperature, and ionic composition. Therefore we conclude that the experiment would be worth repeating, with care taken to secure better supporting data. Suggestions as to how this should be done are made in a longer version of this paper, available from the first author on request. If a future experiment confirms that this method for measuring plasma resistivity at ELF works satisfactorily, then it should be able to measure anomalous as well as normal resistivity.

Acknowledgments. This research was carried out under contract with the Centre National d'Etudes Spatiales. The sensor system was built by SAAB-Scania, Linköping, Sweden, the electronics by Etudes et Equipements Electroniques SA, Toulouse, France, and the test equipment at the LPCE, Orleans, under the responsibility of P. Décréau. We are most grateful to C. G. Fälthammar for inviting us to take part in the Substorm-GEOS program, and we thank the other experimenters for their cooperation, in particular L. Block and G. Marklund with whose instrument our own was combined and who advised us on the interpretation of the data. We most especially appreciated the unfailingly courteous assistance we received from the Project Scientist, the late U. Fahleson. Finally, we thank the reviewers for their helpful comments.

REFERENCES

Fiala, V., and L. R. O. Storey, The response of a double-sphere dipole antenna to VLF electrostatic plasma waves, in *Plasma Waves in Space and in the Laboratory,* edited by J. O. Thomas and B. J. Landmark, Vol. 2, pp. 411–426, Edinburgh University Press, Edinburgh, 1970.

Foster, J. C., C. del Pozo, K. Groves, and J.-P. Saint-Maurice, Radar observations of current driven instabilities in the topside ionosphere, *Geophys. Res. Lett., 15,* 160–163, 1988.

Malingre, M., Rayonnement d'ondes acoustiques ioniques par une source ponctuelle pulsante dans un magnétoplasma chaud, *C. R. Acad. Sci. Paris, Sér. B, 289,* 257–260, 1979.

Malingre, M., Radiation of ion acoustic waves from a solid spherical probe in a warm isotropic plasma, *Radio Sci., 19,* 400–410, 1984.

Marklund, G., L. Block, and P.-A. Lindqvist, Rocket measurements of electric fields, electron density and temperature during different phases of auroral substorms, *Planet. Space Sci., 29,* 249–259, 1981.

Marklund, G., I. Sandahl, and H. Opgenoorth, A study of the dynamics of a discrete auroral arc, *Planet. Space Sci., 30,* 179–197, 1982.

Mishin, E. V., and K. Schlegel, On incoherent scatter lines in aurorae, *J. Geophys. Res., 99,* 11,391–11,399, 1994.

Odéro, D., Possibilités d'utilisation d'une sonde quadripolaire dans la gamme 0–1000 Hz pour mesurer les fréquences de collision des particules chargées dans l'ionosphère, *Ann. Géophys., 28,* 541–574, 1972.

Papadopoulos, K., and T. Coffey, Anomalous resistivity in the auroral plasma, *J. Geophys. Res., 79,* 1558–1561, 1974.

Pfaff, R. F., M. C. Kelley, B. G. Fejer, E. Kudeki, C. W. Carlson, A. Pedersen, and B. Hausler, Electric field and plasma density measurements in the auroral electrojet, *J. Geophys. Res., 89,* 236–244, 1984.

Primdahl, F., Polar ionospheric E-region plasma wave stabilization and electron heating by wave-induced enhancement of the electron collision frequency, *Physica Scripta, 33,* 187–191, 1986.

Rietveld, M. T., P. N. Collis, and J.-P. Saint-Maurice, Naturally enhanced ion acoustic waves in the auroral ionosphere observed with the EISCAT 933-MHz radar, *J. Geophys. Res., 96,* 19,291–19,305, 1991.

Schott, L., Measurement of the plasma potential with ion acoustic waves, *Rev. Sci. Instrum., 51,* 383–384, 1980.

Storey, L. R. O., Mutual-impedance techniques for space plasma measurements, this volume.

Storey, L. R. O., M. P. Aubry, and P. Meyer, A quadripole probe for the study of ionospheric plasma resonances, in *Plasma Waves in Space and in the Laboratory,* Vol. 1, edited by J. O. Thomas and B. J. Landmark, pp. 302–332, Edinburgh University Press, Edinburgh, 1969.

Storey, L. R. O., and M. Malingre, A proposed method for the direct measurement of enhanced resistivity, in *European Programmes on Sounding-Rocket & Balloon Research in the Auroral Zone,* pp. 387–409, European Space Agency, Neuilly, Rep. SP-115, 1976. (Copies are available from the first author on request).

L. R. O. Storey, Quartier Luchène, 84160 Cucuron, France. (e-mail: storey@nssdca.gsfc.nasa.gov)

L. Cairó, MAPMO–UMR 6628, Université d'Orléans, UFR Sciences, BP 6759, 45067 Orléans Cédex 2, France. (email: lcairo@ labomath.univ-orleans.fr)

A Critical Overview of Measurement Techniques of Spacecraft Charging in Space Plasma

Shu T. Lai

Phillips Laboratory, Hanscom AFB, MA 01731

Under certain conditions in space there are physical processes which may cause misinterpretation of spacecraft potential measurements using some very common techniques. The use of long booms to measure the potential difference between a spacecraft and the tip of a boom has at least two deficiencies. First, when the spacecraft potential is high, the tip of the boom may be engulfed by the sheath of the spacecraft. Second, the tip of the boom itself may charge. Even if materials of high secondary emission coefficients are used for the boom tips, charging can still occur when the space plasma is energetic. Another common technique for measuring the energy shift of the incoming electron energy distribution function also has deficiencies. First, if ionization is abundant in the vicinity of the spacecraft (for example, during electron beam emissions), the energy gap resulting from the shift becomes blurred. Second, if the angular momentum of the incoming electrons is significant, the shift should not be attributed to potential energy only. The technique of using Langmuir probes, which are located at short distances from the spacecraft surface, is often deficient in measuring spacecraft potential. It may be too naive to apply either the attractive regime or the repulsive regime of Langmuir. For example, while the probe potential may be negative relative to that of the space plasma, it may be positive relative to that of the spacecraft. In that case, the probe is repelling electrons from the space plasma but attracting secondary electrons from the spacecraft.

1. INTRODUCTION

In designing experiments, one has in mind a set of physical processes expected to occur in the experiments. The instruments so designed may yield good measurements under the conditions being considered. Very often, however, what actually happens in an experiment is complex. There are other physical processes that may have

occurred to affect the measurements. Sometimes the measurements may easily be misinterpreted. One has to identify and take into account these processes in the interpretation of the measurements.

In this paper, we present a critical overview of some very common measurement techniques in spacecraft charging [for reviews, see *Garrett*, 1981; *Whipple*, 1981; *Lai*, 1991a]. We point out that under certain conditions in space, there are physical processes which may cause misinterpretation of spacecraft potential measurements using some very common techniques. Three common techniques are discussed. They are (1) the use of long booms, (2) the energy shift of the charged particle distribution function, and (3) the use of Langmuir probes.

Measurement Techniques in Space Plasmas: Fields
Geophysical Monograph 103

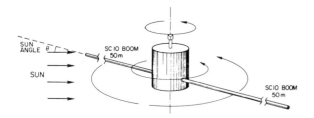

Figure 1. Long booms (SC10) on the SCATHA satellite for measuring spacecraft charging.

2. THE USE OF LONG BOOMS

A very common technique used to measure spacecraft potential ϕ_s is to use long booms [*Aggson, et al.*, 1983]. The booms (Figure 1) are electrically isolated from the spacecraft. The objective is to measure the potential difference, $\Delta\phi$, between the ambient plasma and the spacecraft body. What one actually measures is the potential of the tip of a boom relative to the spacecraft body.

$$\text{Objective}: \quad \Delta\phi = \phi_{\text{plasma}} - \phi_s$$

$$\text{Actual}: \quad \Delta\phi = \phi_{\text{boom}} - \phi_s$$

If the potential, ϕ_{boom}, of the tip of the boom equals nearly the ambient plasma potential, ϕ_{plasma}, the measurement will be a good approximation. However, under certain conditions the boom potential may deviate substantially from that of the ambient plasma. We discuss two possibilities, viz., (1) sheath engulfment of booms, and (2) charging of booms.

2.1 *Sheath Engulfment of Booms*

As the spacecraft potential, ϕ_s, increases, the Coulomb sheath (or shielded sheath) of the spacecraft expands. At high potentials, the sheath may engulf significantly the entire boom. For example, we take the spacecraft sheath potential $\phi(r)$ modeled using the Debye form [*Whipple, et al.*, 1974]:

$$\phi(r) = \phi(0)\,\frac{R}{r+R}\,exp(-r/\lambda_D)$$

where R is the radius of the spacecraft body, and λ_D is the Debye distance. For environments, we assume that the Debye length λ_D of the ambient plasma is about 45m

[*Aggson, et al.*, 1983] and the photoelectron temperature T_{ph} is about 2 eV [*Whipple*, 1981]. The SC10 booms on SCATHA are each 50m long. If the SCATHA spacecraft body is charged to -1 kV, the potential $\phi(r)$ at a distance of r = 50m from the spacecraft body would be -4 volts approximately. In high-potential charging experiments, sheath engulfment of booms is significant (Figure 2).

Sheath engulfment not only renders the potential measurements inaccurate, but also draws current from the booms to the spacecraft body. The current may affect the spacecraft body potential substantially. For example, during electron beam emissions from SCATHA in sunlight on Day 70, the photoelectrons from the SC10 booms flowed down the potential gradient along the booms to the spacecraft body [*Lai, et al.*, 1987; *Lai*, 1994]. As the spacecraft rotated in sunlight, the photoelectrons caused sinusoidal potential variations.

2.2 *Charging of Booms*

To prevent negative charging, the outer surface of the tip of a boom is often made of a material which has high secondary electron emission coefficient δ. When primary electrons have mostly energy, E, for which $\delta(E) > 1$, the number of outgoing secondary electrons will probably exceed the number of primary electrons, thus preventing negative charging. If the flux of secondary electrons exceed that of primary electrons, positive charging may occur. However, it would occur at a low level only, because secondary electrons, which have low energies (typical a few eV), would not escape if the charging level is positive and high.

For example, the outer portion of the SC10 boom on SCATHA is coated with CuBe, which has $\delta_{\text{max}} \sim 4$ at E = 900 eV. Indeed, the outer portion of the SC10 boom is

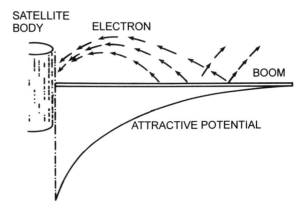

Figure 2. Sheath Engulfment. A boom is engulfed in the potential sheath of the spacecraft body. .

often charged to a few volts positive in sunlight when the ambient plasma is quiet (Figure 3). If a few volts is an acceptable error, the measured potential difference would be a good approximation of that between the ambient plasma and the spacecraft body.

When the ambient plasma is very energetic, the probability of secondary electron emission is less than one, δ (E) < 1. When that occurs, negative charging of the material can occur. For example, when the ambient electrons became energetic after eclipse entrance on Day 114, the CuBe surface of the SC10 boom was charged (Figure 4) to high negative voltages [*Lai*, 1991b]. When a boom tip is charged to high voltages, the measurement $\Delta\phi$ is no longer a good approximation of the potential difference between the ambient plasma and the satellite body. Under such a condition, the measurement technique is not applicable.

3. ENERGY SHIFT IN CHARGED PARTICLE DISTRIBUTION UNCTIONS

Another common method for measuring spacecraft potential is to measure the shift, ΔE, of energy E in the distribution function of electrons or ions. The principle is as follows. The distribution function $f(E)$ of electrons, for example, is measured on the spacecraft surface. If $f(E)$ is Maxwellian, a plot of log[$f(E)$] will be a straight line as a function of E; if $f(E)$ is non-Maxwellian, it will not be. Without charging, the function $f(E)$ will most likely start from E=0.

With charging to a positive potential, ϕ_s, the ambient electrons within the sphere of influence will be attracted towards the spacecraft, resulting in an energy shift (Figure

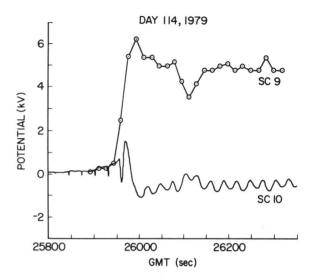

DAY 114, 1979

Figure 4. SC10 and SC9 potential measurements on SCATHA as a function of time in Day 114. The SC10 curve represents the potential of the tip of the boom relative to the satellite body. The SC9 curve is obtained from the shift of the energy distribution functions at every instant of the charged particle spectra. When the boom is suddenly charged to a high negative potential, the SC10 curve reverses sign. The sign of the SC9 curve is unchanged [*Lai*, 1991b].

5) of magnitude $e\phi_s$ in the distribution function $f(E)$. An energy gap from E = 0 to e ϕ_s is identified as the shift. Similarly, if the spacecraft potential is negative, the electron distribution would be shifted by - $e\phi_s$ and the ion distribution by $e\phi_s$. In the former case, there is no gap; nevertheless, a shift of the distribution can often be identified. The magnitude of the energy shift is interpreted as the charging level.

This technique assumes that the lowest energy ambient charged particles are initially at rest. With charging, the attracted species fall towards the spacecraft thereby gaining an energy ϕ_s. Under certain conditions, such as (a) and (b), this assumption does not hold.

(a) During electron beam emissions, there is abundant ionization in the spacecraft sheath [*Lai*, 1992]. The newly created electrons have initially nearly zero energy, but they gain energy as they fall towards the spacecraft. The energy gain is less than $e\phi_s$ if they start from inside the sheath. The electrons's low energies would be in the gap of the shifted distribution and thereby blur the gap.

(b) For a highly charged spacecraft with a sheath large compared with the spacecraft radius, an ambient charged particle may not fall radially towards the spacecraft. Depending on its initial velocity and impact

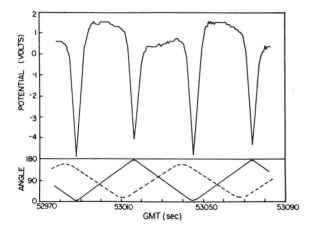

Figure 3. Boom potential on a rotating satellite, SCATHA, in sunlight on a quiet day [*Lai, et al.*, 1986].

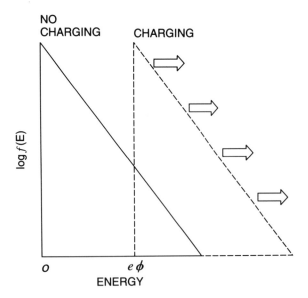

Figure 5. A shift of the energy distribution function of electrons or ions. A Maxwellian distribution is shown.

parameter, it may have substantial angular momentum. When it arrives at the spacecraft, its energy may exceed $e\phi_s$ substantially. To overcome this problem, the instrument should have good angular resolution.

4. LANGMUIR PROBES

Langmuir probes are commonly used in the laboratory and in space. Under certain conditions, their usage have pitfalls, which may be easily overlooked. We discuss two remarks, viz., (1) probe geometry, and (2) non-uniform plasma environment.

4.1 *Probe Geometry*

Mott-Smith and Langmuir [1926] derived the probe formulae for spheres and infinite cylinders only. In the orbit limiting regime, the formulae for the attractive and repulsive species respectively are

$$I(\phi) = I(0)(1 - e\phi/kT)^N$$

$$I(\phi) = I(0)\exp(-e\phi/kT)$$

Here, $I(\phi)$ is the current collected by the probe, ϕ the probe potential, e the elementary charge, T the plasma temperature, and N = 1 for spheres and ½ for infinite cylinders. The formulae can be used not only for Langmuir probes on spacecraft but also for the spacecraft itself.

In practice, the geometry of a Langmuir probe is neither a sphere nor an infinite cylinder. For example, the power N for the SCATHA satellite, which is a short cylinder, turns out to be about 0.774 [*Lai*, 1994]. If one simply uses the spherical probe formula for a short cylinder, for example, the results would be inaccurate.

4.2 *Non-Uniform Plasma Environment*

The Langmuir probe formulae were derived with the assumption that the plasma environment is infinite and uniform. In practice, the plasma may not be infinite or uniform. For examples, the plasma may be partly shaded, the potential distribution may be non-uniform, there may even be potential barriers nearby, and the probe may be both attractive and repulsive to the same species.

To illustrate the last one of the these examples, let us consider a probe on a differentially charged spacecraft. Suppose the probe potential is negative with respect to the ambient plasma. It may also be positive with respect to a nearby surface. While the probe is repelling electrons from the ambient plasma, it may be attracting secondary electrons emitted from the nearby surface. In this case, neither the repulsive formula nor the attractive formula alone is applicable (Figure 6).

5. CONCLUSION

We have given a critical overview on some measurement techniques of spacecraft charging. The use of a long boom for measuring spacecraft potential is good unless (1) when the charging level of the spacecraft is so high that the boom is engulfed by potential sheath of the

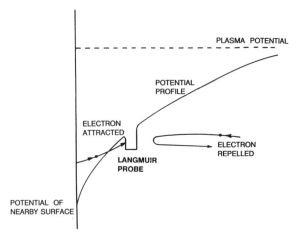

Figure 6. A Langmuir probe repelling electrons while attracting electrons from another source.

spacecraft body, and (2) the boom itself becomes charged. When there is a potential gradient along a boom, electrons may flow down the gradient and affect the spacecraft potential. The method of equating the shift of electron or ion energy distribution function as the spacecraft potential often works unless there is substantial ionization in the sheath. Also this method requires good angular resolution of the incoming charged particles. The original Langmuir probe formulae are available for a sphere or an infinite cylinder only. Also the formulae are derived by assuming an ideal (infinite and uniform) plasma. In practice, the plasma may not be ideal. In certain situations, a probe may be repelling electrons from one source but attracting electrons from another.

The measurement techniques discussed do not alway work perfect. Very often, what happens in an experiment is complex. A technique that works well under certain conditions may not work well under other conditions, in which some physical processes may have occurred to affect the measurements. One has to identify and take into account the processes in the interpretation of the measurements. We advocate the use of multi-instruments in space experiments for better diagnostics and measurements of the various physical processes at work..

REFERENCES

Aggson, T.L., B.G. Ledley, A. Egeland, and I. Katz, Probe measurements of DC electric fields, *ESA-SP-198*, 13-17, 1983..

Garrett, H.B., The charging of spacecraft surfaces, *Rev. Geophys. Space Phys.*, Vol.19, 577-616, 1981.

Lai S.T., H.A. Cohen, T.L. Aggson, and W.J. McNeil, Boom potential of a rotating satellite in sunlight, *J. Geophys. Res.*, vol.91A, 12137-12141, 1986.

Lai S.T., H.A. Cohen, T.L. Aggson, and W.J. McNeil, The effect of photoelectrons on boom-satellite potential difference during electron beam ejection, *J. Geophys. Res.*, vol.92, 12319-123-25, 1987.

Lai, S. T., Spacecraft charging thresholds in single and double Maxwellian space environments, *IEEE TransNucl.Sci.*, Vol.19, 1629-1634, 1991a.

Lai, S.T., Theory and observation of triple-root jump in spacecraft charging, *J. Geophys. Res.*, Vol.96, No.A11, 19269-19282, 1991b.

Lai, S.T., Sheath ionization during electron beam emission from spacecraft, *Physics of Space Plasmas*, Vol.11, 411-420, 1992.

Lai, S.T., An improved Langmuir probe formula for modeling satellite interactions with near geostationary environment, *J. Geophys. Res.*, Vol.99, 459-468, 1994.

Mott-Smith, H.M., and I. Langmuir, The theory of collectors in gaseous discharges, *Phys. Rev.*, Vol.28, 727-763, 1926.

Whipple, E.C., Potentials of surfaces in space, *Rep. Prog. Phys.*, *44*, 1197, 1981.

Whipple E.C., Jr., J.M. Warnock, and R.H. Winckler, Effect of satellite potential in direct ion density measurements through the magnetopause, *J. Geophys. Res.*, *79*, 179, 1974.

Shu T. Lai, Department of the Air Force, Phillips Laboratory, Mail Stop: GPID, Hanscom Air Force Base, MA 01731.

Results from the NRL Floating Probe on SPEAR III: High Time Resolution Measurements of Payload Potential

Carl L. Siefring and Paul Rodriguez

Charged Particle Physics Branch, Plasma Physics Division, Naval Research Laboratory, Washington, DC

We present results from the NRL Floating Probe (FP) which made high time resolution measurements of spacecraft charging and discharging on the Space Power Experiments Aboard Rockets (SPEAR III) payload. SPEAR III was specifically designed to study the physics of spacecraft charging phenomena. We know that spacecraft in the ionosphere can charge to high levels (up to a few kilovolts). Charging occurs naturally in the auroral region due to high-energy streaming electrons or during the operation of active experiments (ion/electron beams or electromagnetic tethers). Charging and discharging events are often impulsive in nature and a method fast enough to track these potential changes will be extremely important. The FP consists of a metallic sphere containing a high-impedance amplifier and a capacitive divider network for scaling large voltages to the range that solid state circuits can handle. The probe can be used for either positive or negative polarity measurements. The 'worst case' time-response is associated with negative charging since the probe must collect ions to stay in contact with the local plasma. Modeling of the FP indicates that it has a time response faster than 1 ms for the entire range of negative charging normally associated with spacecraft (0 to -2 kV). We will discuss the FP design, construction, and theory of operation. SPEAR III also carried an ElectroStatic particle Analyzer (ESA) for monitoring the incoming ion distribution and a comparison of these two measurements is presented.

INTRODUCTION

Spacecraft often acquire large negative electrical potentials; spacecraft charging occurs naturally when passing through the regions of high-energy streaming electrons in an aurora and during active experiments involving electron beams, ion beams, or electromagnetic tethers. Studies performed using the DMSP F6 and F7 [Gussenhoven et al., 1985] and SCATHA [Mullen et al., 1986] satellites indicate that payloads commonly reach hundreds of volts (negative) in the auroral region. Although not as common, satellites in eclipse, in this region, charge to levels below -2 kV (cf., Whipple [1981] and Garrett [1981]). Active experiments also induce spacecraft charging. Ion beam emission will

charge a payload (e.g., Kaufmann et al. [1989] and Olsen et al. [1990]) with maximum levels around a few hundred volts. Electron beam sources are capable of relatively large currents and can also cause high-voltage charging; both positive and negative charging can occur because of ringing [e.g., Borovsky, 1988; Winglee, 1991]. A good example of this phenomenon occurred in the BEAR neutral particle beam experiment where ringing to negative voltages of 300-400 V occurred with just a few milliamps of un-neutralized current [Pongratz et al., 1991].

Large differential potentials can damage sensitive electronic equipment. Arcing between insulators is thought to cause both physical damage to solar arrays [Thiemann et al., 1990], and to induce electromagnetic transients in other components [Metz, 1986]. Such electromagnetic transients can cause command-and-control failures (cf., McPherson and Schober [1976], and Koons et al., [1988]). Also of importance are the effects that spacecraft charging has on space-based measurements. Charging to potentials of even a few volt can seriously effect Langmuir probe and low energy particle measurements. Charging to tens of volts can

Measurement Techniques in Space Plasmas: Fields
Geophysical Monograph 103

saturate electric field probes and kV charging can interfere with energetic particle, X-ray, and optical measurements.

Techniques for Measuring Spacecraft Potential

Electrostatic particle analyzers have historically been successful in measuring payload voltages in the few-hundred volt to few-kV range. The drawback of these measurements is time resolution; the entire energy spectrum of the incoming ions must be sampled to infer the payload voltage. At the lower end of the range (less than ≈300 V) inferring the payload potential from the ion spectrum can be difficult [Mullen et al, 1986]. The use of floating probes is common at lower voltages (0 to 15 V or slightly higher) and these have a relatively rapid time-response which depends on the local plasma conditions. Thus, a floating probe that can measure large voltages offer two advantages. First, in the range from approximately 30 V to a few-hundred volts floating probes provide a measurement that is difficult for electrostatic analyzers. Second, above a few-hundred volts floating probes provide an increase in time resolution over the state-of-the-art particle measurements..

The difficulties in designing a floating probe to be used for measuring large voltages are several. First, the high voltage must be reduced to signal levels that typical electronic circuits can handle and simultaneously a very high input-impedance must be maintained. These two requirements determined the choice of capacitive coupling for our sensor. Second, large payload potentials imply a large ion-sheath around the spacecraft. The FP must be placed outside of this sheath in order to measure the full potential difference between the payload and surrounding plasma. Thus, the required boom length and complexity grow with the maximum voltage of interest. Finally, large voltages can easily damage sensor components and great care must be exercised to protect solid state devices.

THE SPEAR III EXPERIMENT

Recently, experiments such as SPEAR I and III (Space Power Experiments Aboard Rockets) have used 'tether-like' configurations for studying spacecraft charging and *discharging* [Allred et al., 1988; Raitt et al., 1997]. As shown in Figure 1, a large sphere and the payload body are connected to the opposite poles of a capacitor charged to a high voltage. In this configuration, the sphere and the payload 'float' with-respect-to (wrt) the local plasma so that a positive voltage appears on the sphere and a negative voltage appears on the payload body. The high-voltage capacitor on the payload is charged to +10 kV wrt the payload body. This configuration and voltage level was chosen so that in the F-region ionosphere the sphere/payload combination will float (wrt the plasma) with at least -2 kV appearing on the payload and the remaining voltage appearing on the sphere. A resistor across this capacitor allows for a 1 s decay of the voltage. Thus, as the voltage on the capacitor decays the payload voltage will go through the entire range of voltages normally

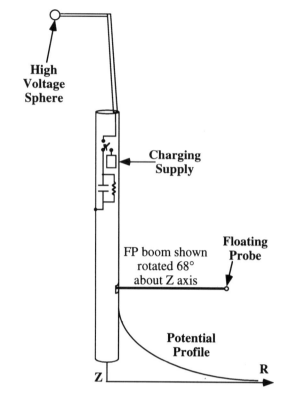

Figure 1. Simplified diagram of the payload charging system for SPEAR III. Payload, high-voltage boom and sphere, and Floating Probe boom and sphere are to scale.

associated with natural charging (0 to -2 kV). During flight, any current flowing through the plasma will shorten the decay time and this decay time was typically measured between 0.3-1.0 sec. The charge and decay cycle was set to repeat every 5 seconds. In this way, the payload was repeatedly charged to several kilovolts (wrt to the local ionosphere), and on each cycle one of several methods was used to discharge (ground) the payload.

We designed and built the NRL Floating Probe (FP) [Siefring et al., 1995] for the expected voltage range relevant to the SPEAR III sounding rocket experiment [Raitt et al., 1997]. The design criteria for this experiment were for a diagnostic instrument capable of measuring payload potentials of -300 Volts with a frequency response up to 1 kHz. The FP easily met these criteria and in fact indications are that the instrument provided good quality measurements up to about -1800 V.

Four devices were studied and tested for there ability to discharge the payload [Raitt et al., 1997]. The devices can be grouped into two broad categories. In the first category are effluent release experiments, either neutral gas or plasma. The spacecraft attitude control thrusters were used to augment the neutral gas release study. In the second category were two electron emitting devices; one a thermionic emitter and the other a field effect emitter.

The SPEAR III program included testing of critical components in a full size, 'mock-up' (as close to flight conditions as possible) configuration in the NASA-Lewis Research Center's Plum Brook Station B2 vacuum facility. The laboratory test of the FP and many of its operational characteristics have been reported on by Siefring et al. [1995]. In this paper we concentrate on the performance of the FP during the SPEAR III flight. Our primary interest, is to compare the FP measurements with data taken an ElectroStatic particle Analyzer (ESA) on the same platform.

HIGH VOLTAGE PROBES

For the SPEAR III experiment the FP requirements were to measure body voltages in the range of 0 V to -300 V with a frequency response of at least 1 kHz. These design goals were determined prior to the SPEAR III flight for two reasons. First, an ElectroStatic particle Analyzer (ESA) was planned that would make measurements of incoming ions in the range from 10 eV to 25 keV. The expected insensitivity of the ESA below ~300 V, suggested that the FP and ESA should both make measurements in this range. Second, a 2.5 m deployable boom was available from a previous mission. Estimates of the spacecraft sheath indicated that with this boom length the probe would be outside of the body-sheath (i.e., in the ambient plasma) at about -500 V.

Calculations of the payload sheath are beyond the scope of this paper and several simulation codes exist for this purpose. In our case, we used numerical simulations similar to those done and published for SPEAR I (cf., Neubert et al.. [1990] and Katz et al.. [1989]). Theses showed that the probe would be located at the -20 V contour for a payload voltage of -500 V.

To theoretically demonstrate the probe performance we consider the following model. During the discharge phase of the cycle the payload potential has gone from -2 kV to below -300 V and, thus, the sheath edge has just passed the probe location. Assuming the probe has not been able to track the local potential variation while inside the body-sheath, it is left with a large negative charge. How long will it take for the probe to relax to the local potential now that it is surrounded by the ionospheric plasma?

To answer this question we consider a simple model and attempt to determine a time-constant for the probe surface by modeling the probe sheath as a parallel combination of C_{sheath} and R_{sheath}. Under these conditions, the probe has a large sheath radius ($r_{sheath} \gg r_{probe}$) and C_{sheath} reduces to approximately the free space capacitance

$$C_{sheath} \cong 4\pi\varepsilon_o r_{probe}. \qquad (1)$$

For our case $r_{probe} \approx 3.2$ cm, and yields $C_{sheath} \approx 4$ pFd

Modeling the sheath resistance is somewhat more involved. Siefring et al. [1995] examines the dynamic sheath resistance of a negatively charged sphere in detail and gives

$$R_{sheath} = \frac{\partial V}{\partial I} = 2.95x10^{-6}(j_{th})^{-1} n^{4/7}\theta^{2/7} r_{probe}^{-6/7} |V|^{1/7}. \qquad (2)$$

where $j_{th} = nqv_{th}$, $v_{th} = (q\theta/2\pi M)^{1/2}$, n is the plasma density, θ the ion temperature (eV) and V the probe potential. Using Eqns. 1 and 2 the time constant for small voltage changes is

$$\tau = R_{sheath}C_{sheath}. \qquad (3)$$

Figure 2 shows plots of the sheath resistance and time constant versus voltage for a range of ionospheric plasmas ($M_{O+} = 16M_p$, $\theta = .01$ eV, and $n = 10^4, 10^5, 10^6 \text{cm}^{-3}$). Notice first that values of R_{sheath} are above few times $10^8 \Omega$ (for n=10^4 cm^{-3} implying that FP input impedance of must be maintained well above this level to insure a good quality measurement over a reasonable range of ionospheric conditions. The time constant plots show that τ is significantly less than 1 ms over the entire voltage range. This differential time constant does not directly answer the previous question (i.e., If the probe is left with a negative charge from when it was inside the payload sheath, how quickly will it dissipate?), but the integrated discharge decay time can not be larger than the maximum value of τ.

This indicates that the FP should be capable of tracking 1 ms changes in spacecraft potentials going to -2 kV (as long as the probe is not contained in the payload body-sheath at the time of the measurement). Thus, not only can the probe provide a 1 kHz measurements in the -300 V case, it should be capable of operation over the entire range of voltages normally associated with natural spacecraft charging with the proviso that the boom length is sufficient to place the sensor in the ambient plasma.

The largest source of error for these calculations results from secondary electron emission. Although there are few studies of oxygen ions bombarding aluminum oxide, indica-

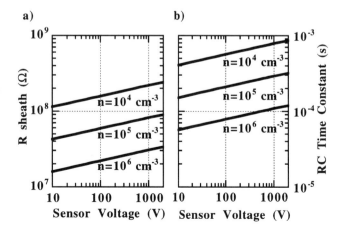

Figure 2. Theoretical probe sheath resistance and time constant for large voltages.

Figure 3. Picture of the Floating Probe sensor.

tions are that it could affect the time-constant significantly at the high end of the voltage range [Neubert et al., 1990; Knudsen and Harris, 1973]. Apparently, a sharp emission-yield threshold exists at ion velocities around 5×10^4 m/s regardless of the ion [Dietz and Sheffield, 1975]. Thus, for oxygen ions secondary emission becomes important at approximately 200 eV and at 460 eV the yield is 50%. Experimenters should be aware that the time response may be significantly faster at voltages above 460 V because of the additional current from secondary emission.

THE FLOATING PROBE SENSOR

The Floating Probe on SPEAR III consisted of three sub-systems: (1) a spherical sensor containing a high impedance amplifier and a capacitive divider network, (2) a deployable boom to position the sensor 2.5 m from the spacecraft, and (3) electronics to condition the signals for output to the spacecraft telemetry. The discussion here is oriented mainly to the spherical sensor.

Figure 3 shows a picture of the FP sensor and figure 4 shows a circuit diagram of the important components. The ratio of the 'pick-up' capacitance to the 'divider' capacitance, C_p/C_d, controls the step-down of the voltage. The input impedance of the sensor is essentially the pick-up capacitance in parallel with the leakage resistance R_1. Thus, if we wished, we could build the sensor with the optimum input-impedance of only stray capacitance (fraction of 1 pFd) and leakage resistance ($\approx 10^{12}$ Ω). The leakage resistance, although large, was found to affect the low-frequency response of the sensor during our testing. The bias resistor R_b (a 100 GΩ glass resistor) serves two purposes; (1) to provide a small current to keep the amplifier biased in the on condition and (2) to produce a slow bleed of charge from C_d to prevent a long lasting offset from being established. The amplifier has an input-resistance specified at 10^{15} Ω, but layout considerations typically limit this to near 10^{12} Ω. A neon

lamp across the input provides over-voltage protection without adversely affecting the input impedance.

The pick-up capacitor was custom made for the sensor and can be seen in figure 3. A wire was affixed near the outside of the circuit board and completely covered with silicon RTV. With this construction the capacitance could be reduced by cutting and removing part of the wire. To increase capacitance the wire and RTV must be removed and the capacitor reconstructed. However, if the wire is made longer than required, then by clipping and measuring the capacitance at a few different lengths, it becomes easy to extrapolate to the desired voltage step-down. The inside surface of the sphere was coated with the same RTV compound to prevent possible arcing to electronic components.

The voltage division and capacitance were measured with standard laboratory equipment by connecting a signal generator to the outer surface of the assembled sensor and measuring the amplitude of the output (step-down) with an oscilloscope. C_d was a standard commercial silver-mica capacitor, of 400 pFd (+/- 5% a component was used but the value was measured to increase accuracy). The flight value of C_p was 2.6 pFd, with voltage division of 154. Although, the sensor is capacitively coupled, the component values allow for measurements down to a fraction of a Hz. To make measurements at the low frequency end, it is important to measure the phase and amplitude characteristics (transfer function) of the sensor. Siefring et al. [1995] shows how to compensate for the effects of the electronics on signals at these low frequencies.

SPEAR III FLIGHT RESULTS

The Floating Probe operated extremely well during the rocket flight. Figures 5, 6 and 7 show three comparisons of the payload potential measurements made by the FP and the ESA. Each figure shows one charging and decay cycle (5 sec). The solid line in these figures are data from the FP

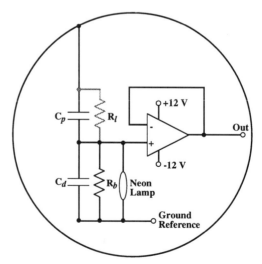

Figure 4. Floating Probe functional circuit diagram.

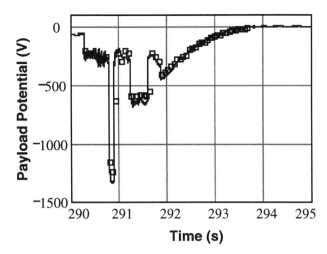

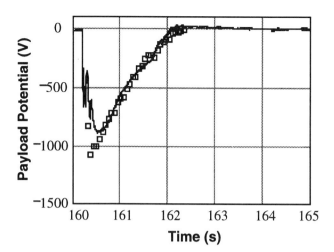

Figure 5. Comparison of payload potential measurements by the FP and the ESA during a pulsed neutral gas release. The squares represent the peak potential of the incoming ions which is inferred to be the payload potential.

Figure 7. Comparison of payload potential measurements by the FP and the ESA. In this case the FP was not able to track the payload potential below ~-800 V because the payload sheath becomes larger than the boom length of 2.5 m.

which have been 'corrected' for the frequency response of the sensor. The boxes are data from the ESA which indicate the measured peak in the incoming ion spectrum.

Figure 5 is a charging cycle were neutral Argon gas is released and breaks down. This break down allows more current flow to the rocket body and, thus, less of the charging capacitor voltage appears on the body [Berg et al., 1995]. The gas was pulsed at two different flow rates. In this case the gas was being released when the high-voltage capacitor was switched at ~290.3 s. We see that both instruments indicate that the payload voltage is held near -200 V to -300 V by the neutral gas. During one of the 'gas-off' phases the payload voltage jumped to ~-1200 V for a short time and was again grounded by a gas release at about 290.9 s. The

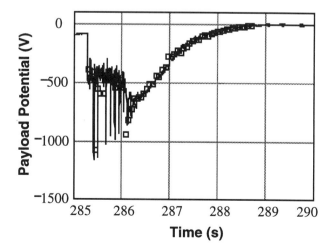

Figure 6. Comparison of payload potential measurements by the FP and the ESA during attitude thruster firings. The plot illustrates the high time resolution capabilities of the FP.

payload voltage bounces up and down several times, as a result of the gas release pulsing, until it reaches about -400 V. It appears that the payload voltage needs to be above 400-500 V to initiate the gas breakdown. The two measurement techniques agree extremely well for this charging cycle.

Figure 6 is a cycle that occurred during an Attitude Control System (ACS) firing. It is known that ACS gas releases can have an effect on spacecraft charging and the comparison of the two systems was a planned part of the experiment. The ACS gas was Nitrogen. The ACS thrusters are pulsed with shorter bursts but have a higher flow rate then the Argon gas release system. In this case the payload voltage bounces around rapidly while the thrusters are firing (285.3 to 286.3 s). The high time resolution advantage of the FP is apparent from the plot.

Figure 7 is a case where the FP does not track the payload potential well. In this case, the grounding technique used was a Field Effect Device (FED). The FED is a solid state device designed to emit electrons on the application of an electric field to very small radius points etched on a substrate [Raitt et al., 1997]. There were nine of these devices contained in TO-5 transistor cans, but some of these devices were damaged during its first grounding cycle. In figure 7, the FED only has an effect on the payload potential for a short time. We expect the payload potential would have reached -2 kV if the FED was not used. What is of interest here is that the FP does not measure the full payload potential. This is because the payload sheath is, at this point, larger than the FP boom length. The ESA and FP measurements start to depart significantly at about -800 V. In cases where we believe that the FP is inside the payload sheath we typically see a break in the exponential character of the payload potential. This break appears at about 168.8 s in figure 7. The voltage level of this break and, thus, the payload sheath size appears to scale properly with plasma density during the flight.

Another advantage of the FP is the ability to detect high frequency phenomena. For SPEAR III, 10 kHz was the highest frequency monitored. Typically, high-frequency oscillations were superimposed on the general trend [Siefring et al., 1995]. In each cycle, the ac signal levels are highest right after the voltage is applied and generally move to lower levels as the payload voltage decreases. Band-limited signals were seen in the 500 Hz to 2 kHz range and indicate organized plasma wave activity. Plasma waves tend to cause turbulence in the local environment and can interfere with current flow to the payload and subsequent neutralization. Future modeling of the plasma wave activity may shed some light on the grounding process.

We have reported on the NRL FP which is a significant advancement in our ability to monitor and study the behavior of spacecraft charged to high-voltages. The problem of spacecraft charging is significant to space research, because of the adverse effects that electrical charging can have on both spacecraft performance and diagnostic instruments. On a more general note, the design and theoretical considerations discussed here have applications to any high-voltage or electric field measurements in space plasmas.

Acknowledgments: The authors would like to thank the numerous employees at Utah State University who supported SPEAR III and D. W. Potter and H. R. Anderson of SAIC for allowing us to use data from their experiment. We owe a great deal to our co-workers at NRL and specifically thank J. A. Antoniades, M. M. Baumback, D. Duncan, J. A. Stracka and David Walker. We also thank M. J. Mandell and others at S³-Maxwell Labs who helped with numerical modeling of payload sheaths. Finally, we thank Principle Scientist W. J. Raitt (USU) for conducting a very well run experiment. The SPEAR III project was jointly supported by BMDO, DNA and NASA. NRL's Floating Probe was funded by DNA and ONR.

REFERENCES

Allred, D. B., J. D. Benson, H. A. Cohen, W. J. Raitt, D. A. Burt, I. Katz, G. A. Jongeward, J. Antoniades, M. Alport, D. Boyd, W. C. Nunnally, W. Dillion, J. Pickett, and R. B. Torbert, The SPEAR-I experiment: High voltage effects on space charging in the ionosphere, *IEEE Trans. on Nucl. Sci., 35, 1386,* 1988.

Berg, G. A., W. J. Raitt, D. C. Thompson, B. E. Gilchrist, N. B. Myers, P. Rodriguez, C. L. Siefring, H. R. Anderson and D. W. Potter, Overview of the effects of neutral gas releases on high-voltage sounding rocket platforms, *Adv. Space Res., 15, 12, 83,* 1995.

Borovsky, J. E., The dynamic sheath: Objects coupling to plasmas on electron-plasma-frequency time scales, *Phys. Fluids, 31,* 1074, 1988.

Dietz, L. A., and J. C. Sheffield, Secondary emission induced by 5-30 keV monatomic ions striking thin oxide films, *J. Appl. Phys., 46,* 4361, 1975.

Garret, H. B., The charging of spacecraft surfaces, *Rev. Geophysics, 19,* 577, 1981.

Gussenhoven, M. S., D. A. Hardy, F. Rich, W. J. Burk, and H. C. Yen, High-level spacecraft charging in the low-altitude polar auroral environment, *J. Geophys. Res., 90,* 11,009, 1985.

Katz, I., G. A. Jongeward, V. A. Davis, M. J. Mandell, R. A. Kuharski, J. R. Lilley, Jr., W. J. Raitt, D. L. Cooke, R. B. Torbert, G. Larson, and D. Rau, Structure of the bipolar plasma sheath generated by SPEAR I, *J. Geophys. Res., 94,* 1450, 1989.

Kaufmann, R. L., D. N. Walker, J. C. Holmes, C. J. Pollock, R. L. Arnoldy, L. J. CaHill and P. M. Kintner, Heavy ion beam-ionosphere interactions: charging and neutralizing the payload, *J. Geophys. Res., 94,* 453, 1989.

Knudsen, W. C. and K. K. Harris, Ion-impact-produced secondary electron emission and its effect on space instrumentation, *J. Geophys. Res., 78,* 1145, 1973.

Koons, H. C., P. F. Mizera, J. L. Roeder and J. F. Fennel. Severe spacecraft-charging event on SCATHA in September 1982, *J. Spacecr. and Rockets, 25,* 239-243, 1988.

McPherson, D. A., and W. R. Schober, Spacecraft charging at high altitudes: The SCATHA satellite program, Spacecraft Charging by Magnetospheric Plasmas, edited by A. Rosen, *Progr. Aeronaut. Astronaut., 47,* 15-30, 1976.

Metz, R. N., Circuit transients due to arcs on a high-voltage solar array, *J. Spacecraft and Rockets, 23,* 499, 1986.

Mullen, E. G, M. S. Gussenhoven, D. A. Hardy, T. G. Aggson, B. G. Ledley, E. Whipple, SCATHA survey of high-level spacecraft charging in sunlight, *J. Geophys. Res., 91,* 1474, 1986.

Neubert, T., M. J. Mandell, S. Sasaki, B. E. Gilchrist, P. M. Banks, P. R. Willimson, W. J. Raitt, N. B. Meyers, K. I. Oyama, and I. Katz, The sheath structure around a negatively charged rocket payload, *J. Geophys. Res., 95,* 6155, 1990.

Olsen, R. C., L. E. Weddle and J. L. Roeder, Plasma wave observations during ion gun experiments, *J. Geophys. Res., 95,* 7759, 1990.

Pongratz, M., D. Walker, M. Baumback, C. Siefring, H. Anderson, D. Potter, Plasma Physics Instrumentation, *BEAR Project Final Report, Volume II: Flight Results and System Evaluation,* G. J. Nunz, A. D. McGuire, P. G. O'Shea, E. B. Barnett, editors, Los Alamos National Laboratory, LA-11737-MS, Vol II, Part I, BEAR-DT-7-2, pp 7-1, 1991.

Raitt, W. J., G. Berg, D. Thompson, A. White, B. Peterson, M. Roosta, M. Jensen, L. Allen, J. Antoniades, P. Rodriguez, C. L. Siefring, H. Anderson, D. Potter, J. Jost, C. Holland, J. Picket, R. Merlino, M. Adrian, N. Grier, N. Poirier, R. Morin, SPEAR III a sounding rocket experiment to study methods of electrically discharging negatively charged space platforms at LEO altitudes, *J. Spacecr. and Roc.,* accepted 1997.

Siefring, C. L., P. Rodriguez, M. M. Baumback, J. A. Antoniades, and D. N. Walker, A method for measuring large changes in the payload voltage of rockets and satellites, *Rev. Sci. Instrum., 66,* (9), 4681, 1995.

Thiemann, H., R. W. Schunk, and K. Bogus, Where do negatively biased solar arrays arc?, *J. Spacecraft and Rockets, 27,* 563, 1990.

Whipple, E. C., Potentials of surfaces in space, *Rep. Prog. Phys., 44,* 1197, 1981.

Winglee, R. M., Simulations of pulsed electron beam injection during active experiments, *J. Geophys. Res., 96,* 1803, 1991.

P. Rodriguez and C. L. Siefring, Code 6755, Plasma Physics Division, Naval Research Laboratory, Washington, DC 20375.

How to Really Measure Low Energy Electrons in Space

Earl E. Scime

Department of Physics, West Virginia University, Morgantown, WV

There is little argument in the space plasma physics community that in-situ, low energy electron measurements are technically challenging. The primary obstacle has been the effects of spacecraft charging on the measured three-dimensional electron velocity space distribution. A successful spacecraft charging correction algorithm used with the three-dimensional electron instrument aboard the Ulysses spacecraft has clarified the role of spacecraft and instrument parameters in the eventual reconstruction of low energy electron distributions. Suggestions for instrument and spacecraft modifications that can minimize spacecraft charging effects are presented in this paper. The emphasis is on designs that lend themselves to robust correction algorithms.

1. INTRODUCTION

The objective of in-situ particle measurements of space plasmas is to provide enough information to fully characterize the plasma particle distributions. The details of both the ion and the electron distributions are needed to understand the growth of instabilities and the partitioning of energy within the plasma. The space environment and typical spacecraft resources, however, can significantly affect the extent to which accurate particle distribution measurements can be obtained. Fundamentally, it is charging of the spacecraft that distorts measurements of the ambient plasma velocity space distributions. As charged particles approach a plasma instrument, their velocities and trajectories are modified by the plasma sheath surrounding the spacecraft. It has been shown [*Scime et. al.*, 1994; *Parker and Whipple*, 1970], that the ambient plasma velocity space distribution can be accurately reconstructed if the plasma sheath structure and the spacecraft potential are known.

The focus of this paper is a discussion of the role played by instrument design, instrument calibration, and spacecraft design in the velocity space distribution reconstruction process. Not all spacecraft shapes or instrument designs are equivalent. For the purposes of discussion, only those techniques that lead to more accurate plasma measurements without a significant increase in instrument resources (mass, power, telemetry) are considered.

To avoid the complications of multiple species and subsonic distributions, discussions will be limited to low energy electrons measured from positively charged spacecraft. It is true that for a positively charged spacecraft, very low energy ions are completely reflected and it is not possible to recover the low energy ion data. In such a case, the spacecraft design must emphasize the complete elimination of the charging of the spacecraft. The instrument modifications and spacecraft designs suggested in this paper for charged spacecraft are also relevant for low energy ion measurements from negatively charged spacecraft, or measurements of low energy particles whose energy exceeds that of a similarly charged spacecraft (e.g., ambient 4 eV electrons measured from a spacecraft charged to -3 V).

2. SPACECRAFT CHARGING EFFECTS

For a plasma in which the electron temperature is greater than a few percent of the ion temperature, the electron flux to the surface of an object immersed in the plasma exceeds the ion flux. In the absence of any other effects, e.g., photoelectron emission or secondary electron emission by

Measurement Techniques in Space Plasmas: Fields
Geophysical Monograph 103

ion or electron impact, the excess electron flux results in a negative floating potential for the object [*Langmuir and Blodgett*, 1924]. This situation occurs only rarely in magnetospheric and heliospheric plasmas. Solar ultraviolet radiation liberates enough photoelectrons from the surface of a typical spacecraft that the photoemission overwhelms the ambient electron flux and the spacecraft floats positive. Because the magnitude of the spacecraft potential is a function of the ambient plasma density, spacecraft illumination (solar ultraviolet level and spacecraft orientation), age of the spacecraft surface, and ambient plasma temperature, a priori calculations of the spacecraft potential are accurate only to within a few volts [*Mandell et al.*, 1978]. When a spacecraft enters the full shadow of celestial body, such as the Earth or the Moon, the photoelectron emission ceases and the spacecraft can charge to large negative potentials [*Rosen*, 1976; *Whipple*, 1981]. Due to their plasma densities and distances from the Sun, different regions of space have different characteristic spacecraft potentials. Figure 1 shows typical spacecraft potentials for the ionosphere, magnetosphere, and heliosphere.

Once it becomes positively charged, a spacecraft will attract negatively charged particles. That the negative charged particles (electrons) will accelerate towards the spacecraft and gain additional kinetic energy equal to the spacecraft potential is obvious. The effects on the details of the measured particle distributions, however, are more subtle. The spacecraft potential distorts the trajectory of the ambient electrons entering the instrument (Figure 2). A positively charged spacecraft will focus ambient electrons. A negatively charged spacecraft will defocus ambient electrons. In the limit of a thin sheath (or a sheath whose

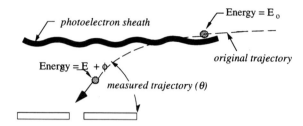

Figure 2. Example of trajectory focusing effect for a positively charged spacecraft attempting to measure ambient, low energy electrons. Measured angle of incidence is θ, while true angle of incidence is θ_O (see Eq. (1)).

equipotential surfaces are parallel to the spacecraft body near the instrument), the relationship between the true incident angle for an electron far from the positively charged spacecraft (θ_O) and the measured angle of incidence (θ) is given by:

$$sin\,\theta_o = \frac{sin\,\theta}{\sqrt{1-U/E_A}}, \qquad (1)$$

where U is the spacecraft potential and E_A is the energy of the electron measured by the instrument [*Scime et al.*, 1994]. For $E_A \approx U$, electrons emitted from the spacecraft body itself ($\theta \approx 90°$) can appear to come from the ambient plasma. The result is an energy-dependent geometric factor for the instrument that is also a function of the local plasma density and solar illumination. Low energy electrons are collected from an enormous field of view and only the higher energy electrons are collected from the intended instrument field of view. It should be added that magnetic field focusing effects should be considered if low energy particle instruments are placed close to high current spacecraft power systems.

This focusing effect has been described by a number of authors [*Garrett*, 1981; *Singh and Baugher*, 1981; *Sojka et al.*, 1984; *Scime et al.*, 1994], but only recently have correction techniques been implemented during routine data analysis [*Comfort et al.*, 1982; *Scime et al.*, 1994]. Left uncorrected, this focusing effect leads to substantial errors in the calculation of the electron density and all the vector moments of the electron distribution, e.g., velocity and pressure tensor. After using a thin sheath (local Debye length small compared to spacecraft scale size) spacecraft charging model that corrects for both the acceleration and focusing of the ambient electrons, the difference between ion and electron density measurements from the Ulysses spacecraft dropped from 60% to less than 0.5% [*Scime et al.*, 1994]. The details of the three-dimensional electron velocity space distribution measured by Ulysses also improved with the thin sheath correction. For example, unless the magnetization of the electrons is systematically

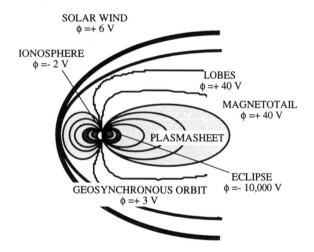

Figure 1. Typical spacecraft potential (ϕ) for different regions of the near-Earth space environment [*Garrett*, 1981; *Scime et al.*, 1994; *Frank et al.*, 1993].

destroyed during an electron gyroperiod (through some type of anisotropic collision effect in the plane perpendicular to the magnetic field), the electron distribution will be isotropic, gyrotropic, in the plane perpendicular to the magnetic field. Before the thin sheath correction (Figure 3a), the measured electron distributions were clearly non-gyrotropic. After the correction, the distributions were appeared remarkably gyrotropic (Figure 3b).

To perform the thin sheath correction, the angular distribution of the ambient electron distribution must be measured. Without measurements from a differential plasma instrument, corrections for sheath focusing effects cannot be performed. Single aperture instruments, such as simple Faraday cups, combine the electron fluxes from different angles of incidence and there is no way to reconstruct the paths through the sheath for individual low energy electrons. Corrections for the overall geometric factor can be estimated [*Scime et al.*, 1994], but the vector moments cannot be accurately repaired.

Accurate measurements of the spacecraft potential are also needed to perform the sheath focusing correction. In many cases, the cloud of photoelectrons surrounding a positively charged spacecraft can be used. Photoelectrons emitted with kinetic energy less than the spacecraft potential are reflected back to the spacecraft and can be detected by onboard plasma instruments. Figure 4 shows a typical electron energy spectrum measured with an electrostatic analyzer aboard the positively charged Ulysses spacecraft. The data in Figure 4 are from a single sampling direction from the spinning spacecraft. As indicated in the figure,

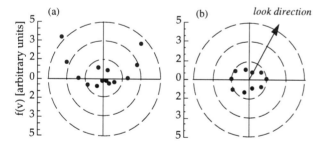

Figure 3. (a) The magnitude, in arbitrary units, of the measured electron distribution function from the Ulysses spacecraft for 12 eV electrons plotted for 14 different viewing directions perpendicular to the local magnetic field. The data has been corrected for the solar wind drift velocity and the acceleration due to the spacecraft potential ($\phi = 6$ V). The dotted circles are lines of constant velocity space density (distribution function magnitude) and are scaled linearly as shown by the axis labels. The measured electron distribution is clearly not isotropic (gyrotropic). (b) The same data corrected for the additional trajectory focusing effects of the spacecraft sheath. Note that those data points projecting back to the spacecraft body (photoelectrons) have been eliminated.

both photoelectrons and ambient electrons are detected by the analyzer. The discontinuity in the slope of the energy spectrum (the two portions of the spectrum have different temperatures) distinguishes the accelerated ambient electrons from the trapped photoelectrons and indicates the spacecraft potential. Although this technique is quite useful, when the ambient electron temperature is low, as in the outer solar system, the discontinuity in the slope of the electron spectrum vanishes. Without some other measurement of the spacecraft potential, correction for charging effects is not possible in the outer heliosphere or on negatively charged spacecraft.

It is the difficulty in correcting for space environmental effects that has lead many researchers to treat electron measurements as unreliable and instead "calibrate" them to agree with ion or plasma wave data once the spacecraft is on orbit (e.g., *Frank et al.* [1993]); ignoring both plasma dependent variations in the responses of different measurement techniques and the careful ground-based calibration of the electron instrument. Since the objective of in-situ electron measurements is to investigate the details of the electron distribution, simply scaling moments of the distribution can obscure non-systematic errors due to variations in spacecraft charging effects. Improper correction of charging effects can result in misinterpretations of plasma parameter gradients. For example, density increases can appear to be density decreases because of the reduced effects of spacecraft charging [*Scime et al.*, 1994].

Unfortunately, mitigating space environmental effects to increase measurement accuracy is not the driving factor in modern instrument design. Future space instruments must be "lighter, smaller, and cheaper." Therefore, new instrument designs must improve accuracy while reducing resource requirements.

3. SPACECRAFT MODIFICATIONS

Without modifying any plasma instruments, a dramatic reduction in the effects of spacecraft charging can be obtained by simply minimizing the charging. This can be accomplished with plasma contactors or ion guns that emit positive ions from the spacecraft. Active spacecraft potential control has been used on a number of spacecraft including SCATHA [*Olsen et al.,* 1990] Geotail [*Schmidt et al.,* 1995] and Polar [*Moore et al.,* 1995]. Active potential control is typically employed to allow the detection of very low energy positive ions that would be reflected away from a positively charged spacecraft [*Olsen et al.,* 1982; *Moore et al.,* 1995]. Recent results from an indium metal based ion emitter aboard the Geotail spacecraft indicate that the spacecraft potential can be maintained between 2 Volts and

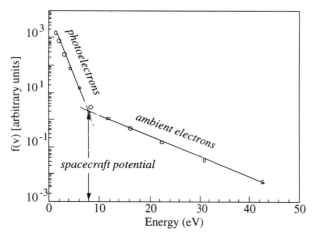

Figure 4. Typical electron energy spectrum that shows distinct ambient and photoelectron populations. The ambient electrons appear at energies above the spacecraft potential.

10 Volts positive in the near-Earth space environment [*Schmidt et al.*, 1995]. Although the Geotail potential control system has been fairly successful, there are significant spacecraft resource penalties for using an emitter or contactor for active potential control. These include: a source of ions is required (adding mass to the spacecraft and increasing power consumption); long term missions such as Voyager, Pioneer, or Ulysses exceed the expected one to two year lifetime of these devices [*Schmidt et al.*, 1993]; contamination of the local ion and electron populations by the emitted ions; the generation of particle beam induced electromagnetic waves [*Olsen et al.*, 1990]; and the effects of plasma contactors on other scientific instruments are not thoroughly understood. If sufficient resources are available and the scientific objectives of the mission are not compromised, active spacecraft potential control can certainly play a crucial role in reducing the adverse effects of the space environment on low energy plasma measurements. However, even when a plasma emitter is used, the residual few volt spacecraft potential can still significantly affect measurements of an ambient particle distribution with a few eV temperature, e.g., the bulk electron population in the heliosphere [*Phillips et al.*, 1993] or the bulk ion population in the plasmasphere [*Moldwin et al.*, 1995].

Another way to improve the measurement process without modifying the plasma instrument itself is to choose a more appropriate spacecraft geometry. Since it is the "view through the sheath" that affects the trajectories of incident particles, spacecraft charging effects can cause identical instruments in the same plasma environment but aboard spacecraft of different shapes to measure the ambient plasma distribution differently. Figure 5 shows the focusing of 9 eV electrons for two different spacecraft. Each

spacecraft is charged to a potential of +7 Volts. In both cases, the electrons appear to have an energy of 16 eV when measured at the spacecraft. The electrons that appear to originate at an angle of 45° with respect to the aperture normal are strongly focused. The difference in the focusing effects for the two spacecraft (planar and spherical) is due entirely to the shape of the spacecraft. Figure 5 shows the advantages of a simple planar spacecraft geometry when it is time to reconstruct the actual ambient particle distribution. These simulations used the NASCAP computer code [*Katz et al.*, 1981; *Mandel et al.*, 1978] to modeling the sheath focusing effects.

Because environmental effects such as shown in Figure 5 are ignored during typical instrument calibration, it is critical to have a complete understanding of the sheath structure near a plasma instrument once it is in space. This is probably best accomplished by a combination of computational and experimental modeling. If the instrument (or a close facsimile) and the nearby spacecraft structure are placed in a carefully designed calibration facility that can simulate the appropriate space plasma environment, the focusing effects of the sheath can be quantitatively assessed and compared to computational models. The models can then be used with confidence during the actual mission and subsequent particle distribution reconstruction process.

Since it is the photoelectron emission current that defines the plasma Debye length near the spacecraft (typical photoelectron emission currents [*Schmidt et al.*, 1995] yield electron densities of approximately 10^7 cm^{-3}), it would not be difficult to construct model spacecraft that maintain the ratio of Debye length to spacecraft structural scale size. In space, the photoelectron dominated Debye length is approximately 0.2 m. Typical low density laboratory plasma Debye lengths are approximately 0.05 m. Thus a one-forth scale model of the relevant spacecraft surfaces and a miniature

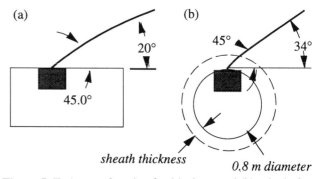

Figure 5. Trajectory focusing for (a) planar and (b) spherical spacecraft geometries. The focusing effect is significantly stronger for the planer style spacecraft. At the instrument aperture, both trajectories appear to originate from an angle of 45° with respect to the aperture normal.

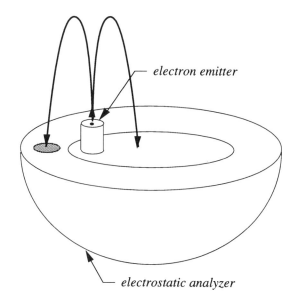

electron emitter

electrostatic analyzer

Figure 6. Schematic of an active electron emitter spacecraft potential measuring device mated to a hemispherical analyzer.

plasma instrument could be used to map the focusing effects of different spacecraft shapes as well as benchmark codes such as NASCAP. Miniature plasma instruments have already been flown on rockets for ionospheric measurements [*Pollock et al.,* 1996].

Having reviewed the effects of the plasma sheath on the ambient particle distribution, it is important to remember that, without an accurate measurement of the spacecraft potential, even a perfect in-situ calibration cannot be used to accurately reconstruct the measured electron velocity space distribution once the spacecraft is in space. Therefore, instruments intended for accurate low energy plasma measurements must be modified to provide accurate measurements of the spacecraft potential.

4. INSTRUMENT DESIGN

Accurate spacecraft potential measurements can be obtained passively or actively from a positively charged spacecraft. The passive approach is to measure the electron distribution with closely spaced energy steps around the spacecraft potential so that photo and ambient electron distributions can be cleanly separated. Regardless of the type of plasma instrument, electrostatic analyzer (e.g., *Bame et al.,* 1992) or Faraday cup (e.g., *Ogilvie et al.,* 1995), the energy steps through which a plasma instrument is scanned are usually logarithmically spaced. This practice permits efficient coverage of a broad energy range. Unfortunately, it also results in relatively coarse coverage in the 0.1 eV to 20 eV range. As shown in Figure 4, finely spaced energy steps

are needed to accurately determine the spacecraft potential. The increased telemetry requirements that would result from additional measurement steps could be eliminated by onboard spacecraft potential calculations. With sufficiently detailed information aboard the spacecraft, it should be possible to only transmit the value of the spacecraft potential and data from the energy steps above it. This would actually reduce the telemetry needs of the final instrument. Such an instrument, the PEACE instrument, was to be flown on the ill-fated Cluster mission [*Johnstone et al.,* 1988].

The active approach is to directly measure the spacecraft potential with an electron emitter. Electrons expelled by an electron emitter aboard a positively charged spacecraft will be reflected back to the spacecraft until the emission energy exceeds the spacecraft potential. By sweeping the energy of the emitted electron beam until the signal from an appropriately positioned detector vanishes, the spacecraft potential can be determined accurately and quickly (see Figure 6). The minuscule emission currents needed for such a measurement would not affect the spacecraft potential. Telemetry requirements of the instrument could be reduced by using the spacecraft potential measurements in the analyzer power supply control circuitry and the data processing system to avoid transmitting, or not even measure, energy steps below the spacecraft potential. The additional mass, power, and volume resources used by the emitter must be weighed against the improved scientific return of a plasma instrument for which the low energy particle distributions can be accurately reconstructed. Inclusion of the emitter and detector assembly will increase the of the electron analyzer. The best placement for an emitter could be determined by a combination of simulation and laboratory experimentation.

5. DISCUSSION

Analysis of low energy electron data from the Ulysses spacecraft indicates that the ambient electron distribution can be accurately reconstructed with an accurate measurement of the spacecraft potential and a thorough understanding of the spatial structure of the sheath in front of the instrument. This information can be integrated with data processing and instrument control systems aboard spacecraft to reduce telemetry requirements and increase measurement speed of future instruments. Control of the spacecraft potential with ion guns or plasma contactors can play an important role in minimizing space environmental effects, but accurate measurements of low energy plasma populations still require accurate spacecraft potential measurements. The spacecraft potential can be determined passively, as is typically done, or actively with the use of an electron emitter aboard the spacecraft. Although the emitter increases the mass and weight of a prospective instrument, the improved scientific reliability of the data may justify employing less capable instruments to stay within mass and power requirements. Finally, laboratory testing of instruments in a plasma filled chamber as a function of spacecraft potential and ambient plasma density are needed to investigate the coupling of ambient plasma parameters and sheath focusing effects while simultaneously benchmarking modeling codes.

The space environment plays an important role in low energy plasma measurements and the need for more detailed distribution measurements will continue to motivate the development of new approaches to minimize space environmental effects. "Really measuring low energy electron distributions in space" requires careful consideration of the impact of spacecraft and instrument designs on the eventual distribution reconstruction process.

Acknowledgments. This work was supported by West Virginia University. The author thanks John Phillips for access to raw electron data from the Ulysses spacecraft.

REFERENCES

Bame, S.J., D.J. McComas, B.L. Barraclough, J.L. Phillips, K.J. Sofaly, J.C. Chavez, B.E. Goldstein, and R. K. Sakurai, The Ulysses solar wind plasma experiment, *Astron. Astrophys. Suppl. Ser., 92,* 237, 1992.

Comfort, R.H., C.R. Baugher, and C.R. Chappell, Use of the Thin Sheath Approximation for Obtaining Ion Temperatures from the ISEE 1 Limited Aperture RPA, *J. Geophys. Res.*, 87, 5109, 1982.

Frank, L.A., W.R. Paterson, and M.G. Kivelson, Galileo observations of the motions of ion and electron plasmas in the magnetotail, *Geophys. Res. Lett.*, 20, 1771, 1993.

Garrett, H.B., The charging of spacecraft surfaces, *Rev. Geophys.,* 19, 577, 1981.

Johnstone, A.D., A. J. Coates, D.S. Hall, B.N. Maehlum, S.J. Schwartz, M. Thomsen, J.D. Winningham, 'PEACE' - A Plasma Electron And Current Experiment, ESA Report #SP-1103, p. 77, 1988.

Katz, I., M.J. Mandell, G.W. Schnuelle, D.E. Parks, and P.G. Steen, Plasma collection by high voltage spacecraft in low Earth orbit, *J. Spacecr. Rockets, 18,* 79, 1981.

Langmuir, I. and K.B. Blodgett, Currents limited by space charge between concentric spheres, *Phys. Rev.*, 24, 49, 1924.

Mandell, M.J., I. Katz, G.W. Schnuelle, P.G. Steen, and J.C. Roche, The decrease in effective photocurrents due to saddle points in electrostatic potentials near differentially charged spacecraft, *IEEE Trans. Nucl. Sci., NS-25,* 1313, 1978.

Moldwin, M.B., M.F. Thomsen, S.J. Bame, D.J. McComas, G.D. Reeves, The fine-scale structure of the outer plasmasphere, *J. Geophys. Res.*, 100, 8021, 1995.

Moore, T.E., C.R. Chappell, M.O. Chandler, S.A. Fields, C.J. Pollock, D.L. Reasoner, D.T. Young, J.L. Burch, N. Eaker, J.H. Waite, D.J. McComas, J.E. Nordholt, M.F. Thomsen, J.J. Berthelier, R. Robson, The Thermal Ion Dynamics Experiment and Plasma Source Instrument, *Space Sci. Rev., 71,* 409, 1995.

Ogilvie, K.W., D.J. Chornay, R.J. Fritzenreiter, F. Hunsaker, J. Keller, J. Lobell, G. Miller, J.D. Scudder, E.C. Sittler, R.B. Torbert, D. Bodet, G. Needell, A.J. Lazarus, J.T. Steinberg, J.H. Tappan, A. Mavretic, and E. Gergin, SWE - A comprehensive plasma instrument for the WIND spacecraft, *Space Sci. Rev., 71,* 55, 1995.

Olsen, R.C., The hidden ion population of the magnetosphere, *J. Geophys. Res.,* 87, 3481, 1982.

Olsen, R.C., L.E. Weddle, and J.L. Roeder, Plasma Wave observations during ion gun experiments, *J. Geophys. Res., 95,* 7759, 1990.

Parker, L.W. and E.C. Whipple, Jr., Theory of Spacecraft Sheath Structure, Potential, and Velocity Effects on Ion Measurements by Traps and Mass Spectrometers, *J. Geophys. Res., 75,* 4720, 1970.

Phillips, J.L., S.J. Bame, J.T. Gosling, D.J. McComas, B.E. Goldstein, and A. Balogh, Solar wind thermal electrons from 1.15 to 5.34 AU: Ulysses observations, *Adv. Space Res., 13,* 647, 1993.

Pollock, C.J., T.E. Moore, M.L. Adrian, P.M. Kintner, R.L. Arnoldy, SCIFER - Cleft region thermal electron distribution functions, *Geophys. Res. Lett., 23,* 1881, 1996.

Rosen, A., Spacecraft charging problems, in *Physics of Solar Planetary Environments*, edited by D.J. Williams, p. 1024, AGU, Washington D. C., 1976.

Schmidt, R., H. Arends, A. Pederson, M. Fehringer, F. Rudenauer, W. Steiger, B.T. Narheim, R. Svenes, K. Kvernsveen, K. Tsuruda, H. Hayakawa, and M. Nakamura, W. Reidler, and K. Tokar, A novel medium-energy ion emitter for active spacecraft potential control, *Rev. Sci. Instrum., 64,* 2293, 1993.

Schmidt, R., H. Arends, A. Pederson, F. Rudenauer, M. Fehringer, B.T. Narheim, R. Svenes, K. Kvernsveen, K. Tsuruda, T. Mukai, H. Hayakawa, and M. Nakamura, Results from active spacecraft potential control on the Geotail spacecraft, *J. Geophys. Res.*, 100, 17253, 1995.

Scime, E.E., J.L. Phillips, and S.J. Bame, Effects of spacecraft potential on three-dimensional electron measurements in the solar wind, *J. Geophys. Res., 99,* 14769, 1994.

Singh, N., and C.R. Baugher, Sheath effects on current collection by particle detectors with narrow acceptance angles, *Space Sci. Instrum., 5,* 295, 1981.

Sojka, J.J., G.L. Wrenn, and J.F.E. Johnson, Pitch angle properties of magnetospheric thermal protons and satellite sheath interference in their observation, *J. Geophys. Res., 89,* 9801, 1984.

Whipple, E.C., Potentials of surfaces in space, *Rep. Prog. Phys., 44,* 1197, 1981.

_____

E. E. Scime, Department of Physics, West Virginia University, Morgantown, WV 26506 (e-mail: Internet escime@wvu.edu)

Imaging Space Plasma With Energetic
Neutral Atoms Without Ionization

K. C. Hsieh and C. C. Curtis

Department of Physics, University of Arizona, Tucson, AZ 85721

In 1951, prior to the Space Age, the existence of energetic neutral hydrogen atoms (as high as 70 keV in energy) in space plasma was discovered. The study of space plasma in the last three decades had been confined to *in situ* measurements of ions until the ingenious interpretation of data gathered by the ion detectors on IMP 7/8 and ISEE 1 under fortuitous conditions was reported in 1985. The year 1990 ushered in a new era in the imaging of space plasma in energetic neutral atoms (ENAs), when the dedicated ENA imager, INCA, was chosen for the Cassini mission to Saturn. We shall discuss some inherent design constraints (such as the interfering EUV and ion fluxes, the relatively low ENA fluxes, and the detection without ionization) and the needed solutions. Based on the optics, the viewing schemes, and spacecraft requirements, we shall review general approaches to imaging ENAs without requiring the ionization of the ENAs prior to detection, thus restricting this review to ENAs of energies > 10 keV or "high-energy" ENAs (HENAs). Combining particle spectrometry with imaging is the main challenge in ENA imaging. We shall review some of the existing designs of ionization-free HENA imagers with discussions on their characteristic features. We hope this review will provoke more innovative designs in the coming years.

1. INTRODUCTION

The investigation of space plasma by way of detecting energetic neutral atoms (ENAs) began with the discovery of energetic hydrogen atoms in aurorae in 1950 (Meinel, 1951). Although much work had been done towards remote sensing space plasma in ENAs since 1969 (*Bernstein et al.*), it was not until 1987, that the first ENA image of a magnetic storm finally caught

the fancy of the space-science community (Roelof, 1987). The idea of using ENAs to obtain global images of space plasma, especially that of Earth's magnetosphere, to answer questions that could not be addressed by statistical analysis of 35 years of *in situ* measurements, is now widely accepted. Only through a global approach, *i. e.* remote sensing with large field of view, could we better understand the core of STP (solar-terrestrial physics). ENA imaging permits study of the ways in which our entire plasma environment -- including the magnetopause, ring current, plasmasphere, auroral zones, plasma sheet, and the ionosphere -- reacts to the changing conditions of the solar wind (Williams, 1990). Using similar techniques,

Measurement Techniques in Space Plasmas: Fields
Geophysical Monograph 103

remote sensing of other planetary magnetospheres, interplanetary particle events, such as transient shocks and co-rotating interaction regions, coronal mass ejecta, and even the very edges of the heliosphere could be studied in ENAs. In the two years since the presentation of this paper in Santa Fe, several noteworthy events in the attempts to study space plasma in ENAs bear mentioning. 1) An ENA imager aboard spacecraft SAC-B was severely crippled following the failure of separating spacecraft SAC-B and HETE from the upper stage of the launcher on 4 November 1996. 2) An instruments on SOHO (Solar & Heliospheric Observer) began the detection of heliospheric ENAs with a limited ENA capability since 12 January 1997. 3) A dedicated ENA imager on Cassini is readied for launch later this year to image the Saturnian magnetosphere beginning in 2002. 4) The first Medium-size Explorer, IMAGE (Imager for Magnetopause to Aurora Global Explorer), which has three ENA imaging instruments passed the confirmation review on 5 March 1997. And 5) magnetospheric ENA flux was detected by ENA-capable HEP on GEOTAIL).

The heightening of activities in the ENA community makes now an exciting time to review the basic concepts of imaging space plasma in ENAs and up-to-date techniques in ENA imaging in all energies. This review will be confined to the imaging of HENAs (high-energy neutral atoms, energies > 10 keV) without ionizing the ENAs first. Hence, the family of ENA instruments originated by Bernstein et al. (1969) at lower energies and later modernized and extended to higher energies by McComas et al. (1991) will not be reviewed here. While HENA remains for ENA >10 keV, the HENA imagers reviewed here are only those which do not employ ionization in the detection of ENAs.

This review shall begin with the general and move to the specific through the following topics:

1. Brief history and significant milestones
2. Characteristics of ENA flux
3. Imaging processes and requirements
4. Instrument design considerations
5. Comparison of some existing designs

We hope more innovative solutions to further improve or even revolutionize the instrumentation and more opportunities to implement them in the pursuit of remote sensing non-radiating space plasma will follow.

2. BRIEF HISTORY

Imaging space plasma in ENAs has its genesis in the sighting of light emitted by ENAs in a radiating space plasma. Meinel (1951) reported the sighting of blue-shifted and broadened hydrogen Hα line (6563Å) in an aurora along the magnetic field line (Fig. 1) and deduced that this Doppler effect must be due to H atoms approaching with a kinetic energy as high as 70 keV. Further observations and calculations by Fan and Schulte (1954) and Fan (1958) gave estimates on the flux of precipitating protons during an auroral break-up. In 1968, the first ENA flux measurement (Fig. 2) by the first dedicated ENA spectrometer on a sounding rocket during an auroral break-up at ~250 km altitude was performed by (Bernstein et al., 1969a,b). Although Dessler and Parker (1959) pointed out the existence and the role of ENAs in the decay of ring currents in sub-storms, this mechanism was not invoked until the detection of protons of energies > 0.25 MeV at equatorial low altitudes (< 600 km) by Moritz (1972). Improved techniques, including the newly developed 2-parameter (time-of-flight and residual-energy) particle identification scheme (Gloeckler and Hsieh, 1979), to study the directional distributions of ENAs in the magnetosphere were proposed (Hsieh et al., 1980) for the OPEN mission in 1980 without success. Techniques for ENA detection continued to be developed, but the space-science community did not warmly embrace the idea of remote sensing a non-radiating space plasma using ENAs until the ingenious interpretation of a set of fortuitous data taken 9 years earlier was reported (Roelof, 1987). Optimism in

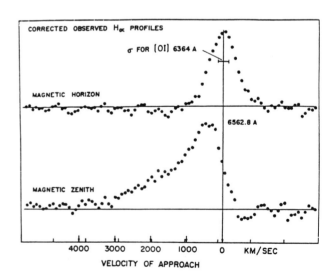

Figure 1. First evidence of ENA in space plasma seen in solar-terrestrial phenomena: Blue-shifted and broadened Hα line along the local magnetic lines of force (magnetic zenith) is compared with the Hα line normal to the field lines (magnetic horizon). Figure taken from Meinel (1951).

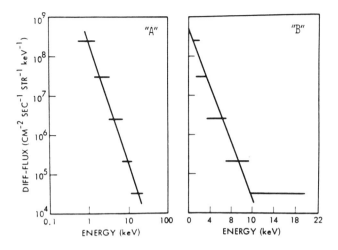

Figure 2. First directly measured ENA spectrum between 1 and 20 keV at ~250 km altitude. A power-law fit of $\propto E^{-3.2}$ ("A") seems better than an exponential law with e-folding of ~1 keV ("B"). Figure taken from Bernstein *et al.* (1969).

sensing space plasma from "directional distributions" to "imaging". What follows is a brief listing of the milestones leading to our present state of development in techniques and current standing in the space-physics community. Conferences in which the topic of HENA was high-lighted are also listed as milestones, because their proceedings contain the most up-to-date reviews on the subject at the time. If any work is left out here, it is either due to the authors' ignorance or for the sake of brevity.

1950-59

First sighting of ENA (up to 70 keV) in blue-shifted hydrogen Hα in aurorae (Meinel, 1951)

First estimate of the proton energy and flux needed to explain the blue shifted auroral spectra (Fan and Schulte, 1954)

Prediction of the production of ENAs in the ring current as means for recovery phase of magnetic storms (Dessler and Parker, 1959)

1960-69

Proposed study of magnetic storm by detection of ENAs (Dessler *et al.*, 1961)

First dedicated ENA instrument and spectrum (1 - 20 keV) (Bernstein *et al.*, 1969a,b)

1970-79

First reports of probable HENA flux from the ring current (Moritz, 1972; Hovestadt *et al.*, 1972; Mizera and Black, 1973)

Laboratory testing of time-of-flight (TOF) and residual energy mass spectrometers for space physics reported (Gloeckler and Hsieh, 1979)

1980-89

Proposal HELENA to detect ENAs on OPEN (Hsieh *et al.*, 1980)

Review of ENAs in precipitation (Tinsley, 1981)

First report of emission of energetic neutral particles from Jupiter and Saturn (Kirsch *et al.*, 1981a,b)

Detection of HENA from the ring current on IMP 7/8 and ISEE-1 (Roelof *et al.*, 1985)

Development of dedicated instrument on Ulysses for the *in situ* detection of penetrating interstellar neutral helium atoms (Rosenbauer *et al.*, 1984)

Development of ENA instruments on Relikt for the detection of neutral solar wind and geomagnetospheric ENAs (Gruntman and Morozov, 1982)

First image of ring current in HENA (Roelof, 1987)

Yosemite Conference, 1988 (*Solar System Plasma Physics*, AGU, 1989)

1990-97

INCA selected for Cassini, 1990 (Mitchell *et al.*, 1993)

Solar Wind Seven in Goslar, 1991 (*Solar Wind Seven*, Pergamon, 1992)

Adoption of an ENA option on SOHO, 1991 (Hovestadt *et al.*, 1995)

NASA mission study of the Inner Magnetospheric Imager (IMI) began in 1991

ENA imaging ring current on CRRES, 1991 (Voss *et al.*, 1993)

Proposal to fly ISENA on SAC-B to Italian Space Agency (ASI), 1992

SPIE Symposium in San Diego, 19 - 24 July, 1992 (*Instrumentation for Magnetospheric Imagery*, Proc. SPIE Intl. Soc. Op. Eng. 1744, 1992)

COSPAR Symposium in Washington, D. C., 1992 (*Advances in Space Research*, 13(6), 1993)

Report of detection of penetration interstellar neutral helium on Ulysses (Witte *et al.*, 1993)

NASA's IMI mission understudy reduced to Magnetospheric Imager (MI) mission, 1994 (Armstrong and Johnson, 1995)

Launch of ASTRID carrying ENA imager, 1995 (Barabash *et al.*, 1996)

Chapman Conference in Santa Fe, 1995 (This volume)

ISENA on SAC-B approved by ASI, September 1995

SOHO with ENA capability launched, 2 December 1995

Pegasus failed to deploy SAC-B in orbit, 4 November 1996

Dedicated ENA, EUV, FUV and radio imaging mission IMAGER selected for first Medium-size Explorer, 25 April 1996

Report of unambiguous detection of energetic neutral atom flux outside of Earth's magnetosphere (Wilken *et al.*, 1997)

The above chronology shows a long lead time in the development towards fruition. Activities have intensified dramatically since 1990 as maturation in the required techniques and opportunities for applications began to converge. Relevant events occurred between Santa Fe and this revision are added and critical references updated.

3. CHARACTERISTICS OF ENA FLUXES

In space plasmas, such as the ring current in the magnetosphere, transient or co-rotating shocks in the heliosphere, and unmodulated cosmic rays at or beyond the edges of the heliosphere, singly-charged ions may become ENAs by charge exchange with atoms in the respective ambient neutral gas -- the geocorona in the magnetosphere (Rairden *et al.*, 1986), penetrating interstellar medium in the heliosphere (Blum and Fahr, 1970; Bertaux and Blamont, 1971; Thomas and Krassa, 1971; Lallement *et al.*, 1992; 1996), and local interstellar medium at the termination shock of the solar wind (Baranov and Malama, 1996; Équemerais *et al.*, 1995; Lallement *et al.*, 1996). Due to the large internuclear distances during charge exchange, negligible energy and momentum are transferred in these interactions; hence, the ENAs preserve the energy spectrum of the original ion population, only modified by the energy dependence in the charge-exchange cross sections. Unaffected by magnetic fields, the ENAs travel in ballistic trajectories and some of them reach distant ENA instruments poised to sample the non-

radiating space plasma in remote and often not so easily accessible regions.

The capability to detect ENAs with good mass, energy and directional resolutions constitutes the basis of ENA imaging. Imaging in ENA overcomes the difficulty inherent in *in situ* measurements, *i.e.* distinguishing spatial variations from temporal changes, thus enabling the global study of interconnectedness of space plasma in different regions over large distances and under diverse conditions. As an example, the global dynamics of the entire magnetosphere, from magnetopause to ring current, plasmasphere, auroral zones, and the plasma sheet, can be studied in the context of changing conditions in the solar wind and events in the solar corona (Williams, 1990). Similar techniques may be used to study the magnetospheres of other planets *e.g.*, (Mitchell *et al.*, 1993), and interplanetary particle populations, such as those in transient shocks and co-rotating interaction regions, coronal mass ejecta, and even the very edges of the heliosphere (Hsieh *et al.*, 1992a,b; 1993; Roelof, 1992; Hovestadt *et al.*, 1995; Hilchenbach *et al.*, 1996)

Figure 3 illustrates the geometry of remote sensing a space plasma *via* ENAs. For an observer $\mathbf{I}$ at $\mathbf{x_o}$, the ENA flux of the ith species of mass (M_i) and energy (E) coming from the direction of the unit vector $\mathbf{s}(\varphi, \theta)$ is

$$j_i(\mathbf{s}, E; t) = \int r^{-2} j_{i,+1}(\mathbf{x}, E; t) \sum_j \left[\sigma_{ij}(E) \cdot n_j(\mathbf{x}) \right. \\ \left. exp\left[-D(\mathbf{s}', E)\right] d\tau \right. \tag{1}$$

typically, in units of $(cm^2 \cdot sr \cdot s \cdot keV)^{-1}$, where $j_{i,+1}$ is the differential flux of the singly-charged ions of the ith species with velocities pointing along $\mathbf{s}'$ at the time of charge exchange, σ_{ij} is the charge-exchange cross-section for ions of the ith species incident on a neutral target of the jth species, and n_j is the number density of the jth species in the neutral gas. The product $r^{-2} d\tau$ reduces to a column element dl subtending a solid-angle element $d\Omega$ at $\mathbf{x_o}$ along unit vector $\mathbf{s}$, the line of sight. The exponential term represents the extinction of ENAs due to re-ionization by photons, electrons, ions, and atoms along the way. The term D is itself a line integral from $\mathbf{x}$ to $\mathbf{I}$ along the unit vector $\mathbf{s}'$;

$$D(\mathbf{s}', E) = \int \left[\frac{\beta(\mathbf{x})}{v} + \sum_k \sigma_{ik}(E) \cdot n_k(\mathbf{x}) \right] dl$$

(2)

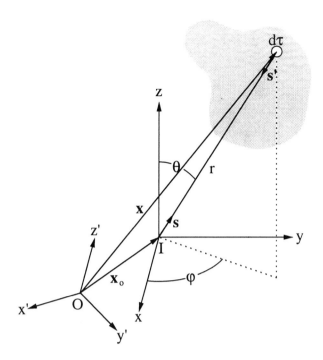

Figure 3. Coordinate system used for Eq. (1): The imager I is located at $\mathbf{x}_o$ and receiving ENA flux from volume element $d\tau$ at $\mathbf{x}$. The shaded area represents the space plasma under investigation. The unprimed coordinate system is fixed to either the imager or the spacecraft.

where $\beta\,(\mathbf{x})$ is the photoionization rate, ν the speed of the ENA, and σ_κ the ionization cross section for collisions with the kth target species, including electrons, whose number densities are n_k. A comprehensive compilation of the cross sections is found in Shih (1993). For most cases, D is sufficiently small so that the exponential term in Eq. (1) can be approximated by $[1 - D]$.

The above consideration highlights three inherent features of ENA flux, $j_i(\mathbf{s}, E; t)$, that will affect the design of any ENA imager.

(1) *Steep energy spectra*: The parent-ion spectra, $j_{i,+1}$, typically decrease with increasing energy, as has been observed in all space plasma . The cross sections for charge exchange, σ_{ij}, typically also decrease with increasing energy. Jointly they yield an ENA spectrum, j_i, that decreases with increasing energy even more rapidly than $j_{i,+1}$ and σ_{ij}.

(2) *Low flux*: ENA flux at $\mathbf{x}$ is lower than its parent-ion flux by a factor of $\sigma_{ij} \cdot n_j$. While n_j may be as

large as 10^4 cm^{-3}, *e. g.* geocoronal hydrogen at $2\,R_E$ (Earth radii), σ_{ij} , on the other hand, is typically $\ll 10^{-15}$ cm^2. Only integration along the line of sight over the length of the object, *e. g.* of the order of R_E or $\sim 10^9$ cm for near-Earth objects, raises the ENA flux to detectable levels. There are regions of space where ENA fluxes can be detected without significant interference from ions, as in the cases reported by (Kirsch *et al.*, 1981a,b; Roelof, 1987), but the laws of mechanics forbid any observer to tarry there indefinitely. The most recent observation of Wilken *et al.* (1997) gave an upper limit of 7 particles/(cm^2 sr s) in the energy interval 77 - 200 keV.

(3) *Co-existence with EUV fluxes*: Besides energetic ions, the ambient neutral atoms -- ENAs' other parent -- are also sources of concern. These atoms, n_j being their number densities, are effective resonant scatterers of solar radiation, *e. g.* 1216Å from H and 584Å from He (Rairden *et al.*, 1986; Bertaux and Blamont, 1971; Thomas and Krassa, 1971; Lallement *et al.*, 1992; 1996). Therefore, regions of ENA production coincide congenitally with regions of EUV emission; hence, ENA fluxes co-exist with and EUV fluxes.

4. ENA IMAGING PROCESSES & REQUIREMENTS

To acquire $\{j_i(\mathbf{s}, E; t)\}$, a set of histograms in $(M, E, \varphi, \theta; t')$ vector space, is the usual design goal for all ENA imagers; to retrieve $j_{i,+1}(\mathbf{x}, E; t')$ is the observational goal for all ENA investigations; and to elicit values of the set $\{j_{i,+1}(\mathbf{x}, E; t')\}$ under different conditions in terms of plasma physics is the ultimate goal of ENA imaging. Identification of ENAs in (M, E) sub-space is particle spectrometry, already a well developed art in ion detection in space. New to the field is imaging -- the identification of ENAs in (φ, θ) sub-space. The challenge of remote sensing space plasma in ENAs lies in combining mass spectrometry with imaging.

A HENA imager is an imaging mass-spectrometer. The essential elements of ENA imaging is summarized in Fig. 4. In general, fluxes coming from the source and projected unto the Object Plane go through the Optics of the imager **I** and fall on the Image Plane; *e. g.*, a particular element Q of the Object Plane falls on a pixel P on the Image Plane. For this review, "Optics" means structural designs that map the Object Plane onto the Image Plane *without systematically altering the direction of flight of the ENAs*. The accumulation of signals in the Image Plane over time provides the

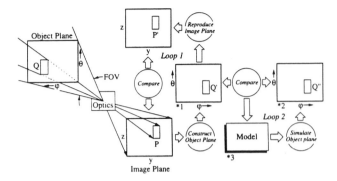

Figure 4. Overview of the essential elements of ENA-imaging: From incident flux to model validation. This review concentrates on instrumentation, which is composed of the FOV, Optics, and the Image Plane. While Loop 1 processes the data to produce the best image, Loop 2 optimizes the model by comparing with the best image.

database from which a Constructed Object Plane (*1) may result after going through image processing (Loop 1). Parallel to this iterative image processing is model validation (Loop 2). The Simulated Object Plane (*2) that best matches the best Constructed Object Plane (*1), which resulted from the best Model of the plasma object (*3), yields the final set { $j_{i,+1}(\mathbf{x}, E; t')$ }, ready for analysis and interpretation.

The first step of imaging is mapping the Object Plane onto the Image Plane. The ideal mapping is a one-to-one transformation, thus obviating Loop 1. An extrapolation from the point of impact on the Image Plane to the point of incidence in some aperture, say, a pinhole, a coded aperture or a position-sensing detector near the aperture, is needed . The low HENA flux usually precludes the use of pinhole. Other aperture configurations, such as coded aperture or apertures defined by collimators, that allow each pixel P to be exposed to only a specific element Q may be acceptable if sufficient flux is accumulated within the acceptable angular and time resolution. There are at least two ways to proceed, or two imaging schemes. To ease further discussion, Fig. 5 represents the principal parts of a HENA imager that does not require the ionization of the ENAs.

A set of plates form the collimator, which is the first stage of the Optics that projects the Object Plane on the Image Plane. Because of the need to suppress ions of masses and energies similar to the HENAs from triggering the detector, and because of the desire for the option to turning on and off the ion rejection capability,

electrostatic field between the collimator plates is preferred to magnetic field.

If P1 is the only position-sensing element of the imager, then P1 *is* the Image Plane and the projection of the Object Plane through the collimator on P1 must be known in order to produce the Constructed Object Plane (*1 in Fig. 4). This is the "one-point" scheme. If each incident ENA needs to be detected twice -- for spectrometric reasons to be discussed later, then P1 needs to be a thin foil to maximize the passage of the ENA to reach P2. Scattering in the thin foil will cause the ENA to impact P2 at point B', instead of B, as indicated in Fig. 5. If P2 is also a position-sensing detector, then the Constructed Object Plane is formed by extrapolating the point of impact on P2, B', to the point of entry on P1, A. This is the "two-point" scheme. (See *Imaging Scheme* below.)

While the "two-point" scheme is inherently two dimensional (2D), even though scattering produces blurring, the "one-point" scheme is one-dimensional (1D), since the collimator plates can only define a distribution of exposure pattern in the dimension along the line P1 in the page of the figure. To extend to two dimensions, one could introduce a 1D coded aperture

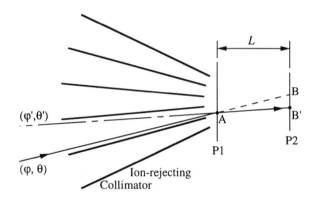

Figure 5. A schematic representation of a HENA imager that does not require ionization of the HENAs. The collimator plates serve to define the FOV of the imager as well as to suppress the entry of ions in the same mass and energy ranges of the expected HENAs. P1 could be a position-sensing element. A second plane P2 separated by a distance L can provide a second signal for the same event to provide a two-parameter measurement of each incident HENA. Scattering in P1 causes the point of impact on P2 to shift from B to some other point B'. The orientation of the imager with respect to the spin axis in a scanning mode determines whether the position sensing detectors are one dimensional or two.

somewhere between the end of the collimator and the Image Plane P1 or have the imager spin about an axis parallel to P1 in the plane of the page. (See *Viewing Modes* below.)

The confidence and precision which could be attached to the best Simulated Object Plane and Model resulting from Loop 2 depend on the resolution of the HENA imager, but the techniques in model validation are independent of the instrumentation and we refer to the excellent works of (Chase and Roelof, 1995; Moore *et al.*, 1995) and several others in this volume.

To summarize, the basic processes in ENA imaging presented above translate into the following requirements on HENA imagers, in addition to the appropriate mass (M) and energy (E) resolutions: 1) minimum energy threshold to take advantage of the steep spectrum, 2) maximum sensitivity to HENAs while effectively rejecting ions, 3) maximum EUV rejection, and 4) optimum use of the Image Plane with maximum spatial resolution.

5. INSTRUMENT DESIGN CONSIDERATIONS

Unlike imagers for ENAs, typically < 30 keV, that require first ionizing the incident ENA by interaction with a specially treated surface or a thin foil or a beam of electrons, followed by an ion analyzer that bends the particle's trajectory by an electric and/or magnetic field, e.g. (Bernstein *et al.*, 1969; McComas *et al.*, 1991; McComas and Funsten, 1997), the HENA imagers under considerations do not require ionization of the ENAs, thus permitting rectilinear trajectories between Object Plane and Image Plane. The elimination of the need for ionization raises the sensitivity of the HENA imagers, but the absence of bent trajectories exposes the imaging detector to direct EUV illumination. Maximum EUV rejection and minimum blurring are, unfortunately, in direct conflict, as we shall see.

Consider now the obvious approaches, techniques, and concerns, seven in all, from which the HENA imager designers could choose to meet the above demands and cover the desired domains in two sub-spaces (M, E) and (φ, θ).

5.1 *Viewing Modes*

In the (φ, θ) sub-space, a 2D image may be gotten by either staring at or scanning the object. The choice depends on the HENA imager's vantage point and the object's location and dimensions relative to the imager, and the time required for flux accumulation. If the plasma object is distributed around a planet and

enclosed by the orbit of the imager, *e. g.* imaging the inner magnetosphere of a planet from a highly elliptical orbit, then Kepler's "equal-area" law suggests staring at the object from the apoapsis. The span of (φ, θ), the field of view (FOV) of the HENA imager, should then be no larger than necessary to circumscribe the object, thus, defining the 2D Image Plane (Fig. 4). If, on the other hand, the orbit is inscribed by the object, *e. g.*, a low Earth-orbiter looking at the ring current, then the φ-range of the FOV can be greatly reduced by letting the spacecraft's spin about its z-axis sweep out the full 2^1 in φ. The scanning mode needs only a 1D image plane in $z(\theta)$, which significantly simplifies the sensor technology and the subsequent image processing. The reduced duty cycle, however, must be weighed against the characteristic time of the plasma object.

5.2 *Imaging Scheme*

After choosing between scanning or staring, one must decide on how to determine the arrival direction of an HENA. The choices to date are between "one-point" and "two-point" schemes (see Fig. 5). The "one-point" scheme requires only one position-sensitive detector. The geometry of the aperture projects specific elements of the FOV, *i. e.* the elements in the Object Plane, on the imaging detector at P1, the Image Plane. This corresponds to the element-to-pixel mapping in Fig. 4. The ambiguity in P1, *e. g.* point A (in Fig. 5) being illuminated by a range of directions contained in two adjacent collimator channels, can be resolved by statistical analysis under the constraint of the FOV of each pixel. An example is given below. The "two-point" scheme requires two position-sensing detectors, P1 and P2, to determine the arrival direction of each HENA by locating two points along the trajectory (points A and B' in Fig. 5) and then taking the line connecting the two points to represent the arrival direction. The locations of the two points have, in most cases, independent two-dimensional uncertainties due to scattering in P1. The required deconvolution is error-prone, especially at low energies.

Consider an example of the "one point" scheme. The collimator plates can only define a distribution of exposure pattern in the dimension along the line P1 in the page of the figure. As shown in Fig. 5, point A, the ENA's point of entry on P1, could be associated with a range of directions defined by two adjacent collimator channels. This means adjacent pixels may have overlapping FOVs, hence, overlapping elements in the Object Plane (Fig. 4). This "over sampling" would provide an image of lower angular resolution but with a better time resolution than a pin-hole imager. When the accumulated events are statistically significant, from either longer exposure or larger geometrical factor per pixel, over-sampling can yield angular resolutions better than the FOV of each pixel by appropriate deconvolution techniques in Loop 1 (Fig. 4). An example of reconstructing the intensity distribution from "over sampling" in 1D using a nonlinear iterative technique is shown in Fig. 6 (Curtis and Hsieh, 1989). It shows that a well-defined geometrical relation between the apertures and the pixels can provide angular resolutions finer than the FOV of each pixel. Besides the simplicity in mechanics and electronics, the "one-point" is free from blurring due to scattering in P1.

The choice between "one point" and "two point" schemes is also tied to the type of mass spectrometry, to be discussed later. The differences in the two

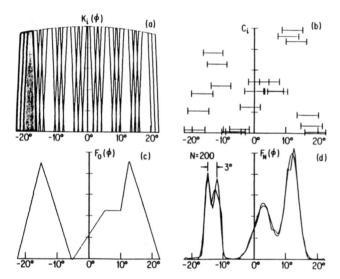

Figure 6. An example of deconvolution of a set of over-lapping one-dimensional pixels. (a) Response curves, $K_i(\varphi)$, of 21 overlapping pixels defined by the collimator sectors spanning 45° in φ. The shaded area represents the curve for the left-most pixel. (b) Over-sampled accumulated counts, C_i, of the pixels in arbitrary units derived from four Gaussian distributions. (c) "Guess" distribution based on the apparent distribution shown in (b) is used to start the iterations. (d) The resulting distribution after 200 iterations is compared with the source distribution that generated the accumulated counts in (b). The resolution is excellent compared to the angular-width of each pixel, when the event accumulation is statistically significant. Figure is taken from Curtis and Hsieh (1989).

approaches are best understood when comparing specific designs.

5.3 Collecting power

The low HENA flux mandates large geometrical factors. There is a distinction between geometrical factor and collecting power of an imager. The often quoted geometrical factor, G, of an HENA imager is the sum of all the geometrical factors, each defined by a pair of ion-rejecting collimator plates. If all the collimator channels are identical and point in the same direction, *i. e.* parallel channels, then G is the number of channels times the geometrical factor per channel; the larger the number of such channels (assuming proportionally larger detectors) the larger is G. The overall FOV (solid angle Ω), however, does not increase with the number of parallel channels, but equals that of a single channel. In this case, the detected fluence from within the FOV is directly proportional to G. On the other hand, if all the identical channels point in different directions, then Ω expands without increasing G. The geometrical factor in a given direction is only that of a single channel. The detected fluence from a given direction is not directly proportional to G, but only the geometrical factor of that single channel. One, therefore, must consider the figure of merit, G/Ω, the *collecting power* of an imager instead of G. For example, a large G derived from a large Ω is less effective than a somewhat smaller G with a much smaller Ω, especially, if the object subtends a solid angle smaller than Ω. More precisely, it is the differential geometrical factor $\dfrac{dG}{d\Omega}$ as a function of pointing direction (φ, θ) that tells how efficient an imager is in collecting data (Sullivan, 1971).

5.4 Mass spectrometry

Scientific objectives and properties of the plasma objects determine the needed resolution in the (M, E) sub-space. If only one species, say, H, is expected to dominate the ENA flux, then a single measurement of E would suffice. If both M and E identification are required, then a two-parameter measurement of each incident HENA must be performed. In addition, a two-parameter measurement also provides an effective discrimination against random noise, such as that induced by copious EUV photons, by a coincidence requirement. In terms of high mass- and energy-resolution and low random noise, the most effective two-parameter measurement technique is the time-of-flight *plus* pulse-height analysis (TOF/PH) developed and used in space ion mass spectrometry (Gloeckler and Hsieh, 1979; Gloeckler, 1990). Since a TOF analysis requires a "start" and a "stop" for each valid event, it already incorporates a double-coincidence measurement. The additional pulse-height (PH) analysis provides a triple-coincidence capability to reduce random noise even more effectively (Hsieh and Curtis, 1989). The "two-point" imaging scheme discussed earlier goes well with a two-parameter spectrometer; *e. g.*, when the "start" (at P1 in Fig. 5) and either the "stop" or the PH detectors (at P2 in Fig. 5) are position sensitive.

5.5 *The PH detector*

In one-parameter measurements, the pulse-height (PH) detector does both event selection and particle identification. In two-parameter measurements, the PH detector complements TOF analysis, not only in resolving masses of the incident particles, but also greatly reducing random noises by requiring triple coincidence for each event. The choices for PH detectors to date are MCPs operating at saturation and with position-sensing anodes (Mitchell *et al.*, 1993; 1996) and pixellated solid-state detectors (PSSDs) (Voss *et al.*, 1993, 1997; Hovestadt *et al.*, 1995), for they can also serve as the Image Plane. When position-sensitive anodes are used in conjunction with MCPs, spatial resolution can be continuous and better than 1 mm, *e. g.* see Siegmund *et al.* (1986) and Carlson and McFadden (1997). PSSDs, on the other hand, restricted by the pixel size, typically > 1 mm in each dimension, and the gap between pixels, typically ~0.1 mm, obviously offer a coarser spatial resolution. The energy threshold for particle detection of MCPs is typically ~1 keV, while that of SSDs is typically >10 keV, due to the conducting layer, dead-layer and detector-amplifier noise. Nevertheless, PSSDs have other advantages over MCPs. 1) The PSSD is immune to EUV, if each pixel has its own amplifier, but the MCP is not. Consequently, with a PSSD a much thinner foil can be used to lower the threshold and scattering, and there is no need for a second foil to protect a large-area imaging and PH-analyzing MCP. 2) The PSSD has excellent PH resolution ($\delta E < 10$ keV by cooling the SSD or appropriate choice of the coupling field-effect transistor), but the MCP pulse-height distribution is highly statistical and sensitive to the gain. 3) The PSSD performance is stable, but gain-shifts in MCPs require in-flight maintenance. 4) The bias for the PSSD is < 100 V, while the MCP stack requires > 3 kV. 5) Micro-electronics for PSSD are already in use. The availability of specific PSSDs may be at times a problem, but technically, choosing between MCPs and PSSDs will always be a challenge to the designer. A combination of both types of PH-detectors should be considered.

5.6 *Integrity of the collimator*

The collimator defines the FOV of a HENA imager. The preference for using electric over magnetic fields to reject incident ions below certain energies also makes the collimator plates electrodes that maintain the needed electric field. Proper mechanical design and choice of material can prevent unwanted particles from entering the imager by either penetration or scattering. Micro-serration of the surface of collimator plates is essential, but needs be supplemented by baffles to further suppress scattering. Thickening the plates also reduces particle penetration. Caution must also be exercised to prevent discharge between adjacent plates at opposite potentials. The use of strong fields for ion rejection must not lead to electron-photon cascades, in which secondary electrons produced on negatively biased plate surfaces are accelerated towards positively biased plate surfaces to produce X-ray photons, which in turn may strike the opposite plate at high negative potential, thus producing more photo-electrons, *etc*. To provide the electric field between the plates, using two high-voltage power supplies, one positive and one negative, instead of one of a single polarity, would lower the absolute potential on each supply and would allow operation at a narrower energy window, even if one supply fails in flight. In all cases, the entire collimator assembly must be shielded from the ambient space-plasma potential.

5.7 *Redundancy*

The fear of failure in flight can be partially assuaged by having redundancies in critical parts. Having two sensor heads, even looking in different directions, yields a better chance of partial success. Having individual signal chains and lower bias voltage, a PSSD has a better chance of partial success than a single pair or stack of large-area MCPs operating at saturation. For long duration missions, some redundancy would be reassuring.

The above seven design considerations are general and current. They are best understood by examining some of the existing designs.

6. SOME EXISTING HENA IMAGER DESIGNS

To date -- March 1997, five instruments with HENA capabilities, two of which were dedicated HENA imagers, have been placed into orbit around Earth. One instrument with limited HENA capability oscillates around the Langrangean point L1; and one dedicated HENA imager is ready to go to Saturn later this year. These HENA instruments in their launch sequence are IMS-HI on CRRES, HEP on GEOTAIL (Doke *et al.*, 1664), PIPPI on ASTRID, SEPS on POLAR, HSTOF on SOHO, ISENA on SAC-B, and INCA on Cassini, as listed in Table 1. Of these instruments, PIPPI (Barabash *et al.*, 1996), ISENA (Orsini *et al.*, 1996) -- an adaptation of the RAPID ion spectrometer on the ill fated CLUSTER (Wilken *et al.*, 1993), and INCA (Mitchell *et al.*, 1996) are separately described in this volume, therefore, their specific features will be cited only for comparison. We choose to present only IMS-HI and HSTOF in some detail. Although they are not dedicated HENA imagers, they have all the required features of HENA imagers. They represent several contrasting aspects -- IMS-HI was on a geomagnetospheric mission, used magnetic fields to reject incident ions, and had a one-parameter particle identifier, while HSTOF is on a heliospheric mission, uses electric fields to reject incident ions, and has a two-parameter mass spectrometer.

The authors apologize for not including HEP on GEOTAIL (Doke and al., 1664) and SEPS on POLAR (Voss *et al.*, 1993) in this review, due to the authors' inadequate preparation.

6.1 *IMS-HI on CRRES*

Fig. 7 is a schematic representation of the magnetic ion-mass-spectrometer IMS-HI on CRRES. (Voss *et al.*, 1993). The 17-cm long 7-kG magnetic analyzer sweeps all charged particles < 50 MeV amu e^{-2} away from the No. 7 SSD. This dedicated HENA channel (10 -1500 keV) makes IMS-HI the first HENA imager in Earth orbit (~1.1 x 6.5 R_E). The single-parameter measurement of the particle energy precludes any mass identification. With the energy window of the cooled (-50°C) SSD set between 10 and 1500 keV, only neutrals, *including* x-ray photons, in this energy window can be detected. The small geometrical factor (10^{-2} cm^2 sr), the short integration time (1 µs), and the relatively high threshold (10 keV) easily EUV photons (10.25 eV for H 1216Å and 41 eV for He$^+$ 304Å) either singly or collectively (in pile-up), from triggering the SSD, even in a 10 kR environment of the

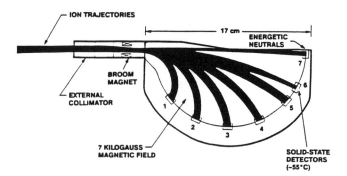

Figure 7. Schematics of IMS-HI flown on CRRES. Particles undeflected by the 17-cm and 7 kG magnet are considered ENAs and are detected and PH-analyzed by SSD No. 7. The other 6 SSDs are designated for protons between 30 keV and 10 MeV. The cooled (-50°) SSDs had a noise level <10 keV. Figure taken from Voss *et al.* (1992).

geocorona. Small-area SSD's immunity to EUV is also exploited in PIPPI. The noise due to penetrating ions, say, cosmic rays, is suppressed by an anti-coincidence SSD placed immediately behind the HENA-detecting SSD, as is also done in ISENA . The angular resolution of IMS-HI is just its FOV. The orbit of CRRES is enclosed by the ring current, so IMS-HI uses the spacecraft spin and the orbital motion to scan the ring current with a time resolution restricted by its small FOV and the duty cycle set by the spacecraft spin.

6.2 *HSTOF of CELIAS on SOHO*

Fig. 8 is a schematic representation of HSTOF of CELIAS on SOHO, an extension of STOF (Hovestadt *et al.*, 1995). While STOF (the portion with curved collimators) faces the Sun and analyzes the charge-states of superthermal ions in interplanetary shocks between 20 and 1,000 keV/amu, HSTOF (the portion with straight collimator plates) has its optical axis at 37° away from the Sun-probe line in the ecliptic and extends the energy range to 2 MeV/amu. Although not a dedicated HENA imager, HSTOF has all the features of one using TOF/PH analysis, like INCA and ISENA -- an ion-rejecting collimator and an EUV-blocker. The straight parallel plates of its collimator, when biased at ±1.295 kV, reject ions of E/q < 100 keV/e with increasing efficiency for ions of lower energies, thus allowing the detection of HENAs < 100 keV. The EUV blocker, also the "start"-signal generator for the TOF analyzer, is a Si/Lexan/C layered foil of respective thickness 5, 4, and 1 µg cm^{-2}, identical to those in ISENA and INCA. The foil suppresses the H Lyα

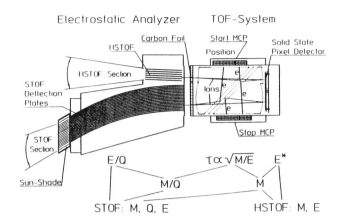

Figure 8. Schematics of HSTOF and STOF of CELIAS on SOHO. HSTOF, with straight ion-deflecting collimator plates, has the ENA capability. HSTOF has its own foil (Si/Lexan/C layered foil), own portions of the START and STOP MCPs, and SSD, which has 64 pixels. Figure taken from Hovestadt *et al.* (1995).

photons from triggering the MCPs of the TOF analyzer either directly, after few reflections, or by their photo-electrons from either the foil or the SSD surface. HSTOF uses a 64-pixel PSSD (32 cm^2 of sensitive area) for its PH detector, as compared to INCA's pair of 100 cm^2 MCPs. The small size of the pixels (0.5 cm^2 each) and the relatively high threshold (45 keV or 3x the noise) render the PSSD EUV-immune, thus, eliminating the additional foil needed for the large-area MCP in INCA. The high threshold of the PSSD, however, eliminates mass identification of HENAs < 50 keV, since only TOF analysis is performed. For particles > 50 keV, HSTOF's position-sensitive "start" MCP determines the exit point on the foil and the PSSD the impact point on it for each incident particle. Although not used for imaging, as in INCA, this capability is used to determine the TOF path-length of each particle -- thus removing uncertainty due to the ±2° spread in φ and ±17° in θ of the FOV, plus the energy- and mass-dependent scattering in the foil -- to refine HSTOF's mass resolution, sufficient to resolve He isotopes. This windfall HENA capability, though limited, makes HSTOF the first HENA imager to sense remotely ions accelerated in transient and co-rotating shocks in interplanetary space and unmodulated cosmic rays (galactic and anomalous) from the edges of the heliosphere (Hsieh *et al.*, 1992a,b; 1993; Roelof, 1992; Hilchenbach *et al.*, 1996), as HSTOF sweeps out each year a 34°-wide band of the heliosphere about the ecliptic with its 4° x 34° FOV.

6.3 *Further comparison*

Unless severely constrained by spacecraft resources, e. g. PIPPI on ASTRID, a two-parameter measurement to populate the 2D (M, E) sub-space is preferred. HSTOF, ISENA and INCA, all adopted the 2-parameter technique of TOF/PH analysis, not only for the resolution in mass and energy, but especially for the effectiveness of triple coincidence in reducing the background noise. These instruments differ in the choice of collecting power, PH detector, imaging scheme, and electron mirrors . We now discuss these choices in some detail. A summary of the characteristics of these instruments is shown in Table 1.

6.3.1 *Imaging scheme.* The choices mentioned are "one-point" and "two-point" schemes (Fig. 5); e. g., PIPPI and ISENA use the former and INCA the latter. For ISENA, the imaging detector is the "start"-MCP with its anode divided into four pixels; each corresponds to one FOV defined by the collimator plates and the solid-state detector D1 (Orsini *et al.*, 1996). Valid events belonging to each FOV go through different portions of the foil (P1 in Fig. 5), from which secondary electrons are guided by the electron mirror to the designated pixels of the "start"-MCP. If the electron mirror performs a perfect one-to-one mapping from the foil to the pixels, then the angular resolutions of this approach coincide with the respective FOVs. *The scattering of HENAs in the foil will not affect the angular resolution*, because the shift from the entry point to the exit point on the thin foil is microscopic; scattering of ENA in the foil will, however, seriously affect the efficiency of detection as some of the scattered HENAs may not fall on D1.

The "two-point" scheme determines the arrival direction of each HENA by locating the exit point on the foil (point A in Fig. 5) and the landing on the PH detector (point B' in Fig. 5), then taking the line connecting the two points for arrival direction. As in "one-point" schemes, the precision in locating the exit point on the foil depends on the effectiveness of mapping the foil onto the position-sensing "start"-MCP by the electron mirror. *The location of landing, however, is affected by the scattering in the foil.* Recent laboratory measurements by Ho *et al.*, (1997) showed that the HWHM (half-width at half maximum) scattering angle in the Si/Lexan/C foil depends on M and E of the particles; for 10 keV H$^+$, 25 keV He$^+$, and 110 keV O$^+$ the HWHM scattering angle in the Si/Lexan/C foil used in HSTOF on SOHO is 10°. The effect of scattering on imaging is especially severe

TABLE 1   Characteristis of Selected HENA Instruments[1]

Instrument Name	PIPPI-SSD	ISENA	INCA	HSTOF
Redundancy	Yes	No	No	No
Physical dimension (cm³)	3 x 14 x 14	11 x 16 x 24	30 x 35 x 49	18 x 18 x 39
Detectors	14 (1.5 x .5 cm²) SSD	2 pairs 1.8 cm² MCP 1 (1.5 x 0.5 cm²) SSD Above set per module	1pair 2 x 10 cm² MCP 1pair 5 x 10 cm² MCP 1pair 10 x 10 cm² MCP	2 pairs 9 x 7 cm² MCP 10.2 x 3.4 cm² PSSD (64 pixels)
Front foil	No	Si/Lexan/C	Si/Lexan/C	Si/Lexan/C
TOF path length (cm)	0	3.4 to 3.9	5 to 10	17.5 to 18.3
Geometrical factor, G (cm² sr)	0.035	8 x 0.02	2.5	0.22
Field of view, Ω	5° x 322°	60° x 3° & 3° x 60°	120° x 90°	34° x 4°
G / Ω (cm²)	0.07	1.5	0.76	4.8
Angular resolution	25° x 5°	15° x 3° (Highly idealized)	12° x 12° O@100 keV² 6° x 6° H@ 20 keV²	34° x 4° (FOV)
HENA energy range (keV)	13 - 140	20 - 150	20 - 500	50 - 80
Energy resolution	8 steps	8 steps	**	10%
Mass range (amu)	NA	1 - 16	1 - 16	1 - 16
Mass resolution	No	H, He, O	**	H, He, O
Detection efficiency (%)	**	**	**	**

[1]The Authors apologize for not inlcuding HEP on GEOTAIL, MIS-HI on CCRES, and SEPS on POLAR due to lack of preparation.

[2]Based on measurement on Si/Lexan/C foil (Ho et al., 1997).

** See Barabash et al. (1997), Orsini et al. (1997), Mitchell et al. (1997), and Hovestadt et al. (1996), respectively.

when the "start" position is only one-dimensional and the distance between the foil and the imaging PH detector is large, e. g., in INCA. The combined uncertainties in locating the two points significantly distort the arrival direction and hence blur the image. Furthermore, the two-point scheme does demand more electronics and data handling.

6.3.2 *Front foil.* An optimal front foil, one that effectively reduces the EUV flux, minimally attenuates and scatters ENAs, and efficiently generates secondary electrons from penetrating ENAs to trigger the "start" detector, is essential for all HENA imagers that use TOF analysis. The search for such a foil began in 1979 (Hsieh *et al.*, 1980) and continues still, e. g., Keath *et al.* (1989); Hsieh *et al.*, (1991); Sandel *et al.* (1993); Ho *et al.* (1997). The ingenious EUV filters with submicron structured channels (Gruntman, 1991; 1995) are ideal for uni-directional ENA fluxes, but could be restrictive for multiple arrival directions. The front foil protects all the detectors of an HENA imager that are either sensitive to EUV photons or EUV photo-electrons. Using an EUV-immune imaging PH detector that looks out of the aperture, e. g. SSD in HSTOF and ISENA, enhances the possibility of finding a foil that attenuates EUV effectively and with less ENA scattering. The search for a replacement for the Si in the Si/Lexan/C layered foil is presently underway for the HENA imager on IMAGE.

6.3.3 *Electron mirrors.* All TOF analyzers need electron mirrors to guide the secondary electrons from the foil to the "start"-MCP and from the front surface of the PH detector to the "stop"-MCP. The arrangements in HSTOF and ISENA are similar, both simple and isochronous (Wilken and Stüdemann, 1984), but the grids that provide the field configuration are obstacles to the HENAs on their way to the PH detector and the electrons from their sources to their respective MCPs. The obstruction produced by the grids is multiplicative in the number of grids, which significantly reduces the efficiency of a HENA imager. INCA has an improved design that replaces the grids with a few fine wires; but the location and biasing of the wires are critical and the spread in electron's flight time affects the TOF resolution (Mitchell *et al.*, 1996). Minimizing the number of grids and removing altogether the burden of position-mapping from the electron mirrors would be one way to simplify the design.

Sensitivity and Calibration: The combined workings of the collimator, foils, electron mirrors, and detectors, that make up the spectrometer determine the overall detection efficiency. Multiplying it by the geometrical factor gives the overall sensitivity. This overall sensitivity determines how particle counts are finally converted to the flux, $j_i(s, E; t)$, for analysis. This critical parameter of an ENA imager can only be gotten through well conducted calibration of imager's ion-rejection efficiency as a function of (M, E) and (φ, θ). This demanding task is not always done adequately due to last minute rushes. Computer simulation is not a sufficient substitute.

CONCLUSION

We have reviewed the general concept of HENA imaging, designs of HENA imagers that do not require the ionization of ENAs for detection, and cited some specific examples. We have identified some of the crucial areas of concern in the design of HENA imagers. The accelerated activities in HENA imaging space plasma in recent years have been encouraging, and we hope that 1) brighter minds will bring forth more innovative techniques to provide even better approaches to visualize space plasma in different regions under varying conditions; and 2) stronger support will offer more instrument development and flight opportunities to apply the best techniques to better understand the global dynamics of our plasma environment that *in situ* measurements alone cannot provide.

Acknowledgments. We are grateful to have been given the opportunity by the organizers of this Chapman Conference to review the topic of HENA imaging. We especially thank M. A. Gruntman, M. Hilchenbach, B. Klecker, T. E. Moore, E. C. Roelof, and H. D. Voss for technical information and helpful discussions.

REFERENCES

Armstrong, T. P., C. L. Johnson, "Magnetosphere Imager Science Definition Team Interim Report" *NASA Reference Publication 1378* (Marshall Space Flight Center, NASA, 1995).

Barabash, S., R. Lundin, O. Norberg, in *Measurement Techniques for Space Plasma* J. E. Borovsky, R. F. Pfaff, D. T. Young, Eds. (AGU, Washington, DC, 1997).

Baranov, V. B., Y. G. Malama, Axisymmetric self-consistent model of the solar wind interaction with the LISM: basic results and possible ways of development, *Space Science Review* 78, 305-316 (1996).

Bernstein, W., G. T. Inouye, N. L. Sanders, R. L. Wax, Measurements of precipitated 1-20 keV protons and electrons during a breakup aurora, *Journal of Geophysical Research* 74, 3601 (1969).

Bernstein, W., R. L. Wax, N. L. Sanders, G. T. Inouye, in *Small rocket instrumentation techniques* . (North-Holland, Amsterdam, 1969) pp. 224-231.

Bertaux, J. L., J. E. Blamont, Evidence for a source of an extraterrestrial hydrogen Lyman Alpha emission: the interstellar wind, *Astronomy and Astrophysics* 11, 200 (1971).

Blum, P. W., H. J. Fahr, Interaction between interstellar hydrogen and the solar wind, *Astronomy and Astrophysics* 4, 280 (1970).

Carlson, C. W., J. P. McFadden, in *Measurement Techniques for Space Plasma* J. E. Borovsky, R. F. Pfaff, D. T. Young, Eds. (AGU, Washington, DC, 1997).

Chase, C. J., E. C. Roelof, Extracting evolving structures from global magnetospheric images via model fitting and video visualization, *Johns Hopkins APL technical Digest* 16, 111-122 (1995).

Curtis, C. C., K. C. Hsieh, in *Solar System Plasma Physics* J. H. Waite, Jr., J. L. Burch, R. L. Moore, Eds. (AGU, Washington DC, 1989) pp. 247-251.

Dessler, A. J., W. W. Hanson, E. N. Parker, Formation of the geomagnetic storm main-phase ring current, *Journal of Geophysical Research* 66, 3631 (1961).

Dessler, A. J., E. N. Parker, Hydromagnetic theory of geomagnetic storm, *Journal of Geophysical Research* 64, 2239 (1959).

Doke, T., et al., The energetic particle spectrometer HEP onboard the GEOTAIL spacecraft, *Journal of Geomagnetism & Geoelectricity* 46, 713 (1664).

Équemerais, E., R. B. Sandel, R. Lallement, J. L. Bertaux, A new source of Lyman a emission detected by the Voyager UVS: heliospheric or galactic origin, *Astronomy and Astrophysics* 299, 249-257 (1995).

Fan, C. Y., Time variation of the intensity of auroral hydrogen emission and the magnetic disturbance, *Astrophysical Journal* 128, 420 (1958).

Fan, C. Y., D. H. Schulte, Variations in the auroral spectrum, *Astrophysical Journal* 120, 563 (1954).

Gloeckler, G., Ion composition measurements techniques for space plasma, *Review of Scientific Instruments* 61, 3613 (1990).

Gloeckler, G., K. C. Hsieh, Time-of-flight techniques for particle identification at energies from 2 to 400 keV/nucleon, *Nuclear Instruments and Methods* 165, 537 (1979).

Gruntman, M. A., Submicron structures: promising filters in EUV - a review, O. H. W. Siegmund, R. E. Rathschield, Eds., EUV, X-Ray, and Gamma-Ray Instrumentation for Astronomy, San Diego, CA (American Optical Society, 1991).

Gruntman, M. A., Extreme-ultraviolet radiation filtering by freestanding transmission gratings, *Applied Optics* 34, 5732-5737 (1995).

Gruntman, M. A., V. A. Morozov, H atom detection and energy analysis by use of thin foils and TOF technique, *Journal of Physics E: Scientific Instruments* 15, 1356-1358 (1982).

Hilchenbach, M., et al., Detection of Energetic Neutral Hydrogen Atoms of Heliospheric Origin, *EOS, Transaction AGU* 77, F557 (1996).

Ho, G. C., et al., Scattering of H+, He+ and O+ below 120 keV in a Si/Lexan/C foil, *Nuclear Instrument and Methods* (1997).

Hovestadt, D., B. Häusler, M. Scholer, Observation of energetic particles at very low altitudes near the geomagnetic equator, *Physical Review Letters* 28, 1340 (1972).

Hovestadt, D., et al., CELIAS - charge, element and isotope analysis system for SOHO, *Solar Physics* 162, 441-481 (1995).

Hsieh, K. C., C. C. Curtis, in *Solar System Plasma Physics* J. H. Waite, Jr., J. L. Burch, R. L. Moore, Eds. (AGU, Washington DC, 1989) pp. 159-164.

Hsieh, K. C., et al., "HELENA: a proposal to perform an in situ investigation of neutrals in geospace and the heliosphere on PPL & IPL of the OPEN mission" (University of Arizona, 1980).

Hsieh, K. C., E. Keppler, G. Schmidtke, Extreme ultra-violet induced forward photoemission from the carbon foils, *Journal of Applied Physics* 4, 2242-2246 (1980).

Hsieh, K. C., B. R. Sandel, V. A. Drake, R. S. King, H Lyman a transmittance of thin C and Si/C foils for keV particle detectors, *Nuclear Instruments and Methods in Physics Research* B61, 187-193 (1991).

Hsieh, K. C., K. L. Shih, J. R. Jokipii, S. Grzedzielski, Probing the heliosphere with energetic hydrogen atoms, *Astrophysical Journal* 393, 756-763 (1992).

Hsieh, K. C., C. C. Curtis, C. Y. Fan, M. A. Gruntman, Techniques for the remote sensing of space plasma in the heliosphere via energetic neutral atoms: a review, in *Solar Wind Seven* E. Marsch, R. Schwenn, Eds. (Pergamon, Oxford, 1992) pp. 357.

Hsieh, K. C., K. L. Shih, J. R. Jokipii, M. A. Gruntman, Sensing the solar-wind termination shock from Earth's orbit, in *Solar Wind Seven* E. Marsch, R. Schwenn, Eds. (Pergamon, Oxford, 1992) pp. 365.

Hsieh, K. C., M. A. Gruntman, Viewing the outer heliosphere

in energetic neutral atoms, Advances in Space Research **13**(6) 131-139 (1993).

Keath, E. P., *et al.*, in *Solar System Plasma Physics* J. H. Waite, Jr., J. L. Burch, R. L. Moore, Eds. (AGU, Washington DC, 1989) pp. 165.

Kirsch, E., S. M. Krimigis, W.-H. Ip, G. Gloeckler, X-ray and energetic neutral particle emission from Saturn's magnetosphere, *Nature* 292, 718 (1981a).

Kirsch, E., S. M. Krimigis, J. W. Kohl, E. P. Keath, Upper limits for x-ray and energetic neutral particle emission from Jupiter: Voyager 1 results, *Physical Review Letters* 8, 169 (1981b).

Lallement, R., J. L. Bertaux, B. R. Sandel, E. Chassefière, in *Solar Wind Seven* E. Marsch, R. Schwenn, Eds. (Pergamon, Oxford, 1992) pp. 209-212.

Lallement, R., J. L. Linsky, J. Lequeux, V. B. Baranov, Physical and chemical characteristics of the ISM inside and outside the heliosphere, *Space Science Review* **78**, 200-304 (1996).

McComas, D. J., B. L. Barraclough, R. C. Elphic, I. H. O. Funsten, M. F. Thomsen, Magnetospheric imaging with low energy neutral atoms, *Proceedings of the National Academy of Sciences, U. S. A.* (1991).

McComas, D. J., H. O. Funsten, in *Measurement Techniques for Space Plasma* J. E. Borovsky, R. F. Pfaff, D. T. Young, Eds. (AGU, Washington, DC, 1997).

Meinel, A. B., Doppler-shifted auroral hydrogen emission, *Astrophysical Journal* 113, 50 (1951).

Mitchell, D. G., *et al.*, INCA: the ion neutral camera for energetic neutral atom imaging of the Saturnian magnetosphere, *Optical Engineering* 32, 3096-3101 (1993).

Mitchell, D. G., *et al.*, in *Measurement Techniques for Space Plasma* J. E. Borovsky, R. F. Pfaff, D. T. Young, Eds. (AGU, Washington, DC, 1997).

Mizera, P. F., J. B. Black, Observation of ring current protons at low altitudes, *Journal of Geophysical Research* 78, 1058-1062 (1973).

Moore, T. E., M.-C. Fok, J. D. Perez, J. P. Keady, in *Cross-Scale Coupling in Space Plasma* J. L. Horwitz, N. Singh, Eds. (AGU, Washington, DC, 1995), vol. 93, pp. 37.

Moritz, J., Energetic protons at low equatorial altitudes: a newly discovered radiation belt phenomenon and its explanation, *Zeitschrift der Geophysik* 38, 701 (1972).

Orsini, S., *et al.*, in *Measurement Techniques for Space Plasma* J. E. Borovsky, R. F. Pfaff, D. T. Young, Eds. (AGU, Washington, DC, 1997).

Rairden, R. L., L. A. Frank, J. D. Craven, Geocoronal imaging with Dynamic Explorer, *Journal of Geophysical Research* 91, 13613 (1986).

Roelof, E. C., Energetic neutral atom image of a storm-time ring current, *Geophysical Research Letters* 14, 652 (1987).

Roelof, E. C., Imaging heliospheric shocks using energetic neutral atoms, in *Solar Wind Seven* E. Marsch, R. Schwenn, Eds. (Pergamon, Oxford, 1992).

Roelof, E. C., D. G. Mitchell, D. J. Williams, Energetic neutral atoms (E ~ 50 keV) from the ring current: IMP 7/8 and ISEE-1, *Journal of Geophysical Research* 90, 10991-11008 (1985).

Rosenbauer, H., *et al.*, "The ISPM interstellar neutral-gas experiment" *ESA SP-1050* (European Space Agency, 1984).

Sandel, B. R., V. A. Drake, A. L. Broadfoot, K. C. Hsieh, C. C. Curtis, Imaging extreme ultraviolet photons and energetic neutral atoms: a common approach, *Remote Sensing Review* 8, 147-188 (1993).

Siegmund, O. H. W., M. Lampton, J. Bixler, S. Chakrabarti, J. Vallerga, S. Bowyer, and R. F. Manila, Wedge and strip image readout systems for photon-counting detectors in space astronomy, Journal of the Optical Society of America A, 3, 2139-2145 (1986).

Shih, K.-L., Ph. D., University of Arizona (1993).

Sullivan, J. D., Geometrical factor and directional response of single and multi-element particle telescopes, *Nuclear Instrument and Methods* 95, 5-11 (1971).

Thomas, G. E., R. F. Krassa, OGO 5 measurements of the Lyman Alpha sky background, *Astronomy and Astrophysics* 11, 218 (1971).

Tinsley, B. A:, Neutral atom precipitation - a review, *Journal of Atmospheric and Terrestrial Physics* 43, 617-632 (1981).

Voss, H. D., J. R. Kilner, R. A. Baraze, J. Mobilia, in *Measurement Techniques for Space Plasma* J. E. Borovsky, R. F. Pfaff, D. T. Young, Eds. (AGU, Washington, DC, 1997).

Voss, H. D., J. Mobilia, H. L. Collin, W. L. Imhof, Satellite observations and instrumentation for measuring energetic neutral atoms, *Optical Engineering* 32, 3083-3089 (1993).

Wilken, B., *et al.*, in *Cluster: mission, payload and supporting activities* W. R. Burke, Ed. (European Space Agency, Paris, 1993), vol. ESA SP-1159, pp. 185.

Wilken, B., *et al.*, Energetic neutral atoms in the outer magnetosphere: An upper flux limit obtained with the HEP-LD spectrometer on board GEOTAIL, *Geophysical research Letters* 24, 111-114 (1997).

Wilken, B., W. Stüdemann, A compact time-of-flight mass spectrometer with electrostatic mirrors, *Nuclear Instruments and Methods* 222, 587-600 (1984).

Williams, D. J., in *Magnetospheric Physics* B. Hultqvist, C. G. Fälthammer, Eds. (Plenum, New York, 1990) pp. 83.

Witte, M., H. Rosenbauer, M. Banaszkiewicz, H. Fahr, The Ulysses neutral gas experiment: determination of the velocity and temperature of the interstellar helium, *Advances in Space Research* 13, 121-130 (1993).

Neutral Atom Imaging: UV Rejection Techniques

H. O. Funsten and D. J. McComas

Space and Atmospheric Sciences Group, Los Alamos National Laboratory, Los Alamos, New Mexico

M. A. Gruntman

Department of Aerospace Engineering, University of Southern California, Los Angeles, California

Neutral atom imaging promises to provide a detailed picture of the global structure and dynamics of the magnetosphere. The neutral atom fluxes are characteristically small, and techniques to image neutral atoms must be highly sensitive to neutral atoms while rejecting the enormous background UV flux. Several imaging techniques that exploit a unique difference in properties of neutral atoms and UV are being developed. Limitations of the fundamental physics of each of these techniques, which are studied and compared here, have resulted in neutral atom imagers optimized for different energy regimes. The emerging technique utilizing a transmission grating promises to span a broad energy range with one instrument.

1. INTRODUCTION: THE UV DILEMMA

An emerging technique for global magnetospheric imaging is remote detection of magnetospheric plasma ions, predominantly H^+, He^+ and O^+, that are neutralized by charge exchange with geocoronal neutral species [e.g., *Williams et al.*, 1992]. By imaging these neutral atoms from a remote location, a source plasma can be imaged and its global properties can be derived.

The primary technical challenge of imaging neutral atoms in space is accurately measuring their velocity and mass while separating them from the tremendous UV background that enters the instrument. UV photoelectrons, which can trigger microchannel plate (MCP) detectors and are emitted from foils used for coincidence and time-of-flight measurements [*Hsieh et al., 1980*], can introduce an exceedingly large noise background in the neutral atom measurement.

The dominant UV flux is H Ly-α (1216 Å), for which the solar flux is ~4×10^{11} photons cm^{-2} s^{-1} [*Rottman et al.*, 1982], and the terrestrial dayglow flux is >25 kR (2.5×10^{10} photons cm^{-2} s^{-1}) [*Chakrabarti et al.*, 1983]. While the EUV flux may be much less than this, the MCP detection efficiency of EUV photons is ~10% at 300-600 Å [*Fraser*, 1982] compared to 1-2% at 1216 Å [*Fraser*, 1982; *Taylor et al*, 1983], so EUV may contribute significantly to noise in a neutral atom imager. Due to the low neutral atom fluxes, either direct UV attenuation or separation of the neutral atoms from the UV is crucial for neutral atom detection.

The techniques utilized to remove the UV background from the neutral atom measurement has resulted in a general partitioning of neutral atoms into the following approximate energy range classifications: energetic neutral atoms (ENAs) with $E \geq 30$ keV, low energy neutral atoms (LENAs) having 1 keV $\leq E \leq 30$ keV, and very low energy neutral atoms (VLENAs) with $E < 1$ keV. In comparing imaging techniques, we attempt to elucidate and assess the key mechanisms that define their utility by asking two questions: (1) what is the physics behind the technique used to distinguish neutral atoms from the UV background, and (2) what are the limitations on imaging performance that result from this technique?

Measurement Techniques in Space Plasmas: Fields
Geophysical Monograph 103

We note two common features of neutral atom imagers that significantly impact their design. First, neutral atom fluxes are anticipated to be relatively small, so imagers typically require both a large geometric factor and a coincidence scheme to minimize the noise. Second, the collimator of a neutral atom imager on a spinning spacecraft can maintain high azimuthal resolution independent of the imaging technique, so image degradation induced by the imaging technique is typically limited to the polar direction.

2. ENA IMAGING: UV BLOCKING FOILS

ENA imagers employ a thick ($\sim$10-15 μg cm^{-2}) foil that attenuates a large fraction of the UV but allows neutral atoms with sufficient energy to pass [e.g., *Mitchell et al.,1993, 1997*]. Ignoring multiple reflections in the foil, the UV transmittance T decays exponentially with increasing foil thickness t according to $T = A \exp(-\mu t)$ where μ is the mass absorption coefficient and t is the areal mass density (e.g., units of μg cm^{-2}). The constant A, which depends on the index of refraction [*Hsieh et al.*, 1991], is independent of foil thickness and does not greatly vary between different foil materials. Experiments by Hsieh *et al.* [1991] and Drake *et al.* [1992] using C foils and Si/C and Al/C composite foils indicate that $\mu \approx 0.6$ cm$^2\mu g^{-1}$ for C, $\mu \approx 0.78$ cm$^2\mu g^{-1}$ for Si, and $\mu \approx 0.37$ cm$^2\mu g^{-1}$ for Al (observed in the foil as Al$_2$O$_3$) for incident Ly-α. Although variations in the microstructure of the foil material resulted in wide variations of μ, Si appears to have the lowest Ly-α transmission per unit areal mass density.

While a thicker foil exponentially reduces the UV flux in the imager, it increases both angular scattering of ENAs, which degrades the imaging resolution, and ENA energy loss, which degrades the speed or energy resolution and can stop ENAs in the foil. For H projectiles at tens of keV, the magnitude of angular scattering, which results primarily from Coulomb interactions with target nuclei and scales as E^{-1}, cannot be inferred from energy loss in the foil, which primarily depends on interactions with target electrons and scales as $E^{1/2}$ for velocities less than the Bohr velocity ($\sim$25 keV/amu). Of these two effects, angular scattering is more important since (a) image resolution should be more important than energy or speed resolution, and (b) energy loss effects should only be observed at low energies in which angular scattering is quite large. We briefly examine energy loss and study angular scattering in detail.

2.1 Energy Loss of ENAs in the Foil

Figure 1 shows the stopping power (energy loss per unit areal mass density) of H projectiles in C, Si, Lexan and Al$_2$O$_3$ as a function of projectile energy based on results from the TRIM computer code [*Ziegler et al.*, 1985]. The results indicate that the energy loss of H is largest in C and least in Al$_2$O$_3$ and Lexan. The results for He and O in these foil materials is qualitatively the same. Therefore, Si would

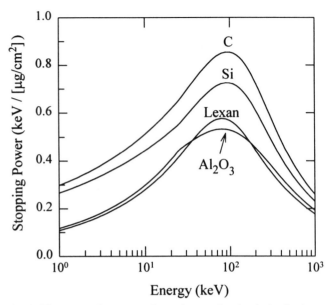

Fig. 1. The energy loss per unit areal mass density derived using the TRIM computer code [*Ziegler et al.*, 1985] is shown for H projectiles in C, Si, and Lexan. For foils of equal areal mass density, the energy loss of ENAs is largest for C foils.

be a preferential foil material relative to C based on energy loss measurements alone. When comparing different foil materials in thickness units of areal mass density, Fig. 1 shows that projectile energy loss does not scale with the foil atomic number Z_2 [*Hsieh et al.*, 1991; *Drake et al.*, 1992] or with the areal mass density [*Mitchell et al.*, 1993].

2.2 Angular Scattering of ENAs in the Foil

The expression for the scattering halfwidth (HWHM) $\psi_{1/2}$ $\propto Z_1 Z_2 / (Z_1^{2/3} + Z_2^{2/3})^{1/2}$ from *Högberg et al.* [1970] (Z is atomic number and the subscripts 1 and 2 refer to the projectile and target, respectively) has been used to justify a larger $\psi_{1/2}$ with increasing Z_2 [*Keath et al.*, 1989]. However, this expression was originally utilized by *Högberg et al.* [1970] to show the variation of $\psi_{1/2}$ with Z_1 assuming a linear dependence of $\psi_{1/2}$ on the foil thickness t. If we do assume this linear dependence and use foil thickness units of μg cm^{-2}, then the full expression from the base equations of *Meyer* [1971] is actually $\psi_{1/2} \propto Z_1 Z_2 / m_2 (Z_1^{2/3} + Z_2^{2/3})^{1/2}$ where m_2 is the atomic mass of the foil material. Since $m_2 \approx 2Z_2$, this dependence reduces to $\psi_{1/2} \propto Z_1 / (Z_1^{2/3} + Z_2^{2/3})^{1/2}$, which infers that $\psi_{1/2}$ actually *decreases* with increasing Z_2 for foils having the same areal mass density. Furthermore, the assumption that $\psi_{1/2}$ is linearly dependent on t is only valid for a "reduced" foil thickness less than $\sim$3 [*Högberg et al.*, 1970]. For a reduced thickness greater than 3, the theoretical model of *Meyer* [1971] infers multiple random collisions so that $\psi_{1/2} \propto t^{1/2}$, which is generally valid (and of particular interest) for foils used in ENA imagers.

Based on *Meyer* [1971],we quantify angular scattering of neutral atoms with energy E that transit a foil using

$$E\psi_{1/2} = k_F (X) \qquad (1)$$

where $k_F(X)$ is a foil constant that depends on the particular combination of foil and neutral atom species X [*Funsten et al.*, 1993]. As the angle of incidence relative to the foil surface normal, the foil thickness that the ENA traverses, and therefore the angular scattering, increase. The lines in Fig. 2 show k_F derived from the theory of *Meyer* [1971] for H, He and O projectiles at normal incidence as a function of the foil thickness t for C and Si foils. For $t > 6$ μg cm^{-2}, for which $\psi_{1/2}$ is nonlinearly dependent on t, k_F is slightly larger for Si foils than for C foils. Also shown are values of k_F derived using the TRIM code [*Ziegler et al.*, 1985] for 50 keV H transiting C and Si foils. The TRIM results indicate that k_F is always slightly larger for Si foils than for C foils. For a foil thickness of ~10 μg cm^{-2}, the results using *Meyer* [1971] and TRIM both indicate that angular scattering is slightly less in a C foil.

For example, TRIM calculations yield $k_F(H) \approx 60$ keV-deg and $k_F(O) \approx 510$ keV-deg for a composite Si/Lexan/C (5.5/5/1 μg cm^{-2}) foil that will be flown on the Ion Neutral Camera (INCA) on the Cassini mission [*Mitchell et al.*, 1997]. At 10 keV, this results in $\psi_{1/2} \approx 6^0$ for H and $\psi_{1/2} \approx 51^0$ for O. Empirical results from Mitchell *et al.* [1997] show $k_F(H) \approx 110$ keV-deg and $k_F(O) \approx 2100$ keV-deg, although these measurements also included effects due to the entrance slit width.

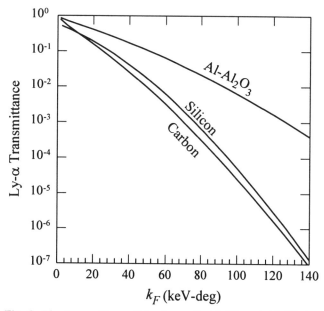

Fig. 3. The transmittance T is shown for C, Si, and Al (likely Al$_2$O$_3$) foils as a function of the foil constant k_F. Based on these results, Si is a superior foil for ENA imaging due to its lower value of k_F for a particular T.

Since angular scattering is more important than energy loss, foil optimization is a trade-off between maximizing UV attenuation and minimizing angular scattering. Figure 3 shows the variation of Ly-α transmission with the foil constant k_F for S, Si, and Al-Al$_2$O$_3$ foils, where μ and A were derived from the data of *Drake et al.* [1992]. The coefficient A is 0.6 for C, 0.96 for SiO (based on an index of refraction of 0.68), and 0.94 for Al (based on an index of refraction of 0.61). The results show that Si is slightly better than C and far superior to Al due to its smaller value of k_F.

We note that the mass absorption coefficient is strongly a function of UV wavelength [*Powell et al.*, 1990], and the optimal foil must meet the requirements of efficient EUV suppression. Some materials are poor attenuators in the spectral range, for example, carbon at wavelengths < 400 Å. Also, composite foils may include one layer, e.g., Lexan, for structural support [*Mitchell et al.*, 1997].

3. LENA IMAGING: CHARGE MODIFICATION BY TRANSIT THROUGH AN ULTRATHIN FOIL

The first attempt to detect LENAs in space utilized a carbon foil to convert the neutral atoms to positive ions and an electrostatic analyzer to sweep them away from the ambient UV into a detector [*Bernstein et al.*, 1969]. This technique has been further developed to include velocity, mass, and coincidence measurements [*McComas et al.*, 1991, 1997]. The key feature of a LENA detector is the ultrathin (e.g., ~1 μg cm^{-2}) charge conversion foil through which LENAs pass and can exit as an ion.

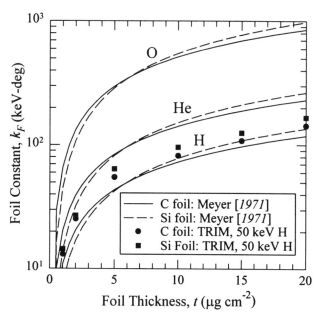

Fig. 2. The foil constant, $k_F = E\psi_{1/2}$, is shown as a function of the foil thickness for C and Si foils. The lines are based on the theory of *Meyer* [1971], and the points are computed from the TRIM computer code [*Ziegler et al.*, 1985].

As with foils used in ENA imagers, significant angular scattering of LENAs in the ultrathin conversion foil can result in degraded image resolution. The measured foil constants of nominal 0.5 µg cm^{-2} carbon foils are $k_F(H) \approx$ 12.6 keV-deg, $k_F(He) \approx$ 34 keV-deg, and $k_F(O) \approx$ 133 keV-deg [Funsten et al., 1993]. These are 5-10 times less than the ENA foil constant, enabling greatly enhanced imaging resolution for $E < 25$ keV.

A LENA imager sensitivity is proportional to the probability that a LENA exits the conversion foil as an ion. The positive ion fraction increases with increasing energy; for example, the H$^+$ fraction ranges from 6% at 1 keV to 42% at 30 keV [Funsten et al., 1993]. Alternately, our measurements show that the negative ion fraction depends on the projectile affinity level and reaches a maximum at a speed corresponding to ~4 keV/amu. Detection of these negative ions enables enhanced measurement of oxygen LENAs, since the O$^-$ fraction, which is 18% at 30 keV, is substantially more abundant than O$^+$ [Funsten et al., 1993].

Energy loss in the foil is negligible, since it is usually smaller than the energy passband of the electrostatic analyzer [Funsten et al., 1995]. Therefore, the lower energy threshold for LENA imaging is ~0.8 keV because of a low ionization probability of LENAs exiting the foil and degraded polar resolution from large angular scattering. For $E > 30$ keV, the positive ionization probability is high (>42%) and angular scattering is small ($\psi_{1/2} < 0.5^0$). Here, the challenge is construction of an electrostatic energy analyzer with a large enough gap to retain a large instrument geometric factor while successfully deflecting these higher energy LENAs.

4. VLENA IMAGING: CHARGE MODIFICATION BY SURFACE REFLECTION

A promising technique for VLENA detection involves conversion of VLENAs to negative ions by their reflection from a low work function surface and subsequent electrostatic deflection from the ambient UV into a detector section [e.g., Gruntman, 1993; Wurz et al., 1995, 1997]. In this technique, formation of a negative ion relies on tunneling of an electron in the valence band of the reflection surface to the neutral projectile affinity level. This process is most efficient when the work function of the reflection surface is small and when the projectile energy is several hundred eV [van Os et al., 1988; Los and Geerlings, 1990]. Interestingly, even though He has a closed shell configuration, metastable (1s2s2p ^{4}P^0) He$^-$ can be formed and was observed as 0.14% of 10 keV He reflected from a Na surface [Schneider et al., 1984].

An ideal conversion surface has a low work function, a low vapor pressure, and is non-reactive to maintain surface purity. Candidate materials tabulated in Gruntman [1993] are highly reactive, oxidize rapidly, and therefore require stringent quality control and, possibly, periodic surface

conditioning to maintain surface integrity. One optimal conversion surface appears to be Ba or BaO, which has a relatively low work function resulting in conversion efficiencies of up to 4% [van Os et al., 1988], a vapor pressure of ~10^{-17} torr, and is the least reactive of the listed materials.

Recent experiments performed for verification of the VLENA imaging technique using incident H$_2^+$ indicate that the negative ion yield might approach 14% for a Cs/W surface [Wurz et al., 1995] and 5.5% for polycrystalline diamond which may act as a negative electron affinity material [Wurz et al., 1997]. Interpretation of these results as detection of metastable H$_2^-$, whose existence is doubtful [Bae et al., 1984], requires experimental confirmation. The diamond surface, which exhibited a high conversion efficiency without surface conditioning, is extremely promising for imaging VLENAs.

An important parameter governing the geometric factor of this technique is the absolute conversion yield, which we define as the product $P_R P^-$ where P_R is the reflection probability and P^- is the probability that a reflected projectile is a negative ion. An irregular surface with surface defects, adsorbates, and other contamination can cause strong nonspecular reflection out of the imager optics or penetration of the projectile into the surface, where it is lost. These effects, which act to decrease P_R, are enhanced at energies greater than several hundred eV as the planar surface potential caused by the collective effect of surface atoms becomes localized around the surface atoms [Oen, 1983]. We know of no studies that have quantified P_R for the energy range and surfaces of interest.

One technical challenge with implementing this technique is maintenance of an atomically smooth, low work function conversion surface over a large area, which scales as $A\sec(\alpha)$, where A is the instrument aperture area and α is the angle between the planes of the aperture and foil.

5. IMAGING WITH FREESTANDING GOLD GRATINGS

An emerging technique for neutral atom imaging uses freestanding transmission gratings that utilize the grating's waveguide effect to efficiently damp the UV, while the open slits, which are aligned with the collimator plates, allow neutral atoms to pass [Scime et al., 1995; Gruntman, 1995; McComas et al., 1997]. In this technique, UV is blocked without disturbing the incident neutral atom velocity, resulting in high image resolution and imaging over a broad energy range.

Figure 4 shows experimental and theoretical results for a grating with a thickness $t = 500$ nm, period $p = 200$ nm, slit width $w = 62$ nm, and transmission $T_S = 0.44$ of the grating support structure. Based on the grating geometry, the neutral atom transmission at normal incidence is $T_S w/p = 14\%$. The symbols and dashed line show excellent

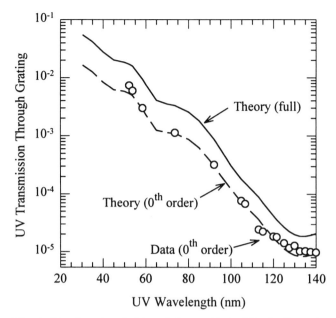

Fig. 4. The data (symbols) and theory (dashed line) of zeroth-order diffraction wavelength-dependent UV transmission through a grating show excellent agreement. The summation over all diffraction orders (solid line), which is obtained by theory, predicts actual grating performance.

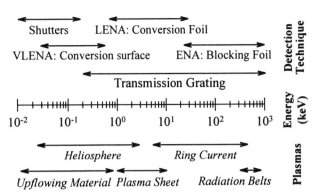

Fig. 5. The comparative utility of neutral atom imaging techniques based on sensitivity and imaging resolution is shown over the neutral atom energy range.

atom imaging using transmission grating promises to span a broad energy range.

Acknowledgments. The authors thank Don Mitchell for his assistance in the ENA foil specifications. At Los Alamos, this work was performed under the auspices of the United States Department of Energy.

agreement for zeroth-order diffraction. The solid line, which is a theoretical result with all diffraction orders included, is representative of the grating performance for unpolarized light [*Gruntman, 1997*] and follows a general exponential decrease of transmittance with increasing wavelength from 0.06 at 304 Å to 4×10^{-5} at 1216 Å. Note that while the EUV flux may be small compared to H Ly-α, both the EUV transmission and MCP detection efficiency are comparatively much larger and must be considered.

When used with a coincidence measurement [*McComas et al.*, 1997], the minimum energy of this technique depends on the neutral atom transmission probability through an ultrathin foil and secondary electron emission from the foil. This has been demonstrated at 0.6 keV [*Gruntman and Morozov, 1982*] and should be possible for neutral atom energies down to several hundred eV [*Funsten et al., 1994*].

Figure 5 compares the applicable energy ranges of the imaging techniques discussed above that are derived from quantitative arguments of the imaging resolution and detection sensitivity associated with each technique used to separate neutral atoms from the ubiquitous UV. Also shown is a shutter concept [*Funsten et al., 1995*] for VLENA detection. Overlap exists between the VLENA detection technique using a conversion surface, LENA detection using an ultrathin conversion foil, and ENA detection using a thick blocking foil. Together, these imaging techniques cover a wide spectrum of terrestrial and extraterrestrial neutral atom sources. The emerging technique of neutral

REFERENCES

Bae, Y.K., M.J. Coggliola, and J.R. Peterson, A search for H_2^-, H_3^- and other metastable ions, in *Production and Neutralization of Negative Ions and Beams*, American Institute of Physics, New York, 90-95, 1984.

Bernstein, W., R.L. Wax, N.L. Sanders, and G.T. Inouye, An energy spectrometer for energetic (1-25 keV) neutral hydrogen atoms, in *Small Rocket Instrumentation Techniques*, North Holland, Amsterdam, 224-231, 1969.

Chakrabarti, S., F. Paresce, S. Bowyer, R. Kimble, and S. Kumar, The extreme ultraviolet day airglow, *J. Geophys. Res.*, *88*, 4898-4904, 1983.

Drake, V.A., B.R. Sandel, D.G. Jenkins, and K.C. Hsieh, H Lyα transmission of thin foils of C, Si/C, and Al/C for keV particle detectors, *Proc. SPIE*, *1744*, 148-160, 1992.

Fraser, G.W., The soft x-ray quantum detection efficiency of microchannel plates, *Nucl. Instrum. Meth.*, *195*, 523-538, 1982.

Funsten, H.O., D.J. McComas, and B.L. Barraclough, Ultrathin foils used for low-energy neutral atom imaging of the terrestrial magnetosphere, *Opt. Eng.*, *32*, 3090-3095, 1993.

Funsten, H.O., D.J. McComas, and E.E. Scime, Comparative study of low energy neutral atom imaging techniques, *Opt. Eng.*, *33*, 349-356, 1994.

Funsten, H.O., D.J. McComas, and E.E. Scime, Low-energy neutral-atom imaging techniques for remote observations of the magnetosphere, *J. Spacecraft and Rockets*, *32*, 899-904, 1995.

Gruntman, M.A., and V.A. Morozov, H atom detection and energy analysis by use of thin foils and TOF technique, *J. Phys. E*, *15*, 1356-1358, 1982.

Gruntman, M.A., A new technique for *in situ* measurement of the composition of neutral gas in interplanetary space, *Plant. Space Sci.*, *41*, 307-319, 1993.

Gruntman, M.A., Extreme ultraviolet radiation filtering by freestanding transmission gratings, *Appl. Opt.*, *34*, 5732-5737, 1995.

Gruntman, M.A., Transmission grating filtering of 52-140 nm radiation, *Appl. Opt.*, *36*, in press, 1997.

Högberg, G., H. Norden, and H.G. Berry, Angular distribution of ions scattered in thin carbon foils, *Nucl. Instrum. Meth.*, *90*, 283-288, 1990.

Hsieh, K.C., E. Keppler, and G. Schmidtke, Extreme ultraviolet induced forward photoemission from thin carbon foils, *J. Appl. Phys.*, *5*, 2242-2246, 1980.

Hsieh, K.C., B.R. Sandel, V.A. Drake, and R.S. King, H Lyman α transmittance of thin C and Si/C foils for keV particle detectors, *Nucl. Instrum. Meth.*, *B61*, 187-193, 1991.

Keath, E.P., G.B. Andrews, A.F. Cheng, S.M. Krimigis, B.H. Mauk, D.G. Mitchell, and D.J. Williams, Instrumentation for energetic neutral atom imaging of magnetospheres, in *Solar System Plasma Physics, Geophysical Monograph Ser.*, vol. 54, eds. J.H. Waite, J.L. Burch, and R.L. Moore, pp. 165-170, AGU, Washington, D.C., 1989.

Los, J., and J.J.C. Geerlings, Charge exchange in atom-surface collisions, *Phys. Rep.*, *190*, 133-190, 1990.

McComas, D.J., B.L. Barraclough, R.C. Elphic, H.O. Funsten and M.F. Thomsen, Magnetospheric imaging with low-energy neutral atoms, *Proc. Natl. Acad. Sci. USA*, *88*, 9598-9602, 1991.

McComas, D.J., H.O. Funsten, and E.E. Scime, Advances in low energy neutral atom imaging, these proceedings, 1997.

Meyer, L., Plural and multiple scattering of low-energy heavy particles in solids, *Phys. Stat. Sol.*, *44*, 253-268, 1971.

Mitchell, D.G., S.M. Krimigis, A.F. Cheng, S.E. Jaskulek, E.P. Keath, B.H. Mauk, R.W. McEntire, E.C. Roelof, C.E. Schlemm, B.E. Tossman and D.J. Williams, The imaging neutral camera for the Cassini mission to Saturn and Titan, these proceedings, 1997.

Mitchell, D.G., A.F. Cheng, S.M. Krimigis, E.P. Keath, S.E. Jaskulek, B.H. Mauk, R.W. McEntire, E.C. Roelof, D.J. Williams, K.C. Hsieh, and V.A. Drake, INCA: The ion neutral camera for energetic neutral atom imaging of the Saturnian magnetosphere, *Opt. Eng.*, *32*, 3096-3101, 1993.

Oen, O.S., Universal shadow expressions for an atom in an ion beam, *Surf. Sci.*, *131*, L407-L411, 1983.

Powell, F.R., P.W. Vedder, J.F. Lindblom, and S.F. Powell, Thin film filter performance for extreme ultraviolet and x-ray applications, *Opt. Eng.*, *29*, 614-624, 1990.

Rottman, G.J., C.A. Barth, R.J. Thomas, G.H. Mount, G.M. Lawrence, D.W. Rusch, R.W. Sanders, G.E. Thomas, and J. London, Solar spectral irradiance, 120 to 190 nm, October 13, 1981-January 3, 1982, *Geophys. Res. Lett.*, *9*, 587-590, 1982.

Schneider, P.J., W. Eckstein, and H. Verbeek, Trajectory effects in the negative charge state fraction of ^{3}He and ^{4}He reflected from a solid target, *Nucl. Instrum. Meth.*, *B2*, 525-530, 1984.

Scime, E.E., E.H. Anderson, D.J. McComas, and M.L. Schattenburg, Extreme-ultraviolet polarization and filtering with gold transmission gratings, *Appl. Opt.*, *34*, 648-654, 1995.

Taylor, R.C., M.C. Hettrick, and R.F. Manila, Maximizing the quantum efficiency of microchannel plate detectors: The collection of photoelectrons from the interchannel web using an electric field, *Rev. Sci. Instrum.*, *54*, 171-176, 1983.

van Os, C.F.A., P.W. van Amersfoort, and J. Los, Negative ion formation at a barium surface exposed to an intense positive-hydrogen ion beam, *J. Appl. Phys.*, *64*, 3862-3973, 1988.

Williams, D.J., E.C. Roeloff, and D.G. Mitchell, Global magnetospheric imaging, *Rev. Geophys.*, *30*, 183-208, 1992.

Wurz, P., M.R. Aellig, P. Bochsler, A.G. Ghielmetti, E.G. Shelley, S.A. Fuselier, F. Herrero, M.F. Smith, and T.S. Stephen, Neutral atom imaging mass spectrograph, *Opt. Eng.*, *34*, 2365-2376, 1995.

Wurz, P., R. Schletti, and M.R. Aellig, Hydrogen and oxygen negative ion production by surface ionization using diamond surfaces, *Surf. Sci.*, *373*, 56-66, 1997.

Ziegler, J.F., J.P. Biersack, and U. Littemark, "Transport of Ions in Matter (TRIM) Computer Code," The Stopping an Range of Ions in Solids, pp. 202-263, Pergamon, New York, 1985.

Energetic Neutral Atom Imager on the Swedish Microsatellite Astrid

S. Barabash, O. Norberg, R. Lundin, S. Olsen, K. Lundin, P. C:son Brandt

Swedish Institute of Space Physics, Kiruna, Sweden

E. C. Roelof, C. J. Chase, B. H. Mauk

The Johns Hopkins University / Applied Physics Laboratory, Laurel, Maryland

H. Koskinen, J. Rynö

Finnish Meteorological Institute, Helsinki, Finland

The Swedish microsatellite ASTRID was launched by a Russian Cosmos rocket on January 24, 1995 into a 1000 km circular orbit with 83° inclination. Besides the main objective of technological demonstration, imaging of energetic neutral atoms (ENAs) was attempted. The imager detected ENA in the energy range 0.1 - 140 keV utilizing two different techniques. Neutrals of the energy 13 - 140 keV were recorded by 14 solid state detectors with the total field of view $5° \times 322°$. For half a spin (~1.5 s) of the ASTRID spacecraft, almost all of space was covered with an angular resolution $2.5° \times 25°$. Less energetic neutrals of ~0.1 - 70 keV were converted on a graphite target into secondary particles which then were detected by a microchannel plate with 32 anodes. A fraction of primary neutrals was directly reflected towards the sensor. This technique provided the total ENA flux with an angular resolution $4.6° \times 11.5°$. The instrument weight is 3.13 kg. Successful operation of the instrument during the first 5 weeks of the mission provided the first ENA images of the ring current at low altitudes.

1. INTRODUCTION

Any energetic plasma immersed in a neutral gas background emits energetic neutral atoms (ENAs) generated by the charge - exchange process. ENAs are not affected by electromagnetic fields and propagate essentially rectilinear like photons, and a direction-responsive neutral particle detector can image the emitting region. The hot plasma of the ring current interacting with the hydrogen exosphere is the main source of ENA in the Earth's magnetosphere. High energy (> 10 keV) neutral atoms

(HENAs) emitted from the ring current have been detected by charged particle detectors at several occasions [*Roelof et al.*, 1985; *Voss et al.*, 1993; *Lui et al.*, 1997; *Wilken et al.*, 1997], and yet never by a specifically designed ENA instrument.

The basic problems in ENA measurements are relatively low ENA flux (< 10^3 cm^{-2}sr^{-1}s^{-1}keV^{-1}, for 10 - 30 keV) and an ever-present flux of UV photons (>10^8 cm^{-2}sr^{-1}s^{-1}), which make detection of low energy (< 20 keV) neutral atoms (LENAs) quite challenging. Consequently, except for the aforementioned energetic charged particle detectors, very few experiments have been so far performed to measure LENA in space. We mention the most recent one. *Witte et al.* [1992] utilized a secondary ion conversion technique with a LiF target to successfully perform imaging of interstellar LENAs onboard the Ulysses spacecraft. Note that the instrument was even able to detect ENAs originated in the Jovian magnetosphere [*Witte at al.*, 1993].

Measurement Techniques in Space Plasmas: Fields
Geophysical Monograph 103

More comprehensive reviews of ENA instrumentation can be found in *McEntire and Mitchell* [1989], *Hsieh et al.* [1991], *Funsten et al.* [1994], and *Barabash,* [1995]. The present report describes the basic design, some calibrations and tests, and initial results of the ENA imager PIPPI (Prelude In Planetary Particle Imaging) which have been flown on the Swedish microsatellite Astrid.

The Astrid microsatellite [*Grahn and Rathsman,* 1995] was launched into a 1000 km polar orbit on January 24, 1995 from the Russian cosmodrom Plesetsk as a piggyback passenger on a Kosmos-3M rocket. Astrid is 10 times smaller in scale than the Swedish satellite Freja and weights 27 kg. In the launch configuration the dimensions of the satellite are approximately 0.45 × 0.45 × 0.29 m. The satellite is spin-stabilized with the spin axis pointing towards the Sun. A typical spin period is 3 - 4 s.

Astrid carries three scientific instruments with the total mass of 4.08 kg. The main instrument was an ENA imager PIPPI. Besides PIPPI, there are two supporting instruments onboard. These are an electron spectrometer and UV photometers (for more details, see *Norberg et al.,* 1995).

2. THE ENA IMAGER PIPPI

PIPPI is almost identical to the Neutral Particle Imager (NPI), which is a part of the ASPERA-C experiment on the Russian Mars - 96 spacecraft and consists of two sensor heads, PIPPI-SSD, PIPPI-MCP, and an electronics unit. The former sensor head is dedicated to measuring HENAs in the energy range 13 - 140 keV using SSDs. The latter measures ~ 0.1 - 70 keV LENAs by means of the reflection and conversion of primary neutrals into secondary particles (electrons and ions) followed by detection with a MCP. Mechanically the instrument is one unit. The cross section of the instrument is given in Figure 1.

In the PIPPI-SSD charged particles are removed by a two deck electrostatic deflection system consisting of 4 disks (Figure 1). Neutrals are detected by 14 Si solid state detectors located on two decks in order to increase the

geometrical factor of each sensor. Eight plastic spokes between each pair of the disks divide the 2π field of view into 8 collimators with an aperture of $5° \times 30°$ each. The upper and lower levels of the sensors are turned with respect to each other by 22.5° to provide an angular resolution of $\approx 25°$. For one 180° turn around an axis in the plane of the instrument, in practice for half a satellite spin period, almost the entire 4π space is covered by all sensors. Two sensors pointing towards the Sun are obscured to avoid direct solar light from reaching them. One of these detectors is completely blocked and is used to estimate the electronics noise and penetrating radiation. The other is shielded by two layers of the 0.6 μm aluminium foil to detect only particles with energies above 140 keV. It has a separate counter and monitors the energetic particles which can pass the deflector system. The 16 detectors are sampled simultaneously. Each pulse from the 15 SSD detectors (except the background sensor with the foil) is discriminated in 8 levels to give the energy spectrum. The direction is given by the sensor number. The direction - energy matrix of 15 × 8 elements are accumulated during 31.25 ms. When the high voltage for the electrostatic deflector is off, the PIPPI-SSD measures ions and electrons as an ordinary energetic particle detector.

The PIPPI-MCP deflection system includes only one level and due to the higher gap between the deflector plates removes charged particles only up to 70 keV. The space between the disks is divided into 32 sectors by plastic spokes forming 32 collimators with an aperture of $9° \times 18°$ each. The sector pointing towards the Sun is blocked. Neutrals passing through the deflection system hit a 32 sided cone target with a grazing (70°) angle of incidence. The interaction with the target results in the secondary particle production or reflection of the primary neutrals. All particles leaving the target block are detected by a MCP stack with 32 anodes. The signal from the MCP gives the direction of the primary incoming neutral. The MCP can operate in either electron mode with positive bias or in ion mode with negative bias. In order to improve the angular resolution and collimate the secondary particles, 32 separating walls are attached to the target block forming a star-like structure. This configuration allows the entering particles to experience multiple reflections and reach MCP. Like the PIPPI-SSD, the PIPPI-MCP covers almost 4π in half a satellite spin period and produces an image of the LENA distribution in the form of an azimuth × elevation matrix. The direction vector of 32 elements is read out once per 31.25 ms. A summary of the instrument characteristics is given in Table 1.

3. CALIBRATIONS AND TESTS

The Astrid project had an extremely short time schedule and the PIPPI instrument had to be designed and manufactured within only 13 months only. To deliver the instrument in time we had to limit calibrations and tests to

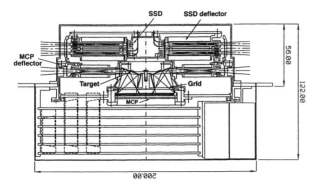

Figure 1. Cross section view of the PIPPI instrument with principal components.

TABLE 1. The PIPPI ENA Imager Characteristics

Parameters	Values
PIPPI-SSD	
Energy range (neutrals)	13 - 140 keV
Energy resolution	8 steps
Angular resolution (FWHM)	$2.5° \times 25°$
Aperture per sensor	$5° \times 30°$
Full field of view	$5° \times 322°$
Azimuthal sectors	16
	(2 background sectors)
Geometrical factor	3.5×10^{-2} cm^{2}sr
Geometrical factor per sector	2.5×10^{-3} cm^{2}sr
PIPPI-MCP	
Energy range	~0.1 - 70 keV
Energy resolution	No
Angular resolution (FWHM)	$4.6° \times 11.5°$
Aperture per sensor	$9° \times 18°$
Full field of view	$9° \times 344°$
Azimuthal sectors	32
	(1 background sector)
Geometrical factor	7.8×10^{-2} cm^{2}sr
Geometrical factor per sector	2.5×10^{-3} cm^{2}sr
Efficiency @ 6 keV Ar beam	$\approx 0.5\%$
TM budget	78 kbps (high mode)
	4.9 kbps (low mode)
Power	4.0 W
Mass	3.13 kg

the most crucial ones and to do them in the simplest way. We have, for instance, simplified the energy level calibrations of PIPPI-SSD by using a Ba-133 monoenergetic (conversion) electron source which provides the discrete electron and gamma - ray spectrum.

The electrostatic deflector is an entirely new element and experience was lacked in designing such a system. Moreover, very limited literature on this subject is available. To our knowledge, only two experimental studies of an electrostatic deflector have been published so far [*Keath et al.,* 1989; *Wilken et al.,* 1997]. The simplest electrostatic deflector is a pair of parallel plates of length L, separation distance D, and having a voltage V between the plates. Such a system rejects all particles of charge q with energy less then the cut-off energy, E_c, which is given by the simple formula [*McEntire and Mitchell* , 1989]

$$E_c = q V \left(1 + \left(\frac{L}{4D} \right)^2 \right) \qquad (1)$$

The PIPPI deflector plates are disks but the above formula with L equal to the difference between the outer and inner radii is still valid to give the minimum cut-off energy. To optimize the PIPPI-SSD deflector design we

have chosen $L = 47$ mm and $D = 2$ mm, i.e., $E_c / qV = 35.5$ keV / kV. To reduce forward scattering of particles inside the deflector, the serrations are machined in the inner disk surfaces according to the design by *Keath et al.* [1989]. In addition, the deflector surface is blackened by copper sulphide.

We have experimentally checked the deflector formula (1) with the deflector in the flight configuration but with a channel electron multiplier (CEM) as a particle detector. The charge particle flux was simulated by an electron beam. For each beam energy, the dependence of the CEM response on the potential between the plates was obtained. The CEM response was given by integral counts for the +7° / -6° sweep through the beam. The potential V varied from 0 to 2000 V. The first potential when the CEM signal did not drop any further with increasing V was taken as the deflector potential providing the cut-off for the given beam energy. Figure 2 presents the dependence of the cut-off energy on the deflector potentials. There is a reasonable agreement with the theoretical dependence $E_c / qV = 35.5$.

Another important characteristic of the deflection system is the transmittance. We define the transmittance for the potential V as the ratio of the detector response for the potential V to the detector response for $V = 0$. The PIPPI deflection system was calibrated using a 72 keV electron beam with a 3.8 keV potential between the plates. The transmittance was found to be 4×10^{-4}, in agreement with the results obtained by *Keath et. al.* [1989] for a similar deflector.

One of the most important issues in the PIPPI-MCP development was the choice of a UV absorbing coating for the target. This must satisfy the following conditions listed in order of significance. (1) A combination of the coating and substrate must effectively absorb UV photons. (2) The coating should be readily available and simple to apply. (3)

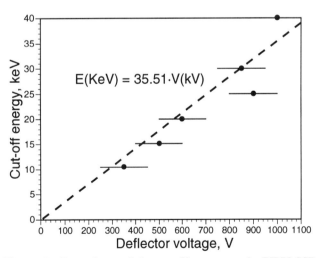

Figure 2. Dependence of the cut-off energy on the PIPPI-SSD deflector potential. A reasonable agreement with the theoretical dependence E_c (keV) = 35.5 V (kV) can be found.

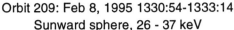

Orbit 209: Feb 8, 1995 1330:54-1333:14
Sunward sphere, 26 - 37 keV

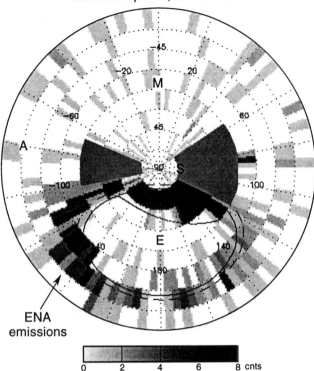

ENA emissions

```
0      2      4      6      8 cnts
```

Figure 3. The ENA image obtained by the PIPPI imager from the polar cap. Figure shows PIPPI-SSD raw counts integrated over 2 min 20 sec when the Astrid spacecraft was at 85° solar-magnetic latitude and 165° longitude. The data are presented in the fish-eye projection of the entire sphere. The ellipse at the bottom part of the images is the Earth's limb. The line in the ellipse area is the terminator. The dashed line near the limb is the exobase at 500 km altitude. *S* and *A* mark the sunward and antisunward directions, *M* the magnetic pole and *E* the nadir. The pixels overexposed by the reflected light, which can reach SSDs at moments when the Sun is close to or in the sensor aperture, are masked. The light reflected by the Earth's surface on the dayside is also seen in the corresponding pixels. The bright pixels near the exobase are HENAs generated by the ring current.

It is desirable to have comparatively high secondary ion and electron yields and reflection efficiency. We have ruled out LiF coating used by *Witte et al.* [1992] due to the complicated mechanical design of our target block and difficulties in maintaining the quality of such surfaces. We chose DAG 213, a resin-base graphite dispersion, an analog to Aquadag, which is a graphite dispersion in water. The DAG 213 photoelectric properties are rather close to those of Aquadag (T. Harley, private communication) and only slightly worse than for LiF [*Grard*, 1973]. DAG 213 is used as a paint for satellite-borne Langmuir probes and has flight heritage. However, DAG 213, as any graphite

compound, has relatively low secondary electron (kinetic) yield. That is the main disadvantage of this material.

In our calibrations we did not define separately the absolute secondary ion and electron yields from coating materials because the scattered primary particles contribute to the instrument response as well. These particle include both ions and neutrals since the charge - state equilibrium in the emerging beam is established very quickly during the interaction with the target (on a depth of few Å). For the PIPPI - MCP final calibrations we, thus, just checked the instrument response against a neutral (ion) beam of known intensity and defined the efficiency for particle detection.

The instrument response, R (s^{-1}), is connected to the particle differential flux, j (E) (cm^{-2}s^{-1}sr^{-1}keV^{-1}) via the equation

$$R = \int_{E_1}^{E_2} C(E)j(E) \, dE \; , \qquad (2)$$

where $C(E)$ is the conversion factor (cm^{2}sr) and the integration is performed over some energy range. The conversion factor can be presented as

$$C(E) = \eta(E) \, G. \qquad (3)$$

Here $\eta(E)$ is the efficiency and G is the geometric factor. Physically, the product of the geometric factor and the incoming differential flux defines the amount of particles per sec entering a detecting unit, the target block with MCP in our case. The efficiency determines what part of these particles is recorded. According to this definition, the geometrical factor is fully defined by the collimator mechanical design and can be readily calculated analytically. The efficiency should be determined in calibrations. To calculate the geometrical factor, we used a simple mathematical model of the instrument. The model predicted the angular responses which agree very well with calibrations and, therefore, is quite accurate. The model gives the geometrical factor 2.5 × 10^{-3} cm^{2}sr / pixel for both PIPPI-SSD and PIPPI-MCP.

Due to shortage of time we did not run the full calibration cycle but defined the instrument response in the ion and electron modes against the 6 keV Ar beam with the retarding grid biased at -100 V and +5 V respectively. The efficiencies turned out to be 0.6% and 0.5%, i.e., the same in the ion and electron modes. This indicates that the particles reflected from the target surface rather than secondaries give the main contribution to the response. In the opposite case, the efficiency in the ion mode would be one order of magnitude less than in the electron mode. Thus reflection is the main process defining the instrument efficiency. Scaling this result to the H beam is not straight forward due to complexity of the processes involved. However, we do plan additional calibrations of the Mars - 96 spare instrument which contains the ENA imager identical to PIPPI - MCP. The LENA sensor head does not have energy resolution and simultaneously measures neutral atoms of all energies. Therefore, the integral efficiency is higher than the one obtained for a fixed energy

In order to check how efficient the target block suppresses UV radiation we have performed a number of tests with the V.03 deuterium lamp with a magnesium fluoride window (*Key and Preston*, 1980). The lamp produces the UV radiation within the wavelength range 115 - 165 nm according to the manufacturer's specification. Approximately 20% of the net irradiance are accounted for by Lyman - α. The net radiance has been evaluated by a simple gold cathode detector (see *Barabash*, 1995, for details). For this particular test the instrument MCP assembly was modified. It included only two microchannel plates instead of three in the flight instrument. The efficiencies were found to be 3×10^{-8} and 6×10^{-8} in the ion mode and electron mode with negative potential on the retarding grid. The MCP UV efficiency is about 1% and, hence, the target UV suppression is about 2×10^5.

4. ENA IMAGE

In the initial stage of the analysis only the data from the PIPPI - SSD sensor head were considered, since they are simpler for interpretation. Over the polar part of the Astrid orbit, up to 70° magnetic latitude, the energetic particle background is low and detection of the ENAs may be performed even with no voltage on the deflection system.

Figure 3 shows PIPPI-SSD raw counts integrated over 2 min 20 sec when the Astrid spacecraft was at 85° solar-magnetic latitude and 165° longitude. The data are presented in the fish-eye projection of the entire sphere, i.e., as an image. In these polar coordinates, the radius is given by the polar angle of each sensor in the frame related to the spacecraft spin axis, and the azimuth angle is the satellite spin angle. The ellipse at the bottom part of the images is the Earth's limb. The line in the ellipse area is the terminator. The dashed line near the limb is the exobase at 500 km altitude. The energy window is 26 - 37 keV. *S* and *A* mark the sunward and antisunward directions, *M* the magnetic pole and *E* the nadir. The pixels overexposed by the reflected light, which can reach SSDs at moments when the Sun is close to or in the sensor aperture, are masked. The light reflected by the Earth's surface on the dayside is also seen in the corresponding pixels. The contamination by photons is unavoidable for the PIPPI design, since neither protective foils nor a thick window on the sensor surfaces were used in order to keep the energy threshold as low as possible. The pixels with low count rate in the center of the image correspond to the blind sensor used to monitor the internal instrument noise and cosmic radiation level. The noise was 0.19 s^{-1} for the energy channel in question. The bright pixels near the exobase are HENAs generated by the ring current. The general structure of the HENA image is as one expects from geometrical considerations, a dawn - dusk

elongated asymmetrical narrow band at low altitudes. Altogether four sequences of HENA images of the ion population at low altitudes in the precipitation region have been identified in the PIPPI-SSD data (*Barabash et al.*, 1997). The images demonstrate clearly the most powerful feature of ENA imaging from near-Polar vantage points; the global dawn - dusk asymmetry in the ring current is instantaneously manifested.

5. SUMMARY

This paper provides a short description of the ENA imager PIPPI which has been flown on the microsatellite Astrid in a low altitude polar orbit of the Earth. The following has been achieved in this first ever satellite-borne ENA experiment .

(1) Unambiguous HENA images of the ring current structures at low altitudes have been obtained.

(2) Several important components of ENA instrumentation such as a deflection system, target block and UV suppressing coating have been developed and tested in our calibration facilities. Now, we are performing analysis of the flight data.

(3) The PIPPI experiment was a part of the Astrid mission which was focused on technological demonstration that microsatellites can be used in performing highly innovative and risky experiments. This demonstration was clearly achieved.

Acknowledgements. The Swedish microsatellite Astrid as well as the PIPPI ENA experiment was financed by grants from the Swedish National Space Board. The Astrid project was initiated, managed and operated by the Swedish Space Corporation. Support to the Astrid project was also provided by the Finnish Meteorological Institute and the Academy of Finland. The efforts of E. C. R., C. J. C., and B. H. M. were supported in part by Grants NAGW-2619 and NAGW-4729 from NASA to the Johns Hopkins University.

REFERENCES

Barabash, S., Satellite observations of the plasma - neutral coupling near Mars and the Earth, Ph. D. thesis, *Sci. Rep. 228,* Swedish Institute of Space Physics, Kiruna, Sweden, 1995.

Barabash, S., P. C:son Brandt, O. Norberg, R. Lundin, E. C. Roelof, C. Chase, B. Mauk, H. Koskinen, Energetic neutral atom imaging by the Astrid microsatellite, *Adv. Space Res.,* in press, 1997.

Funsten, H. O., D. J. McComas, E. E. Scime, Comparative study of low-energy neutral atom imaging techniques, *Optical Engineering, 33,* 349-356, 1994.

Grahn, S., and A. Rathsman, ASTRID: An attempt to make the microsatellite a useful tool for space science, *Proc. the 11th Annual AIAA/USU Conf. on Small Satellites,* Logan, Utah, September, 1995.

Grard, R. J. L., Properties of the satellite photoelectron sheath derived from photoemission laboratory measurements, *J. Geophys. Res., 78,* 2885-2906, 1973.

Hsieh, K. C., C. C. Curtis, C. Y. Fan, and M. A. Gruntman, Techniques for the remote sensing of space plasma in the heliosphere via energetic neutral atoms: a review, in *Solar Wind Seven,* edited by E. Marsch and R. Schwenn, pp. 357-364, Pergamon Press, New York, 1991.

Keath, E. P., G. B. Andrews, A. F. Cheng, S. M. Krimigis, B. H. Mauk, D. G. Mitchell, and D. J. Williams, Instrumentation for energetic neutral atom imaging of magnetospheres, in *Solar System Plasma Physics, Geophys. Monogr. Ser.,* vol. 54, edited by J. H. Waite Jr., J. L. Burch, and T. E. Moore, pp. 165-170, AGU, Washington D. C., 1989.

Key, P. J., and R. C. Preston, Magnesium fluoride windowed deuterium lamps as radiance transfer standards between 115 and 350 nm, J. Phys. E: Sci. Instrum., 13, 867-870, 1980.

Lui, A. T. Y., D. J. Williams, E. C. Roelof, R. W. McEntire, D. G. Mitchell, First composition measurements of energetic neutral atoms, *Geophys. Res. Lett.,* 23, 2641-2644, 1996.

McEntire, R. W., and D. G. Mitchell, Instrumentation for global magnetospheric imaging via energetic neutral atoms, in *Solar System plasma physics, Geophys. Monogr. Ser.,* vol. 54, edited by J. H. Waite Jr., J. L. Burch, and T. E. Moore, pp. 69-80, AGU, Washington, D. C., 1989.

Norberg, O., S. Barabash, I. Sandahl, R. Lundin, H. Lauche, H. Koskinen, P. C:son Brandt, E. Roelof, L. Andersson, U. Eklund, H. Borg, J. Gimholt, K. Lundin, J. Rynö, and S. Olsen, The microsatellite ASTRID, *Proceedings 12th ESA Symposium on Rocket and Balloon Programmes and Related Research,* Lillehamer, Norway, 273-277, 1995.

Roelof, E. C., D. G. Mitchell, and D. J. Williams, Energetic neutral atoms (E ~ 50 keV) from the ring current: IMP 7/8 and ISEE 1, *J. Geophys. Res., 90,* 10,991-11,008, 1985.

Voss, H. D., J. Mobilia, H. L. Collin, and W. L. Imhof, Satellite observations and instrumentation for measuring energetic neutral atoms, *Optical Engineering, 32,* 3083-3089, 1993.

Wilken, B., I. A. Daglis, A. Milillo, S. Orsini, T. Doke, S. Livi, and S. Ullaland, Energetic neutral atoms in the outer magnetosphere: An upper flux limit obtained with the HEP-LD spectrometer on board GEOTAIL, *Geophys. Res. Lett.,* 24, 111-114, 1997.

Witte, M., H. Rosenbauer, E. Keppler, H. Fahr, P. Hemmerich, H. Lauche, A. Loidl, and R. Zwickl, The interstellar neutral-gas experiment on ULYSSES, *Astron. Astrophys. Ser. 92,* 333-348, 1992.

Witte M., H. Rosenbauer, M. Banaszkiewicz, and H. Fahr, The ULYSSES neutral gas experiment: determination of the velocity and temperature of the interstellar neutral helium, *Adv. Space Res., 13,* (6)121-(6)130, 1993.

S. Barabash, P. C:son Brandt, K. Lundin, R. Lundin, O. Norberg, S. Olsen, Swedish Institute of Space Physics, Box 812, S-981 28, Kiruna, Sweden.

E. C. Roelof, C. J. Chase, B. H. Mauk, The Johns Hopkins University / Applied Physics Laboratory, Johns Hopkins Road, Laurel, MD 20723-6099, USA.

H. Koskinen, J. Rynö, Finnish Meteorological Institute, Box 503, SF-00101, Helsinki, Finland.

Imaging Low-Energy (< 1 keV) Neutral Atoms: Ion-Optical Design

Mark F. Smith, D. J. Chornay, J. W. Keller, and F. A. Herrero

Laboratory for Extraterrestrial Physics, NASA/GSFC, Greenbelt MD, USA

M. R. Aellig, P. Bochsler, and P. Wurz

Physikalisches Institut, University of Bern Switzerland

Neutral atom imaging allows plasma populations to be remotely sensed enabling instantaneous images of Earth's magnetosphere and ionosphere to be obtained. The technique has been widely discussed, particularly the imaging of high energy neutrals. Much of the magnetosphere/ionosphere plasma population, however, lies at energies below 1 keV. The most promising development for neutral atom imaging at these low energies is the surface interaction technique, which uses a conversion surface to change the neutral atoms into negative ions. In this paper we discuss the design of such an instrument. We focus on the ion optics required to make such an instrument work and present new laboratory results achieved with a novel ion optic system.

INTRODUCTION

Neutral atom imaging is a technique by which ion populations may be remotely sensed. In the case of Earth, magnetospheric and singly-charged ionospheric ions may charge exchange with the cold, neutral geocorona. The result of such an interaction is to produce a cold ion and a hot neutral. The hot neutral then travels, unaffected by magnetic fields, to the remote imaging position preserving information about the original ion energy and direction of travel (pitch-angle) before the interaction. This technique allows a neutral image to be built-up that can then be inverted to obtain the original ion distribution [e.g., Perez et al., 1995]. The first unambiguous neutral atom observations were from the ISEE-1 plasma instrument which fortuitously detected neutrals [Roelof, 1987]. Since then much work has been published on the technique, including reviews on low energy neutral atom detection [Funsten et al., 1994] and on magnetospheric imaging [Williams et al., 1992].

Many important regions of the magnetosphere and ionosphere are dominated by lower (< 1 keV) energy particles. For example, the cusp and auroral ion outflows, which have energies of typically less than 1 keV, are responsible for populating large fractions of the magnetosphere with ionospheric (O^+) ions [Chappell et al., 1987]. These sources

Measurement Techniques in Space Plasmas: Fields
Geophysical Monograph 103

have been widely studied by *in situ* measurements but the time-dependency and spatial inhomogeneities of these regions have proved difficult to understand. It is thus highly important to study these regions through imaging techniques.

In this paper we briefly review the various techniques available for studying low energy (< 1 keV) neutral atoms. We show that to detect the low fluxes expected from the cusp [Hesse et al., 1993] only one technique, surface conversion is applicable. We then show a new ion optic system which, when used with the surface conversion technique, can provide the necessary energy, angle, and mass resolution to image cusp and auroral ion outflows. Details of the surface conversion technique are given in a companion paper [Aellig et al., this issue]. Finally, we show results from laboratory tests on this system.

INSTRUMENT CONCEPTS

Although various techniques are available to measure neutral atoms, few can measure the low fluxes expected. Those techniques that are sensitive enough separate into two categories. The first are direct techniques where the neutrals themselves are detected using, for example, solid state detectors. Due to the response of these detectors, direct measurements are limited to high energies above many 10's of KeV. In addition, mass resolution is difficult, although not impossible, to achieve. Recently, other direct detection techniques such as using free-standing gold gratings [Scime et al., 1994] have been investigated. At present it is not clear whether these techniques are viable at energies below 1 keV.

The second category of techniques are those which transform the neutrals into ions and then use various electrostatic or magnetic methods to obtain the ions energy, mass, and direction of arrival. The commonly used thin carbon foil will certainly convert the incoming neutrals into ions. The efficiency of this technique is, however, poor at energies below 10 keV and virtually zero at energies below 1 keV; the effects of energy straggling and angle scattering render this method ineffective. These problems led researchers [Herrero and Smith 1992; Gruntman, 1993] to suggest the use of a reflection surface to convert the neutrals into negative ions. This technique has the advantage of high efficiency at energies below 1 keV because of the quasi-specular character of the reflection mechanism down to kinetic energies as small as a few eV. Since then this conversion technique has been widely studied [e.g., Wurz et al., 1995]. Recent designs use a cesiated tungsten surface at high grazing incidence to obtain ion conversion efficiencies in excess of 10% [Wurz et al., 1995]. Details can be found in Aellig et al. [this issue] and will not be discussed further here.

Based on this technique, a complete instrument has been designed [Ghielmetti et al., 1994] and proposed for a number of missions, such as HI-LITE [Smith et al., 1992] and IMAGE (MIDEX proposal, J. Burch PI). In this original design, photon rejection was solved by dragging the ions off the converter surface at angles almost normal to the surface using a strong electric field, and hence out of the light path. An electrostatic lens system was then used to focus the negative ions emitted by the conversion surface so that energy and mass analysis could be undertaken. Here we have, on the basis of work needed for the Polar/Wind missions [Herrero, 1990], kept the ions at grazing incidence after detection and used a light trap and toroidal analyzer to achieve the required photon rejection. This instrument is the subject of this paper.

INSTRUMENT REQUIREMENTS

Instrument design is always driven by the measurement requirements. To measure the low energy ion population that makes up the ion outflow from the ionosphere to the magnetosphere any instrument must be capable of measuring atom fluxes below 10^4 cm^{-2} s^{-1} with an angular resolution of 8° x 8° in each pixel, and good time resolution (better than 5 mins per image) [see Hesse et al., 1993]. These requirements determine the following instrument characteristics:

1. A large geometric factor. Hence, the interaction surface must be large and the instrument must be a spectrograph.
2. Mass analysis needed to separate H and 0
3. Elimination of a large photon flux with a rejection ratio of about 10^7.
4. A field-of-view of at least 90° x 8°
5. Limited energy resolution (e.g., 5 energy bins over the full energy range of few eV to 1 keV)

These requirements, however, do pose some problems for the ion optics. To obtain a large geometric factor, the converter surface must have a nominal diameter of 1 cm or larger and could lead to significant spherical aberration problems. For ion optics of practical size, a 1 cm diameter object is quit large. Typically, the energy analyzers that are used for space plasma experiments, have real object apertures in the range below a few mm. The dynamic energy range we are interested in covering is also very large (few eV to 1 keV). This leads to energy dependent effects, analogous to chromatic aberration effects in visible optics. Consideration also has to be given to UV photon rejection i.e. a good light trap. Closely related are problems with

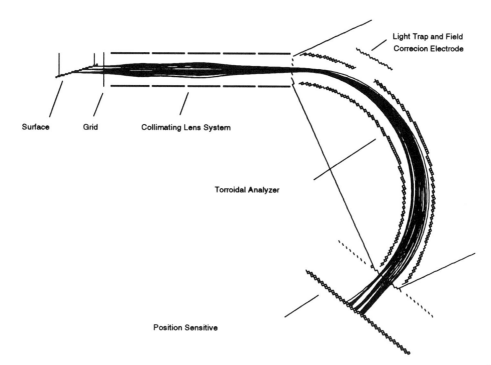

Light Trap and Field
Correcion Electrode

Surface   Grid   Collimating Lens System

Torroidal Analyzer

Position Sensitive

Figure 1. Ray tracing showing three groups of ions having energies 50 eV, 500 eV, and 1 keV, leaving a 1 cm^2 surface. Here the collimating lens elements are planar symmetric, i.e. stretch in and out of the paper. The azimuthal angle corresponds to different impact angles on the surface in the plane of the paper, while the elevation or fan angle is in the plane perpendicular to the paper. A grid at 0 V lies between the source and the entrance to the lens stack thus maintaining a field-free region in the vicinity of the source. The ions are brought to a focus at the entrance of the toroidal analyzer, which then disperses them according to energy, and brings each group to a focus at a different point beyond the exit. The location of the position sensitive detector is a compromise between these points, and is chosen so that the ions leaving the surface may be binned into 4 energy bins, over the range 20-1000 eV. A hole in the outer electrode allows UV photons reflected from the surface to enter a light trap. A potential is applied to a field correction element within the volume of the light trap to maintain uniform equipotentials in the toroidal analyzer. Mass analysis can be achieved by using a conventional time-of-flight system on the back end.

photoelectrons that may be created, then accelerated from surfaces in the lens and analyzer system. One helpful point is that high energy and angular resolution are not required.

OPTICAL DESIGN

Our approach to designing the optical system follows naturally from the above considerations. In order to deal with the chromatic aberration problems associated with the large initial spread in energies, a collimating lens system is used to apply a strong initial acceleration. This reduces the relative energy spread of the ions. Furthermore, since the instrument needs to have a wide field of view in the fan or elevation angle direction (defined in Figure 1) it is preferable to avoid as much bending in the elevation plane as possible. This leads us to employ some form of cylindrical symmetry to this lens system (see Figure 1). We also want to preserve information on the elevation angles at which the ions leave

the surface. Thus, any strong asymmetric fields in the region from the surface to the acceleration lens should be avoided. This leads us to place a high transparency (95%) grounded grid at the entrance to the lens system as a precaution. Finally, for the energy dispersive element, a toroidal analyzer was chosen, since it has the desirable property of focusing in the fan angle plane, in addition to the energy-dispersive focusing properties (Figure 1).

The initial design was also driven in part by the availability of a toroidal electrostatic analyzer, a two dimensional imaging detector, and the desire to restrict ourselves to lens elements that could be easily fabricated. Lenses were chosen to be formed from flat plates, as these tend to have superior optical properties over aperture lenses, for a given volume [Harting and Read, 1976]. Ray tracings were performed to optimize the lens parameters. The toroidal analyzer was the prototype used for the development of the Strahl (electron) detector now onboard the spacecraft as part of the SWE

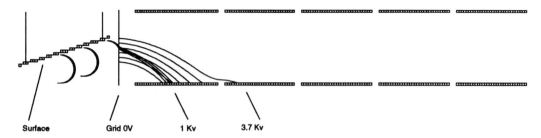

Figure 2. Ray tracing showing the effect of a pair of sweep magnets on undesired low energy (≤ 20 eV) photoelectrons, leaving the conversion surface and the grid. For clarity the magnets are not shown, but are rectangular in shape (1.5 by 2.0 cm) separated by 3.0 cm, and are located above and below the plane on the paper. The field in the region of the grid is 15 Gauss.

experiment The imaging detector together with its associated GSE and display software, were the engineering prototype for the PPA instrument (electron detector), currently onboard the POLAR spacecraft [Scudder, 1990].

The toroid has major and minor radii of 160 mm and 60 mm respectively, with a gap between the electrodes of 12 mm. This leads to a plate factor of approximately 5, i.e. the total voltage between the electrodes needs to be 2/5 the energy of the beam, so that it is brought to a focus near the exit. In order to use this as a dispersive element such that there is an energy spread of approximately 1 keV across the gap, (to cover the range from a few eV to 1 keV for the negative ions leaving the conversion surface,), it is necessary to accelerate all ions by the same amount From a lens chromatic aberration point of view, this potential should be as high as possible, (within the limits of the HV power supply envisioned) while at the same time not so high that we lose the ion energy information at the exit slit of the toroid. This trade-off leads to an applied potential of 6 keV as being acceptable.

A sweep magnet is used in the vicinity of the surface and the first element of the lens system to remove unwanted photoelectrons that may be produced from the surface, and the grid (Figure 2). Without the magnet, the electrons would be indistinguishable, as far as the electrostatic lens system and analyzer are concerned, from the negative ion signal. The presence of the magnet has a barely discernible effect on the ion trajectories shown in Figure 1.

TEST SETUP

For testing purposes a low-efficiency conversion surface was formed from brass shim stock (25 mm wide, by 32 mm high) mounted on an aluminum form. Located in the center was a slit aperture (0.7 mm by 15 mm). The length of the slit was along the energy dispersive direction of the electrostatic

analyzer (ESA). Ions could be fired either straight through the slit in order to determine lens optics and ESA characteristics, or reflected from the surface to emulate the conversion surface.

The whole assembly could be rotated about the center of this surface, along both azimuth and elevation axes. The elevation angle corresponds to different 'fan' angles with an elevation of 58° degrees corresponding to ions traveling straight down the center of the collimating lens system. An azimuth angle of 0° degrees corresponds to ions passing through the slit aperture, and straight down the lens system into the ESA. An azimuthal angle of approximately -28° is used to reflect ions off the surface, with an angle of incidence of 76° into the ESA. Current falling on the surface was monitored and was typically in the range of 100 pA

A 32x32 pixel position sensitive detector with an active region of 20x20 mm was placed at the exit of the ESA. The position sensitive detector has two outputs: an analog signal proportional to the total count rate (no position information), and a two dimensional 'digital snapshot' containing position information.

This work focuses on testing the ion optic train. Full tests of the instrument require a neutral atomic beam with adjustable energy from a few eV to 1 keV. Therefore, future tests will be conducted at the Atomic Oxygen Beam Facility at the University of Denver which is one of the few facilities able to produce well-characterized, low energy, neutral atom beams. For these tests characterization we used positive ions, which the GSFC facility can produce over the range from 50 eV to 30 keV with good optical properties. For direct testing of the optics system the use of positive ions as compared to the negative ions produced from a cesiated surface only requires the swapping the polarities of the electrostatic components. For tests using the brass surface some positive ions will be converted into negative ions, although the efficiency will be low because of the two-electron transfer process required in the conversion.

RESULTS

The angular and energy response of the prototype was characterized in a series of laboratory tests, which are summarized in Figures 3 through 5. In Figure 3 the angular response of the instrument is characterized. Figure 3(a) shows the output obtained from directly illuminating the instrument with a beam of monochromatic (500 eV) positive ions. The instrument has been rotated through three steps, 5° apart in fan angle. Clearly the ion optics described provide

a)

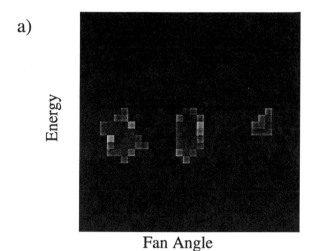

Fan Angle

b)

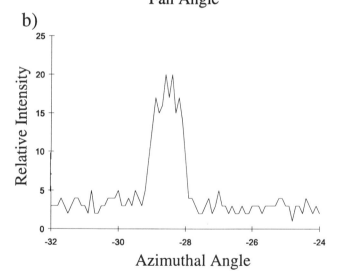

Azimuthal Angle

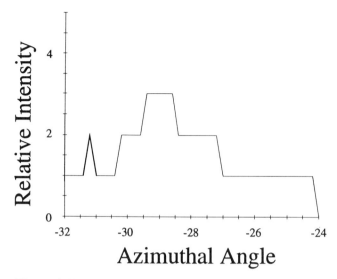

Azimuthal Angle

Figure 4. The azimuthal response of the instrument to a monochromatic beam of positive ions and neutrals fired onto a low efficiency brass conversion surface. As the surface is of low efficiency the number of negative ions coming off of the surface into the optic system is extremely low.

good angular resolution (better than 5°). In addition, the beam falls consistently at one position along the energy dispersion axis for all fan angles. The response of the instrument in the azimuthal angle is shown in Figure 3(b). The measured azimuthal width is less than 2° but this does not include the effect of scattering on the conversion surface. This effect is apparent in Figure 4 which shows data obtained from reflecting the incident ion beam off the brass surface, which is acting as a very low efficiency converter surface. As the input beam is a mixture of ions and a small proportion (< 10%) neutrals both of these populations can be converted to negative ions. The angular response in Figure 4 has been degraded to a value of approximately 3' which is small compared to the expected FWHM scattering on the converter surface of around 5°. This is due to the design of the ion optics which serve to compress the angular width of the scattered ions, leading to the small degradation of the angular resolution.

A measurement of the instrument's response to a 200 eV ion beam is displayed in Figure 5. Although this is an integrated signal of the total flux on the position sensitive detector, we derive an energy resolution of the instrument, $\Delta E/E$, of about 7% using the width in energy of the response in Figure 5 with the spatial extent of the beam measured for a fixed plate voltage (Figure 3(a)). Thus, for an acceleration voltage of 6 kV the energy bin width is about 450 V. This resolution is somewhat less than predicted by the ray tracing calculations, which is partly attributable to limitations in the detector which was not optimized for this application.

Figure 3. Panel (a) shows the instrument response to a monochromatic beam of ions fired directly into the instrument aperture. The instrument was rotated in three, 5° steps. The horizontal axis corresponds to the fan angle of the detector and shows the 3 beams are well-separated. The vertical axis is the direction of energy dispersion. The beams are all thus seen at the same vertical position, as expected. The bottom panel (b) shows the response in the azimuthal plane to an ion beam.

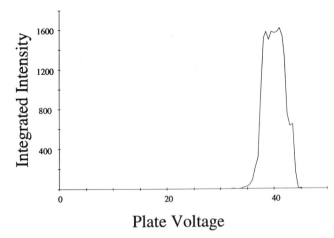

Plate Voltage

Figure 5. Response of the analyzer to a 200 eV ion beam, as a function of the analyzer plate voltage. The ESA plate voltage is scanned while recording the integrated signal from the position sensitive detector. The flat top peak is as expected since we are recording the amplitude of the total signal. At the same time the arrival position of the ions at the position sensitive detector was observed to travel across the field-of-view, as expected. The shoulder on the high energy side of the peak is probably due to ions at grazing incidence scattering off the inner surface of the toroidal analyzer.

SUMMARY

We have presented a new ion optic design which will preserve the distribution of negative ions reflected from a high efficiency converter surface and meet the measurement requirements necessary to measure the cusp and auroral ion outflows. Optimization of the system is currently being investigated.

The instrument design is a candidate for adoption for the IMAGE mission which would provide the first images of the cusp and auroral ion outflow on time scales of 5 min. The data obtained from such an instrument will provide images of the low energy (< 1 keV) ion population which is of the utmost importance because of its role in populating the magnetosphere with ions. The proposed imager will open a new window on our understanding of this phenomena.

Acknowledgments. The authors acknowledge discussion with J. Quinn at the University of New Hampshire, and E. Shelley, A. Ghielmetti and S. Fuselier at Lockheed-Martin.

REFERENCES

Chappell, C. R. , T. E. Moore, and J. H. Waite Jr, The ionosphere as a fully adequate source of plasma for the Earth's magnetosphere, *J. Geophys.* Res., 92, 5896, 1987.

Funsten, H. O, D. J. McComas, E. E. Scime, Comparative study of low-energy neutral atom imaging techniques, *Opt. Eng.* 33, 349,1994.

Ghielmetti, A. G., E. G. Shelley, S. Fuselier, P. Wurz, P Bochsler, F. Herrero, M. F. Smith, T. S. Stephen, Mass Spectrograph for imaging low-energy neutral atoms, *Opt Eng.*, 33, 362, 1994.

Gruntman, M. A., A new technique for *in situ* measurement of the composition of neutral gas in interplanetary space, *Planet. Space Sci.*, 41,307, 1993.

Harting, E., and F. H. Read, Electrostatic Lenses, Elsevier, New York, 1976.

Herrero, F. A., Light-trap design using multiple reflections and solid-angle attenuation: application to a spaceborne electron spectrometer, *Applied Optics*, 31, 5331.

Herrero, F. A., and M. F. Smith, Image of low energy neutral atoms (ILENA): Imaging neutrals from the magnetosphere at energies below 20 keV, Instrumentation for Space Imagery, *SPIE Proc.*, 1744, 1992.

Hesse, M. , M. F. Smith, F. Herrero, A. G. Ghielmetti, E. G. Shelley, P. Wurz, P. Bochsler, D. L. Gallagher, T. E. Moore, and T. S. Stephen, Imaging ion outflow in the high-latitude magnetosphere using low-energy neutral atoms, *SPIE*, 2008,1993.

Perez, J. D., J. P. Keady, T. E. Moore, and M. -C. Fok, Microphysics from Global Images, in 1995 *Cambridge symposium in Theoretical Geoplasma Physics*, ed. T. Chang, in press, 1995.

Roelof, E. C., Energetic neutral atom image of a storm-time ring current, *Geophys. Res.* Lett, 14, 652, 1987

Scudder, J., F. Hunsaker, G. Miller, J. Lobell, T. Zawistowski, K. Ogilvie, J. Keller, D. Chornay, F. Herrero, R. Fitzenreiter, D. Fairfield, J. Needell, D. Bodet, J. Googins, C. Kletzing, R. Torbert, J. Vandiver, R. Bentley, W. Fillius, C. McIlwain, E. Whipple, and A. Korth, HYDRA - A 3-D Electron and Ion Instrument for GGS/Polar, *Space Science Rev.*, 71, 459, 1995

Scime, E. E., H. O. Funsten, D. J. Mc Comas, K. R. Moore, and M. A. Gruntman, Novel low-energy neutral atom imaging technique, *Opt. Eng.*, 33, 357,1994.

Smith, M. F. and the HI-LITE team, The high-latitude and Energetics (M-LITE) Explorer: A mission to investigate ion outflow from the high-latitude ionosphere, *SPIE*, 2008,1992

Williams, D. J., E. C. Roelof, and D. G. Mitchell, Global Magnetospheric Imaging, *Rev. Geophys.*, 30, 183, 1992

Wurz, P., M. R. Aellig, P. Bochsler, A. G. Ghielmetti, E. G. Shelley, S. A. Fuselier, F. Herrero, M. F. Smith, and T. S. Stephen, Neutral Atom Imaging Mass Spectrometer, *Opt. Eng.*, 34, 2365, 1995.

M. R. Aellig, P. Bochsler, and P. Wurz Physikalisches Institut, University of Bern Switzerland

D. J. Chornay, J. W. Keller, and F. A. Herrero, Laboratory for Extraterrestrial Physics, NASA/GSFC, Greenbelt MD, 20771 USA

M. F. Smith, Logica UK Ltd, Wyndham Court, Cobham, Surrey, KT11 3LG England.

Imaging Earth's Magnetosphere: Measuring Energy, Mass, and Direction of Energetic Neutral Atoms with the ISENA Instrument

S. Orsini[1], P. Cerulli-Irelli[1], M. Maggi[1], A. Milillo[1], P. Baldetti[1], G. Bellucci[1], M. Candidi[1], G. Chionchio[1], R. Orfei[1], S. Livi[2], I. A. Daglis[3], B. Wilken[2], W. Güttler[2], C. C. Curtis[4], K. C. Hsieh[4], J. Sabbagh[5], E. Flamini[5] E. C. Roelof[6], C. Chase[6], M. Grande[7]

ISENA (Imaging Spectrometer for Energetic Neutral Atoms) was launched on board the Argentinean-US Satellite SAC-B on 4 November 1996. The 3-axis stabilized SAC-B was to orbit the Earth at 550 km with an inclination of 38°, making it a suitable platform for detecting ENAs produced by charge exchange within Earth's magnetosphere. A failure of the spacecraft to separate from the Pegasus booster has resulted in the spacecraft's assuming an attitude which does not provide sufficient illumination of the solar panels, greatly reducing the likelihood of useful data production. ISENA's sensors are based upon the GEOTAIL/HEP-LD and CLUSTER/RAPID instruments. The instrument is comprised of two identical sensors, each with a field of view of 60° x 12°. The sensors are oriented at 90° to each other, but have parallel optic axes. Each sensor contains an entrance collimator electrically biased to reject charged particles, an entrance foil to suppress H Ly α photons and provide electrons for a start signal, two microchannel plate assemblies associated with a time-of-flight measuring system, a 4-segment anode to measure ENA arrival directions with 15° angular resolution, and a solid state detector to measure the energies of 35 keV - 150 keV ENAs. ISENA has a mass of 6 kg and consumes 6W. The design and operating characteristics of ISENA's sensors and associated electronics are described. Simulation of the predicted ENA signal is discussed.

[1]Istituto di Fisica dello Spazio Interplanetario, Frascati, Italy

[2]Max-Planck Institut für Aeronomie, Katlenburg-Lindau, Germany

[3]Institute of Ionospheric and Space Research, National Observatory of Athens, Greece

[4]Physics Department, University of Arizona, Tucson, Arizona, U.S.A

[5]Agenzia Spaziale Italiana, Roma, Italy

[6]Applied Physics Laboratory, Johns Hopkins University, Laurel, Maryland, U.S.A.

[7]Rutherford Appleton Laboratory, Oxfordshire, United Kingdom

Measurement Techniques in Space Plasmas: Fields
Geophysical Monograph 103
Copyright 1998 by the American Geophysical Union

PROJECT OBJECTIVES

We planned to perform in situ measurements of precipitating energetic neutral atoms (ENA) of energies 35 keV to 150 keV, using the Imaging Spectrometer for Energetic Neutral Atoms (ISENA). The instrument was placed aboard the SAC-B spacecraft, an Argentina-USA joint project (Gulich and White, 1993) and launched on a PEGASUS rocket on 4 November 1996. The spacecraft achieved its planned Earth orbit at 550 km altitude, with 38° inclination to the equator, but the spacecraft failed to separate from the booster. Although the solar panels deployed, the present spacecraft attitude does not permit sufficient illumination of the solar cells to provide the minimum power necessary for operation of the spacecraft. ISENA was to generally point antisunward and observe

ENA created in the inner magnetosphere, the radiation belts, and ring current. With data from ISENA , we hoped to gain an understanding of:

1) the morphology of the current system, e.g. the ring current structure and dynamics as a function of distance from Earth and of geomagnetic parameters like AE and Dst, and its effect on the transport of mass and energy within the magnetosphere, especially during magnetic storms and magnetospheric substorms and in connection with auroral phenomena (e.g.Roelof and Williams, 1988; Daglis and Livi, 1994),

2) the spatial distribution and energy dependence of the ratio O/H as a function of time, to clarify the transport between the magnetosphere and the ionosphere (e.g., Balsiger et al., 1980);

3) requirements for future instrumentation for imaging the magnetosphere via ENA detection..

SENSOR CHARACTERISTICS

There are several types of ENA imagers under development. Barabash et al. (1994) recently put an ENA imager into polar orbit on the ASTRID micro satellite. See Hsieh and Curtis (this volume) for a recent review.

Detecting 35 - 150 keV ENA, ISENA is derived from the SCENIC ion mass spectrometer (Fig 1), developed by the Max-Planck Institut für Aeronomie (MPAe) and currently flying on the GEOTAIL spacecraft (Doke et al.,1994). A similar instrument was set to fly on the ill-fated CLUSTER (Wilken et al.,RAPID: The imaging energetic particle spectrometer on the European CLUSTER mission, submitted to Space Science Reviews, 1995). Biased collimator plates, serated to minimize grazing incidence scattering, reject charged particles and define the 60° x 12° field of view (FOV), with an angular resolution of 15° in the 60° plane. ISENA has 2 identical sensors, oriented with the two 60° FOV planes orthogonal to each other. An ENA of mass M and energy E is identified using

$$M = 2E \frac{(TOF)^2}{D^2}$$

where TOF is the ENA's time of flight between a thin foil and a solid state detector (SSD) located a distance D (3 cm) downstream. The TOF interval begins when secondary electrons ejected from the foil by the ENA arrive at a START detector: it ends with the arrival at the STOP detector of electrons ejected from the front surface of the SSD when it is struck by the ENA. E is derived from the ENA residual energy deposited in the SSD. The foil consists of 5.5 μg/cm² of Si on a 3.5 μg/cm² Lexan substrate, with 1 μg/cm² C on the exit face.. Its Ly α transmission was

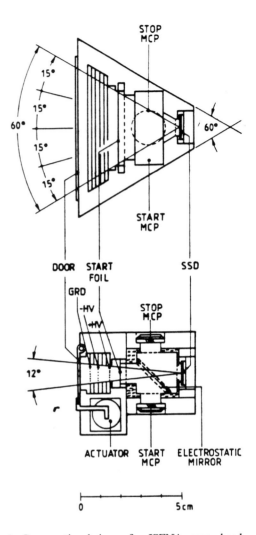

Figure 1. Cross sectional views of an ISENA sensor head

measured at ~10⁻³. Foils of this thickness permit transmission of H with less than 10 keV, albeit with considerable straggling at the lowest energies. The SSD is a 5 x 15 mm Planar Silicon Implanted Ion detector with a 50 nm thick layer of Al deposited on the front surface to reduce sensitivity to scattered Ly α fluxes. The expected operating temperature for the passively cooled SSD was -10 C. After measuring the detector and electronics response to low energy protons, the energy threshold was set at 35keV to minimize noise and the effects of scattering in the foil.

The START and STOP detectors are microchannel plate (MCP) detectors. A 4-element rectangular anode behind the START MCP generates fast START and position signals. Since all STOP signal electrons are emitted by the narrow (5mm width) SSD, no position sensing is required for the STOP detector. Secondary electrons released from the foil or SSD are accelerated to 1 keV and reflected 90° in an isochronous electrostatic mirror before striking the appropriate MCP. Each anode element is mapped to an area on the entrance foil to determine the arrival angle within 15°. Because the foil and front face of the SSD are at ground potential, there is a high positive voltage on the exit surface of both MCP stacks. The MCP output pulses are isolated from this potential and then applied to the timing and position circuitry, whose resolution is 1 ns. Position detection is done using a slower (1 μ s time constant), power-saving charge analysis technique. A valid event is comprised of a START timing signal, a START position signal in one (and only one) of the 4 anode elements, and a STOP timing signal. An energy signal from the SSD is desirable but not essential.. The follow-on electronics encode the event in a 2-bit binary address. ISENA is packaged in an L-shaped box whose longest dimensions are 24 x 16 x 11 cm. It weighs 6 kg, and draws 6 W of power.

THE SIGNAL CONDITIONING UNIT

ISENA contains a Signal Conditioning Unit which prepares analog signals for transfer to the Data Processing Unit (DPU) and executes commands sent by the DPU. A DPU-controlled High Voltage Power Supply (HVPS) generates bias voltages for the MCPs and grids. The Aperture Control (APER CTRL) supplies high voltage to the deflection plates in the entrance collimator. An ENA's energy, START and STOP pulses, and anode identification are transmitted to multiplexers (MUX). TOF intervals are measured by a fast Time-to-Amplitude-Converter (TAC), whose variable amplitude pulse output is stored in a Sample and Hold (S&H) and transmitted through an 8-bit Analog/Digital Converter (ADC) to the DPU. Energy information (a variable amplitude pulse from the SSD) is discriminated in a peak detector and transmitted to the DPU through an 8-bit ADC. A sample rate is also generated by the DPU. A sensor calibration control (E and TOF)) checks instrument performance during flight operations.

THE DATA PROCESSING UNIT

The DPU collects particle data from the sensors, analyzes and organizes the data, and transmits them to the spacecraft on board data handling (OBDH) processor. The DPU has been designed around an Intel 80C186 16-bit microprocessor, with 64kb EEPROM memory for program storage and 512-kb static RAM area. Spacecraft and sensor interfaces have been designed using a Xilinx XC3030 FPGA. Each detected ENA has a 3 element vector: E, direction and TOF. The matrix of particle vectors is analyzed once per minute and compared with calibration data stored in a look-up table (LUT). The resulting 3-D spectra are stored

in the main memory buffer and transferred to the OBDH once per orbit. Any telecommands received are stored in a memory buffer for later execution. The operating software, written using an MS-C compiler and Paradigm locator, is stored in EEPROM. At power-on, the program is copied into the RAM area and executed; it may be downloaded through a telecommand for debugging.

SPECTROMETER OPERATION

ISENA was to be allotted 250 kB of data during each 5-minute transmission to ground. These would occur 4 times daily at consecutive 90 minute intervals. The data contain events and spectra at 1-minute time resolution. The spectra are organized into 512 bins (4 mass bins x 16 energies x 8 angular sectors). Three mass bins contain H, He, O; the 4th contains events (treated as H) for which only a TOF is available. ISENA's 115 keV energy range is divided into 16 log-equidistant energy channels. Encoding each of the 512 bins into an 8-bit logarithm requires 50 kB for the 90 minute period.. The remaining 200 kB will be filled with raw data. During the ~18 h daily period with no data transmission, spectra would be collected over 150° of longitude along the night side orbits and transmitted at the first ground connection. By telecommand it would be possible to select different accumulation times (15 seconds to several minutes) for each energy spectrum, as well as different operation periods for those orbits with no connection to ground. A calibration mode, activated by telecommand, would fully devote the instrument to housekeeping and duty cycle checkouts.

ISENA SAC-B TELEMETRY INTERFACE

The Power Distribution Unit supplies 28 V on dual circuits for redundancy. The Telemetry Processor (also fully redundant) generates clock signals (bit, word, housekeeping minor and major frame) and 16-bit data words, transferred at either 50, 100, or 200 kHz (selectable). Another path for data transfer is the Digital Housekeeping (8 bits), taken from ISENA and stored in the main satellite memory. The Telecommand Processor (fully redundant) generates clock, 16-bit data word and a Pulse Command Actuator (10 ms).

THE GROUND SUPPORT EQUIPMENT

The Ground Support Equipment (GSE) simulates the telemetry and the telecommand lines necessary for receiving data from the instrument, sending the commands to the DPU. GSE hardware consists of a 66 MHz PC equipped with a custom interface board; this generates the requested timing for both commands and data. Telecommands can be transmitted at up to 200 kHz. Data are received by the GSE after an interrupt request from the data telemetry line. Digital words are read from the serial line and stored in the PC memory. The housekeeping header is decoded and stored in a separate location. GSE software is composed of a Microsoft Windows interface together with an IDL package which provides a graphical facility. Some C-language modules which control the interface board are dynamically linked to IDL. This provides an on-line display of the housekeeping data with plots of the instrument characteristics. Off-line data analysis may also be performed.

ANGULAR COVERAGE, MASS AND ENERGY RESOLUTION, AND SENSITIVITY

The sensor system has a total FOV of 12° x 60°, with the 60° interval subdivided into four 15° sectors. The ENA mass resolution is limited by energy straggling in the SSD ($\Delta E/E$), path length differences in the TOF system ($\Delta S/S$), and timing errors ($\Delta T/T$) in the TAC due to amplitude variations in the START and STOP signals. The resolution permits detection/identification of H and He > 35 keV, and O >50 keV. Energy resolution for O in the SSD decreases with energy from 100% at 50 keV to 30% at 150 keV.

The four 15° sectors can each be considered to have a differential geometric factor G_n. Although the individual values of G_n depend on geometry, an estimate to within 20% can be calculated by assuming 4 equal foil segments each having $G_n = 0.0025$ cm^2 sr. This gives a total G = 0.01 cm^2 sr for each sensor; the 2 ISENA sensors combined have G = 0.02 cm^2 sr. The overall detection efficiency ϵ of the TOF/E spectrometer depends on geometric factor G_n, the probability function T_r describing the transmission through the TOF system, the electron emission efficiency of the foil and SSD $\epsilon(S)$, the detection efficiency $\epsilon(T)$ of the TOF-detectors, and the detection efficiency $\epsilon(E)$ of the SSD. The quantities G_n and T_r are instrument parameters while $\epsilon(S)$, $\epsilon(T)$ and $\epsilon(E)$ depend on ENA mass and energy. The probability of observing an ENA-identifying triple coincidence (START, STOP, and E-signals) is proportional to the product $\epsilon(S)$ $\epsilon(T)$ $\epsilon(E)$. The triple-coincidence signal provides significant immunity to background signals. Below the SSD lower energy threshold, only a double coincidence signal (TOF) is available. Folding in the efficiencies, ISENA would detect ENAs at a rate of ~10(s keV)$^{-1}$ for a flux of~10^3(cm^2 s sr keV)$^{-1}$ at energies of ~35 keV.

STRUCTURE AND THERMAL CONTROL

ISENA is placed adjacent to the CUBIC experiment, pointing anti-sunward. The instrument housing is attached to the satellite frame only, maintaining a 1 cm separation from CUBIC. The box is made of an Al alloy (Anticorodal 6061), anodized with ALODINE 1200. The power dissipated by the instrument when switched-on is sufficient for thermal balance. Switched-off, a ~ 4W heater protects the electronics from low temperatures. Two thermistors monitor the temperature inside the box. The experiment uses thermal blankets to insulate ISENA from CUBIC and the satellite. The only exposed surface is blackened with DAG213.

RADIATION EXPOSURE

ISENA would fly across the southern Atlantic anomaly and spend ~1 hr every day in the low altitude extension of the Earth's radiation belts. During these crossings the instrument would absorb energetic protons from the radiation belt. The maximum proton flux at 50 MeV is 2340 protons/(cm^2 s) at geographic latitude -34° and longitude -33°. The ISENA sensors and the electronics are shielded by the instrument box which has 1 mm minimum thickness. A 1 mm aluminum shielding totally shields against protons with energy <10 MeV (ESA PSS-010609, Issue 1, May 1993). Using a polynomial logarithmic interpolation between 1 and 50 MeV, we estimate an integral flux $\Phi(>10$ MeV) = 2987 protons/(cm^2 s). In a zero-order approximation, we consider a semi-infinite aluminum shielding, using the proton planar dose/ fluence values given by ESA PSS-010609. We consider three integral flux intervals: 10 - 50 MeV, 50 - 150 MeV, and > 150 MeV. The maximum dose rate for ISENA (~ 1 hr/day) is 3 10^{-4} rads/s, or 400 rads/year.

PHOTON BACKGROUND

Solar H Ly α radiation scattered in the Earth's exosphere generates geocoronal photons. These photons would produce spurious events in ISENA because they generate accidentally coincident START and STOP signals. At ISENA's altitude, the geocorona zenith emission rate is about 10 kR (Rairden et al., 1986). The instrument would be operational primarily on the night side of each orbit, where the photon flux is considerably reduced. It is realistic to assume an H Lyα of ~5 kR impacting ISENA, i.e.~4.8 x 10^8 photons cm^{-2} sr^{-1} s^{-1}. The ISENA geometric factor for photon response is ~0.2 cm^2 sr (based on the gerometry of the collimator and foil) which is 10 times the geometric factor for ENA. The photoelectron production efficiency at the rear surface of the START foil is about 1%. For a foil with 100% photon transmission, the counting rate in the START or STOP detectors due to these photoelectrons would be ~ 2 x 10^6 photons s^{-1}

Additionally, photons can trigger the MCPs directly. If 20% of the incident photons scattered into the detectors and triggered them with 5% efficiency, an additional 2 x 10^6 counts s^{-1} would be generated, for a total counting rate of about 4 x 10^6 s^{-1}. If the foil reduces the photon flux by a factor F, the counting rate becomes 4 x 10^6 /F s^{-1}. The number of paired signals occurring in a 100 ns TOF window is $(4 \times 10^6/F \times 10^{-7})^2 = 16\ 10^{-2}/F^2$. The probability that the pair sequence is exactly START-STOP is 1/4 of the total number of signal pairs. This yields $4 \times 10^{-2}/F^2$ events / (TOF window). For 10^7 TOF windows/sec, the apparent ENA signal due to photons is ~ 0.4 events/s for F=1000, but rises to ~ 40 events/s for F=100, which is larger than the expected ENA detection rate. It is clear that a foil with a H Lyα transmission of 10^{-3} or less is needed.

ANTICIPATED ENA MEASUREMENTS

Orsini et al. (1994) modelled the ENA flux that might be seen by ISENA. The model assumes a radial dependence of the geocoronal H atom density on the basis of the best fit of the DE-1 geocoronal observations presented by Rairden et al. (1986). The ring current/radiation belts source distribution for ENA production modeling has been derived from the energetic ion measurements made from 1985 to 1987 by the Charge - Energy - Mass (CHEM) spectrometer on board the Charge Composition Explorer (CCE), one of the three Active Magnetospheric Particle Tracer Explorer (AMPTE) spacecraft (Gloeckler et al., 1985). To simulate ENA production at the geographic equator, the source population data set covers the geocentric distance range of 1.25 - 9.75 R_E (binned in 0.25 R_E intervals) and a range of Magnetic Local Time from 00 - 24 MLT. Three different geomagnetic activity levels defined as "low" (AE < 100 nT), "medium" (100 nT < AE < 600 nT) and "high" (AE > 600 nT) are used. H^+ and O^+ are included in the data base. Considering H, the ENA flux intensity F_H [(cm^2 sr s keV)^{-1}] of energy E originating from R and arriving at an observation point R_0 along the direction P (see Fig 3) can be derived using the following formula:

$$F_H(R_0, P, E) = \sigma(E) \int_0^P n_H(R)\, F_{prot}(R, E)\, dp$$

where σ(E) is the charge exchange cross section for ions of energy E in the target gas H, n_H is the number density of neutral gas H, F_{prot} is the proton flux at R with energy E, and P is the total integration path related to R by the expression:

$$|P| = \sqrt{R_0^2 + R^2 - 2 R_0 R \cos(\delta)}$$

where δ, the angle between R_0 and R, is given by:

$$\delta = 180° - \alpha - \arcsin[\sin(\alpha) R_0 / R]$$

with α the angle between R_0 and P.

To identify source ions whose pitch angle is aligned with the instrument look direction, the IGRF modified dipole internal magnetic field model has been used (GEOPACK, M. Peredo and N.A. Tsyganenko, private communication). In most cases the pitch angles are ~ 90°, because we consider ENA sources which are near the magnetic equator. Fig 4 shows the estimated ENA spectra for a vantage point at an altitude of 550 km and MLT = 1930. The contribution of the inner region (R < 3.0 R_E) is dominating, so that the discrimination of the ENA originating from the ring current could be problematic. To resolve this, a geometrical technique has been developed by Milillo et al., (Low-altitude energetic neutral atoms imaging of the inner magnetosphere ..., submitted to J. geophys. Res., 1995) that makes use of measurements made simultaneously along different lines-of-sight which are at 60° to each other in the equatorial plane. Information from the differing integration paths is used as input for a kind of "tomographic" analysis, which gives an estimate of the ENA fluxes generated in the two different magnetospheric regions.

TEAM MEMBER RESPONSIBILITIES

Istituto di Fisica dello Spazio Interplanetario, CNR, Frascati (Roma)
S. Orsini: PI
P. Cerulli-Irelli: Payload manager; Instrument DPU and GSE.
M. Candidi: Payload scientist; Science data analysis.
M. Maggi: Payload technical engineer; Sensor electronics.
P. Baldetti: Payload mechanical engineer; Mechanical box.
G. Bellucci: GSE
R. Orfei: DPU engineering.
G. Chionchio: DPU engineering
A. Milillo: GSE. Scientific data analysis

Agenzia Spaziale Italiana, Roma
E. Flamini: Scientific data analysis and interpretation.
J. Sabbagh: ENA modelling and analysis.

Max-Plank Institut für Aeronomie, Lindau
S. Livi: Leading investigator; Sensor design.
B. Wilken: Sensor electronics; Calibrations.
I. A. Daglis: Scientific data analysis and interpretation.
W. Güttler: Analog electronics

Department of Physics, University of Arizona, Tucson, Arizona
C. C. Curtis: Instrument collimator and foils

K. C. Hsieh: Instrument performance and science data analysis.

Applied Physics Laboratory, J.H.U., Laurel, Maryland
E. C. Roelof: Data analysis and interpretation.
C. Chase: Data analysis and interpretation

Rutherford Laboratories, Oxford, U.K.
M. Grande: Foil calibration

Acknowledgements: The authors would like to thank A. Morbidini and A. Pavoni for their contributions to the instrument hardware. NASA Research Grant NAGW-3635 has supported the participation of C.C. Curtis and K.C. Hsieh in this project

REFERENCES

Balsiger, H., P. Eberhardt, J. Geiss, and D. T. Young, Magnetic storm injection of 0.9 to 16 KeV/e solar and terrestrial ions into the high-altitude magnetosphere, *J. Geophys. Res*, 85, 1645, 1980.

Barabash, S., R. Lundin, O. Norberg, C. Chase, E. Roelof, The Swedish microsatellite ASTRID: a first attempt at global magnetospheric imaging, *EOS*, 75, 44, 546, 1994.

Daglis, I. A., and S. Livi, Potential merits for substorm research from imaging of charge-exchange neutral atoms, *Ann. Geophys.*,13, 505, 1995.

Doke,T., M. Fujii, M. Fujimoto, K. Fujiki, T. Fukui, F. Gleim, W. Güttler, N. Hasabe, T. Hayashi, T. Ito, K. Itsumi, T. Kashiwagi, J. Kikuchi,T. Kohno, S. Kokubun, S.Livi, K. Maezawa, H. Moriya, K. Munakata, H. Murakami, Y. Muraki, H. Nagoshi, A. Nakamoto, K. Nagata, A. Nishida, R. Rathje, T. Shino, H. Sommer, T. Takashima, T. Terasawa, S. Ullaland, W. Weiss, B. Wilken, T. Yamamoto, T. Yamaginachi, and S. Yanagita, The energetic particle spectrometer HEP onboard the GEOTAIL spacecraft , J. *Geomagnetism and Geoelectricity* 46, 713, 1994.

Gloeckler, G., F. H. Ipavich, W. Studemann, B. Wilken, D. C. Hamilton, G. Kremser, D. Hovestadt, F. Gliem, R. A. Lundgren, W. Rieck, E. O. Tums, J. C. Cain, L. S. MaSung, W. Weiss, and H. P. Winterhoff, The charge-energy-mass (CHEM) spectrometer for 0.3 to 300 keV/e ions on the AMPTE/CCE, IEEE Trans. *Geosci. Remote Sens.*, GE-23, 234-240, 1985.

Gulich, M., and C. White, Satelite de Aplicationes Cientificas (SAC-B), International Cooperative Mission, *A.I.A.A.*, 93-4257, 1, 1993.

Hsieh, K.C., and C.C. Curtis, Imaging space plasma with energetic neutral atoms without ionization, this volume.

Orsini, S., I.A. Daglis, M. Candidi, K.C. Hsieh, S. Livi and B. Wilken, Model calculation of energetic neutral atoms precipitating at low altitudes, *J. Geophys Res.*, 99, 13489, 1994.

Rairden, R.L., L.A. Frank and J.D. Craven, Geocoronal imaging with Dynamics Explorer, *J. Geophys. Res.*, 91, 13613, 1986.

Roelof, E.C., and D.J. Williams, The terrestrial ring current: from in situ measurements to global images using energetic neutral atoms, *John Hopkins APL Technical Digest*, JHU/APL, 1988.

Istituto di Fisica dello Spazio Interplanetario, CNR, 00044 Frascati, Italy

Max-Planck Institut für Aeronomie, Katlenburg-Lindau, Germany

Institute of Ionospheric and Space Research, National Observatory of Athens, Greece

Physics Department, University of Arizona, Tucson, AZ 85721 USA

Agenzia Spaziale Italiana, 00100 Roma, Italy

Applied Physics Laboratory, Johns Hopkins University, Laurel, MD 20723 USA

Rutherford Appleton Laboratory, Oxfordshire, OX11 0QX, UK

Advances in Low Energy Neutral Atom Imaging

D.J. McComas[1], H.O. Funsten[1], and E.E. Scime[2]

In the near-Earth space environment, charge exchange continually occurs between the magnetospheric ions and very cold (~ eV) neutral atoms in the geocorona. This process creates both energetic (> tens of keV) and low energy (< tens of keV) neutral atoms (ENAs and LENAs, respectively) that radiate from the magnetosphere. Over the past several years, imaging techniques have been developed for observing these neutrals in space. Such techniques promise to open a new window onto the magnetosphere, providing synoptic measurements and a global view of the magnetospheric plasma and energetic particle environments for the first time. In this paper we describe recent advances in two of the leading techniques developed for LENA imaging: thin foil transmission-based imagers and UV-blocking grating-based imagers. In particular, we will discuss the development and testing of a flight-quality prototype foil-based imager and present an advanced design for a grating-based imager that has just been selected for NASA's IMAGE spacecraft.

1. INTRODUCTION

The current understanding of magnetospheric physics has been built on over three and a half decades of single point *in situ* measurements of the space environs. These observations have provided a large amount of detailed information about specific plasma properties and a statistically derived understanding of the global morphology and interconnections between various magnetospheric regions. Magnetospheric imaging promises to provide the first truly global, simultaneous observations of the structure and dynamics of these plasmas. This new view should prove invaluable for both testing our present understanding of the magnetosphere as well as enabling new discoveries in a manner similar to the early days of space observations [i.e., Williams et al., 1992; Moore et al., 1994].

Low energy neutral atoms (LENAs) [McComas et al., 1991] and energetic neutral atoms (ENAs) [Roelof, 1987] arise from the charge exchange process between cold geocoronal neutral hydrogen and the local plasma and energetic ion populations, respectively. This process produces neutrals which radiate from the charge exchange regions with energies and compositions that directly reflect these charged particle populations. From a remote vantage point, LENA and ENA measurements will provide global line-of-sight integrated images of the optically thin plasma and energetic particle environments.

The fundamental problem with detecting neutral atoms in space is that they must be measured against a very large ultraviolet (UV) background. Ly-α fluxes from the Sun at the Earth's orbit are ~10^{11} cm^{-2} s^{-1} while much more dispersed fluxes of ~10^9 cm^{-2} s^{-1} are observed from scattering of this light off the hydrogen geocorona [Hsieh et al., 1980]. Detectors suitable for measuring neutral atoms, such as microchannel plates and channel electron multipliers, are sensitive to UV radiation with about a 1% efficiency.

Techniques for imaging ENAs typically rely on relatively thick (~1000 Å) foils that serve to suppress the UV [e.g., Williams et al., 1992, and references therein]. Such thick foils, however, are not appropriate for energies less than several tens of keV owing to the large angular scattering, energy straggling, and even complete stopping

[1]Space and Atmospheric Sciences Group, Los Alamos, National Laboratory, Los Alamos, New Mexico
[2]West Virginia University, Morgantown, West Virginia

Measurement Techniques in Space Plasmas: Fields
Geophysical Monograph 103

of particles at sufficiently low energies. These problems required that other approaches to neutral atom imaging be developed in order to provide measurements of the lower energy LENAs [McComas et al., 1991].

LENA imaging techniques envisioned to date can be divided into two fundamentally different types: direct and indirect detection techniques [McComas et al., 1994]. In direct detection, the UV is blocked while the LENAs are transmitted. Structures that have been previously suggested to accomplish this include nuclear track filters [e.g., Gruntman, 1991, and references therein] and free-standing gold gratings [Scime et al., 1994; 1995].

High frequency moving shutters have also been used to provide UV suppression and neutral velocity selection. One type of moving shutter device that has been flown on sounding rockets utilized a set of slotted disks mounted on a rapidly rotating shaft [Moore and Opal, 1975]. While excellent in theory, the engineering realities of flying a high speed rotor make it difficult for long duration space missions. Recently, a more robust technique has been examined which uses sets of apertures, some of which are fixed and others of which are mounted to a high-Q bending-bar structure (similar to a tuning fork) so that they oscillate back and forth [Funsten et al., 1995]. A particular range of LENA speeds is selected by the timing of the apertures and gating of the detector behind the apertures; UV is rejected by making use of the fact that the LENAs travel much more slowly than the speed of light. As long as the oscillations are kept below the elastic limit of the material, such a device should be capable of functioning reliably over very long mission lifetimes.

In contrast to these direct detection techniques, indirect detection relies on modifying the incoming LENAs in some way so that they can be moved away from the path of the UV prior to detection. To the best of our knowledge, all such techniques proposed so far utilize positive or negative ionization of the neutrals and subsequent deflection by electric and/or magnetic fields. For LENAs with energies from several 100s of eV to several 10s of keV, the primary indirect technique is based on transmission through ultra-thin foils [McComas et al., 1991; 1994].

In this paper we briefly review the most mature indirect detection technique for LENAs: that using charge state conversion during transmission through an ultra-thin foil. We then describe its present state of development and discuss, for the first time, a flight-quality LENA prototype development effort presently underway within the Space and Atmospheric Sciences Group at Los Alamos National Laboratory. Finally, we discuss a new, alternate type of direct LENA detection technique that combines specially-developed, free-standing gold gratings with a subset of the components used in the foil-based LENA imager.

2. INDIRECT DETECTION: FOIL-BASED IMAGER

A foil-based LENA imager consists of four basic components: 1) a collimator to remove the ambient ions and electrons and to set the azimuthal field-of-view (FOV); 2) an ultrathin foil to convert a fraction of LENAs to ions; 3) an electrostatic analyzer (ESA) to reject UV and high energy particles and set the imager's energy resolution; and 4) a coincidence or time-of-flight (TOF) detector section. This latter section measures the LENA trajectory (described below) and rejects spurious counts from residual UV scattered through the imager, penetrating radiation, or background detector noise. Finally, TOF provides a direct measurement of the ion's speed and, combined with the energy per charge measurement of the ESA, its mass per charge or species.

The critical component of this type of LENA imager is the ultrathin foil charge state converter. Over the past decade, we have been developing the technology to reliably mount foils with thicknesses <50 Å for space flight applications [e.g., McComas et al., 1990]. When LENAs with energies greater that ~1 keV pass through such foils, they undergo minimal scattering and straggling. They also come to an equilibrium charge state that depends on the speed of the ion, its species, and the foil material. This conversion is independent of input charge state since the particle comes into equilibrium within several tens of angstroms of the entrance surface of the foil [Funsten et al., 1993a; 1994b].

As shown in Figure 1, exit charge state fractions are a function of both particle species and energy [Funsten et al., 1993a; 1993b; 1994]. For H^+, the fraction of positive ions drops from ~40% at 30 keV to ~5% at 1 keV. Hydrogen also forms a negative ion (H^-) after passing through a foil, although the probability remains well below the positive ion yield at all energies. Oxygen, on the other hand, has a higher yield for negative than positive ions due to its high electron affinity. Helium has a positive ion yield that is qualitatively similar to oxygen's but does not form a stable negative ion owing to its full 1s orbital in the ground state.

The LENA imager prototype presently under development at Los Alamos is schematically displayed in Figure 2. This prototype is being designed and built to flight quality specifications in the hope that ultimately a LENA proof-of-principle flight of opportunity can be found for it. The full mechanical design is complete and includes such details as a wax motor actuated reclosable door.

LENAs enter the instrument through the collimator which consists of nine very thin stainless steel sheets 75 mm in radius and spaced 5 mm apart, giving an azimuthal FOV of +/-2°. The plates are alternately grounded and biased at high voltage to sweep out all charged particles below several hundred keV. The nominal 0.5 mg cm^{-2}

carbon conversion foil covers a 20 x 40 mm aperture. These mounted foils have been successfully vibrated to typical flight levels without producing holes or tears. The foil is electrically isolated from the front plate of the instrument and biased with a variable positive high voltage. The electrostatic analyzer comprises two nested hemispherical (180°) analyzer plates with radii of 80 and 120 mm; a commandable positive high voltage is applied to the outer plate. Various combinations of foil and ESA bias voltages set different post-accelerations and energy per charge passbands for the ions.

At the exit of the hemispherical ESA, the transmitted ions have the same polar angular distribution that they did just after passing through the foil at the entrance aperture. The polar angle, or inclination, of an ionized LENA's orbit through the analyzer is determined in the detector section by measuring two points along the particle's trajectory. After exiting from the ESA, the ions pass through another ultra-thin (nominal 0.5 mg cm⁻²) foil. Secondary electrons from this foil are imaged on a 1-dimensional position-sensitive detector to give a "start" position at the foil. LENAs that pass through the detector section are imaged on a separate 1-dimensional "stop" position-sensitive detector. The combination of these two positions identifies the polar angle trajectory of the LENAs. The coincidence measurement between the electrons and ions greatly reduces the noise counting rates, even in high background environments. The TOF in the detector section is derived from the time between the start and stop events while the distance traveled across the

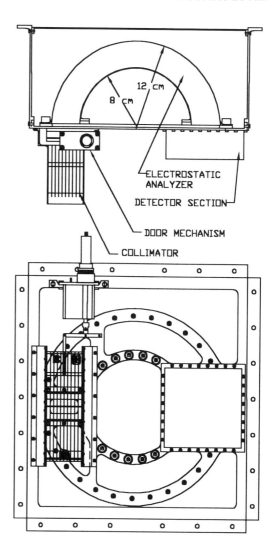

Figure 2. Schematic diagram of the foil-based LENA imager under development.

detector section is obtained from the trajectory information. These yield the LENA velocity, and, when combined with the ESA energy knowledge, provide information about the particles mass, and hence, species.

In order to simplify the mechanical design of this prototype, we detect both the LENAs and their associated secondary electrons using a single microchannel plate (MCP) Z-stack located directly below and parallel to the foil. A further refinement was to minimize the area required to detect the secondary electrons by electrostatically compressing them into the center region of the MCP detector. The detector section is shown in Figure 3 where the electrons are focused and imaged in a region only ~20% as wide as the foil aperture. Electrostatic focusing of the secondary electrons is achieved through a set of bias voltages applied to wires in

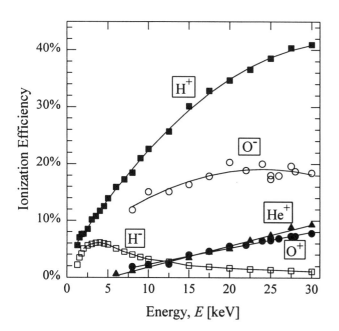

Figure 1. Ionization efficiency for hydrogen, helium, and oxygen neutrals as a function of energy.

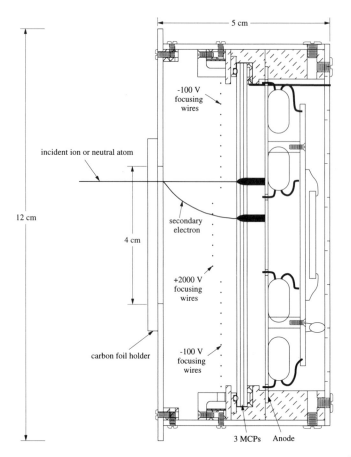

Figure 3. Schematic diagram of the imager detector section. Start (ejected secondary electrons) and stop (LENAs) locations out of the plane of the page provide the unique trajectory information.

front of the MCP stack so that their deflection is in the azimuthal plane and the polar angle location is unchanged.

In our test facility, a positive ion beam is produced in a microwave source with energies from <0.4 keV/q to 50 keV/q. The monoenergetic beam is then mass per charge selected in a 60° bending magnet and subsequently passes through electrostatic deflectors which can provide either fixed or scanning deflections. The beam can be imaged using a standard imaging microchannel plate (IMCP) detector at the entrance of the instrument chamber, which also houses an instrument platform having three axis positioning. Sensor position, internal sensor voltages, ion beam parameters, and IMCP detector images are all managed with an automatic computer control system that has been developed specifically for these purposes at Los Alamos. The entire vacuum system uses oil-free pumps and the instrument chamber pressure is typically ~3x10⁻⁸ Torr during testing.

Figure 4 shows an example of the results from our detector section testing. For the purposes of initial

testing, we temporarily replaced the one-dimensional anode in the LENA prototype detector section with a commercial two-dimensional position sensitive resistive anode encoder. Figure 4 shows the compression of electrons into the central region of the detector. The top panels are greyscale images of the ion distributions striking the MCP while the bottom panels are plots of the integrated total counts as a function of location across the MCP. The left pair of panels show the relatively uniform distribution of ions and electrons produced by electrostatically scanning the incident beam over the entire entrance aperture with the harp wires at ground potential. The harp wire shadows appear as reductions in the ion counts. The right panels show the results with an identical set up, but with +1000 V on the center harp wires and -100 V on the outside ones. Subsequent testing has demonstrated that replacing the harp wires with biased grids covering only the LENA detecting regions of the MCP can also compress the electrons onto the central one fifth of the detector.

In flight configuration, the commercial 2-D encoder anode is replaced with a dual 1-D wedge-type anode. For each of 1) the secondary electron and 2) LENA regions of the MCP, charge from the two ends of the wedge anodes is fed into separate charge sensitive preamplifiers.

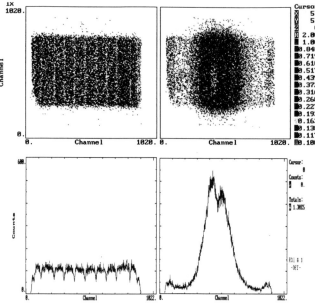

Figure 4. Results of detector section prototype testing showing how the biased wires act to compress and focus electrons from the start foil into a narrow band through the center of the MCP. The top left panel is taken with the harp wires unbiased, while the top right panel shows the result with the voltages on. The bottom panels display the integrated counts in the top panels as a function of location across the MCP.

Outputs from each of the two pairs of channels are then summed and divided by the signal from one side to provide simultaneous position information for both the electron (start) and LENA (stop) signals. The anode outputs also provide the TOF or coincidence measurements.

3. DIRECT DETECTION

The second type of LENA imaging technology that has been under development at Los Alamos over the past several years is based on gold grating technology [Scime et al., 1994; 1995; Funsten et al., 1995]. These free standing microstructures consist of evenly spaced gold bars which, in their present rendition, are ~140 nm wide and ~510 nm tall with an interbar spacing of ~60 nm to give a period of ~200 nm. The slots between the bars, which are smaller than or comparable to the wavelength of the UV light, primarily attenuate the UV by acting as waveguides. LENA transmission through existing gratings has been measured at ~10-15%.

Unlike our previously suggested designs for grating-based imagers [Scime et al., 1994; Funsten et al., 1995], which used sets of crossed gratings to achieve sufficient UV suppression, this paper introduces a significant improvement which utilizes single gratings. This allows LENAs to be detected simultaneously over a broad range of incident polar angles with a large effective aperture. This design can be implemented in many ways; one simple rendition, shown in Figure 5, is achieved by using the collimator and detector sections of the foil-based

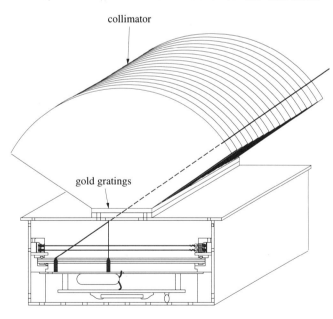

Figure 5. Schematic diagram of an implementation of a grating-based imager that utilizes the front (collimator) and back ends of the foil-based imager described above.

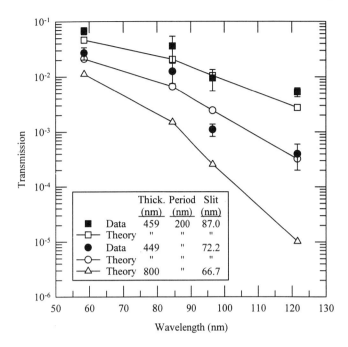

Figure 6. Plots of measured (solid symbols) and calculated (open symbols and lines) EUV transmission through gold gratings as functions of wavelength.

imager previously described. In place of the ESA is a gold grating with its bars oriented parallel to the collimator plates; the polar directions of the LENAs are determined directly in the detector section, just as in the foil-based imager design.

Figure 6 shows UV transmission as a function of wavelength. Measurements from two test gratings are indicated by the solid circles and squares; the calculated transmissions for these gratings, shown with the open symbols and lines, match the observations well. In the bottom curve, the calculations are extended to predict the transmission through thicker gratings that may be available in the near future. Such gratings should be capable of achieving attenuations for Ly-α of $>10^5$.

The background or noise counting rates in such a LENA instrument can be easily estimated from the combination of this transmission factor (presently ~5×10^{-4}), the foil's UV transmission efficiency (<0.1), the MCP detection efficiency for EUV (~0.01), the solar or geocoronal-scattered EUV flux, and the aperture area. For a geocoronal EUV flux of 2×10^9 cm^{-2} s^{-1}, the background singles counting rates would be <1 kHz for each square centimeter of open aperture area. Since coincidence measurements can easily handle many kHz of uncorrelated background counts, the present grating technology is sufficient to produce a grating-based LENA imager and incremental improvements expected in the near term in this technology will make the capabilities of grating-based imagers even better.

4. DISCUSSION

LENA imaging should provide the next major step forward in magnetospheric research by enabling synoptic measurement of the dynamic magnetospheric plasma populations. These unprecedented observations will make simultaneous measurement of the various plasma regions throughout the inner magnetosphere possible for the first time. In this paper we described advances in both direct and indirect LENA detection techniques. In particular, we describe a foil-based LENA imager prototype that is being developed as a protoflight unit, to flight quality design standards, in the hope that we may be able to find a flight of opportunity for it.

In addition, we describe for the first time, a completely new and advanced design for a direct detection LENA imager. This design uses a single gold grating to attenuate the UV radiation by many decades while still transmitting the LENAs with reasonable efficiency. We have examined a number of different LENA imager designs using this new grating technology; one particularly simple rendition combines the gratings with the collimator and detector section for the foil-based imager described above. There are a number of advantages of the grating-based imager technology. These include a wide range of simultaneously accepted energies leading to a nearly continuous duty cycle; a compact, low mass design; and a large geometric factor per unit volume of the instrument. Because of these advantages, our grating-based imager has been selected and is presently under flight development to measure neutral atoms with energies in the one to 10s of keV energy range for NASA's IMAGE spacecraft.

Acknowledgements. We thank Rudy Abeyta, Phil Barker, Danny Everett, Bill Spurgeon, Paul Stigell, and Steve Storms for help in developing these sensors. This work was carried out under the auspices of the United States Department of Energy.

REFERENCES

Funsten, H.O., B.L. Barraclough, and D.J. McComas, Shell effects observed in exit charge state distributions of 1-30 keV atomic projectiles transiting ultra-thin carbon foils, *Nuc. Inst. and Meth.*, *B80/81*, 49-52, 1993a.

Funsten, H.O., D.J. McComas, and B.L. Barraclough, Ultrathin foils used for low energy neutral atom imaging of planetary magnetospheres, *Optical Engineering, 32*, 3090-3095, 1993b.

Funsten, H.O., B.L. Barraclough, and D.J. McComas, Interaction of slow H, H_2, and H_3 in thin foils, *Nuc. Inst. Meth. in Phys. Res. B, 20*, 24-28, 1994.

Funsten, H.O., D.J. McComas, and E.E. Scime, Low energy neutral atom imaging for remote observations of the magnetosphere, *J. Spacecraft and Rockets, 32*, 899-904, 1995.

Gruntman, M.A., Submicron structures: promising filters in EUV - a review, *Proc. SPIE Internl. Symp.*, 1549, 385, 1991.

Hsieh, K.C., E. Keppler, and G. Schmidtke, Extreme ultraviolet induced forward photoemission from thin carbon foils, *J. Appl. Phys.*, 51, 2242-2246, 1980.

McComas, D.J., and J.E. Nordholt, New approach to 3-D, high sensitivity, high mass resolution space plasma composition measurements, *Rev. Sci. Inst., 61*, 3095-3097, 1990.

McComas, D.J., B.L. Barraclough, R.C. Elphic, H.O. Funsten III, and M.F. Thomsen, Magnetospheric imaging with low energy neutral atoms, *Proceedings of the National Academy of Sciences, USA, 88*, 9589-9602, 1991.

McComas, D.J., H.O. Funsten, J.T. Gosling, K.R. Moore, E.E. Scime, and M.F. Thomsen, Fundamentals of low energy neutral atom imaging, *Optical Engineering, 33*, 335-341, 1994.

Moore, J.H., Jr. and C.B. Opal, A slotted disk velocity selector for the detection of energetic atoms above the atmosphere, *Space Sci. Inst., 1*, 337-386, 1975.

Moore, K.R., E.E. Scime, H.O. Funsten, D.J. McComas, and M.F. Thomsen, Low energy neutral atom emission from the Earth's magnetosphere, *Optical Engineering, 33*, 342-348, 1994.

Roelof, E.C., Energetic neutral atom image of a storm-time ring current, *Geophys. Res. Lett.*, 14, 652-655, 1987.

Scime, E.E., H.O. Funsten, D.J. McComas, K.R. Moore, and M.A. Gruntman, A novel low energy neutral atom imaging technique, *Optical Engineering, 33*, 357-361, 1994.

Scime, E.E., D.J. McComas, E.H. Anderson, and M.L. Schattenburg, Extreme ultraviolet polarization and filtering with gold transmission gratings, *Applied Optics, 34*, 648-654, 1995.

Williams, D.J., E.C. Roelof, and D.G. Mitchell, Global magnetospheric imaging, *Rev. Geophys., 30*, 183-208, 1992.

H.O. Funsten and D.J. McComas Space and Atmospheric Sciences Group, Los Alamos National Laboratory, Los Alamos, NM 87545.

E.E. Scime, Department of Physics, West Virginia University, Morgantown, WV 26506.

The Imaging Neutral Camera for the Cassini Mission to Saturn and Titan

D G Mitchell, S. M. Krimigis, A. F. Cheng, S. E. Jaskulek, E. P. Keath, B. H. Mauk, R. W. McEntire, E. C. Roelof, C. E. Schlemm, B. E. Tossman, and D. J. Williams

Applied Physics Laboratory Johns Hopkins University

The INCA sensor is the first Energetic Neutral Atom (ENA) imager funded for flight by NASA. It is a part of the Cassini mission to Saturn, where it will be well suited to monitoring the global dynamics of the Saturn-Titan magnetospheric system throughout the orbital tour. The investigative requirement is to perform remote sensing of the magnetospheric energetic (E ≥ 20 keV) plasma ions by detecting and imaging charge exchange neutrals, created when magnetospheric ions capture electrons from ambient neutral gas. In fact, this instrument will work to somewhat lower energies. Escaping charge exchange neutrals, detected by the Voyager-1 spacecraft outside Saturn's magnetosphere, can be used like photons to form images of the emitting regions, as has been done at Earth. Since Cassini is 3-axis oriented, INCA is designed as a 2-D imager with a field of view of 90 by 120 degrees. The technique involves sensing the position of the ENA as it penetrates an entrance foil and again on the back-plane microchannel plate, thereby establishing the ENA's trajectory and time-of-flight (TOF). Along with rough composition determined by pulse-height analysis, the sensor produces images of the hot plasma interaction with the cold ambient neutral gas as a function of species and energy, from ~20 keV to several MeV. A large geometric factor (~2.5 cm²-sr) allows sufficient sensitivity to obtain statistically significant images in ~1 to 30 minutes, for most conditions and locations. We will discuss several of the design details unique to this instrument, and will also review some of the its limitations as well as modifications required for imaging in the Earth magnetospheric environment.

INTRODUCTION

Charge exchange interactions between energetic trapped ions and cold ambient neutral exospheric gasses result in the creation of ENA's in the magnetospheres of Earth, Jupiter, Saturn, and possibly Uranus and Neptune. These ENA's then escape the magnetospheric systems since they are no longer trapped, and a detector using straight-path optics may be used to form global images of the emission regions, and thus of the magnetospheric system. A comprehensive description of the principles behind this approach, as well as the modeling techniques required to make it useful, are given by Roelof [*Roelof, 1987*].

The Ion Neutral CAmera (INCA) head on the Cassini mission Magnetospheric IMaging Instrument (MIMI) is a neutral particle detection system for the Cassini Orbiter spacecraft which will perform global imaging measurements to study the overall configuration and dynamics of Saturn's magnetosphere and its interactions with the solar wind, Saturn's atmosphere, Titan, and the icy satellites. The investigative approach is to perform remote sensing of the magnetospheric energetic (E ≥ 20 keV) ion plasmas by detecting and imaging charge exchange neutrals, created when singly-charged magnetospheric ions capture electrons from ambient neutral gas. The escaping charge exchange neutrals, detected by the Voyager-1 spacecraft outside Saturn's magnetosphere,

Measurement Techniques in Space Plasmas: Fields
Geophysical Monograph 103

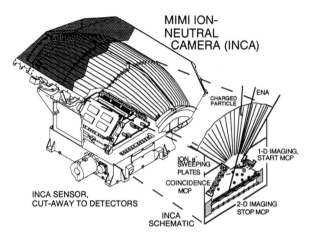

Table 1. INCA characteristics

Measurement	Capability
Energy range	<10 keV - 8 MeV/nuc.
Velocity resolution	50 km/sec (1ns TOF)
Composition	H, O, Heavies
Field Of View	120° x 90°
Angular coverage	.76π sr (3π, if spinning)
Angular resolution	~4° x 4°, high energy
(scattering in foils at low E)	>4° x >4°, low energy
Time resolution	0.1 sec., PHA events
	85 sec, low resolution
	6 min., high resolution
	23 min., full sky
G x ε (cm²-sr)	~2.5 for O, ~2.2 for H
G x ε for 4° x 4°. pixel	.0075, tapered at edges
Dynamic range	~10⁷

Figure 1. Engineering drawing of the INCA head, with section schematic. Serrated charged collimator plates sweep charged particles out of aperture. ENA penetrate front foil covering entrance slit, producing secondary electrons for start pulse and 1-D position. ENA travels to back foil and 2-D imaging MCP, and back-scattered secondary electrons provide coincidence pulse. Dots indicate the locations of wire electrodes for secondary electron steering.

can be used as if they were photons to form images of the emitting regions, as has been done at Earth [*Roelof, 1987; Roelof et al., 1985*]. The magnetospheric imaging will be complemented by the first in-situ measurements at Saturn of 3-D particle distribution functions and comprehensive ion composition, including molecules and charge state (E ≥ 10 keV/e), provided by the University of Maryland CHarge Energy and Mass Spectrometer (CHEMS) head of the MIMI investigation; and total ion and electron flux measurements by the Max Planck Institut für Aeronomie Low Energy Magnetospheric Measurements System (LEMMS). The combination of in-situ measurements with global images, together with analysis and interpretation techniques, will yield a global assessment of magnetospheric structure and dynamics, including those of a) magnetospheric ring currents and hot plasma populations, b) magnetic field distortions, c) electric field configuration, d) particle injection boundaries associated with magnetic storms and substorms, and e) the connection of the magnetosphere to ionospheric altitudes.

1. THE INCA SENSOR

INCA is a large ~2.5 cm²-sr (G·ε, where G is the geometric factor, and ε is the efficiency) time-of-flight (TOF) detector that analyzes separately the composition and direction of motion of energetic neutral atoms. Elements of the detector assembly can be seen in Figure 1. Sensor characteristics are summarized in Table 1. The entrance includes a serrated-plate fan for charged particle deflection,

with a field-of-view (FOV) of 90° in the spin direction (the azimuthal direction) by 120° in the direction perpendicular to the spin plane (elevation), centered on the spin plane.

1.1 Deflector

The magnesium deflector plates are serrated to inhibit forward scattering of incident particles. Commandable potentials of up to ±6 kV are applied to alternate plates to sweep energetic charged particles with energies ≤ 500 keV/e into the plate walls (excluding them from the detector). The shielding effectiveness of the collimator has been measured in the laboratory (Figure 2; also in [*Keath et al., 1989*]) and is adequate to permit neutral particle imaging below the sweeping energy even while the instrument is within moderate magnetospheric plasma environments.

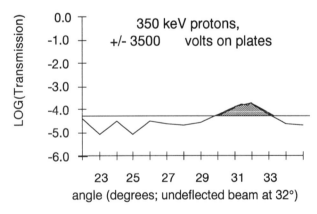

Figure 2. Accelerator test of deflector ion sweeping at an intermediate voltage, indicating charged particle rejection at better than 4 orders of magnitude. The peak with shading is in the position of the undeflected beam, and results from energetic neutrals created via charge exchange of energetic ions in the beam tube.

1.2 Measurement Technique

Incoming neutral particles encounter and penetrate a thin foil, producing secondary electrons (Figure 1). The secondary electrons are first accelerated perpendicular to the foil by the E-field which is locally normal to the equipotential surface of the foil, and then steered electrostatically (using a combination of wires and shaped electrodes at fixed potentials, Figure 3) onto a side start microchannel plate (MCP) with 1-D position sensitive anode. A start time taken from a capacitively coupled electrode of the start MCP is generated by this event. The original incident particle, after some angular scattering in the front foil, continues through the instrument, striking a second foil just in front of the stop microchannel plate assembly. Secondary electrons produced on the exit surface of the imaging foil are accelerated into the stop microchannel plate and 2-D imaging anode, mapping the position of impact and registering the stop time for the TOF measurement. In addition to this TOF and trajectory measurement, secondary electrons produced as the ENA enters the back foil are electrostatically accelerated and guided to the side coincidence MCP (see Fig. 1). The electron travel time for these back-scattered electrons is constrained to < 40 ns by the steering potentials. The

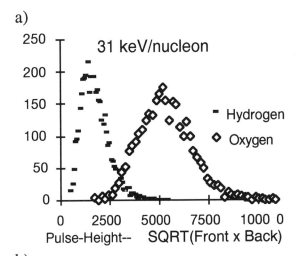

a)

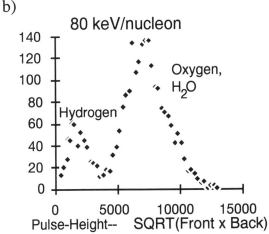

b)

Figure 4. Separation of oxygen and hydrogen using MCP pulse-height signatures. a) 31 keV O and H measurements, overplotted. b) Accelerator run of 80 keV/nuc H_2O, in which some of the molecules split in the beam tube. The separate H and O peaks are clear.

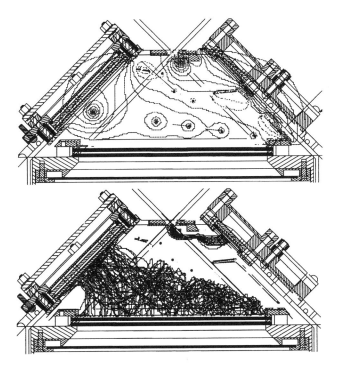

Figure 3. a) Secondary electron steering potentials. b) Secondary electron trajectories: Start e⁻ sweep from front foil to start MCP, with ~500 ps dispersion, coincidence e⁻ oscillate between back foil and potential barrier, reaching coincidence MCP in < 40ns.

pulse generated in the coincidence plate is used in the valid event logic circuitry as a coincidence check on the measurement, further reducing the probability that uncorrelated background will be falsely counted as a neutral.

The number of secondary electrons produced in each foil is dependent on the atomic number of the neutral atom; for the two most common neutral atoms expected, oxygen has been found to produce ~8 to 16 times the number of secondaries that hydrogen will produce (Figure 4). Exploiting this phenomenon, we record the pulse-height of the microchannel plate signal and determine the species based on that measurement. The measured positions of the penetration of the foils determines the particle arrival direction and its path length within the instrument; the TOF associated with this path length (and thus velocity) combined with the species determination based on pulse-height analysis determine particle energy. Although the species determination is poorly resolved, the neutral fluxes at Saturn are expected to consist almost exclusively of hydrogen and oxygen, and these will be relatively well determined.

1.3 Foils and UV Sensitivity

Electrons are also produced in the foils by photoionization by vacuum ultraviolet light EUV and FUV, predominantly in the Ly-α line at Saturn. The foil must therefore be chosen thick enough to attenuate the Ly-α to an acceptable level, so that the position and TOF circuitry is not swamped by false counts. The TOF measurement is made by recording a start signal followed by a stop signal some time later. Any single photon can produce a start or a stop, but not both. As long as the time between a false Ly-α start and a separate false Ly-α stop is long compared with the maximum valid TOF period (~100 nsec), this background will not produce false events. The electronic design assumes nominal upper limits for Ly-α start events of $\leq 5 \times 10^4$ secondary electrons per second, and for stop events of ≤ 1000 secondary electrons per second. The peak Ly-α flux at Saturn (excluding the sun itself) is ~1 kR [Broadfoot et al., 1981], which corresponds to 5.6×10^8 photons/sec incident on the front foil (which has a ~7 cm^2-sr geometry factor). The secondary electron production efficiency for Ly-α photons is ~1% [Hsieh et al., 1980]. A very thick foil would eliminate Ly-α entirely; however, the minimum energy particle which can penetrate the foil and the angular scattering of particles within the foil both increase with foil thickness. Therefore, it is desirable to select the thinnest foil possible, consistent with the adequate suppression of Ly-α.

Transmittance of various thin foils for H Ly-α has been studied [Keath et al., 1989; Hsieh et al., 1980; Hsieh et al., 1991; Powell et al., 1990; Powell, 1993]. Materials tested in those studies include pure carbon foils, aluminum on carbon, silicon on carbon, Lexan, aluminized Lexan, Parylene, and others. The material found to have the least particle stopping power (and least scattering) for a given Ly-α transmittance is silicon on carbon. For example, 7 μg cm^{-2} Si on 7 μg cm^{-2} C has a Ly-α transmittance [Hsieh et al., 1991] of 10^{-3} and in terms of particle scattering is 25% thinner than a pure C foil having the same transmittance. An additional factor influencing the foil design is strength. In general, polymer foils such as Lexan or Parylene are considerably stronger and less brittle than carbon or silicon. We therefore have designed a foil of 5.5 μg cm^{-2} Si, 5μg cm^{-2} Lexan, and 1.0 μg cm^{-2} C with a measured 1.0×10^{-3} Ly-α transmittance to reach a conservative design goal of $\leq 6 \times 10^3$ Ly-α induced secondary electrons per second when the brightest portion of the Saturnian atmosphere/exosphere is in view. This foil is mounted on a 70 line/inch stainless mesh (82% transmission). The stop foil, composed of Lexan and carbon, will reduce the Ly-α flux to the back, 2-D imaging MCP by about another factor of 100. With the sun in view at Saturn, because the quantum efficiencies for secondary electron production are so high at EUV wavelengths, no practical foil composition-thickness combination could be identified to sufficiently suppress it. We have therefore included a recloseable shutter, designed such that when the sun is nominally in view, the shutter will reduce the field of view and the geometry factor enough to allow operation.

The scattering in the foils, along with the angular spread introduced in the electron optics, has been measured during accelerator tests of the complete engineering model INCA sensor. Representative angular standard deviations are: at 31 keV/nucleon, $\sigma_\theta = 4.5°$, $\sigma_\phi = 3.1°$ for hydrogen, $\sigma_\theta = 2.5°$, $\sigma_\phi = 3.6°$ for oxygen; at 80 keV/nuc, $\sigma_\theta = 2.6°$, $\sigma_\phi = 1.5°$ for hydrogen, $\sigma_\theta = 3.1°$, $\sigma_\phi = 2.7°$ for oxygen; at 5 keV/nuc, $\sigma_\theta = 9.3°$, $\sigma_\phi = 6.5°$ for oxygen (no measurement has yet been made at low hydrogen energies); at 200 keV, $\sigma_\theta = 3.4°$, $\sigma_\phi = 1.2°$ for hydrogen. The ϕ resolution is dominated by scattering in the front foil and the entrance slit width in all cases. The θ resolution is dominated by the scattering in the front foil and by the electron optics for the front foil imaging. The latter effect is dominant at high hydrogen energies, where the number of secondary electrons produced on average is ~1.

1.4 Sensitivity and Background Rejection

The INCA head relies upon valid time of flight (separate start and stop) measurements, as well as coincidence pulses to differentiate background from foreground events. FUV will produce secondary electrons, but not correlated start, stop, and coincidence pulses. In order to produce false events that appear valid, an FUV start must be followed within the valid time windows by FUV (or other) stop and coincidence pulses. Other sources of false signals may come from high energy penetrating > 2 MeV energetic electron fluxes, which should only become a significant background inside L = 5 to 6, as well as ion fluxes above the rejection cut-off energy of the deflector deflection plates. The latter will be discriminated against by their TOF signatures (i.e., too fast to be valid neutrals below the deflector cut-off), although we expect to be able to image neutrals up to ~6 MeV when the ambient energetic ion flux above the cut-off energy is sufficiently low. The TOF and coincidence technique for identifying valid events has been very successful in our experience with the MEPA experiment on the AMPTE CCE spacecraft. The INCA geometry factor is ~2.5 cm^2 sr. The expected environment will give foreground count rates ranging from ~10^3/s for the inner Saturnian magnetosphere down to ~1/s at great distances, corresponding to unusually high apoapsis and quiet conditions.

The system has an intrinsic window for valid events of ~100ns, based on the maximum valid TOF for a ~50 keV oxygen (which is the slowest valid ENA expected to be analyzed), and the coincidence window is ~40ns. The background rate for false valid events from uncorrelated background rates will therefore be $\sim 1 \times 10^{-7} \cdot \sim 4 \times 10^{-8} \cdot R_{start} \cdot R_{stop} \cdot R_{coincidence}$, where R_{start} (R_{stop}, $R_{coincidence}$) is the uncorrelated singles rate due to penetrating background and FUV on the start (stop, coincidence) microchannel plate. Thus the false valid event rate is $\sim 4 \times 10^{-15} \cdot R_{start} \cdot R_{stop} \cdot R_{coincidence}$, assuming the accidental rates are uncorrelated. For the FUV generated background rates expected with the design foil thicknesses, i.e. $R_{start} = R_{coincidence} = $ ~5600 counts/sec and $R_{stop} = $ ~56 counts/sec (see discussion in section 1.3, above), the maximum false coincidence rate attributable to FUV will be $\sim 7 \times 10^{-6}$ events per second in regions with low foreground rates. If a penetrator hits both MCP's, and produces a correlated pulse pair, the probability of a false coincidence event being registered goes up. However, by requiring a minimum TOF of about 1.5ns (corresponding to a ~6 MeV proton TOF), we will discriminate against some of these very fast particles. Using the coincidence requirements, background rates from penetrators (cosmic rays and magnetospheric energetic particles) combined with uncorrelated FUV rates are expected to be <0.001 per second at apoapsis, increasing as Cassini enters the penetrating > 2 MeV electron flux in the inner magnetosphere.

2. PROCESSING

The analog electronics amplifies the MCP detector outputs, processes the signals, converts them to digital words, and sends them to Data Processing Unit (DPU) for analysis. This analysis derives the mass, direction, and count rates of the particles entering the sensors. The DPU then increments the appropriate rate elements in the counting memory, and stores the mass, velocity, and direction data in a FIFO buffer. The processor uses the data in this buffer to generate both the image planes and high resolution event words. The DPU also performs telemetry formatting, command processing, control I/O, and alarmed housekeeping functions.

Figure 5 shows a block diagram of the INCA analog processing electronics. The INCA sensor generates an MCP "start" signal, an MCP "stop" signal, and MCP position signals. The timing signals will be processed in the Time-to-Amplitude Converter (TAC) circuitry over the range of ~2 to ~100 nsec with total system resolution of ~1.5 nsec. The TAC processing time through the A/D converter is ~10 μsec. Position measurements are derived from the "start" and "stop" MCP anode signals. The

timing and angle information from a valid event is latched into an event register and stored until the DPU is ready to process it.

2.1 High Voltage Processing

The INCA sensor will contain a total of three MCP HV supplies, one to bias the start MCP, one to bias the coincidence MCP's, and the third for the stop MCP. The electron steering elements use voltages provided by these same supplies. INCA will also have similar high voltage supplies to bias the charged particle deflection plates. These supplies will be commandable, such that the deflection plate voltages can be varied between low voltage and about 6 kV.

3. INCA HERITAGE

The INCA sensor has substantial similarities to detector heads that we have already flown on NASA spacecraft and are now developing for flight. These include the highly successful Medium Energy Particle Analyzer (MEPA) flown on the CCE spacecraft of the AMPTE program [*McEntire et al., 1985*], the EPD experiment launched on the Galileo spacecraft to Jupiter, and the EPIC experiment recently launched on the Geotail spacecraft of the ISTP program.

However, there are aspects of the proposed head that have not been incorporated into our previously flown particle analyzers: (i) electrostatic plates to sweep out charged particles, (ii) imaging anodes used together with the large microchannel plates to perform spatial imaging. These technology areas have received substantial attention in our laboratory studies via the NASA Innovative Research Program (IRP) NAGW-865, 1986-1989. Several results from this work have already been published [*Keath et al., 1989; McEntire and Mitchell, 1989*].

4. FUTURE POSSIBILITIES

The INCA ENA imager, while designed specifically for the Saturn environment, is well suited for application at Earth. The Earth's magnetosphere has been shown to be an excellent candidate for both ENA and LENA imaging [*Roelof et al., 1985; Roelof, 1987; McEntire and Mitchell, 1989; Funsten et al., 1994; McComas et al., 1992, 1994, 1996*], and as at Saturn, ENA imaging will require as much sensitivity as resources will permit. INCA combines a very large geometry factor (~2.5 cm^2-sr) with very good angular resolution (~4 degrees). Simulations of images in the Earth environment, using ISEE 1 measurements of neutral fluxes as input, predict that INCA would obtain images with peak counts of ~2,000/pixel in 1 minute, for 4° x 4° pixel resolution under moderately active (Dst ~-100 nT) conditions. The front foil characteristics will have to

INCA Block Diagram

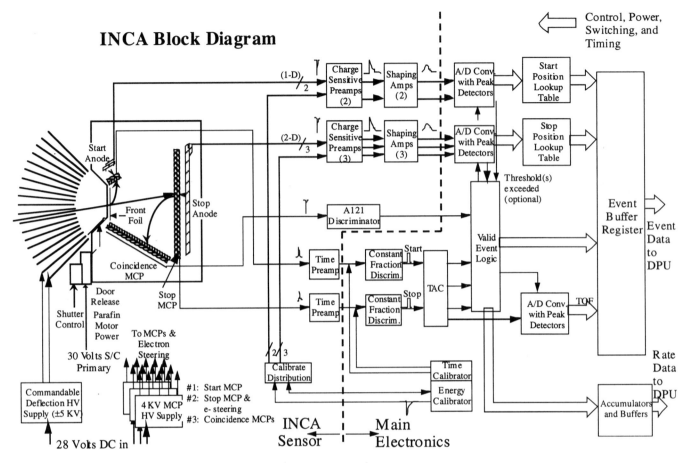

Figure 5. Block diagram of the INCA sensor analog electronics and pre-DPU main electronics.

be modified, consistent with FUV and EUV intensities emanating from the Earth's atmosphere and plasmasphere, but these modifications are minor. Other conceptual designs for ENA imaging are discussed by McEntire and Mitchell [*McEntire and Mitchell, 1989*].

Acknowledgements. DGM acknowledges the contributions of K. C. Hsieh, V. A. Drake, and F. Powell to the foil design. This work was supported in part by NASA Grants NAGW-865 for Innovative Research and NAGW-1862 for Planetary Instrument Definition and Development, as well as NASA under Task I of Contract N00024-97-C-8119 to the Navy.

REFERENCES

Broadfoot, A. L., B. R. Sandel, D. E. Shemansky, J. B. Holberg, G. R. Smith, D. F. Strobel, J. C. McConnell, S. Kumar, D. M. Hunten, S. K. Atreya, T. M. Donahue, H. W. Moos, J. L. Bertaux, J. E. Blamont, R. B. Pomphrey, and S. Linick, "Extreme ultraviolet observations from Voyager 1 encounter with Saturn," *Science, 212,* 206-211, 1981

Funsten, H.O., D.J. McComas, K.R. Moore, E.E Scime, and M.F. Thomsen, "Imaging of magnetospheric dynamics using low energy neutral atom detection," *Solar System Plasmas in Space and Time, Geophysical Monograph 84,* 275-282, 1994.

Hsieh, K. C., B. R. Sandel, V. A. Drake, and R. S. King, "H Lyman α transmittance of thin C and Si/C foils for keV particle detectors," *Nuc. Inst. Meth., B61,* 187-193, 1991

Hsieh, K. C., E. Keppler, and G. Schmidtke, "Extreme ultraviolet induced forward photoemission from thin carbon foils," *J. Appl. Phys., 51,* 2242-2246, 1980

Keath, E. P., G. B. Andrews, A. F. Cheng, S. M. Krimigis, B. H. Mauk, D. G. Mitchell and D. J. Williams, "Instrumentation for energetic neutral atom imaging of magnetospheres," *Solar System Plasma Physics, Geophysical Monograph 54,* J. H. Waite, Jr., J. L. Burch, and R. L. Moore, Ed., 165-170, American Geophysical Union, 1989.

McComas, D.J., H.O. Funsten, and E.E. Scime, "Advances in Low Energy Neutral Atom Imaging," *this publication,* 1996.

McComas, D.J., H.O. Funsten, J.T. Gosling, K.R. Moore, and M.F. Thomsen, "Low energy neutral atom imaging", *Instrumentation for Magnetospheric Imagery, SPIE Proc., V. 1744,* 40-50, 1992.

McComas, D.J., H.O. Funsten, J.T. Gosling, K.R. Moore, E.E. Scime, and M.F. Thomsen, "Fundamentals of low energy neutral atom imaging," *Optical Engineering, 33,* 335-341, 1994.

McEntire, R. W., and D. G. Mitchell, "Instrumentation for global magnetospheric imaging via energetic neutral atoms," *Solar System Plasma Physics*, p 69, J. H. Waite, Jr., J. L. Burch, and R. L. Moore, Ed., *Geophysical Monograph 54*, American Geophysical Union, 1989.

McEntire, R. W., E. P. Keath, D. E. Fort, A. T. Y. Lui, and S. M. Krimigis, "The medium energy particle analyzer (MEPA) on the AMPTE CCE spacecraft," *IEEE Trans. Geosci. Remote Sensing, GE-23,* 230-233, 1985.

Powell, F., "Care and feeding of soft x-ray and extreme ultraviolet filters," Luxel Corporation, Friday Harbor, WA 98250, 1993

Powell, F., P. W. Vedder, J. F. Linblom, and S. F. Powell, "Thin film performance for extreme ultraviolet and x-ray applications," Opt. Eng. 29, 614-624, 1990.

Roelof, E. C., "Energetic neutral atom image of a storm-time ring current," *Geophys. Res. Lett., 14,* 652-655, 1987.

Roelof, E. C., D. G. Mitchell, and D. J. Williams, "Energetic neutral atoms (E~50 keV) from the ring current: IMP 7/8 and ISEE-1," *J. Geophys. Res., 90,* 10991-11008, 1985.

Applied Physics Laboratory Johns Hopkins University, Johns Hopkins Road, Laurel, MD 20723-6099

Surface Ionization with Cesiated Converters for Space Applications

M.R. Aellig[1], P. Wurz[1], R. Schletti[1], P. Bochsler[1], A.G. Ghielmetti[2],

E.G. Shelley[2], S.A. Fuselier[2], J.M. Quinn[3], F. Herrero[4], M.F. Smith[4]

Neutral particle imaging can be used for remote sensing of magnetospheric plasmas. Due to the low fluxes of neutral particles and the very transient nature of many phenomena in such environments, a highly sensitive detection method is required. Neutral particles in the energy range between 10eV and 1keV have not previously been accessible to a mass, energy and angle analysis. Surface ionization, a well-established laboratory technique, can efficiently convert neutral particles in this energy range into negative ions to be analyzed with mass spectrographs. This article describes surface ionization with low work function surfaces as a method and discusses its applicability in spaceborne instrumentation.

1. INTRODUCTION

One of the main objectives of space physics is to understand the Earth's magnetosphere, its interaction with the ionosphere, and the energy input by solar radiation (photon and particle flux). After decades of in situ plasma measurements, which provided point measurements, new techniques are needed to provide maps of plasma distributions by remote sensing. In addition to the imaging capability, high temporal resolution, and thus high sensitivity, is necessary, since many magnetospheric processes are transient.

Although neutral atom imaging is a relatively new technique [*McEntire and Mitchell, 1989*] a few applications have been reported [*Hsieh et al., 1992*]. For energies larger than 10keV, adaptations of standard instruments for measuring energetic ions can be used. *Hsieh and Curtis [1997]* give a review of the status of this instrumentation. However, the energy range below 1keV has been inaccessible to direct measurements so far. For standard mass spectrometry, neutral particles have to be ionized before being analyzed and detected. Surface ionization has been identified as the only ionization technique which has the potential for sufficient ionization yield in the energy range of 10eV to 1keV within the constraints for space instrumentation [*Ghielmetti et al., 1994; Wurz et al., 1995*].

In the past 15 years new surface ionization techniques have been developed for application in fusion plasma research. With these techniques ionization efficiencies up to 67% in the energy range from several eV to about 1keV [*Van Wunnik et al., 1983, Geerlings et al., 1985*] have been achieved. Surface ionization makes use of low work function (WF) surfaces for converting neutral particles to negative ions by resonant charge exchange after reflection from a converter surface. This technique introduces new demands on the design of instruments and requires the development of new analyzer elements with matched ion optical properties. An instrument satisfying these demands has been described by *Ghielmetti et al. [1994]* and by *Wurz et al. [1995]*. Alternate concepts that combine surface ionization techniques with a spectrometer have been described by *Herrero and Smith [1992]* and by *Gruntman [1991]*.

[1]Physikalisches Institut, University of Bern, Switzerland
[2]Lockheed Martin Palo Alto Research Lab, Palo Alto, CA
[3]SSC Morse Hall, University of New Hampshire, Durham, NH
[4]Lab f.Extraterrestrial Physics, NASA/GSFC, Greenbelt, MD

Measurement Techniques in Space Plasmas: Fields
Geophysical Monograph 103
Copyright 1998 by the American Geophysical Union

2. SURFACE IONIZATION

An extensive review of charge exchange in atom-surface collisions is given by *Los and Geerlings* [*1990*]. To describe the concept of charge exchange, a one-dimensional model is used. As the neutral particle moves towards a metal surface, its affinity level is energetically lowered by an image-charge potential resulting from the interaction between the atom and the solid. Affinity states lying below the Fermi level of the metal can be populated by resonant electron tunneling from the metal, but the electrons can also tunnel back to the metal if empty states are available. The finite lifetime of an electron in the affinity state leads to an energy broadening via the Heisenberg uncertainty principle. The lowering and broadening of the affinity level is illustrated schematically by *Van Amersfoort et al.* [*1985*] in their Figure 1. Close to the surface an equilibrium of bi-directional tunneling electrons is established. Thus the charge state of the incoming particles does not matter. Incoming positive ions are neutralized by resonant and Auger type electron transfer [*Hagstrum, 1954*]. If an equilibrium situation is considered, the charge state of a particle at a certain distance from the surface is given by the overlap of the broadened and lowered affinity state and the metal's conduction band. A simple model [*Overbosch et al., 1980*] describes the charge exchange efficiency in a non-equilibrium situation with the assumption that the charge state is frozen at the so-called freezing point due to the decreasing resonance width. Several parameters are important for the charge exchange efficiency. These are the electron affinity of the incoming particle, the WF of the surface, the electron density of the surface, and the velocity and angle of the impinging particles.

The difference between the WF and the electron affinity is the crucial parameter determining the efficiency, since it is a measure for the overlap between the affinity state and the conduction band. For this reason, surface ionization works well for atoms with high electron affinities, e.g., H, C, and O atoms (0.75, 1.27, and 1.46eV). Measurements of O^- and C^- have been reported by *Van Pinxteren et al.* [*1989*]. The production of singly charged negative ions has also been observed for species with low negative electron affinities such as He [*Verbeek et al., 1984*], which forms a metastable negative ion.

The velocity of the particle is important for the charge exchange efficiency because it determines the interaction time of the particle with the surface. In the limit of very high velocities the resonant charge transfer is too slow to populate the affinity level and thus the charge exchange efficiency goes to zero. At very low energies few negative ions are observed since the electron in the affinity level has sufficient time to tunnel to an empty state in the metal on the outgoing

trajectory. For H atoms reflected off a Cs/W(110) converter the conversion efficiency shows a broad maximum around an energy of 100eV normal to the surface [*Van Os et al., 1988*]. When increasing the normal velocity component to achieve a higher charge exchange efficiency by choosing a smaller impact angle one has to account for inferior reflection properties as the angular broadening of the reflected beam increases.

Low WF surfaces are typically generated by coating a metal surface with a thin layer of an electropositive metal. This causes an additional surface dipole layer which reduces the WF markedly [*Lang, 1971*]. An extensively studied combination is Cs/W(110), used either with 0.6 monolayer or a full monolayer of Cs [*Van Wunnik et al., 1983; Geerlings et al., 1985; Amersfoort et al., 1985*]. The WF shows a minimum of 1.45eV for the submonolayer coverage and equals the value of bulk Cs (2.15eV) for the full monolayer, which is far below the WF of the bare W substrate of 5.25eV.

3. EXPERIMENTAL SETUP

We built an experiment using an ion beam to test various conversion surfaces for their suitability for application on a space platform. The aim of this experiment was to show that surface ionization works under moderately good vacuum conditions and thus to qualify this technique for application in an outgassing environment in space.

The experiment consists of an ion source, a beam guiding system, a sample chamber with surface reconditioning units and a detection unit. These parts are in a vacuum chamber pumped by a turbomolecular pump. After bake-out, a pressure of 5×10^{-8}mbar is achieved. A schematic of the experimental setup is shown in Figure 1. The ion beam is formed in an electron impact ion source. From a similar ion source we know for H_2 that H^+ and H_3^+ are present only at a percent level relative to H_2^+ in the ion beam [*Ghielmetti et al., 1983*]. After acceleration the ions are deflected with a cylindrical energy analyzer (resolution of 1% FWHM) on the entrance aperture of the sample chamber. Two collimators limit the beam divergence at the sample surface to 1^0. By rotating the conversion surface, the impact angle α of the ion beam relative to the surface normal ranges between 90^0 and 0^0. The reflected beam is analyzed with a 2D position-sensitive MCP detector. A retarding potential analyzer (RPA) with three grids is mounted in front of the detector, which floats on an adjustable high voltage. The detector unit, including the RPA, is shielded electrostatically. It can be rotated independently of the converter surface around the same axis. The outer grids of the RPA are grounded to shield the inner grid, which has a permanent bias to suppress positive ions. Their fraction in the reflected beam is

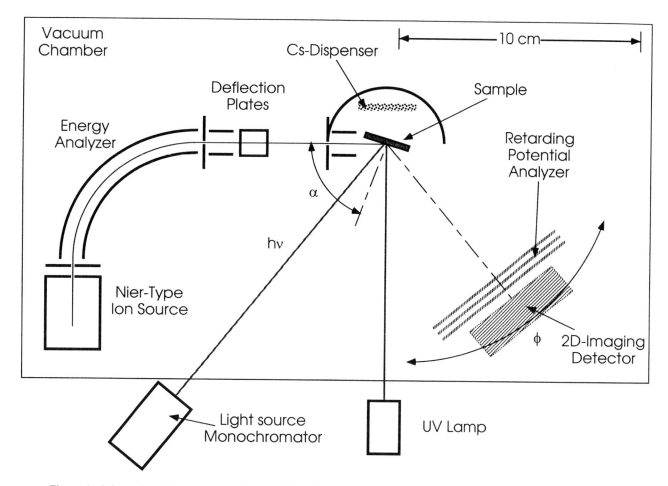

Figure 1. Schematics of the experimental setup. The primary ion beam impinges on the sample surface at an adjustable angle α relative to the surface normal. The reflected beam is detetcted with a 2D MCP detector that has a retarding potential analyzer mounted on its front. The work function is measured photoelectrically with monochromatic light of ajustable wavelength coupled into the chamber with fiber optics.

negligible at these energies, however, parts of the primary beam could hit the detector at impact angles α close to 90^0. To sweep out negatively charged particles from the reflected beam, a negative bias is applied to the detector. By measuring the intensity of the entire reflected beam, and its neutral component (by sweeping out the negatively charged ions), the ionization efficiency η of the surface can be calculated from

$$\eta = 1 - \frac{N_0}{N_{tot}} \qquad (1)$$

where N_0 denotes the counts of the neutral beam within a given detector area and time interval, and N_{tot} the counts of the total beam within the same area and time interval. To this simple equation corrections are applied accounting for the energy dependence of the detection probability, for the different detection efficiencies for neutral particles and for negative ions, and for the effects of contaminants in the ion

beam resulting from ionized background gas [*Schletti, 1996*]. From the 2D-images, the conversion efficiency can be measured at different reflection angles. In addition to the conversion efficiency, angular scattering is measured.

The conversion unit consists of a W(110) single crystal serving as substrate on which an accurately controlled amount of Cs is deposited from a dispenser by evaporation. It takes 45s to deposit one monolayer of Cs based on manufacturer's specifications and our own measurements. To clean the surface, it is heated resistively up to 1600K for a few seconds.

With a fiber-optical system mounted on the vacuum chamber the surface is illuminated with monochromatic light of adjustable wavelength. By measuring the photoelectric current the WF of the converter surface is determined.

To test the UV response of the surface, it can be exposed to very intense ultraviolet radiation in the wavelength range from 145 to 185nm produced by a QUANTATEC Xe lamp.

4. THE APPLICABILITY OF SURFACE IONIZATION IN SPACE INSTRUMENTATION

To assess the applicability of surface ionization with cesiated surfaces to space instrumentation, four important issues have to be investigated:

I) The degradation of the converter due to residual gas.
II) The degradation of the converter due to intense UV radiation.
III) Cs migration in the instrument.
IV) Particle reflection properties of the converter surface.

4.1. Converter Degradation due to Residual Gas

We did not perform this study under stringent UHV conditions because our explicit aim is to show the applicability of surface ionization under less favourable circumstances as experienced in space instrumentation. Our apparatus has no device to assess the atomical cleanliness of the W substrate and the Cs layer, thus lacking precise information about surface contamination. This is similar to the situation where an instrument would be in space and surface contamination could not be assessed either.

Figure 2 shows the measured negative ion fraction in the reflected beam for H_2^+ primary ions with an energy of 225eV impinging on the surface under an angle of $\alpha=82^0$. The W(110) substrate was covered with a full monolayer of Cs. High ionization yields are obtained with this configuration. On the inbound path the incoming positive molecular ions are efficiently neutralized and dissociated [*Van Toledo et al., 1992*]. Thus the charge state before reflection has no influence on the final charge state in this energy range.

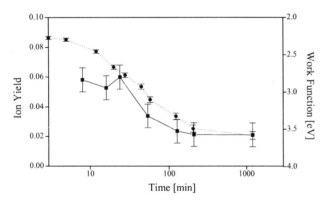

Figure 2. Measured negative ion fraction in the reflected beam (solid line) for H_2^+ impinging under 82^0 at an energy of 225eV on a W(110) surface covered with a full monolayer of Cs. The molecular ions are efficiently neutralized and dissociated on the inbound path. The decrease in ionization efficiency is caused by the increasing WF (dashed line) due to the adsorption of residual gas (background pressure 10^{-7}mbar).

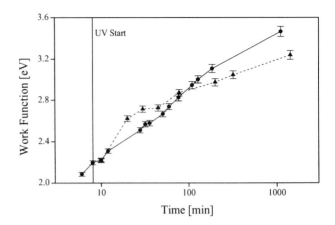

Figure 3. Time dependence of the WF with (triangles) and without (circles) UV exposure. Shortly after surface preparation the UV exposure increases the WF slightly, but on long term the UV exposed surface shows lower WF. The Cs overlayer is basically unaffected by the very intense UV.

Schneider et al. [*1982*] have observed no difference in the final charge state for D and D^+ as primary particles.

A few minutes after surface preparation, we measured an ionization efficiency of 6%. The efficiency decreases with time due to the adsorption of background gas on the converter surface, which increases the WF. It is very important that even 24 hours after the surface preparation the conversion efficiency, while significantly decreased, was still observable in spite of the relatively high pressure of 10^{-7}mbar. Another test run was performed to confirm that the residual gas affects the converter surface: Leaking O_2 into the chamber to deliberately increase the pressure increases the WF while lowering the ionization efficiency. The adsorption of electronegative constituents will cause an increase of the WF, whereas hydrogen will lead primarily to a decrease of the resonance width [*Amersfoort et al., 1986*]. Both mechanisms reduce the conversion efficiency of the surface. If the ambient pressure is reduced by one order of magnitude the characteristic time of operation would increase by the same factor. Such an improvement is possible for a space instrument so that a regeneration of the converter surface would not be necessary more frequently than every ten days.

The functional dependence of the conversion efficiency on the WF is also displayed in Figure 2. The close correlation between the WF and the ionization efficiency can be used for a reliable in-flight diagnostics of the instrument. By means of the WF measurement and a thorough pre-flight calibration determining both ionization and reflection probabilities, the absolute neutral particle flux can be determined in space accurately at any time. In a spaceborne sensor the WF will be measured by illuminating the converter surface with light emitting diodes. The measured photocurrent depends strongly on the WF of the surface [*Schletti, 1996*], and thus the WF can be retrieved accurately.

4.2. Ultraviolet Response of the Cs/W(110) Surface

To test the UV response of the Cs/W(110) surface, it was illuminated with a UV lamp for extended periods and the WF, as a measure of the conversion efficiency, was monitored. Our lamp produces a continuous spectrum of UV photons with lower energy than the Lα photons which dominate geocoronal UV light. Since the photon energy of our lamp by far exceeds the relevant binding energies of Cs on the substrate and of contaminants, our conclusions are qualitatively applicable. The results of one run with UV exposure and one without UV illumination are shown in Figure 3. For short times the WF is slightly higher when the surface is exposed to UV light. This is due to photodesorption of background gas from the walls of the vacuum chamber, which considerably increases the pressure in the chamber. On longer timescales, the UV irradiated surface has lower WF. This means that contaminating adsorbates on the converter surface are removed by photodesorption. However, the Cs overlayer is not affected by UV light as one can conclude from the fact that the WF evolves qualitatively the same way with and without UV exposure. The converter surface in the test stand is irradiated with 2×10^{13}ph s^{-1}cm^{-2} (manufacturer spec.). In space, a Lα photon flux from the geocorona on the converter surface of 2×10^{7}ph s^{-1}cm^{-2} is expected [*Meier and Mange, 1973*], assuming a field of view of the instrument of $8^0 \times 90^0$. We conclude that the conversion surface is only slightly affected by UV exposure and that photodesorption of the Cs overlayer can be neglected.

4.3. Thermal Desorption of Cesium

It is well-known that Cs with its low melting point of 302K is a reactive and volatile element. This, however, is not a problem in our application because we are using only one monolayer at a time. One monolayer Cs on W(110) corresponds to a surface density of 5.5×10^{14}atoms cm^{-2}. For thin layers the heat of desorption is a strong function of the Cs coverage [*Ehlers and Leugn, 1984*]. For the Cs/W(110) surface the heat of desorption increases from 0.82eV per atom, which is the value for bulk Cs, to more than 3eV for almost zero coverage. If there is more than one monolayer of Cs on the substrate, the excess Cs will quickly evaporate (evaporation of one monolayer of bulk Cs takes of the order of seconds at room temperature) and as soon as one monolayer is reached the evaporation rate decreases expo-

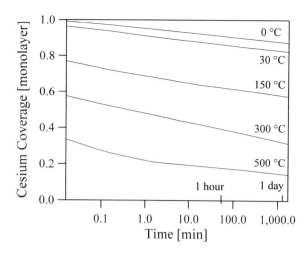

Figure 4. Calculated thermal desorption of a Cs overlayer on W(110) for different temperatures when starting with a full monolayer at t=0. Because the heat of desorption increases strongly with decreasing Cs coverage, evaporation is very slow compared with bulk Cs. Desorption heat data of *Ehlers and Leugn* [*1984*] have been used for this calculation.

nentially. The combination of these two facts lets us conclude that we know the Cs layer thickness with sufficient precision to be one monolayer, which is confirmed by our WF measurements. The cesium coverage shown in Figure 4 has been calculated using experimental data of the heat of desorption for a Cs/W(110) [*Ehlers and Leugn, 1984*] surface without any contamination by assuming a first order desorption process. We conclude from Figure 4 that thermal desorption is not a problem at room temperature. The surface is regenerated by thermally desorbing the Cs and the adsorbed contaminants and then coating the substrate with Cs again. The Cs within the dispenser is bound in a metal salt and starts to evaporate at around 250°C, so no contamination of the instrument will take place during normal operation from either the Cs dispenser or from the converter surface. Contamination of internal surfaces of an instrument due to Cs evaporation during reconditioning, which would be performed an estimated 10 to 100 times during a mission, would be below one monolayer of Cs. With the low vapor pressure for submonolayers, migration of Cs within the instrument is minimal. In our test stand, the sensitive MCP detector is about 12cm away from the Cs dispenser, and after one and a half year of operation, no decrease in performance was observed.

4.4. Reflection Properties

The use of converter surfaces in mass- and energy-analyzing instruments requires sophisticated ion optics, because the ions start from the converter surface with considerable

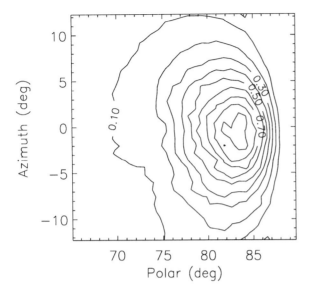

Figure 5. Angular scattering of a 400eV H_2^+ beam impinging at 80^0 on a W(110) surface coated with one monolayer of Cs. The beam profile is normalized to its maximum intensity.

angular and energy spread. Thus the reflection properties of the surface are relevant to the resolution of the instrument. Simulations with the MARLOWE code [*Robinson, 1989*] showed that atomically flat surfaces produce less angular scattering than atomically rough surfaces [*Aellig, 1995*]. This can be observed when comparing a cesiated W(110) substrate coated with either a full or a half monolayer of Cs. Since the latter does not form a closed surface, it exhibits stronger angular scattering. Together with the more complicated preparation this makes the submonolayer system less preferable. The particle reflection coefficient is also an important parameter, because it is proportional to the sensitivity of the instrument. In this respect also the fully Cs covered surface is more preferable than the half-covered. However, even for the half-covered surface a particle reflection coefficient of 0.92 was simulated for 300eV atomic hydrogen impinging at 75^0 relative to the surface normal. Figure 5 displays the measured angular scattering in polar and azimuthal direction for a H_2^+ beam impinging on a Cs coated W(110) surface at an angle of 80^0 with an energy of 400eV. The angular scattering (FWHM of 11^0 in azimuthal direction and 8^0 in polar direction) is consistent with the results obtained with the MARLOWE simulations [*Aellig, 1995*].

5. CONCLUSIONS

We discussed the crucial aspects of the Cs/W(110) conversion system for a space application in detail and demonstrated its feasibility for space instrumentation. We showed

that surface conversion into negative ions can be performed despite contamination due to residual background gas and that the converter surface does not degrade upon intense UV irradiation. Other candidate surfaces for conversion systems are currently under investigation. Future goals are not only higher efficiency, but also a longer standing time of the surface between reconditioning and the search for converter systems which do not need the application of a surface coating, and therefore considerably increase the ease of use.

Acknowledgments. This work was supported by the Swiss National Science Foundation and by Lockheed Martin Missiles and Space Independent Research.

REFERENCES

Aellig, M.R., Master's Thesis, University of Bern, Bern, Switzerland, 1995.

van Amersfoort, P.W., Geerlings, J.J.C., Kwakman, L.F.Tz., Hershcovitch, A., Granneman, E.H.A., and Los, J., *J. Appl. Phys. 58*, 3566-3572, 1985.

van Amersfoort, P.W., Geerlings, J.J.C., Rodink, R., Grannemann, E.H.A., and Los, J., *J. Appl. Phys. 59*, 241-248, 1986.

Ehlers, K.W., and Leugn, K.N., 0094-243X/84/1110227-12, American Institute of Physics, 227-238, 1984.

McEntire, R.W., and Mitchell, D.G., in *Solar System Plasma Physics*, edited by J.H. Waite Jr., J.L. Burch, and R.L. Moore, 69-80, AGU Monograph 54, 1989.

Geerlings, J.J.C., van Amersfoort, P.W., Kwakman, L.F.Tz., Granneman, E.H.A., and Los, J., *Surf. Sci. 157*, 151-161, 1985.

Ghielmetti, A.G., Shelley, E.G., Fuselier, S., Wurz, P., Bochsler, P., Herrero, F., Smith, M.F., Stephen, T., *Opt. Eng. 33*, 362-370, 1994.

Gruntman, M.A., *Planet. Space Sci. 41(4)*, 307-319, 1991.

Hagsturm, H.D., *Phys. Rev. 96*, 336-365, 1954.

Herrero, F.A., and Smith, M.F., in *Instrumentation for Space Imagery, SPIE proceedings Vol. 1744*, 32-39, 1992.

Hsieh, K.C., and Curtis, C.C., in *Measurment Techniques for Space Plasmas*, edited by J. Borovsky, R. Pfaff, and D. Young, AGU Monograph, 1997.

Hsieh, K.C., Curtis, C.C., Fan, C.Y., and Gruntman, M.A., in *Remote Sensing of Plasmas 357*, 357-364, 1992.

Lang, N.D., *Phys. Rev. B4*, 4234-4244, 1971.

Los, J., and Geerlings, J.J.C., *Phys. Reports 190*, 133-190, 1990.

Meier, R.R., and Mange, P., *Planet. Space Sci. 21*, 309-327, 1973.

van Os, C.F.A., van Amersfoort, P.W., and Los, J., *J. Appl. Phys. 64*, 3863-3873, 1988.

Overbosch, E.G., Rasser, B., Tenner, A.D., and Los, J., *Surf. Sci. 92*, 310-324, 1980.

van Pinxteren, H.M., van Os, C.F.A., Heeren, R.M.A., Rodnik, R., Geerlings, J.J.C., and Los, J., *Europhys. Lett. 10(8)*, 715-719, 1989.

Robinson, M.T., *Phys. Rev. B4*, 10717-10726, 1989.

Schletti, R., Master's Thesis, University of Bern, Bern, Switzerland, 1996.

Schneider, P.J., Eckstein, W., and Verbeek, H., *Nucl. Instr. and Meth. 194*, 387-390, 1982.

van Toledo, W., van Buuren, R., Donné, A.J.H., and de Kluiver, H., *Rev. Sci. Instrum. 63(4)*, 2223-2231, 1992.

Verbeek, H., Eckstein, W., and Schneider, P.J., *Proc. of the Third International Symposium on Production and Neutralization of Negative Ion Beams,* American Institute of Physics, 273-280, 1984.

van Wunnik, J.N.M., Geerlings, J.J.C., and Los, J., *Surf. Sci. 131*, 1-16, 1983.

Wurz, P., Aellig, M.R., Bochsler, P., Ghielmetti, A.G., Shelley, E.G., Fuselier, S.A., Herrero, F., Smith, M.F., and Stephen, T.S., *Opt. Eng. 34*, 2365-2376, 1995.

M.R. Aellig, P. Bochsler, R. Schletti, and P. Wurz, Physikalisches Institut, University of Bern, Sidlerstrasse 5, CH-3012 Bern, Switzerland.

S.A. Fuselier, A.G. Ghielmetti, and E.G. Shelley, Lockheed Martin Palo Alto Research Laboratory, 3251 Hanover St., Palo Alto, CA 94304, USA.

J.M. Quinn, SSC Morse Hall, University of New Hampshire, Durham, NH 03824, USA.

F. Herrero, M.F. Smith, Laboratory for Extraterrestrial Physics, NASA/GSFC, Greenbelt, MD 20771, USA.

Neutral Atom Imaging of the Plasma Sheet: Fluxes and Instrument Requirements

Michael Hesse

NASA/Goddard Space Flight Center, Greenbelt, MD 20771

Joachim Birn

Los Alamos National Laboratory, Los Alamos, New Mexico

We use the results of three-dimensional self-consistent MHD simulations of magnetotail dynamics to provide a time-varying source density and temperature as a model of the plasma sheet proton population. Assuming Maxwellian distribution functions, we can then derive energy dependent proton fluxes as a function of position and time. Assuming observed radial distributions of exospheric hydrogen neutrals and fitting the energy dependence of the proton-hydrogen cross-section, we can calculate the neutral hydrogen fluxes generated by charge exchange between the two species. We then line-of-sight integrate these resulting fluxes to yield images such as would be observable by suitably equipped spacecraft missions. By investigating separate energy ranges, we can derive time and spatial resolution, and sensitivity requirements for observing instrumentation.

INTRODUCTION

Among the various magnetospheric regions, the plasma sheet is one of the most dynamical. It reflects through temperature, density, composition, and geometry changes the dynamical evolution of the largest section of the magnetosphere, the magnetotail. These changes are most commonly associated with the substorm process responsible for the dissipation of the majority of the energy previously extracted from the solar wind [e.g., *Baker et al.*, 1995].

Typical changes of the plasma sheet structure during magnetospheric substorms are manifold. During the substorm growth phase, commonly associated with storage of energy in the magnetosphere, the plasma sheet is often observed to thin in the region roughly delineated by geosynchronous orbit as an inner boundary, and more than $10 R_E$ radial distance as an outer boundary [e.g., *Sanny et al.*, 1994]. This thinning is followed by a longitudinal

expansion [e.g., *Hones et al.*, 1984], plasma heating [e.g., *Huang et al.*, 1992], and often compositional changes during the so-called "expansive phase" [e.g., *Lennartson*, 1992; *Shelley*, 1986]. These effects can often be localized in the meridional direction, and expand during the evolution.

Models based on the interpretation of spacecraft data have suffered from the fact that almost all relevant investigations, save auroral imaging, have necessarily been based on single point measurements alone. The absence of a simultaneous global overview of the dynamics of the plasma sheet has thus contributed to the prevalent problems to uniquely determine which mechanism is responsible for the onset of substorm dynamics [e.g., *Hesse*, 1995]. An example of the usefulness of global imaging can also be found in laboratory experiments [e.g., *Yur et al.*, 1995].

A promising step in the direction of global imaging of the plasma sheet might be accomplished by the novel technique of neutral atom imaging. Neutral atom imaging utilizes the flux of neutral atoms which is generated by charge exchange collisions between neutral exospheric hydrogen atoms, and magnetospheric ions [e.g., *McComas et al.*, 1991]. As an example, the imaging of oxygen ions

Measurement Techniques in Space Plasmas: Fields
Geophysical Monograph 103

originating in the cusp/cleft ion fountains has recently been investigated by *Hesse et al.* [1994]. In the case of the plasma sheet, the most abundant ion species is protons, but oxygen as well as helium ions can also constitute significant fractions of the ion population at times. These ions undergo charge exchange collisions with cold exospheric hydrogen, after which the ionized hydrogen remains attached to the magnetic flux tube while the neutralized plasma sheet particle follows a ballistic trajectory influenced only by gravity. These particles can, in principle, be detected remotely by spacecraft equipped with suitable instrumentation. Fluxes obtained at a given spacecraft location are then given by line-of-sight integration of the local charge exchange effects. A global two-dimensional image can be obtained by variation of the observation angle, and it is often possible to deconvolve the resulting image to gain information about the three-dimensional structure of the original ion population [e.g., *Roelof et al.*, 1985].

For the design of such a mission, prior knowledge of the magnitude of the expected fluxes and the angular variation of the flux intensity is quite important. In this paper, we provide a first step in this direction. Since protons are the most abundant population in the plasma sheet, we restrict this investigation to neutral hydrogen produced by charge exchange from plasma sheet protons. In order to calculate the expected fluxes and their angular variations, we use three-dimensional MHD simulations of magnetotail dynamics to provide the proton density and temperature required as input into the model. Assuming Maxwellian distributions, and using observed radial density variations of the exospheric hydrogen density in conjunction with measured charge exchange cross-sections, we can then calculate the local charge exchange rates, and perform line-of-sight integrations to obtain the resulting fluxes at a spacecraft location. In section 2, we will briefly discuss the MHD model. In section 3, we will present the used neutral hydrogen density, the charge exchange cross sections, and list the integrations performed. Results of this procedure as a function of energy will be discussed in section 4, and section 5 will sum up the results with an outlook and interpretation.

MHD MODEL

We initiate the MHD simulation with a three-dimensional numerically derived equilibrium. The initial magnetic field configuration, derived from a Tsyganenko 89 [*Tsyganenko*, 1989] magnetic field for Kp=0, is relaxed numerically by a friction-type approach similar to the method used by *Hesse and Birn* [1993]. The equilibrium covers the magnetotail region $-5 \geq x \geq -65\, R_E$, $-10 \leq y,z \leq 10\, R_E$. In the following, we assume a normalization of $B_0 = 300 nT$ for the magnetic field, $v_A = 1400\ km/s$ for the velocity, and $L=1\ R_E$ for the length scale. Together with

the assumption that the plasma sheet ions are protons, derived normalizations are: $n_0 = 22\ cm^{-3}$ for the density, $kT_0 = 20 keV$ for the temperature, $p_0 = 70\ nPa$ for the pressure, and $t_0 = 4.3\ s$ for the time. To model typical observed temperature variations along the tail [e.g., *Lui and Hamilton*, 1992], we impose a temperature variation with pressure of the form

$$T(p) = \frac{\tau}{1 + 0.0125\,\tau / p}$$

with   (1)

$$\tau(p) = 0.5\left[1 + 2\left(\frac{p}{0.08}\right)^2\right]$$

This choice reproduces qualitatively the temperature variations measured by *Lui and Hamilton* [1992] (see also Figure 3).

The equilibrium is then subjected to driving electric field boundary conditions in qualitatively the same fashion as recently analyzed by *Birn and Hesse* [1995], resulting in the formation of a thin current sheet. The onset of instability requires dissipation. It is initiated here by anomalous resistivity which arises if the local current density exceeds a certain threshold (of 1.5 in the above units). This resistivity is assumed to be of the form

$$\eta = \eta_0 \begin{cases} 10^{-3} & \text{if } j \leq 1.5 \\ 0.25\,(j-1.5)^{0.5} + 10^{-3} & \text{if } 1.5 < j < 2.5 \\ 1 & \text{otherwise} \end{cases} \quad (2)$$

The dynamical evolution initiates once the current density exceeds $j=1.5$ locally. The resulting magnetic field evolution in the noon-midnight meridional plane, together with plasma flow vectors, is shown in Figure 1. The figure demonstrates that reconnection starts around $t=100$, and causes the formation of a plasmoid tailward of the reconnection region, and magnetic field dipolarization earthward of it. Similar effects have been reported earlier [e.g., *Birn*, 1984; *Hesse and Birn*, 1991; *Birn and Hesse*, 1995]. Of particular interest here is the time evolution of the density and temperatures in the plasma sheet. They are shown, in the equatorial plane, in Figures 2 and 3, respectively. The figures demonstrate that density and temperature changes are localized around $y=0$, roughly in the interval $-3 \leq y \leq 3$. These changes are caused by the fast progression of substorm effects in the midnight region. Earthward plasma transport leads to enhanced densities, and, by virtue of adiabatic compression, also to temperature enhancements. Changes in density and temperature extend to the inner boundary at $x=0$, corresponding to $x=-5\ R_E$ in GSM coordinates. In the following, we will address the question of how such signatures can be recovered in neutral atom images obtained by a virtual spacecraft mission.

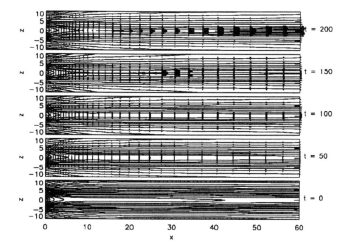

Figure 1. Magnetic field evolution in the midnight meridional plane in the MHD model. At about $t=100$, reconnection causes plasmoid formation and the dipolarization of the magnetic field in the inner section of the model. The arrows indicate flow vectors.

CALCULATION OF NEUTRAL HYDROGEN FLUXES

The total neutral hydrogen flux J_n in the energy band $E_0 \leq E \leq E_1$ is given by the line-of-sight integral [e.g., *McComas et al.*, 1991]

$$J_n = \int dL \int_{E_0}^{E_1} dE \sigma(E) j_p(E) n_H \qquad (3)$$

Here L denotes a coordinate along the line-of-sight vector, σ is the energy dependent charge exchange cross-section for H-p collisions, n_H represents the exospheric cold neutral hydrogen density, and j_p is the energy dependent differential proton flux.

In order to calculate J_n, models are needed for the latter three quantities. For the charge exchange cross-section, we perform a fit to the measurements by *Fite et al.* [1962]. Our best fit is given by

$$s = 10^{-16} cm^2 \begin{cases} 16.2 - 8.51 \log(E/keV) & for\ E < 30\ keV \\ 1.35 \times 10^5 (E/keV)^{-3.05} & for\ E \geq 30\ keV \end{cases} \qquad (4)$$

The quality of the fit is shown in Figure 4. The fit was used for the entire energy range of the study, as it well represents the measurement of *Fite et al.* [1962]. The exospheric hydrogen density is given by a fit to the data of *Rairden et al.* [1986]. We find for an excellent fit in the region under consideration

$$n_H = 2.3 \times 10^4 (x/R_E)^{-3.3} cm^{-3} \qquad (5)$$

The plasma sheet proton flux needs to be determined from the MHD model. For this purpose, we assume for simplicity, that the proton distribution functions are Maxwellian distribution functions, with the density and temperatures given by the MHD model. Thus the distribution function is given by

$$f(E) = n \left(\frac{m_p}{2\pi kT} \right)^{1.5} \exp(-E/kT) \qquad (6)$$

From (6), the flux j_p in units of the inverse of the product of energy, area, time, and solid angle, can be derived as

$$j_p(E) = \frac{2E}{m_p^2} n \left(\frac{m_p}{2\pi kT} \right)^{1.5} \exp(-E/kT) \qquad (7)$$

Inserting the appropriate normalizations, the energy integral in (3) can be evaluated numerically if the density n and temperature T are taken from the MHD simulations. The resulting flux per unit length can then be line-of-sight integrated in the MHD model. In the following section, we will discuss results of this procedure for four different energy bands.

INTEGRATION RESULTS

For the purpose of this investigation, we choose a satellite location at the GSM coordinates $x=-5\ R_E$, $y=0$, and $z=5\ R_E$. We then plot all results of the line-of-sight integrations as functions of look angle, schematically illustrated in Figure 5. Thus cross-tail (α) and down-tail (β) look angles of 0 degrees correspond to looking straight down toward the equatorial plane from the satellite. A down-tail look direction of 45 degrees corresponds to looking toward the $x=-10\ R_E$ line in the equatorial plane, whereas cross-tail look directions of -45 and 45 degrees correspond to looking toward the $y=5\ R_E$ and $y=-5\ R_E$ lines in the equatorial plane, respectively. The angular grid size employed for this study is 100x100.

We split our investigation into four parts by choosing four energy bands. For the lowest energy band, we choose $10eV \leq E \leq 1keV$. The second energy band comprises $1keV \leq E \leq 10keV$, followed by $10keV \leq E \leq 30keV$. Finally, the fourth band ranges from $30keV$ to $50keV$.

The results of the integrations of (3) for all four energy channels are shown in Plate 1. Plate 1 displays the results of the flux integrations in units of cm^{-2}s^{-1}sr^{-1} for the initial configuration and for the final result at $t=200$, corresponding to about 15mins later. The first obvious result from Plate 1 is that the angular distributions of the fluxes at the later times are quite different for different energies. For the very low range ($10eV-1keV$) we find a significant reduction of the flux in the central region around

Densities

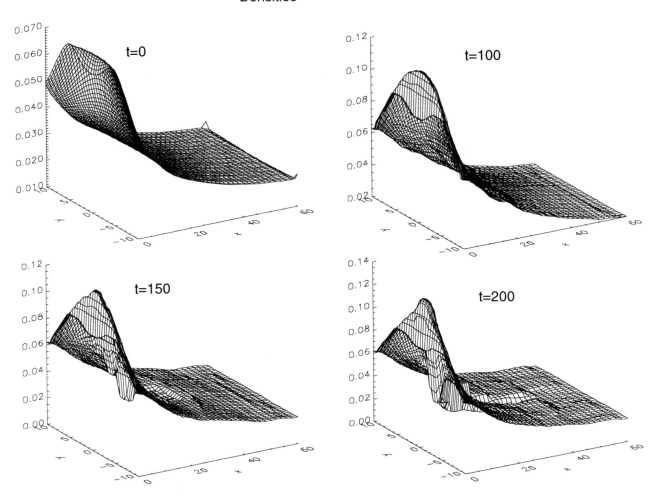

Figure 2. Density variation in the equatorial plane during the simulation.

$\alpha=0^o$, and $\beta=10^o$. In the low energy range (1-$10keV$ some albeit not very substantial flux enhancements are found in the same region, accompanied by a more significant increase (by about 20%) at the inner edge (i.e., $\beta=0^o$) with a larger spread across the tail.

The situation changes appreciably if higher energies are considered. In either case, we find large flux enhancements quite localized around $\alpha=0^o$ with a spread $\delta\alpha \approx 20$, extending from $\beta=0^o$ to about $\beta=30^o$. The flux levels increase by more than 40% in the 10-$30keV$ range, and by about 50% in the highest energy channel. These changes, as well as the partial flux reductions in the lower energy ranges can be understood by considering the temperature distributions shown in Figure 3. Using the normalization for the temperature $kT_0=20keV$, we find typical temperatures in the region inside of $x=5$ (corresponding to a distance of 10 R_E from the Earth), of about 10-$30keV$. The increase in temperatures during the evolution, ranging from

a factor of about 1.2 at the inner edge of the simulation region to about a factor of 2 further downtail, effectively shifts the maximum fluxes to higher energies. Thus we find a flux reduction in the low energy channels, despite the density enhancements (see Figure 2), and a pronounced flux enhancement at the higher energies.

Of high relevance for practical applications are the actual flux levels in the different energy bands. Table 1 lists the maximum fluxes obtained from each energy channel, either at $t=0$, or $t=200$. Also listed in table 1 are estimates of flux levels which need to be observable in order to distinguish the evolution in the images. Inspecting table 1, one finds very low fluxes in the low energy band, as would be expected from the thermal energies of the plasma sheet plasma. It appears that extraordinary sensitivities would be required to compile an image of sufficient angular resolution in a sufficiently small time interval (see below) for this energy range. The three higher energy channels are

TEMPERATURES

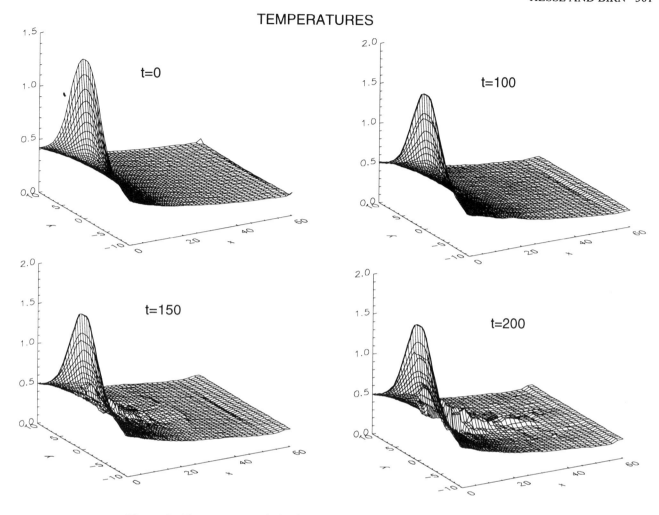

Figure 3. Temperature variation in the equatorial plane during the simulation.

much more promising, in particular in the 10-30 keV range. Here we find that instruments able to measure fluxes of a few 100cm^{-2}s^{-1}sr^{-1} during a suitable time frame with a reasonable resolution should be able to perform adequately.

A closely related issue for a spacecraft mission is the required angular resolution for the image. Note that we are working here for the angular size of features in ideal images. After finding those we can conclude that an image with at least that resolution, including counting statistics, should be able to distinguish the feature. By inspection of plate 1, we find that typical gradient scale lengths are of the order of several degrees. Thus it appears that pixel sizes of about 5x5 degrees, or even 10x10 degrees might be sufficient to provide enough of an overview of the dynamical evolution. Either of these resolutions is, however, insufficient to investigate highly localized phenomena such as a possible onset region. The short time-scale of candidate processes, usually seconds to about a minute [e.g., *Lui*, 1991] on the other hand, renders their

detectability by neutral atom imaging extremely difficult, if not impossible.

Finally, the question of required time resolution needs to be addressed. Typical substorm time scales range from seconds, related to magnetic field fluctuations during dipolarization, to about an hour, the extent of an entire substorm cycle. As noted above, it is probably impossible to achieve time and angular resolutions adequate to identify the onset mechanism, but the overall topology of, e.g., the substorm expansion phase dynamics could be studied with integration times of about 5 minutes.

SUMMARY AND DISCUSSION

We have used three-dimensional MHD simulations of magnetotail dynamics as the basis of an investigation of the viability of neutral atom imaging of the inner plasma sheet. The plasma sheet proton density and temperature were assumed to equal the plasma density and temperature

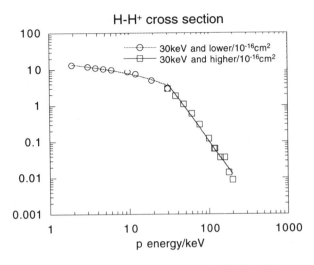

Figure 4. Fit to the measured [*Fite et al.*, 1962] p-H+ cross-sections as adopted in this investigation.

of the MHD simulation. The MHD model was initialized with a realistic magnetotail configuration, covering the range from -5 to -65 R_E in the GSM x direction, and -10 to 10 R_E in both GSM y, and z directions. The plasma pressure-magnetic tension equilibrium was calculated numerically by means of a relaxation method [*Hesse and Birn*, 1993]. The temperature and density were then initialized following typical observed radial profiles [*Lui and Hamilton*, 1992].

The MHD equations were integrated in a form similar to *Birn and Hesse* [1995], and the simulation led to quite similar results. For the purpose of the charge exchange calculations, we assumed that the proton distribution functions were simple Maxwellians, with zero bulk flow velocity. Using this assumption, we could calculate from the MHD densities and temperatures energy and position dependent fluxes of protons. In order to derive neutral hydrogen fluxes generated from plasma sheet protons by charge exchange with the Earth's exosphere, we numerically fitted the energy dependence of the charge exchange cross section for neutral hydrogen-proton collisions, and the experimentally determined radial density dependence of the cold exospheric atomic hydrogen.

The product of these three quantities was then integrated first over four ranges of energy, and then along the line-of-sight from a spacecraft position assumed at x=-5 R_E, y=0, and z=5 R_E. The energy channels chosen were comprising the very low energy range (*0.01-1keV*), the low energy range (*1-10keV*), the high energy range (*10-30keV*), and the very high energy range (*30-50keV*). Using our model, we found that the best results were found in the three higher energy ranges, in particular the high energy range -- a fact that is not surprising in view of the average plasma sheet temperature inside of a radial distance of 10 R_E. In the lowest energy channels we found a significant reduction of flux during substorm expansion, due to the associated plasma sheet heating, whereas the higher energy channels indicated clear flux enhancements. All of these effects were seen in a limited angular region, corresponding to a limited local time sector in the actual MHD simulations.

The visibility of any substorm associated changes appeared limited to radial distances inside of 10 R_E, primarily due to the steep drop-off of the exospheric neutral density with radial distance. In reality, nonspherically symmetric neutral hydrogen densities might slightly extend the observable region in the tailward direction (see below). Using our model, we then proceeded to estimate angular resolution requirements for spacecraft missions to successfully image the plasma sheet in the region under investigation. Based on the flux maps, we estimated that typically, for the distances considered here, a pixel size of some 5x5 degree should be sufficient to resolve the most significant gradients. Clearly, for orbits closer to the equatorial plane large pixel sizes might be permissible, whereas higher inclination orbits might demand smaller pixel sizes.

Looking at the actual energy dependent flux levels, we found that there should be ample neutral hydrogen flux in the energy range from about *10-30keV*. The lower, and higher energy channels offer also quite promising flux levels. Imaging in the very low energy range, however, will most likely be quite difficult due to the low fluxes resulting from the high plasma sheet temperature. Generally, we do not believe that flux levels in any energy range are sufficient to image highly localized (~1x1 degree) and very fast ($\delta t \approx 1s$) phenomena, unless extremely sensitive instruments are developed. It is very conceivable,

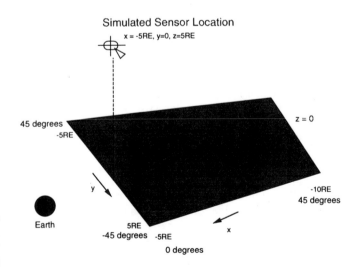

Figure 5. Illustration of the assumed satellite position and the viewing angles.

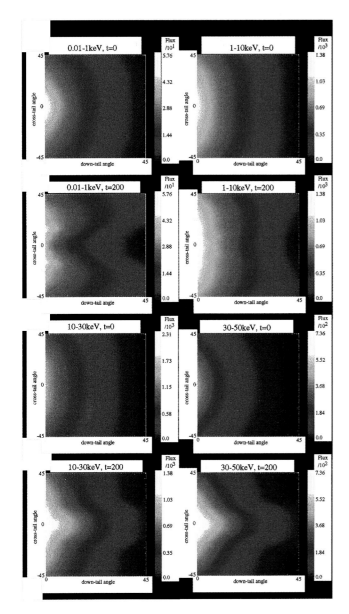

Plate 1. Neutral hydrogen fluxes at the assumed spacecraft location for the initial and final times of the simulation and all four energy channels.

Table 1. Maximum fluxes calculated for each energy channel and estimates of minimum fluxes required for a successful observation.

energy \ time	measurement requirement	maximum flux
0.01 - 1.0keV	2.0×10^1	6.0×10^1
1.0 - 10 keV	3.0×10^2	1.4×10^3
10 - 30 keV	5.0×10^2	2.3×10^3
30 - 50 keV	1.5×10^2	7.4×10^2

unit: $cm^{-2} s^{-1} ster^{-1}$

fluxes. Since, to our knowledge, no improved model exists to-date, we just point out that in actuality fluxes might somewhat higher than suggested by our model.

Another significant simplification lies in our assumption of Maxwellian distribution functions with vanishing bulk flow velocities. More complicated distributions, with cold cores or hot tails, would increase fluxes in some, and decrease fluxes in other energy ranges. Thus it is, for example, possible that even in the very low energy range meaningful measurements might be possible if a sufficiently dense cold core population is present. Similarly, fast plasma flow, as often seen in bursty bulk flows [e.g., *Angelopoulos et al.*, 1992] would lead to enhanced fluxes in energy ranges corresponding to the associated velocities, if the line-of-sight direction is anti-parallel to the velocity direction. Finally, the assumed initial radial temperature distribution might not be representative for all substorm events. In case of a lower initial temperature, much more pronounced signatures would be found in the corresponding energy range. It is quite unlikely, however, that the temperature would be low enough to render the very low energy channel viable.

During this investigation, we have assumed that the MHD density corresponded to a proton population alone. In reality, admixtures of helium and, in particular, oxygen are often found during the substorm evolution. The presence of these heavier ions would reduce the proton and thus the neutral hydrogen fluxes. Unfortunately, the oxygen ions themselves would most likely not lead to observable neutral oxygen fluxes, since, due to the large oxygen mass, associated fluxes would most likely be too small.

In summary, we have presented results of a study focusing on the viability of neutral atom imaging as a tool to study the structure and dynamics of the inner plasma sheet. In view of its limitations we understand our study as a first step toward a more complete investigation of the topic. It nevertheless, however, indicates that neutral atom imaging might constitute a new very promising tool to,

however, that valuable global information on time scales of a few minutes can be obtained from neutral atom imaging in a suitable energy range.

Since all of our results are derived from a model, a discussion of the limitations of the model, and of possible consequences for neutral atom imaging is in order. Beginning with the exospheric hydrogen densities derived from the Rairden et al. [1986] model, we note that deviations from the spherical symmetry of the former, by, e.g., the solar photon pressure, could lead to higher densities in the magnetotail region, and thus to higher for the first time, obtain global snapshots of a large section of the outer magnetosphere. Due to the possible angular resolutions, the resulting images should then also help to resolve questions related to the substorm onset mechanism.

REFERENCES

Angelopoulos, V., W. Baumjohann, C. F. Kennel, F. V. Coroniti, M. G. Kivelson, R. Pellat, R. J. Walker, H. Lühr and G. Paschmann, Bursty bulk flows in the inner central plasma sheet, *J. Geophys. Res.*, **97**, 4027, 1992.

Baker, D. N., T. I. Pulkkinen, M. Hesse, and R. L. McPherron, A Quantitative Assessment of Energy Storage and Release in the Earth's Magnetotail, submitted to *J. Geophys. Res.*, 1995.

Birn, J., E. W. Hones, Jr., Three-Dimensional Computer Modelling of Dynamic Reconnection in the Geomagnetic Tail, *J. Geophys. Res.*, **86**, 6802, 1981.

Birn, J., and M. Hesse, Details of current disruption and diversion in simulations of magnetotail dynamics, submitted to *J. Geophys. Res.*, 1995.

Fite, W. L., A. C. H. Smith, and R. F. Stebbings, Charge transfer in collisions involving symmetric and asymmetric resonances, *Proc. Roy. Soc. London*, A268, **527**, 1962.

Hesse, M., The magnetotail's role in magnetospheric dynamics: Engine or exhaust pipe, US National Report to International Union of Geodesy and Geophysics 1991-1994, *Rev. Geophys. Suppl.*, **675**, 1995.

Hesse, M., and J. Birn, On dipolarization and its relation to the substorm current wedge, *J. Geophys. Res.*, **96**, 19417, 1991.

Hesse, M., and J. Birn, Three-dimensional magnetotail equilibria by numerical relaxation techniques, *J.Geophys. Res.*, **98**, 3973, 1993.

Hesse, M., M. F. Smith, F. Herrero, A. G. Ghielmetti, E. G. Shelley, P. Wurz, P. Bochsler, D. L. Gallagher, T. E. Moore, and T. Stephen, Imaging ion outflow in the high latitude magnetosphere using low-energy neutral atoms, *Optical Engineering*, **32**, 3153, 1993.

Hones, E. W., Jr., T. Pytte, and H. I. West, Jr., Associations of geomagnetic activity with plasma sheet thinning and expansion: A statistical study, *J. Geophys. Res.*, **89**, 5471, 1984.

Huang, C. Y., L. A. Frank, G. Rostoker, J. Fennell and D. G. Mitchell, Nonadiabatic heating of the central plasma sheet at substorm onset, *J. Geophys. Res.*, **97**, 1481, 1992.

Lennartsson, O. W., A scenario for solar wind penetration of the Earth's magnetic tail based on ion composition data from the ISEE-1 spacecraft, *J. Geophys. Res.*, **97**, 19221, 1992.

Lui, A. T. Y., A synthesis of magnetospheric substorm models, *J. Geophys. Res.*, **96**, 1849, 1991.

Lui, A. T. Y., and D. C. Hamilton, Radial profiles of quiet time magnetospheric parameters, *J. Geophys. Res.*, **97**, 19325, 1992.

McComas, D. J., B. L. Barraclough, R. C. Elphic, H. O. Funsten III, and M. F. Thomsen, Magnetospheric imaging with low-energy neutral atoms, *Proc. Natl. Acad. Sci.*, **88**, 9598, 1991.

Rairden, R. L., L. A. Frank, and J. D. Craven, Geocoronal imaging with Dynamics Explorer, *J. Geophys. Res*, **91**, 13613, 1986.

Roelof, E. C., D. G. Mitchell, and D. J. Williams, Energetic neutral atoms (E~50keV) from the ring current: IMP 7/8 and ISEE 1, *J. Geophys. Res.*, **90**, 10991, 1985.

Sanny, J., R. L. McPherron, C. T. Russell, T. I. Pulkkinen, and A.Nishida, Growth-phase thinning of the near-Earth current sheet during the CDAW 6 substorm, *J. Geophys. Res.*, **99**, 5805, 1994.

Shelley, E. G., Magnetospheric energetic ions from the Earth's ionosphere, *Adv. Space Res.*, **6**, 121, 1986.

Smith, P. H., and N. K. Bewtra, Charge exchange lifetimes for ring current ions, *Space Sci. Rev.*, **22**, 301, 1978.

Tsyganenko, N. A., A magnetospheric magnetic field model with a warped tail current sheet, *Planet. Space Sci.*, **37**, 5, 1989.

Yur, G., H. U. Rahman, J. Birn, F. J. Wessel, and S. Minami, Laboratory facility for magnetospheric simulation, *J. Geophys. Res.*, **100**, 23727, 1995.

A Fully Integrated Micro-Magnetometer/Microspacecraft for Multipoint Measurements: The Free-Flyer Magnetometer

R. Goldstein, M. Boehm, E. Cutting, E. Fossum, H. Javadi,

L. M. Miller, B. Pain, J. E. Randolph, P. R. Turner

Jet Propulsion Laboratory, California Institute of Technology, Pasadena, CA

K. Lynch

Institute for the Study of Earth Oceans and Space, University of New Hampshire, Durham, NH

Traditionally, the design of a spacecraft and its payload proceed along relatively independent paths, with little attempt at producing an optimized "whole". The need for optimization becomes more critical as the spacecraft becomes smaller, since interactions between subsystems become more severe. However, to achieve such optimization, the spacecraft + instruments must be treated as a complete system. This paper describes the results and status of an ongoing design study to understand the problems and trade space of fully integrating an instrument into a micro-spacecraft. The example chosen for the study is a miniature ("chip" sized) magnetometer, under development at the Jet Propulsion Laboratory, using the electron tunneling effect in silicon. The micro-spacecraft is then built around the very specific needs of the sensor. The fully integrated electronics include circuits for the sensor, signal processing, and telemetry. As an application, a set of these autonomous, expendable "Free-flyer Magnetometers" would be ejected from a parent spacecraft to measure both temporal and spatial structure of the magnetic field in selected regions of the magnetosphere during the approximately 1 hour life of the free-flyers' battery power. The basic concept will be tested on a suborbital flight ("Enstrophy"), approved for early 1999 launch.

1. INTRODUCTION

The Free-Flyer Magnetometer (FFM) program at the Jet Propulsion Laboratory has pursued dual goals: 1) develop a process for the design and fabrication of a highly integrated, as nearly monolithic as possible, highly miniaturized, autonomous spacecraft which 2) can provide scientifically meaningful measurements. In addition, an eventual goal of the FFM program is to show that most of the free-flyer itself can be mass produced in the manner of integrated circuit chips, greatly reducing their cost.

Attempts at miniaturizing most space instruments (whether counting photons or ions) eventually run into an entrance aperture problem; no matter how small the other components can be made, a useful signal to noise ratio limits how small the front end of the instrument can be made. However, magnetic field sensors are unique in that regard since they have no "aperture", although in a sense, they are measuring the magnetic flux. Hence we have chosen magnetic field measurements as the basis of the micro-

Measurement Techniques in Space Plasmas: Fields
Geophysical Monograph 103

spacecraft. To allow the possibility of a monolithic structure, we have been developing a silicon-based sensor using the electron tunneling effect [*Miller et al.*, 1996]. We have also investigated other sensor possibilities, but the tunneling sensor appears to satisfy overall requirements best. Details of this sensor, which is still in an early development stage, are given below.

The scientific application of the FFM is to provide the capability of multipoint vector measurements of the magnetic field in regions of the magnetosphere or ionosphere. This capability would allow determination of source currents, and the separation of spatial and temporal effects. These have both long been goals in the magnetospheric community and are goals of the Cluster as well as Grand Tour Cluster Missions. (See for example, "Cluster: mission, payload and supporting activities", ESA SP-1159, European Space Agency, March 1993.) Whereas these two missions employ relatively small numbers of spacecraft, widely separated, the FFM can provide dozens of measurement points at a reasonable cost. We are currently targeted to be able to cover scale sizes from ~100m to the order of a few tens of km. Hence the application would be to investigate small scale structures such as in boundary regions and current filaments. See, for example, *Lynch et al.* [1996].

The overall concept of the free-flyer magnetometer is as follows. A host spacecraft contains an ejector mechanism which, on command, selects a free-flyer (FF) from a storage magazine and ejects it with a relative speed of a few m/s. The FF sensor measures the magnetic field and telemeters the data back to the host for storage, processing, and later transmission to the ground. The measurements continue for as long as the FF's battery lasts. A sufficient number of FFs are carried on the host to allow multiple ejection events. These events can be initiated either by on board stored, time-tagged command, by real time ground command, or autonomously when, for example, measurements on the host meet some predetermined criteria for an ejection condition.

This paper is a report of work in progress. We describe the overall concept of the FFM study, give a summary of results of work to date, and discuss briefly some of the open issues yet to be tackled.

2. FREE-FLYER MAGNETOMETER ARCHITECTURE

The free-flyer magnetometer system can conveniently be divided into three functionally separate systems: 1) the free-flyer itself, 2) the host flight system, and 3) the host spacecraft system. Their relationship and component parts are shown schematically in the block diagram of Fig. 1. Although separate components are shown to emphasize functionality, many of these in fact will be physically combined, in keeping with our goal of producing a highly integrated system.

In order to bound the design, we have assumed an FF ejection speed of 2-5 m/s from the host spacecraft, a measurement duration of 1 hr for each FF, and a maximum range of 10-20 km from the host spacecraft. We also took the host spacecraft to be in a high inclination orbit for magnetic field measurements in the polar cusp region.

2.1. *The Free-Flyer*

The free-flyer consists of a set of sensors to provide three-axis measurements, sensor drive electronics, logic and data storage electronics, a telemetry transmitter and antenna, and a battery and power processing electronics. These components are shown in the "Free-flyer" box on the left of Fig. 1. The heart of the free-flyer is the sensor, for which we plan ultimately to use an electron tunneling effect device. Since this sensor is fabricated in silicon (see below), it allows the possibility of combining it with much of the FF electronics into a monolithic structure. This is one of the goals of the program, as mentioned above.

The FF is disk-shaped,~ 6 cm dia by 2 cm thick, with mass ~ 0.25 kg. It is spun up on ejection to maintain its inertial orientation in space. The ejection process also turns the circuitry on. The required power level (~few hundred mW) is kept low by sharing power alternately between the measurement electronics and the transmitter. This technique also minimizes interference between the transmitter electronics and the sensor. Transmission is via a patch antenna on the base of the FF. The carrier of each FF is coded to distinguish the signal of one from another. Preliminary studies have shown that thermal control of the FFs can be accommodated by using appropriate coatings.

The measurement duration (and therefore spatial range), size and weight of the FF, and the battery size are interrelated. We have not done any detailed trade studies, but obviously, increasing the measurement time in order to increase the range, for measurements in the magnetosheath, for example, would require more stored energy, and hence a larger FF.

2.2 *The Host Flight System*

The host flight system consists of those elements on the host spacecraft which directly interface with the free-flyers: the ejector and FF storage, the FF signal receiver and antenna, and a data processing subsystem. Several concepts for FFM ejection, such as by spring loaded mechanisms, are under consideration. In each of these, the disk is spun up as it is ejected. Analysis indicates that less than a few hundred RPM is sufficient to provide the needed inertial stability.

2.3 *The Host Spacecraft System*

The host spacecraft system consists of the components necessary to "service" the flight system just described. This includes power, command, and data busses. To allow for missions of opportunity, in which the flight system is a piggyback payload on a spacecraft not specifically designed

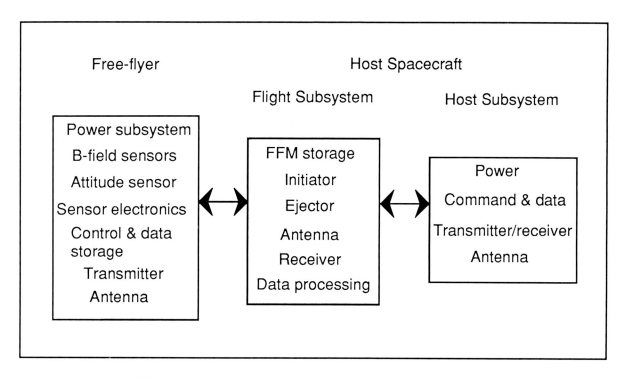

Figure 1. System block diagram of Free-flyer Magnetometer system.

for it, these interfaces should be kept simple. The flight system will therefore be designed to accommodate standard spacecraft interfaces.

3. THE ELECTRON TUNNELING SENSOR MAGNETOMER

The tunneling sensor represents a very new approach for magnetic field measurements. This "chip-size", all silicon device is based on the quantum mechanical electron-tunneling effect which has been used in scanning-tunneling microscopy since 1982 [*Binnig and Roher*, 1982]. The magnetometer sensor exploits the tunneling effect in an arrangement that nulls the deflection of a thin, flexible silicon membrane caused by the Lorentz force resulting from the ambient field to be measured, and a current carrying element on the membrane. The tunneling effect itself can be described with reference to Fig. 2. The tunneling tip is typically several microns in radius, with a gap spacing between the tip and the opposing surface of ~10 Å. The tunneling current between the tip and the surface depends exponentially on that distance. This current therefore provides a sensitive measure of the gap distance. By use of a flexible membrane for the surface, this current then becomes a sensitive measure of the membrane displacement, and therefore the force which produced the displacement. Further, electrically rebalancing this force provides a nulling method of force measurement [*Kenny et al.*, 1994]. For the arrange-

ment in the Figure, the device would be sensitive to components of magnetic field normal to the page, as indicated by the cross at the membrane. Wide dynamic range (>100 dB), mÅ displacement sensitivity, wide bandwidth operation (>10 kHz), robust designs (continuous operation in excess of 3 years in ambient conditions have been demonstrated in the electron tunneling transducer) making this device ideal for use in a low-mass, low-power, highly sensitive μmagnetometer. The drive electronics for the sensor is relatively simple, and is also shown schematically in Fig. 2. The gap drive (labeled "Hi V") is ~ 50 - 100 V.

The device is fabricated by microelectromechanical systems ("MEMS") technology, and is currently under development by the JPL Microdevices Laboratory [*Miller et al., 1996*]. Results from a proof-of-concept μmagnetometer (Prototype I) are tabulated in Table I. The dominant noise mechanism in the tunneling transducer is a 1/f noise. Therefore, the operating point of the device is "pushed" to a higher frequency by biasing the loop with an AC drive current. The response to a DC field is therefore shifted to the drive frequency of the bias current. A typical spectral response is shown in Figure 3, to both a DC and an AC (10 Hz) field, while biasing the μmagnetometer with a 196 Hz current through a single loop. Operating in this mode, a noise-equivalent B-field, NEB, of 7 μT/√Hz has been demonstrated in good agreement with the theoretical value of 3 μT/√Hz. The power consumption is dominated by the bias current through the wire loop, which was not optimized in this experiment. Good linear response was demonstrated for

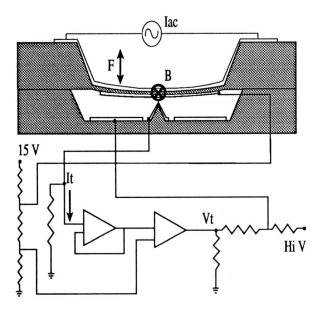

Figure 2. Schematic of bulk micromachined tunneling magnetometer and drive electronics. A Lorentz force, F, is generated by the cross product of the current I_{AC} and the magnetic field, B. The Lorentz force is monitored by the tunneling transducer which generates an electron tunneling current, I_t, which is exponentially related to the membrane displacement. The tunneling voltage, V_t, is the output signal of the device.

both AC and DC fields. Linearity varied from 0.7% FSO in the AC field measurement to 2% FSO in the DC field measurement. Figure 4 illustrates the results for the DC case.

Based on these results, projected performance for a next generation device (Prototype II) is also shown in Table 1. This device, currently in development, uses a loop of 54 turns for the current element in order to increase the sensitivity to the magnetic field.

TABLE 1. Device Performance Comparison between Prototypes I and II.		
Parameter	Prototype I (Demonstrated)	Prototype II (Projected)
NEB (nT/√Hz)	7	10
Bias Frequency (Hz)	200	1000
Number of Loops	1	54
Membrane Dimension (mm)	2.5	3.3
Bias Current (mA)	10	28
Noise (μV/√Hz)	35	10
Responsivity (V/T)	3	610
Power (mW)	50	100

4. MISSION CONSIDERATIONS

The mission analysis performed to date has been directed toward understanding some of the characteristics of how the FFs disperse relative to the host. The result of one such study is shown in Fig. 5, for the trajectories of 8 FFs ejected in 45° intervals at 2m/s relative to the host at an altitude of 6600 km. The ordinate is along the radial direction, while the abscissa is horizontal and in the plane of an 80° orbit. The points are plotted at 30 min. intervals. For this example, the FFs therefore cover a region of about 5 to 10 km around the host in an hour after ejection. Similar calculations for any arbitrary host orbit and ejection velocity are straightforward.

5. CURRENT STATUS

As indicated in the introduction, this is a report on work in progress. Current efforts proceed along three lines: 1) the sensor, 2) system issues, and 3) mission analysis. Some aspects of the tunneling sensor work were summarized above. A proof of concept test of the FFM system has been approved for a suborbital flight in early 1999. This "Enstrophy" mission (K. Lynch, PI) will launch from Poker Flat, Alaska, on a Black Brant 10 class rocket. The scientific objective of the mission is to understand the fine structure of electric currents in the nightside auroral region. In this mission four FFMs will be spun up and ejected from the spinning rocket near its ~1000 km apogee. A brief description of the FFM performance for this mission follows.

Since development of the tunneling sensor is not advanced sufficiently to be included in this flight, the FFMs for Enstrophy will incorporate a 3-axis miniature fluxgate magne-

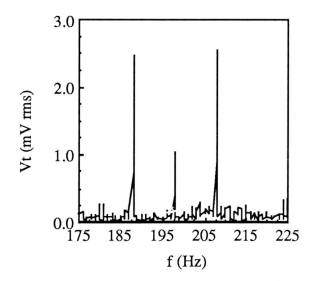

Figure 3. Output spectral response of a tunneling μ-magnetometer to DC (peak at f = 196 Hz) and AC (peaks at f = 196 ± 10 Hz) magnetic fields.

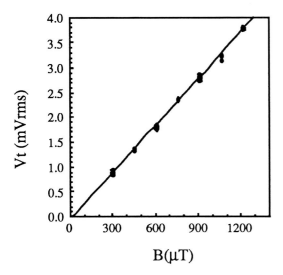

Figure 4. Output response of tunneling μ-magnetometer to DC magnetic field. Bias current, I = 10 mA at 400 Hz. Non-linearity < 2 % FSO.

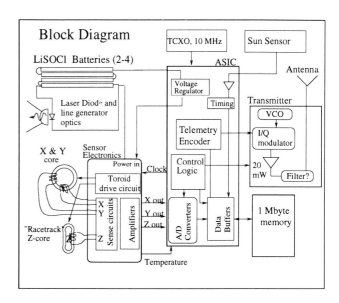

Figure 6. Schematic block diagram of the FFM to be flown on the Enstrophy sub-orbital mission.

tometer sensor, manufactured by Applied Physics Systems, Inc., Mountain View, CA. The sensor has a range of $\pm$ 65536 nT, to be digitized to 1nT resolution, at 7 ms time resolution. The sensor noise level is ~ 0.5 nT/$\sqrt{Hz}$. The FFM also contains a temperature controlled oscillator for accurate timing, ~1 Mbyte memory (for store and forward of the data), A/D and data flow control (ASIC or hybrid chip), LiSOCl batteries, and a telemetry transmitter and patch antenna. Downlink from the Binary Phase Shift Keying (BPSK) transmitter on each FFM from near apogee of the

flight is at S-band directly to the Poker Flat ground station. This link is capable of up to 100 kbps. A schematic block diagram of this FFM is shown in Fig. 6.

6. SUMMARY

We have demonstrated the feasibility of the concept of launching sets of miniature, autonomous "sciencecraft" to measure the small scale structure of the magnetospheric or ionospheric magnetic field. Such fleets will allow separation of temporal and spatial variations of the field and determination of current distributions. The Free-Flyer Magnetometer program at the Jet Propulsion Laboratory involves the development of miniature sensors along with the architecture for these FFMs. In particular, the Laboratory is developing a magnetometer sensor based on the electron tunneling effect in silicon, which will allow a high level of integration with its associated electronics. FFMs using miniature fluxgate magnetometers, as an intermediate step, will be flown as a proof of concept payload on the "Enstrophy" suborbital flight in early 1999.

Acknowledgments: Portions of this work were performed at the Jet Propulsion Laboratory, California Institute of Technology, Pasadena, CA under contract with the National Aeronautics and Space Administration.

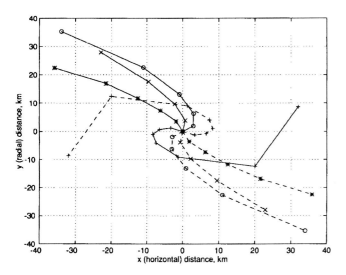

Figure 5. Dispersal pattern of eight free-flyers ejected 2 m/s in 45° intervals for a release altitude of 6600 km. Positions are shown at 30 min. intervals in the plane of the 80° inclination orbit of the host.

REFERENCES

Binnig, G. and H. Roher, Scanning tunneling microscopy, *IBM J. Res. Dev., 30,* 355-369, 1982.

Kenny, T. W., et al., Widebandwidth electromechanical actuators for tunneling displacement transducers, *J. Microelectromech. Sys., 3,* 97-104, 1994.

Lynch, K. A., R. L. Arnoldy, J. Bonnell, and P. M. Kintner, The AMICIST auroral sounding rocket - a comparison of transverse ion acceleration methods, *Geophys. Res. Lett., 23,* 3293-3296, 1996.

Miller, L. M., J. A. J. A. Podosek, E. Kruglick, T. W. Kenny, J. A. Kovacich, and W. J. Kaiser, A μ-magnetometer based on electron tunneling, *Proc. Electro Mechanical Systems*, 467-471, 1996.

_____

R. Goldstein, M. Boehm, E. Cutting, E. Fossum, H. Javadi, L. M. Miller, B. Pain, J. E. Randolph, P. R. Turner, Jet Propulsion Laboratory, Mail Stop 169-506, 4800 Oak Grove Dr., Pasadena, CA 91109.

K. Lynch, Institute for the Study of Earth Oceans and Space, University of New Hampshire, Durham, NH 03824.

Measuring Magnetic Field Gradients from Four Point Vector Measurements in Space

Krishan K. Khurana, Larry Kepko, and Margaret G. Kivelson

Institute of Geophysics and Planetary Physics, University of California, Los Angeles, CA

The determination of all nine first order spatial gradients requires the knowledge of differences between the magnetic field measurements made simultaneously at (at least) four different points in space. As the first order differences between measurements are typically very small compared to the background field, even small errors resulting from an inadequate knowledge of the orientations, zero levels and scale factors of the magnetometer sensors affect the calculation of field gradients disproportionately and must be corrected with high accuracy. Other sources of error include the noise introduced in the data by propagating waves and pulsations and errors in the knowledge of inter-spacecraft spacing. We will discuss how each of these errors can be minimized by improved data processing techniques and in-flight calibration of the measurements. We first introduce an intra-spacecraft calibration technique suitable for spinning spacecraft which relies on the fact that errors in many of the calibration parameters generate monochromatic signals in the despun data at the first two harmonics of the spin frequency. We develop equations that relate the power in the two harmonics to the calibration parameters. In another procedure called inter-calibration use is made of the fact that $\nabla \cdot B$ is zero everywhere and $\nabla \times B$ is vanishingly small in certain regions of the magnetosphere. Calibration parameters are sought that minimize measured $\nabla \cdot B$ and $\nabla \times B$ in those regions.

INTRODUCTION

Direct measurement of spatial magnetic gradients in the neighborhood of the Earth is an objective of the Cluster mission. As the spatial gradients are typically very small, they cannot be obtained directly from gradiometer-type measurements made from a single spacecraft and can only be determined from first differences (the differences in the values of a field component measured simultaneously at several locations). Even for spacecraft separated by distances of several thousand kilometers, the first differences are expected to be of the order of one nT in most of the

magnetosphere. The spatial gradients are therefore easily corrupted by problems like (1) instrument-related measurement errors, (2) spacecraft-related artifacts, e.g. interference from other subsystems and (3) sources external to the spacecraft like temporal variations caused by propagating waves and oscillations. As problems caused by sources (2) and (3) are in general not systematic in nature, they can be easily dealt with by a careful data refinement strategy as discussed later. The instrument related sources of errors stem from an inadequate knowledge of the zero levels, orientations and the scale factors of the magnetometer sensors. In order to improve upon the laboratory-determined values of these quantities (called calibration parameters in this work), further "onboard calibrations" are required. We show below that the problem of minimizing measurement errors can be divided into two parts. In a procedure called intra-calibration, use is made of the fact that

Measurement Techniques in Space Plasmas: Fields
Geophysical Monograph 103

for spinning magnetometers, errors in many of the calibration parameters generate coherent signals at the first two harmonics of the spin frequency in the despun data. Intra-calibration procedures seek parameters that minimize these harmonics in the calibrated data (see e.g., *Kivelson et al.,* [1992] or *Farrell et al* [1995]). In this work, we shall follow a new Fourier transform-based procedure [*Kepko et al.,* 1996]. In a second procedure called inter-calibration use is made of the fact that $\nabla \cdot \boldsymbol{B}$ is zero everywhere and $\nabla \times \boldsymbol{B}$ is vanishingly small in certain regions of the magnetosphere [*Khurana et al.,* 1996]. If the data have not been properly inter-calibrated, they yield non-zero values for $\nabla \cdot \boldsymbol{B}$ and $\nabla \times \boldsymbol{B}$ in those regions. Correct calibration parameters are determined by requiring that the final data set must yield values of $\nabla \cdot \boldsymbol{B}$ and $\nabla \times \boldsymbol{B}$ close to zero.

Twelve parameters are required to relate the output of a near-orthogonal sensor triad to the actual value of the field. These could be thought of as the nine elements of a coupling matrix ($\mathbf{C}$) that scales and reorients the sensor data and three offsets that correct for the zero levels of the sensors. If $\mathbf{B}_i$ and $\mathbf{B}_s$ denote the measurements in the spinning spacecraft and the sensor coordinates, respectively, then

$$\mathbf{B}_i = \mathbf{C}_{ij} \times \left(\mathbf{B}_{S_j} - \mathbf{O}_j \right) \tag{1}$$

INTRA-CALIBRATION

We shall first relate the calibration parameters to the elements of the above coupling matrix and show that eight of the twelve parameters can be determined from intra-calibration. Then we will outline a procedure that uses data from all four of the spacecraft to infer the remaining calibration parameters. Let z denote the spin axis of the triad and let x-y lie in the spin plane. If θ is the elevation angle that a sensor makes with respect to the spin axis and φ is the azimuthal angle the projection of the sensor makes with the x-axis in the x-y plane, then the measurements made by the sensors can be related to the magnetic field expressed in spacecraft coordinates by

$$\mathbf{B}_{S_i} - \mathbf{O}_i = \mathbf{G}_i \left(\sin\theta_i \cos\varphi_i \quad \sin\theta_i \sin\varphi_i \quad \cos\theta_i \right) \mathbf{B} \tag{2}$$

where G_1, G_2 and G_3 denote the gain factors of the three sensors [*Kepko et al.,* 1996]. Equation (3) can be simplified by measuring the elevation and azimuthal angles of the sensors with respect to the nearest spacecraft coordinate axis, that is

$$\begin{array}{lll} \theta_1 = 90 - \Delta\theta_1 & \theta_2 = 90 - \Delta\theta_2 & \theta_3 = \Delta\theta_3 \\ \varphi_1 = \Delta\varphi_1 & \varphi_2 = 90 + \Delta\varphi_2 & \varphi_3 = \varphi_3 \end{array} \tag{3}$$

where the elevation angles $\Delta\theta_1$ and $\Delta\theta_2$ are measured from the x-y plane, $\Delta\theta_3$ is measured from the z axis, $\Delta\varphi_1$ and φ_3 are measured from the x axis, and $\Delta\varphi_2$ is measured from the y axis. Notice that with the exception of φ_3,

all of the angles are expected to be small. Using the small angle approximation we get

$$\begin{pmatrix} B_{S_1} - O_1 \\ B_{S_2} - O_2 \\ B_{S_3} - O_3 \end{pmatrix} = \begin{pmatrix} G_1 & G_1\Delta\varphi_1 & G_1\Delta\theta_1 \\ -G_1(\Delta\varphi_1 + \Delta\varphi_{21}) & G_1 + \Delta G_{21} & G_1\Delta\theta_2 \\ G_3\Delta\theta_3 \cos\varphi_3 & G_3\Delta\theta_3 \sin\varphi_3 & G_3 \end{pmatrix} \mathbf{B} \tag{4}$$

The gain factors in the above equation are numbers close to unity. We further define ΔG_{21} and $\Delta\varphi_{21}$ by

$$G_2 = G_1 + \Delta G_{21} \quad \text{and} \quad \Delta\varphi_2 = \Delta\varphi_1 + \Delta\varphi_{21} \tag{5}$$

Notice that $\Delta\varphi_{21} = \Delta\varphi_2 - \Delta\varphi_1 = 90° - (\varphi_2 - \varphi_1)$ is a measure of the non-orthogonality of the spin plane sensors and ΔG_{21} is a measure of the non-equality of their gains. Substituting (5) into (4) and ignoring second order terms we get

$$\begin{pmatrix} B_{S_1} - O_1 \\ B_{S_2} - O_2 \\ B_{S_3} - O_3 \end{pmatrix} = \begin{pmatrix} G_1 & G_1\Delta\varphi_1 & G_1\Delta\theta_1 \\ -G_1(\Delta\varphi_1 + \Delta\varphi_{21}) & G_1 + \Delta G_{21} & G_1\Delta\theta_2 \\ G_3\Delta\theta_3 \cos\varphi_3 & G_3\Delta\theta_3 \sin\varphi_3 & G_3 \end{pmatrix} \mathbf{B} \tag{6}$$

By multiplying both sides of (6) with the despin matrix

$$D = \begin{pmatrix} \cos\omega t & -\sin\omega t & 0 \\ \sin\omega t & \cos\omega t & 0 \\ 0 & 0 & 1 \end{pmatrix} \tag{7}$$

we obtain the despun vector data, $(B_{X'}, B_{Y'}, B_{Z'})$

$$\begin{aligned} B_{X'} = {} & G_1 B_H \left(\cos\psi + \Delta\varphi_2 \sin\psi \right) \\ & + G_1 \frac{B_H}{2} \left(\Delta G'_{21} \cos\psi - \Delta\varphi_2 \sin\psi \right) \\ & + \cos\omega t\, G_1 \left(B_z\Delta\theta_1 + O'_1 \right) \\ & + \sin\omega t\, G_1 \left(-B_z\Delta\theta_2 - O'_2 \right) \\ & + \cos 2\omega t\, G_1 \frac{B_H}{2} \left(-\Delta G'_{21} \cos\psi - \Delta\varphi_{21} \sin\psi \right) \\ & + \sin 2\omega t\, G_1 \frac{B_H}{2} \left(-\Delta G'_{21} \sin\psi + \Delta\varphi_{21} \cos\psi \right) \end{aligned} \tag{8a}$$

$$\begin{aligned} B_{Y'} = {} & G_1 B_H \left(\sin\psi - \Delta\varphi_2 \cos\psi \right) \\ & + G_1 \frac{B_H}{2} \left(\Delta G'_{21} \sin\psi + \Delta\varphi_2 \cos\psi \right) \\ & + \cos\omega t\, G_1 \left(B_z\Delta\theta_2 + O'_2 \right) \\ & + \sin\omega t\, G_1 \left(B_z\Delta\theta_1 + O'_1 \right) \\ & + \cos 2\omega t\, G_1 \frac{B_H}{2} \left(\Delta G'_{21} \sin\psi - \Delta\varphi_{21} \cos\psi \right) \\ & + \sin 2\omega t\, G_1 \frac{B_H}{2} \left(-\Delta G'_{21} \cos\psi - \Delta\varphi_{21} \sin\psi \right) \end{aligned} \tag{8b}$$

$$\begin{aligned} B_{Z'} = {} & G_3 \left(B_z + O'_3 \right) \\ & + \cos\omega t\, G_3 B_H \left(\cos\varphi_3\Delta\theta_3 \cos\psi + \sin\varphi_3\Delta\theta_3 \sin\psi \right) \\ & + \sin\omega t\, G_3 B_H \left(-\sin\varphi_3\Delta\theta_3 \cos\psi + \cos\varphi_3\Delta\theta_3 \sin\psi \right) \end{aligned} \tag{8c}$$

where $\psi = \tan^{-1}(B_y/B_x)$ and B_H is the magnitude of the field in the spin-plane. We have also defined:

$$O'_3 = \frac{O_3}{G_3} \quad O'_1 = \frac{O_1}{G_1} \quad O'_2 = \frac{O_2}{G_2} \quad \text{and} \quad \Delta G'_{21} = \frac{\Delta G_{21}}{G_1} \quad (9)$$

Equations (8a) through (8c) form the basis of the calibration procedure. They relate the calibration parameters to the amplitudes of the first and second spin harmonics in the despun data. It is seen that the elevation of S_1 and S_2 out of the spin plane ($\Delta\theta_1$ and $\Delta\theta_2$) and the offsets (O_1 and O_2) in the spin-plane sensors produce first harmonics in $B_{X'}$ and $B_{Y'}$ (through their association with the $\sin\omega t$ and $\cos\omega t$ terms). The non-orthogonality of the spin plane sensors ($\Delta\varphi_{21}$) and the mismatch of the gains (ΔG_{21}) generate signals at the second harmonic in $B_{X'}$ and $B_{Y'}$. The elevation ($\Delta\theta_3$) of the spin axis sensor from the z axis generates a first harmonic in $B_{Z'}$. Three of the calibration parameters (G_3, φ_1, O_3) are not associated with the two harmonics and thus do not produce coherent spin related signals in the despun data. The calibration parameter G_1 does appear in the first and second harmonic terms but always in association with other calibration parameters. From now on we will assume that fairly reliable values of these four parameters are available from ground calibrations. In the section on inter-calibration, we will describe how these parameters can be refined further.

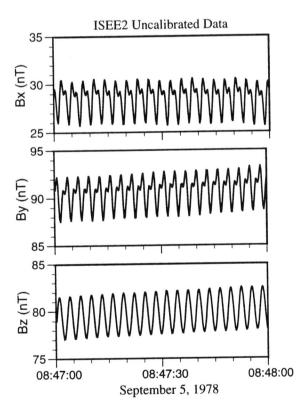

Figure 1. An example of a data segment which was despun without calibration. The data were obtained by ISEE-2 in the inner magnetosphere. Figure reproduced from *Kepko et al.* [1996].

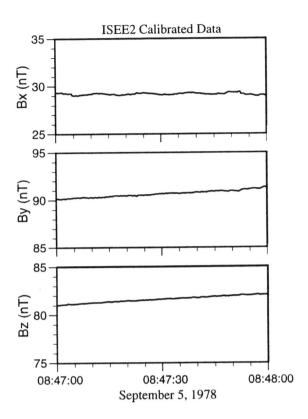

Figure 2. The same segment of data despun after calibration. Notice that the spin harmonics in all three of the components have been reduced to the background level of the noise. Figure reproduced from *Kepko et al.* [1996].

Thus eight calibration parameters can be determined by analyzing the power in the first and second harmonics of the despun data. *Kepko et al.* [1996] describe a Fourier transform method to calculate the power in the spin harmonics and show that equations (8 a-c) reduce to a set of linear equations in the frequency domain which can be solved for the eight calibration parameters.

The technique described above has been successfully applied on data obtained from the ISEE 1 and 2 and the Galileo spacecraft. Here we present results from a calibration performed on an ISEE 2 data set from September 5, 1978 during an inbound part of its orbit through the inner magnetosphere. In Figure 1 we show a segment of the despun uncalibrated data which contain power at the first and second spin harmonics. The same segment despun after calibration is plotted in Figure 2. The calibrated data do not show enhanced power at the spin harmonics.

INTER-CALIBRATION

Once the intra-calibrations have been performed, measurements from the four spacecraft must be compared in order to remove the remaining spurious differences among them. This is the process we call inter-calibration. The intra-calibrated data from all four spacecraft are despun and rotated into a global geophysical coordinate system. This rotation requires accurate knowledge of the orientation of the spacecraft but the information may not be available with the required precision of better than $0.1°$. Therefore, we shall assume that the data are properly orthogonalized but their absolute orientations are in error. Then for each spacecraft six unknowns remain (three Euler angle rotations to orient the independent orthogonal coordinate system in space, two gains to correct the gain mismatch of the spin axis and spin plane sensors, and the offset of the spin axis sensor). In order to complete the inter-calibration, one needs to determine only the *relative* angles, *relative* gains and *relative* offsets between a "mother" and three "daughter" spacecraft. This means that only 6x3 = 18 quantities are needed to complete the inter-calibration. After inter-calibration, the effective orientations, gains and offsets of the data sets from the three daughter spacecraft would be identical to those of the mother spacecraft.

The inter-calibration method uses an error propagation scheme for determining the calibration parameters. Miscalibrated magnetic field data introduce errors into the nine first order gradients. By using Taylor series expansions of the measurements around a point enclosed by the tetrad, we analytically propagate errors from the field values into the gradients. The errors are further propagated into $\nabla \cdot \boldsymbol{B}$ and $\nabla \times \boldsymbol{B}$. Calibration parameters are sought that minimize the measured $\nabla \cdot \boldsymbol{B}$ and $\nabla \times \boldsymbol{B}$ in those regions of space where no electric currents are known to be present.

BASIC EQUATIONS

The four vector measurements $\mathbf{B}_i=(B_{xi}, B_{yi}, B_{zi})$ obtained at the vertices of a tetrahedron are related to the field $\mathbf{B}_0=(B_{x0}, B_{y0}, B_{z0})$ at the center of the tetrahedron by

$$\mathbf{B}_i = \mathbf{B}_0 + \left(\mathbf{dR}_{i0} \cdot \nabla \right) \times \mathbf{B} \qquad (10)$$

where $\mathbf{dR}_{i0} = (dx_{i0}, dy_{i0}, dz_{i0})$ is the distance of the (i+1)'th spacecraft from point 0 at the center. Differencing the field of spacecraft 1 (referred to as the mother spacecraft) from those of the other three spacecraft gives

$$B_{x_{i+1,1}} = \left(\mathbf{dR}_{i+1,1} \cdot \nabla_i \right) B_x \qquad (11)$$

where $B_{x_{i+1,1}}$ is the difference between the x components of the i'th and the mother spacecraft and $\mathbf{dR}_{i+1,1}$ are the distances between the (i+1)'th and the mother spacecraft. Equation (11) can be inverted to obtain

$$\nabla_i B_x = A_{i,j} \times B_{x_{j+1,1}} \qquad (12)$$

where the matrix A_{ij} is the inverse of the distance matrix $(\mathbf{dR}_{i1})$ of equation (11). Similar equations are obtained for B_y and B_z and solved for their respective gradients.

The magnetic field measurements are often inaccurate because six of the calibration parameters are not known accurately. As a result there may be large errors in the first differences. The first differences of the imperfect measurements (denoted by double primes), will satisfy a generalization of equation (12):

$$\nabla_i B_x + \delta_{xi} = A_{i,j} \times B''_{x_{j+1,1}} \qquad (13)$$

where δ_{xx}, δ_{xy} and δ_{xz} are the errors in the x, y and z derivatives of B_x. Similar equations can be written for the derivatives of B_y and B_z components.

ANALYTICAL ERROR PROPAGATION

We will now introduce errors of the types discussed above into the magnetic field measurements and see how they propagate into the inferred gradients. As the calibration parameters that we seek are expected to be small in magnitude, the analysis will be carried out to first order only. Let α, β and γ denote the three Euler angle rotations around the principal axes that would rotate the data from the global coordinate system into the local coordinates of a misoriented magnetometer. Let O_z be the offset, ε_z be the fractional correction to the gain of the spin aligned sensor and $\varepsilon_x = \varepsilon_y = \varepsilon_H$ be the fractional gain correction of the spin plane sensors. Then the measured field is given by

$$B''_x = B_x + \gamma B_y - \beta B_z + \varepsilon_H B_x$$
$$B''_y = B_y - \gamma B_x + \alpha B_z + \varepsilon_H B_y \qquad (14)$$
$$B''_z = B_z + \beta B_x - \alpha B_y + \varepsilon_z B_z + O_z$$

Substituting equation (14) into (13) we obtain

$$\nabla_i B_x + \delta_{xi} = A_{ij} \times M_j \qquad (15a)$$

for the x component, where

$$M_j = \left(B_{x_{j+1,1}} + \gamma_{j+1,1} B_{y_{j+1}} - \beta_{j+1,1} B_{z_{j+1}} + \varepsilon_{H_{j+1,1}} B_{x_{j+1}} \right) (15b)$$

and the subscript $j+1,1$ on a quantity implies that the difference between the (j+1)'th and the mother spacecraft are to be taken. For example $B_{x_{j+1,1}} = B_{x_{j+1}} - B_{x_1}$ and $\gamma_{j+1,1} = \gamma_{j+1} - \gamma_1$.

Subtracting equation (12) from equation (15), we obtain

$$\delta_{xi} = A_{ij} \times M_j \qquad (16)$$

The above equation relates the errors in the three gradients of B_x to the inverse of the distance matrix and the calibration parameters. Similar equations can be written for the gradients of B_y and B_z.

FORMULATING THE LEAST SQUARES EQUATIONS

The measured (denoted by subscript m) gradients are defined as

$$\left(\nabla_i \cdot B_x \right)_m = A_{ij} \times \left(B_{x_{j+1,i}} \right)_m \qquad (17)$$

and similar equations for the gradients of B_y and B_z. These equations are used to compute $(\nabla \times B)_m$ and $(\nabla \cdot B)_m$.

The analytical $\nabla \times B$ and $\nabla \cdot B$ (denoted by subscript a) are given by

$$(\nabla \times B)_{xa} = \delta_{zy} - \delta_{yz}, \qquad (\nabla \times B)_{ya} = \delta_{xz} - \delta_{zx}$$
$$(\nabla \times B)_{za} = \delta_{yx} - \delta_{xy}, \qquad (\nabla \cdot B)_a = \delta_{xx} + \delta_{yy} + \delta_{zz} \qquad (18)$$

in regions where the background $\nabla \times B$ is assumed to be zero. It can be seen that non-zero values of $\nabla \times B$ and $\nabla \cdot B$ are caused by the errors in the calibration parameters. The difference between the measured and the analytical $\nabla \times B$ and $\nabla \cdot B$ can then be minimized in the least squares sense to obtain the calibration parameters [*Khurana et al.*, 1996].

A TEST OF THE TECHNIQUE

As no four-spacecraft data are yet available, we created a simulated database by using the *Tsyganenko* [1989] model in regions of the magnetosphere where background currents are known to be small. We assumed several geometries of the Cluster tetrahedron and used inter-spacecraft distances similar to those expected for the Cluster mission (from 200 km to 6400 km, (see *Dunlop* [1990], *Dunlop et al.* [1990]) with an average inter-spacecraft distance $\approx$ 1000 km. Next, the data were corrupted by introducing small errors in the calibration parameters. The corrupted synthetic data were then used in the iteration procedure described above to recover the calibration parameters. The parameters were recovered with a remarkable precision (error in angles < 0.05 degrees, error in gain factors < .001, and error in offsets < 0.1 nT).

In Figure 3a we show the y components of $\nabla \times B$ at X = -15 R_E calculated from *Tsyganenko* [1989] model by using a Cluster type tetrad. As expected, the current density maximizes at the center of the magnetotail plasma sheet. Next, small calibration errors are introduced for three of the magnetometers in the tetrad. Figure 3b shows the y component of $\nabla \times B$ inferred from the corrupted measurements. The modest errors introduced in the calibration parameters produced large errors in the measured currents. The errors are not random in nature and their effect cannot be removed by simple averaging. Next, the measurements were improved by applying corrections obtained from the procedure. The y component of $\nabla \times B$ was calculated

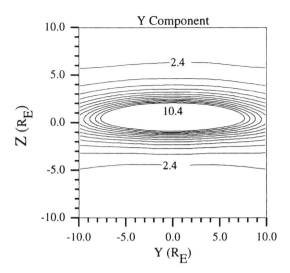

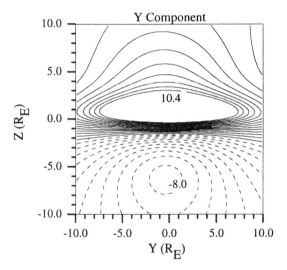

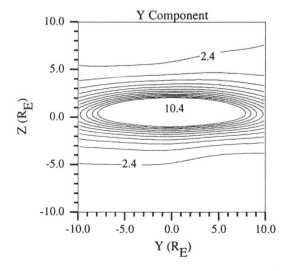

Figure 3. The contour plots of the y component of $\nabla \times B$ (expressed in nT/R_E) calculated from   *a*. uncorrupted magnetic field generated from *Tsyganenko* [1989] model in the Y-Z cross-section of the magnetotail at X = -15 R_E.   *b*. From the corrupted magnetic field data and *c*. From the inter-calibrated magnetic field data.

from the re-calibrated data set and as shown by Figure 3c, is very close to its true value plotted in Figure 3a.

DISCUSSION

In this section we will discuss other sources of errors in the calculation of spatial gradients and how their effects can be minimized. To begin with, a minor, but mostly unpredictable source of error in the first differences is the imperfect knowledge of inter-spacecraft positions. Equation (11) shows that the errors in the first order differences would be at the same level as are the percentage errors in the knowledge of inter-spacecraft distances (which are estimated to be at a few percent level at this time for the Cluster mission). We, therefore, do not expect errors from this source to significantly impact the calculation of gradients. The presence of errors in the inter-spacecraft distances can be detected by comparing the observed value of the total time derivative of B (i.e. time derivative of B in the spacecraft rest frame) with its expected value which can be estimated from:

$$d\mathbf{B}/dt = \left(u_{s/c} \cdot \nabla\right)\mathbf{B} \qquad (22)$$

under the assumption that $\partial\mathbf{B}/\partial t$ is small compared to the RHS of equation (22). Errors in the inter-spacecraft distances produce errors in the calculated spatial gradients and therefore the calculated and the expected values of the total time derivative would be different.

Additional errors in the calculated spatial gradients arise from low frequency magnetic noise produced by other spacecraft subsystems and instruments. This problem is best avoided by adhering to an effective magnetic cleanliness and compensation program on the spacecraft. Long magnetometer booms help by reducing the impact of noise sources on the magnetometer. Detailed modeling of magnetic sources is also highly recommended. These models can then be used to reduce the effect of magnetic sources on the inboard data.

Another type of error arises from propagating waves and pulsations that introduce temporal-spatial gradients which can mask the steady spatial gradients. If time series of sufficiently long duration are available from all four spacecraft, it would be possible to distinguish between the steady state and fluctuating components of the spatial derivatives and filtering techniques can be devised to separate the two effects [*Neubauer and Glassmeier, 1990*]. In general, the presence of errors in the calculated spatial gradients produces non-zero values of $\nabla \cdot B$. We, therefore recommend that $\nabla \cdot B$ calculated from the inter-calibrated data should be monitored regularly to assess the quality of the inter-calibrations and the accuracy of the spatial gradients.

Finally, we would like to comment on the sensitivity of the measurements on the configuration of the spacecraft tetrahedron. As discussed by *Dunlop et al.* [1990], when all four spacecraft lie in a plane or along a straight line (collapsed tetrahedron), gradients can be obtained with confidence only in certain directions. The Q parameter introduced by *Dunlop et al.* describes the quality of the tetrahedron configuration when determining the spatial gradients. We recommend that only those data sets be included in calibration and four spacecraft analysis which have a Q value of at least 3.2.

Acknowledgments. This work was supported by the National Aeronautics and Space Administration Division of Space Physics under grant NAG5-1167. UCLA-IGPP publication number 4602.

REFERENCES

Dunlop, M. W., "Review of the Cluster orbit and separation strategy: consequences for measurements", *Proceedings of an International Workshop on Space Plasma Physics Investigations by Cluster and Regatta*, Graz, Austria, ESA SP-306, 1990.

Dunlop, M. W., A. Balogh, D. J. Southwood, R. C. Elphic, K.-H Glassier and F. M. Neubauer, "Configurational sensitivity of multipoint magnetic field measurements", *Proceedings of an International Workshop on Space Plasma Physics Investigations by Cluster and Regatta*, Graz, Austria, ESA SP-306, 1990.

Farrell, W. M., R.F. Thompson, R. P. Lepping, and J. B. Byrnes, "A method of calibrating magnetometers on a spinning spacecraft", *IEEE Trans. Magnetics*, 31, 966, 1995.

Neubauer, F. M. and K.-H. Glassmeier, "Use of an array of satellites as a wave telescope", *J. Geophys. Res.*, 95, 19,115, 1990.

Kepko, E. L., K. K. Khurana, M. G. Kivelson, "Accurate determination of magnetic field gradients from four point vector measurements: 1. Use of natural constraints on vector data obtained from a single spinning spacecraft", IEEE Transactions on Magnetics, *32*, 377. 1996.

Khurana, K. K., E. L. Kepko, M. G. Kivelson, and R. C. Elphic, "Accurate calculation of magnetic field gradients from four point vector measurements: 2. Use of natural constraints on vector data obtained from four spinning spacecraft", IEEE Transactions on Magnetics, *32*, 5193, 1996.

Kivelson, M. G., K. K. Khurana, J. D. Means, C. T. Russell, and R. C. Snare, "The Galileo magnetometer investigations", *Space Science Rev.*, *60*, 357, 1992.

Tsyganenko, N. A., "A magnetospheric model with a warped tail current sheet", *Planet. Space Sci.*, 37, 5-20, 1989.

Krishan Khurana, Larry Kepko, Margaret Kivelson, Institute of Geophysics and Planetary Physics. University of California at Los Angeles, Los Angeles, CA 90096-1567. USA.

Ionospheric Multi-Point Measurements Using Tethered Satellite Sensors

B. E. Gilchrist[1], R. A. Heelis[2], and W. J. Raitt[3]

Many scientific questions concerning the distribution of electromagnetic fields and plasma structures in the ionosphere require measurements over relatively small temporal and spatial scales with as little ambiguity as possible. It is also often necessary to differentiate several geophysical parameters between horizontal and vertical gradients unambiguously. The availability of multiple tethered satellites or sensors, so-called "pearls-on-a-string," may make the necessary measurements practical. In this report we provide two examples of scientific questions which could benefit from such measurements (1) high-latitude magnetospheric-ionospheric coupling; and, (2) plasma structure impact on large and small-scale electrodynamics. Space tether state-of-the-art and special technical considerations addressing mission lifetime, sensor pointing, and multi-stream telemetry are reviewed.

1. INTRODUCTION

The space tethered satellite concept has been under serious development for more than a decade. Four orbital demonstration and science flights have been successfully flown from unmanned launch vehicles and two flights have also been conducted from the space shuttle. In addition, numerous sub-orbital tethered rocket flights have been conducted successfully starting in the early 1980's and continuing to this day [James and Rumbold, 1995].

Here, we discuss possible benefits, technical capabilities, and challenges in conducting magnetospheric-ionospheric-thermospheric-mesospheric (MITM) measurements using multiple sensors which are constrained to "fly in formation" because of an interconnecting long tether (Figure 1). Such tethered instrument arrays operated from small and medium sized spacecraft, the space shuttle, or the space station, appear able to address essential elements of scientific importance.

At a 1994 University of Michigan summer workshop on the use of space tethers for MITM research [Gilchrist, et al., 1995a] it was determined that from a measurement perspective, a viable tether capability would enable (1) vertical sampling of small-scale particle and field structures, (2) access to large scale horizontal sampling at fixed altitudes below 180 km, (3) a remote sensing multi-look capability, and, (4) the ability to conduct controlled cause-and-effect ionospheric experiments. The merit of using space tethers for multi-point measurements was also recognized as part of NASA's Space Physics Division Strategy-Implementation Study - 1995-2010 where it was noted that "...the unique and important nature of the scientific results [using space tethers for multi-point measurements] would be very strong."

Viable space tether systems have now seen their first flight demonstrations. Three Small Expendable Deployable System (SEDS) missions have successfully flown as secondary payloads on Delta II boosters (SEDS-1 - March, 1993; PMG - May, 1993; and, SEDS-2 - March, 1994). These missions demonstrated the feasibility of simple, low-cost, automated deployment and stabilization of long tethered satellite systems (at least to 20 km) which do not require recovery (Figure 2). The same SEDS deployment system was recently used on a U.S. Navy mission called Tether Physics and Survivability (TiPS) which, as of this

[1]Space Physics Research Laboratory, University of Michigan, Ann Arbor, MI
[2]Center for Space Sciences, University of Texas at Dallas, Richardson, TX
[3]Center for Atmospheric and Space Science, Utah State University, Logan, UT

Measurement Techniques in Space Plasmas: Fields
Geophysical Monograph 103

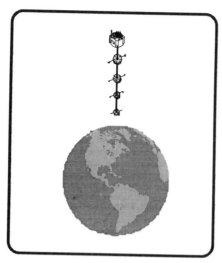

Figure 1. The concept of using a space tether instrument string ("pearls-on-a-string") from a space platform to achieve multi-point and simultaneous measurements along the local vertical is one of the exciting techniques which can be practically achieved in the 1990s.

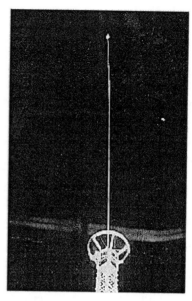

Figure 3. Photo as seen from the shuttle aft flight deck of the TSS-1 tethered satellite. Both TSS-1 and TSS-1R tether dynamics showed a high degree of stability and predictability.

writing has been in orbit for nearly one year flying a 4 km long tether system at 1000 km altitude.

The first Tethered Satellite System (TSS-1) missions were flown on board the space shuttle in 1992 and 1996 (Figure 3). The TSS was developed to provide a reusable facility with advanced capabilities of deploying complex satellites requiring dynamic stability control of long, gravity-gradient stabilized tethers and then recovering them to the Space Shuttle. Both the TSS-1 1992 and TSS-1R 1996 tether missions ended prematurely due to problems unrelated to fundamental tether dynamical issues. For TSS-1, a mechanical blockage in its deployer mecha-

nism limited deployment to 267 m. However, the system was found to be highly stable in proximity to the Orbiter which was of considerable concern for safety reasons. The TSS-1R mission deployed to a distance of approximately 19.7 km before suffering a tether break due to a high-voltage electrical arc between the tether conductor and Orbiter electrical ground. Prior to the break the tether dynamics was behaving as planned, the deployment system was working nominally, and electrodynamic tether measurements were highly successful. High-voltage electrical arcs are not of concern for multi-point applications discussed here as non-conducting tethers are preferred due to their lower mass and volume.

This paper is organized as follows. Section 2 provides some examples of scientific questions which motivate the need for multi-point measurements. Section 3 describes example mission scenarios. Section 4 provides more detailed description of technological considerations as well as relevant efforts to improve upon tether state-of-the-art.

2. SCIENCE MOTIVATION

Knowledge of the MITM system has thus far been accumulated largely from global synoptic data sets gathered from single spacecraft, ground based facilities, and sounding rockets. There are advantages and disadvantages to all these observational techniques. Spacecraft in circular orbits are able to readily diagnose horizontal variations in longitude and/or latitude, but fail in the event of significant altitude variations which become mixed with and in some cases unresolvable from horizontal variations. Therefore,

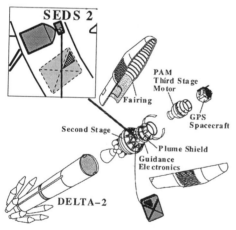

Figure 2. The three initial SEDS tether missions have flown as secondary payloads on the second stage of a Delta-2 launch vehicle. The hardware can easily be configured for a dedicated launch vehicle.

in order to unravel questions related to the horizontal and vertical variations of the region that may change over time scales of a few hours or less and spatial scales of a few 100 km or less, the use of tethered observational platforms will be essential.

A number of scientific challenges can directly benefit from tethered spacecraft systems to advance our under- standing. To illustrate, below we briefly discuss two compelling MITM science examples which require the use of tethered satellite systems to make significant advances.

Magnetospheric-Ionospheric Coupling: Energy Dissipation and Configuration of 3-D High Latitude Current Systems

The importance of simultaneous, multi-point measure- ments in the Earth's magnetosphere is well recognized and was the driving factor behind the ISTP multi-satellite CLUSTER initiative [1993]. Only through such measure- ment techniques is it possible to truly distinguish funda- mental mechanisms. A similar need exists to address the rich nature of fundamental, yet unexplored, high latitude physics in the lower thermosphere and auroral coupling regimes as well as their intimate ties to the more tenuous and difficult-to-characterize magnetosphere.

The closure path for very intense magnetospheric current systems is highly structured, both horizontally and verti- cally, as well as dynamic. These currents flow along the earth's magnetic field lines to and from the magnetosphere and close across relatively narrow vertical slabs in the ionosphere. The altitude distribution over which the hori- zontal ionospheric currents flow depends on a number of factors including (1) the energy distribution of precipi- tating particles, (2) the solar zenith angle, (3) the horizontal scale size of the convection electric fields and magnetic field-aligned current distribution, and, (4) the temporal behavior of these electrodynamic drivers. Many of these factors are thought to be inter-related. In general, all other parameters being equal, the larger the scale size of the driving electrodynamics, the lower the altitude of horizon- tal current closure [Heelis and Vickrey, 1990]. It should be emphasized that the largest altitude gradients for large and medium spatial scales exists in the region below 180 km where substantial horizontal currents can flow.

The closure pattern for magnetospheric currents in the ionosphere involves a number of complex issues including the conductivity profile, the scale size of the driving electrodynamics, and the neutral wind patterns. The situa- tion becomes even more complex when temporal dynamics are also considered in the system. It appears that the modeling work is presently ahead of the observational data base necessary to verify or dispute the model validity. A tethered satellite mission making simultaneous, multiple altitude measurements of electric and magnetic fields and neutral wind vectors would address and resolve outstand- ing questions about the electrodynamics of the entire ionosphere, thermosphere, and magnetosphere system.

Effects of Plasma Structures on Large and Small Scale Electrodynamics

There have been many observations of plasma layers in the bottom-side of both the F and E regions which are thought to be formed by shears in the meridional and zonal wind fields of the region. To date, little information is available on the horizontal extent of these intermediate and sporadic layers other than from knowledge of the tidal or gravity wave wind fields which may be responsible for their formation. This information is essential for determin- ing the electrodynamic consequences of the layers, but requires the kinds of measurements only easily achieved with space tethers to unambiguously resolve relevant processes.

Bottom-side ionospheric layers may have a significant effect on the flux tube integrated conductivity depending on the altitude of their formation. The winds that produce the layers must have relatively small vertical wavelengths and thus drive local current loops in the lower F and E- regions. Winds with larger vertical wavelengths drive current systems that are dramatically modified by the local conductivity gradients and in turn the large scale convec- tive motions of the ionospheric plasma are modified. The nature of the modification depends on the latitude and longitude extent of the conductivity modification. Observa- tions of ionospheric concentration from an orbiting space- craft suffer from an interpretive ambiguity. Variations along the spacecraft track may be signatures of ionospheric layers or simply the passage of the spacecraft through vertical variations in the height of the main F-peak. Such an ambiguity could be removed by simultaneous measure- ment of the ion concentration from tethered platforms spaced 5 km to 20 km apart. Then, measurements of the ion concentration and the neutral and ion drift velocities would completely resolve questions surrounding the dynamics and electrodynamics of the layers.

Ion density gradients of the magnitudes produced by neutral winds may be the seat of other smaller scale struc- turing mechanisms. In the collisional regime of the bottom-side F-region, small scale structures are most likely to be formed at high latitudes by imposed structure in the electric field from the magnetosphere. However, the mapping efficiency of such fields is highly dependent on the altitude profile of the ionospheric number density, which is in turn dependent on the neutral wind field. Thus, the spectrum of irregularities has a complex dependency on altitude which requires simultaneous measurements at different altitudes to resolve. Again, the use of multiple tethered platforms will be invaluable in furthering our understanding of spatial structures and their effects on radio propagation paths.

3. MISSION EXAMPLES

A variety of scientific mission scenarios can be considered which range from very simple to advanced and long lived. To illustrate the possible range of missions, we describe example missions to the lower thermosphere using the space shuttle, tethered multi-point free-flyer satellites, and use of the space station.

Exploration of the Lower Thermosphere from the Space Shuttle

NASA is currently evaluating the technical feasibility of a shuttle-based downward deployed tether mission to place a satellite in the lower thermosphere near 130 km using a 100 km tether. Such a mission could provide more observational time within the span of one week than was planned for the entire 2 year TIMED mission before in-situ science was eliminated. At the conclusion of a recent engineering evaluation, it was estimated that such a mission would be similar in cost to Small Explorer (SMEX) missions [Tether-based Investigation of the Ionosphere and Lower Thermosphere (TIILT), R. Heelis, University of Texas-Dallas, 1996]. A scaled-up SEDS type deployer or the original TSS deployment hardware are both capable of 100 km deployment, although payload recovery is not an option with the SEDS deployer. With the addition of a smaller version of the original SEDS deployer on the main tethered satellite, it is also possible to consider a second, smaller sensor payload deployed below the main satellite to a few kilometers distance. With the reboost capability of the shuttle, an extended multi-point mission to low altitudes for a week or possibly longer appears feasible and rich with reward. Such a mission obtains first ever constant altitude measurements of horizontal and vertical gradients on a global scale. A high inclination orbit of 57 degrees (maximum for the space shuttle due to launch safety constraints) would be preferred.

Tethered Multi-point Free-Flyer Satellites

Using existing tether technology and flight proven hardware, it is possible to define important multi-point missions based on 2 to 3 sensor platforms ("pearls") or more. This includes both simple, short-duration missions (few days) as well as more sophisticated, longer duration MITM missions (few months to one year or longer). Relatively simple missions with two point measurements and limited instrumentation are possible to address basic questions about horizontal and vertical plasma structures [Gilchrist, et al., 1995c]. With the addition of more sophisticated subsystems and instrumentation it is possible to include plasma and neutral constituent instrumentation which requires ram direction pointing/knowledge. Two or three "pearl" missions of this class would be the first to give highest priority to MITM measurements for which the

tether provides a unique capability. These missions, for example, will address the measurement of horizontal and vertical gradients of parameters in the ionosphere/thermosphere using passive and active measurement techniques. The separation of the payloads would likely be in the 1-20 km range. Mission duration would depend on mission goals, initial orbit, and tether lifetime issues.

Several missions of interest to the MITM community have recently been proposed: (1) A Space Physics New Mission Concept definition study recently won NASA approval for a 3-sensor tethered system with up to 10 km separation between platforms intended to address MITM coupling processes [Multiple Tethered Satellites for Ionospheric Studies (MTSIS), R. Heelis, University of Texas-Dallas, 1996]. (2) The Canadian BICEPS mission [James, et al., 1995] which will undertake passive and active radioscientific investigations using bistatic tethered platforms separated over distances ranging from 10 m to 10 km. Both active radio and passive two point measurements will be attempted in coordination with ground stations studying a broad range of questions in equatorial and polar ionospheric physics which benefit from bistatic measurements. (3) The AMPAS mission [Neubert, et al., 1995] is an orbital auroral electron beam experiment which takes advantage of a tethered payload concept to provide a short separation distance between an active emitting platform and sensitive optical observation platform. Its primary objective is to utilize a variable energy electron beam (1-10 keV) to probe from below field aligned electric field acceleration structures. It would utilize sensitive optical signatures of reflected beam particles as seen below in the atmosphere as a large detection screen. (4) A Russian mission is being developed to fly a 20 km long, multicomponent tether experiment on a future ALMAZ unmanned automated space station. The plan calls for two tethered spacecraft to be separated by 20 km at 1 km and 21 km below the ALMAZ spacecraft. Synchronous tethered geophysical measurements of electric, magnetic, and gravitational fields will be accomplished [Zaicev, et al., 1995]. (5) A proposal for a 100 km class mission is under investigation in Japan [Oyama, et al., 1992]. The so-called TSS-J mission would deploy a 100 km tether with instrument packages placed every 25 km along the tether and be launched on a Japanese large rocket.

MITM Tether Missions from the Space Stations

The International Space Station offers a large, stable platform to incorporate a tether facility which could support MITM measurement objectives requiring a longer exposure than is available with the Space Shuttle, yet which can benefit by the orbital mass, power, and communication capabilities offered by the Space Station. The form of a space station based system is hard to predict without further study. However, it is likely that the tether will be required to carry multiple platforms, which will have the capability of varying the separation from the Station and will be able to deploy up or down, the latter direction reaching lower thermospheric altitudes below 150 km. Such a system could also support technological goals onboard the space station such as space station reboost, limited duration electrical power enhancement, sample-return, and even trash removal. A long-term facility on the Space Station could also be designed with the capability of exchanging tether material by a reel interchange.

4. TECHNOLOGICAL CONSIDERATIONS

While early flight demonstrations have shown that relatively simple automated systems are capable of deploying and stabilizing long tethers with probes, there are additional factors which must be addressed for dedicated MITM tether missions. The tether itself becomes an important aspect to be considered for operational phases. Its survivability in the space environment and additional drag must be considered as part of mission design. Tether dynamical inputs to sensor platforms must also be understood and accounted for in the design of pointing systems. Finally, the requirement for simultaneous telemetry from multiple sensor platforms requires additional sophistication in the data handling system.

Long Lifetime Tether Systems

Long lifetime tether systems require consideration of three aspects: (1) survivability in a micrometeoroid and debris environment, (2) increased atmospheric drag, and (3) effects of atomic Oxygen flux on the tether material. The first two aspects, survivability and drag, have opposing design requirements when a simple tether of uniform cross section is utilized. The latter, implies constraints on material selection.

All missions to-date have used a tether consisting of a single bundle of strands. For such a tether, the minimum tether diameter is determined by considerations of survivability from micrometeoroid and debris damage rather than tensile strength. This has lead to strength safety margins of 80 or higher for missions designed to last only several days [Gilchrist, et al., 1995b]. However, as tether diameter increases, atmospheric drag on the system also increases. For example, a 20 km tether with only a 1 mm diameter still represents a total cross-section of 20 m^2, a significant contributor in assessing overall atmospheric drag effects. For tethers covering multiple atmospheric scale heights, such as required to reach the lower thermosphere using a 100 km tether, it is not surprising that the tether drag contribution from the lowest portions of the tether is the most important. To estimate total drag, F_{drrg}, the following expression can be used:

$$F_{drag} = \frac{1}{2} \int \rho \, C_d V_{rel}^2 \, W \, dz$$

where ρ is the atmospheric particle mass density which is a function of altitude, C_d is the coefficient of drag with typical value of 2.2 [Penzo and Ammann, 1989], V_{rel} is the satellite velocity relative to the corotating atmosphere, W is the effective diameter of the tether, and the integral is taken along the vertical altitude spanned by the tether. Atmospheric density variations and winds [Killeen, et al., 1993] therefore represent a possible perturbation to tether dynamics which may need to be considered.

Drag can be reduced while maintaining adequate micrometeoroid survival probability by various approaches. The most straightforward is to taper the tether diameter in steps along its length. A new and particularly exciting concept, however, is to use widely spaced cross-connected tether strands. It uses redundancy to significantly improve micrometeoroid survivability while lowering both drag (smaller cross-sectional area) and total mass at the expense of a somewhat lower tensile strength safety margin. This approach is based on the fact that micrometeoroid flux rapidly decreases for increasing diameter. One example of such a configuration is shown in Figure 4 and is designed to survive a multitude of hits by micrometeoroids without severing [Hoyt and Forward, 1995]. With such tether construction, mission durations of well over a year become practical (Figure 5).

For higher altitude applications where drag is not as significant, a simple technique for producing large effective diameter ("Fat") tethers to improve micrometeoroid survivability has been developed. Its first use was part of the U.S. Navy's TiPS mission which is still in orbit as of this writing. The technique uses a low weight acrylic yarn with a spectra outer braid to increase the diameter of a spectra tether after deployment with a moderate mass penalty using a SEDS deployer. The increased diameter improves the calculated survivability of the tether to about 8 kilometer-years [J. Carroll, Tether Applications, personal communications].

Sensor Attitude Pointing

A tether system introduces tether unique dynamical responses which must be considered in any attitude pointing and sensing system. These are: tether-satellite pendulous motions and satellite yawing motions (Figure 6). Although, spaceflight missions to date have shown that these inputs are well behaved and manageable.

All tether systems can expect to have a libration of the overall tether system along the local vertical which is described as an angle from local vertical. The SEDS-2 20 km tether system, with its automated deployment, is estimated to have achieved a maximum 4° in-plane libration angle [Glaese, 1995]. Initial investigations of three sensor multipoint tether system dynamics have also been conducted showing fundamental stable libration characteristics [Niles, et al., 1995; Lorenzini, et al., 1995]. Roll and pitching motions can also be present in the tethered sensor platform.

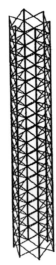

Figure 4. Section of a multistranded tether called a "Hoytether." (Courtesy of R. Forward)

However, for the SEDS experiments it was found that attitude oscillations of the tethered end-mass tended to dampen with time [Cosmo, 1995]. This was attributed in-part to a passive damping system near the end of the tether. Yawing motions from stored energy in tether twists (due to manufacturing imperfections and natural twisting motions experienced on orbit) will act to provide tether unique torque in the yaw direction of the tethered satellite. This introduced a rotation of approximately $35°$ s^{-1} on the SEDS-2 25 kg end-mass [Cosmo, 1995]. In general, however, it should be completely feasible to minimize or eliminate such yaw torques by determining the minimum tether torque point after deployment and then stabilizing around it.

Multi-point Telemetry Requirements

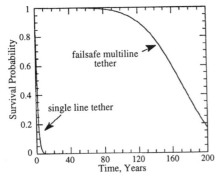

Figure 5. Lifetime comparison of equal-weight single line and fail-safe multiline tethers for a low-load mission.

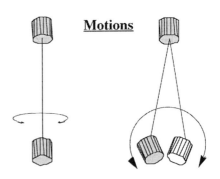

Satellite Pendulum
and Yawing Motion

Figure 6. A space tether can couple its own dynamics into a spacecraft platform including pendulous and yawing motions, but can also act to damp such motions.

Multi-point measurements necessarily introduce a requirement for handling simultaneous, widely separated multiple telemetry streams. Two basic approaches can be considered: (1) independent telemetry of data from each measurement platform directly to the ground, or (2) transmittal of data to a single (main) platform which combines and relays a single data stream to a ground facility. The former avoids the complexity of ingesting and processing multiple data streams and is less sensitive to single point failures. The latter, on the other hand, reduces telemetry power and/or antenna gain requirements for all but the main platform since the distance between the satellites should always be smaller than a space-to-Earth link.

5. CONCLUSION

It is generally recognized that there are many phenomena, often those of greatest interest and controversy, which can not be resolved from the data of a single spacecraft. Multiple tether based sensing platforms "flying" in formation represents a new capability for space research which surpasses single satellite measurements by providing continuous measurement of vertical gradients. Focused campaigns using a multi-point tether system and drawing upon continuous ground based observations and global models offer the best opportunity for addressing complex physical phenomena. Here, we have provided examples of fundamental science questions which can be addressed, presented example mission profiles, and discussed technological considerations to provide the reader an opportunity to better understand the potential for space tethers in MITM research.

REFERENCES

CLUSTER: 1993, Mission, Payload and Supporting Activities, ESA SP - 159.
Cosmo, M. L., E. C. Lorenzini and G. E. Gullahorn, Acceleration

levels and dynamic noise on SEDS end-mass, paper presented at *Fourth International Conference on Tethers in Space*, Washington, D.C., 1995.

Gilchrist, B. E., R. Heelis, W. J. Raitt, C. Rupp, H. G. James, C. Bonifazi, K.-I. Oyama, G. Wood, L. M. Brace and G. R. Carignan, Space tethers for ionospheric-thermospheric-mesospheric science - Report on the 1994 International Summer Workshop, Ann Arbor, MI, paper presented at *Fourth International Conference on Tethers in Space*, Washington, D.C., 1995a.

Gilchrist, B. E., R. A. Heelis, W. J. Raitt and C. Rupp, 1994 International Summer Workshop on Space Tethers for Ionospheric-Thermospheric-Mesospheric Science, Space Physics Research Laboratory, Ann Arbor, MI, 1995b.

Gilchrist, B. E., P. L. Niles, J. Dodds, B. C. Kennedy and C. C. Rupp, AIRSATT - Atmospheric/Ionospheric Research Satellite using Advanced Tether Technology, paper presented at *Fourth International Conference on Tethers in Space*, Washington, D.C., 1995c.

Glaese, J. R., A comparison of SEDS-2 flight and dynamics simulation results, paper presented at *Fourth International Conference on Tethers in Space*, Washington, D.C., 1995.

Heelis, R. A. and J. F. Vickrey, Magnetic field-aligned coupling effects on ionospheric plasma structure, *J. Geophys. Res.*, 95, 7995-8008, 1990.

Hoyt, R. P. and R. L. Forward, Failsafe multistrand tether SEDS technology, paper presented at *Fourth International Conference on Tethers in Space*, Washington, D.C., 1995.

James, H. G., and J. G. Rumbold, The OEDIPUS-C sounding rocket experiment, paper presented at *Fourth International Conference on Tethers in Space*, Washington, D.C., 1995.

James, H. G., A..W. Yau, and G. Tye, Space research in tthe BICEPS experiment, paper presented at *Fourth International Conference on Tethers in Space*, Washington, D.C., 1995.

Killeen, T. L., A. G. Burns, R. M. Johnson and F. A. Marcos, Modeling and prediction of density changes and winds affecting spacecraft trajectories, *Environmental Effects on Spacecraft Positioning and Trajectories Geophysical Monogragh 73, IUGG*, 13, 83-109, 1993.

Lorenzini, E. C., M. L. Cosmo, M. D. Grossi, K. Chance and J. L. Davis, Tethered multi-probe for thermospheric research, paper presented at *Fourth International Conference on Tethers in Space*, Washington, D.C., 1995.

Neubert, T., B. Gilchrist and E. Ungstrup, AMPAS-a new active experiment mission, *Adv Space Res.*, 15, (12)3-(12)12, 1995.

Niles, P. L., B. E. Gilchrist and J. N. Estes, Dual tethered satellite systems for space physics research, paper presented at *Fourth International Conference on Tethers in Space*, Washington, D.C., 1995.

Oyama, K.-I., S. Sasaki, N. Muranaka, N. Kawashima, M. Ogasawara, N. Kato and T. Orii, Feasibility study of a tethered satellite system, paper presented at *Proceedings of the 18th International Symposium on Space Technology*, Koagoshima, 1992.

Penzo, P. A. and P. W. Ammann, Tethers in Space Handbook - Second Edition, NASA, Washington, DC, 1989.

Shen, J. S., W. E. Swartz and D. T. Farley, Ionization layers in the nighttime E-region valley above Arecibo, *J. Geophys. Res.*, 81, 5517-5526, 1976.

Zaicev, E. A., V. A. Modestov, V. A. Ponjukhov, A. E. Reznikov and G. V. Bashilov, Orbital tethered experimental diagnostic gradiometric system on board the ALMAZ-1B SS for Earth remote sensing, paper presented at *International Round Table on Tethers in Space*, ESTEC, Noordwijk, The Netherlands, 1995.

Gilchrist, B. E., Space Physics Research Lab., Dept. of Atmospheric, Oceanic and Space Sciences, The University of Michigan, 2455 Hayward, Ann Arbor, MI 48105-2143.

Heelis, R. A., Center for Space Science, University of Texas at Dallas, P. O. Box 830-688, Richardson, TX 75083.

Raitt, W. J., Center for Atmospheric and Space Science, Physics Department, Utah State University, UMC-34, Logan, UT 84322-4415.